DESIGN OF
SMART POWER
GRID RENEWABLE
ENERGY SYSTEMS

DESIGN OF SMART POWER GRID RENEWABLE ENERGY SYSTEMS

Second Edition

ALI KEYHANI

IEEE PRESS

WILEY

Library of Congress Cataloging-in-Publication Data:

Names: Keyhani, Ali, 1942- editor.
Title: Design of smart power grid renewable energy systems / Ali Keyhani.
Description: Second edition. | Hoboken, New Jersey : John Wiley & Sons, Inc., [2017]
 | Includes bibliographical references and index.
Identifiers: LCCN 2016009194 | ISBN 9781118978771 (cloth)
Subjects: LCSH: Smart power grids–Design and construction. | Electric power
 systems–Automatic control. | Distributed generation of electric power–Computer simulation.
 | Renewable energy sources.
Classification: LCC TK1007 .K49 2016 | DDC 621.319/1–dc23
LC record available at http://lccn.loc.gov/2016009194

Printed in the United States of America

V10005086_101018

I dedicate this book to my parents,
Dr. Mohammed Hossein Keyhani
and
Mrs. Batool Haddad

CONTENTS

PREFACE

Sustainable energy production and the efficient utilization of available energy resources, thereby reducing or eliminating our carbon footprint, is one of our greatest challenges in the twenty-first century. This is a particularly perplexing problem for those of us in the discipline of electrical engineering. This book addresses the problem of sustainable energy production as part of the design of microgrid and smart power grid renewable energy systems.

Today, the Internet offers vast resources for engineering students; it is our job as teachers to provide a well-defined learning path for utilizing these resources. We should also challenge our students with problems that attract their imaginations. This book addresses this task by providing a systems approach to the global application of the presented concepts in sustainable green energy production, as well as analytical tools to aid in the design of renewable microgrids.

In each chapter, I present a key engineering problem, and then formulate a mathematical model of the problem, followed by a simulation test bed in MATLAB, highlighting solution steps. A number of solved examples are presented, while problems designed to challenge the student are given at the end of each chapter. Related references are also provided at the end of each chapter. This book is accompanied by a companion website (www.wiley.com/go/smartpowergrid2e), which includes a Solution Manual and PowerPoint lecture notes with animation that can be adapted and changed as instructors deem necessary for their presentation styles. Solutions to the homework problems presented at the end of each chapter are also included in the Solution Manual. A prerequisite for the book is a basic understanding of electric circuits. The book presents a historical perspective of energy use; an analysis of fossil fuel use is provided through a series of calculations of human's carbon footprints in relation to fossil fuel consumption or that of a single household appliance. The book integrates and presents three areas of electrical engineering:

design of smart, efficient photovoltaic (PV), how to compute the energy yield of photovoltaic modules and the angle of inclination for modules with respect to their position to the sun for maximum energy yields, and wind microgrids. The book builds its foundation on the design of distributed generating system and the design of PV generating plants by introducing design-efficient, smart residential PV microgrid, including energy monitoring systems, smart devices, building load estimation, and load classification and real-time pricing. The book presents basic concepts of phasor systems, three-phase systems, transformers, loads, DC/DC converters, DC/AC inverters, and AC/DC rectifiers, which are all integrated into the design of microgrids for renewable energy as part of bulk interconnected power grids. The focus is on the utilization of DC/AC inverters as a three-terminal element of power systems for the integration of renewable energy sources; MATLAB simulations of PWM inverters are also provided. Topics covered are the basic system concepts of sensing, measurement, integrated communications, and smart meters; real-time pricing; cyber-control of smart grids; high green energy penetration into the bulk interconnected power grids; intermittent generation sources; and the electricity market and the basic modeling and operation of synchronous generator operations, the limit of power flow on transmission lines, power flow problems, load factor calculations and their impact on the operation of smart grids, real-time pricing, and microgrids, power grid bus admittance and bus impedance as well as a power flow analysis of microgrids as part of interconnected bulk power systems. In the final part of the book, the Newton formulation of power flow, the Newton–Raphson solution of a power flow problem, and the fast decoupled solution for power flow studies and short circuit calculations are presented.

This book provides the fundamental concepts of power grid integration with microgrids using green energy sources, which are on the technology road map of virtually all nations. The design of smart microgrids is the driver for the modernization of infrastructure using green energy sources, sensor technology, computer technology, and communication systems.

ALI KEYHANI

January 2015
Berkeley, California

ACKNOWLEDGMENTS

Over the years, many graduate and undergraduate students have contributed to the material presented in this book, in particular, the following students: Chris Zuccarelli, an undergraduate student; Abir Chatterjee, Ehsan Dadash-nialehi, and Paloma Sodhi of IIT, India; Vefa Karakasli of Istanbul Technical University, Turkey; and Adel El Shahat Lotfy Ahmed, Department of Electrical Power and Machines Engineering, Faculty of Engineering, Zagazig University, Egypt. Finally, I acknowledge the support of Professor Joel Johnson, Chair, and Mr. Edwin Lim of the Department of Electrical and Computer Engineering, Ohio State University, for software assistance.

I would also like to thank Simone Taylor, Director, Editorial Development.

ABOUT THE COMPANION WEBSITE

This book is accompanied by a companion website:

www.wiley.com/go/smartpowergrid2e

The website includes:

- Solution Manual for instructors
- PowerPoint presentations for instructors

CHAPTER 1

GLOBAL WARMING AND MITIGATION

1.1 INTRODUCTION—MOTIVATION

The world electric energy production is 65% thermal (coal, gas, and oil), 22% hydro, 12% nuclear, and 1% renewable. The 1% renewable energy is produced by 3% solar, 24% wind, 29% geothermal, and 4% biomass.[1] The trend of renewable energy production is alarming low. The United States and China have the largest energy consumption. The two countries' combined energy productions are 31% and consumptions are 41% of the world's total energy in 1999 as reported by International Energy Agency (IEA).[2] Sustainable energy production and the efficient utilization of available energy resources, thereby reducing or eliminating our carbon footprint, is one of our greatest challenges in the twenty-first century. This is a particularly perplexing problem for those of us in the discipline of engineering and sciences. This book addresses the problem of sustainable electric energy production as part of the design of building efficient microgrids and distributed generation and smart power grid renewable energy systems.

1.2 FOSSIL FUEL

It is estimated that fossil fuels—oil, natural gas, and coal—were produced 300–370 million years ago.[1] Over millions of years, the decomposition of the flora and fauna remains that lived in the world's oceans produced the first oil. As the oceans receded, these remains were covered by layers of sand and earth, and were subjected to severe climate changes: the Ice Age, volcanic eruption, and

Design of Smart Power Grid Renewable Energy Systems, Second Edition. Ali Keyhani.
© 2017 John Wiley & Sons, Inc. Published 2017 by John Wiley & Sons, Inc.
Companion website: www.wiley.com/go/smartpowergrid2e

drought burying them even deeper in the earth's crust and closer to the earth's core. From the earth's core intense heat and pressure, the remains essentially were boiled into oil. If you check the word "petroleum" in a dictionary, you will find it means "rock oil" or "oil from the earth."

The ancient Sumerians, Assyrians, Persians, and Babylonians found oil at the bank of the Karun and Euphrates rivers as it seeped above ground. Historically, humans have used oil for many purposes. The ancient Persians and Egyptians used liquid oil as a medicine for wounds. The Zoroastrians of Iran made their fire temples on top of percolating oil from the ground.[1] Native Americans used oil to seal their canoes.[1]

In fact, although our formally recorded history of humanity's energy use is limited, we can project the impact of energy on early civilizations from artifacts and monuments. The legacy of our oldest societies and their use of wood, wood charcoal, wind, and water power can be seen in the pyramids of Egypt, the Parthenon in Greece, the Persepolis in Iran, the Great Wall of China, and the Taj Mahal in India.[3]

1.3 ENERGY USE AND INDUSTRIALIZATION

The first energy source was wood. Then coal replaced wood, and oil began to replace some of the coal usage to the point that oil now supplies most of the energy needs around the world.

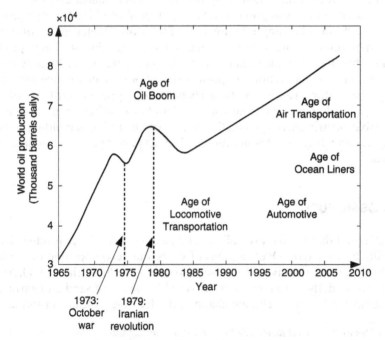

Figure 1.1   The world's oil production (consumption) from 1965–2000 and estimated from 2005–2009.[4,5]

Since the Industrial Revolution, humans have been using coal. Since 1800, for approximately 200 years, humans have been using oil. Sometime in the past, human societies left nomadic way of life, developed agriculture to produce food. Humans used wood and wood charcoal for cooking food. Recorded history shows that human societies have been using wood energy for 5000 years out of 100,000 years living on earth. Similarly, modern human societies have been using oil for 200 years out of 5000 years of recorded history. In the near future, the human societies would exhaust the oil reserve. Oil is not renewable; it cannot be made again. The oil and gas should be conserved. Figure 1.1 depicts the plot of oil production from 1965 to 2000 and estimated from 2005 to 2009[4,5] and the impact of oil on industrialization.

1.4   NEW OIL BOOM–HYDRAULIC FRACTURING (FRACKING)

The new technology in oil and gas extraction from wells deep in the earth rock crust is known as hydraulic fracturing.[6] Fracturing of the rock crust of earth is accomplished by directing pressurized mixture of water with sand and chemicals into crust of earth below the water lines. This technique of oil and gas extraction is known as fracking. Fracking is the method used in wells for shale gas, tight gas, tight oil, and coal seam gas[6] hard rock wells. Fracking has raised environmental concerns. Water contamination, air quality, and migration of chemicals to the ground surface have become a major concern of environmental groups. Fracking has reversed the decline of oil and gas production in United States and has made United States self-sufficient in oil and gas. Fracking methods have become a high charged political issue.[6–9]

1.5   NUCLEAR ENERGY

The US departments of energy estimates worldwide uranium resources are generally considered to be sufficient for at least several decades. The amount of energy contained in a mass of hydrocarbon fuel such as gasoline is substantially lower than in a much less mass of nuclear fuel. This higher density of nuclear fission makes it an important source of energy; however, the fusion process causes additional radioactive waste products. The radioactive products will remain for a long time giving rise to a nuclear waste problem. The counterbalance to a low carbon footprint of fission as an energy source is the concern about radioactive nuclear waste accumulation and the potential for nuclear destruction in a politically unstable world.

1.6   GLOBAL WARMING

Figure 1.2 depicts the process of solar radiation incident energy and reflected energy from the earth's surface and the earth atmosphere. Greenhouse gases in

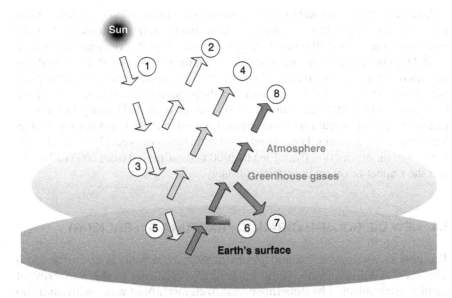

Figure 1.2   The effects of sun radiation on the surface of the earth.

the earth's atmosphere emit and absorb radiation.[10–12] This radiation is within the thermal infrared range. Since the burning of fossil fuel and the start of the Industrial Revolution, the carbon dioxide in the atmosphere has substantially increased as shown in Figures 1.3 and 1.4. The greenhouse gasses are primarily carbon dioxide, carbon monoxide, ozone, a large amount of water vapor, and a number of other gases. Within the atmosphere of earth greenhouse gasses are trapped.

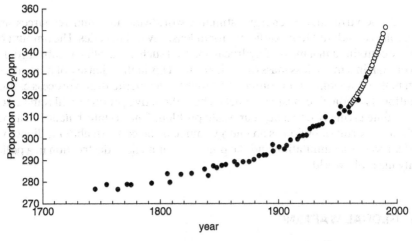

Figure 1.3   The production of CO_2 since 1700.[13]

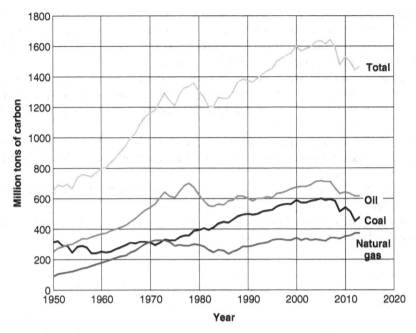

Figure 1.4 The production of CO_2 since 1950.[13]

The solar radiation incident energy as depicted by circle 1 is emitted from the sun and its energy is approximated as 343 W/m^2. Some of the solar radiation, depicted by circles 2 and 4, is reflected from the earth's surface and the earth's atmosphere. The total reflected solar radiation is approximated as 103 W/m^2. Approximately 240 W/m^2 of solar radiation, depicted by circle 3, penetrates through the earth's atmosphere. About half of the solar radiation as depicted by circle 5, approximately 168 W/m^2, is absorbed by the earth's surface. This radiation is depicted by circle 6 and it is converted into heat energy. This process generates infrared radiation in the form of emission of a long wave back to earth. A portion of the infrared radiation is absorbed. Then, it is reemitted by the greenhouse molecules trapped in the earth's atmosphere. Circle 7 represents the infrared radiation. Finally, some of the infrared radiation (circle 8), passes through the atmosphere and into space. As the use of fossil fuel is accelerated, the carbon dioxide in the earth's atmosphere is also accelerated. The growth of carbon dioxide in the atmosphere is shown in parts per million in Figure 1.3.

The World Meteorological Organization (WMO)[13] is the international body for monitoring climate change. The WMO has clearly stated the potential environmental and socioeconomic consequences for the world economy if the current trend continues. In this respect, global warming is an engineering problem, not a moral crusade. Until the world takes serious steps to reduce carbon footprints, pollution and the perilous deterioration of earth's environment will continue.

The CO_2 emission into the atmosphere has peaked during the last 100 years. If concentrated efforts are made to reduce the CO_2 emission and it is reduced over the next few hundred years to a lower level, the earth's temperature will still continue to rise, however, and then stabilize.

Reduction of CO_2 will reduce its impact on the earth's atmosphere; nevertheless, the existing CO_2 in the atmosphere will continue to raise the earth's temperature by a few tenths of a degree. The earth's surface temperature will stabilize over a few centuries.

The rise in temperature due to trapped CO_2 in the earth's atmosphere will impact the thermal expansion of oceans. Consequently, the sea level will rise due to melting of ice.

As the ice sheets continue to melt due to rising temperatures over the next few centuries, the sea level will also continue to rise.

The US contribution is shown in Figure 1.5 which depicts the contributions of various fossil fuel types to CO_2.

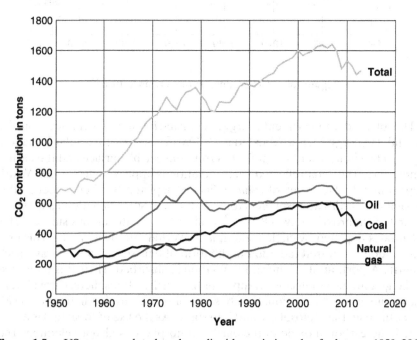

Figure 1.5   US energy-related carbon dioxide emissions by fuel type, 1950–2012, with projection for 2013.[13] Source: Compiled by Earth Policy Institute with 1950–1993 from "Carbon Dioxide Emissions from Energy Consumption by Source, 1949–2011," Table 11.1 in U.S. Department of Energy (DOE), Energy Information Administration (EIA), "Annual Energy Review," available at www.eia.gov/totalenergy/data/annual/showtext.cfm?t=ptb1101, updated September 27, 2012; 1994–2013 from "U.S. Macroeconomic Indicators and CO_2 Emissions," Table 9a in DOE, EIA, "Short-Term Energy Outlook," available at www.eia.gov/forecasts/steo/tables/?tableNumber=5}, updated September 10, 2013. Note: Emissions figures are in million tons of carbon; for tons of CO_2, multiply by 44/12.

1.7 ESTIMATION OF FUTURE CO$_2$

The rapidly growing electrification of developing world and their reliance on fossil fuel vehicles increases, the demand for energy has turned developed nations to mass production of fossil fuels for power and energy. The subsequent burning of these resources has increased the amount of carbon dioxide in the earth's atmosphere. Figure 1.3 shows the measured proportion of CO$_2$ in the atmosphere over the last few centuries. Table 1.1 presents the CO$_2$

TABLE 1.1   Annual Production of CO$_2$

Year	CO$_2$ (ppm)	Year	CO$_2$ (ppm)
1745	277	1930	306
1754	279	1933	307
1764	277	1937	308
1772	279	1940	310
1774	279	1945	308.5
1793	279	1952	311
1805	284	1954	315
1808	280	1959	312
1817	283	1960	314
1826	284	1961	312
1837	287	1962	316
1840	283	1963	317
1845	288	1964	318
1847	287	1965	320
1850	288	1967	320.5
1854	289	1968	321.5
1860	290	1969	322
1865	288	1970	323
1870	289	1971	324
1875	290	1972	327
1880	291	1974	328.5
1882	292	1976	330
1886	292	1977	330.5
1893	294	1978	332
1894	298	1979	333
1900	296	1980	335
1904	295	1981	337
1906	297	1982	339
1910	299	1983	340
1916	300	1984	341
1917	300	1985	342.5
1919	301	1986	344
1920	301.5	1987	345
1921	301	1988	347
1923	305	1989	349
1926	306	1990	352

measurements for each year recorded in this figure. With the accelerated increase in transportation and electricity generation since the 1900s, the burning of more fossil fuels has caused an exponential rising trend in the carbon dioxide concentration.

Using Table 1.1 data, an exponential equation is used to estimate the carbon dioxide concentration for future years if energy production trends are unaltered. This equation was found by fitting an exponential best fit line to the data in Figure 1.3 and Table 1.1 using Microsoft Excel. The equation is

$$\text{Concentration} = 1.853^* e^{(0.0146^* x)} + 277 \tag{1.1}$$

Here, x is the number of years since 1745 and "Concentration" is the proportion of carbon dioxide in parts per million. Equation (1.1) was plotted against the available data in Figure 1.6 and also projected out to the year 2100 to estimate the CO_2 concentration for the year 2100 in future. The estimated carbon dioxide proportion in the year 2100, if current trends continue, is about 610 parts per million, which is nearly double the current proportion of carbon dioxide in the atmosphere. Such data can show the importance of reducing the global carbon footprint by investing in clean energy technologies and more efficient power generation.

If this projection in carbon dioxide concentration would become a reality, the increased amount of CO_2 in the earth's atmosphere would have disastrous effects. More carbon dioxide, as a major greenhouse gas, would trap more of

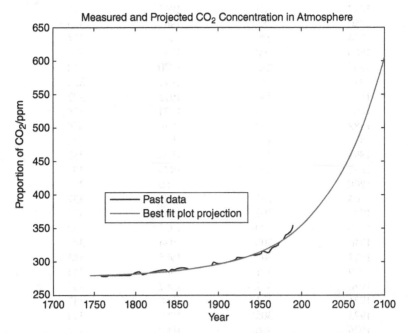

Figure 1.6   Recorded and estimated annual production of CO_2.

the sun's energy in the earth's atmosphere. This would increase the greenhouse effect and rise the earth's average temperature. This would speed up the melting of the polar ice caps and the raising of sea levels. Global warming will greatly change the planet's weather patterns and ocean currents and reduce the amount of habitable land for a growing population. A dramatic increase in carbon dioxide concentration, as projected by the presented data, would have far-reaching effects on the planet's climate, ecosystem, and atmosphere.

According to Intergovernmental Panel on Climate Change, IPCC (https://www.ipcc.ch/report/ar5/wg1/) report of 2013 climate change and global warming due the man-made carbon foot prints states that "the globally averaged combined land and ocean surface temperature data as calculated by a linear trend, show a warming of 0.85 [0.65 to 1.06] °C3, over the period 1880 to 2012, when multiple independently produced datasets exist." (http://www.climatechange2013.org/images/report/WG1AR5_SPM_FINAL.pdf)

The IPCC recommendation states that "Mitigation is a human intervention to reduce the sources or enhance the sinks of greenhouse gases." Mitigation, together with adaptation to climate change, contributes to the objective expressed in Article 2 of the United Nations Framework Convention on Climate Change (UNFCCC): *The ultimate objective of this Convention and any related legal instruments that* the Conference of the Parties may *adopt is to achieve, in accordance with the relevant provisions of the Convention, stabilization of greenhouse gas concentrations in the atmosphere at a level that would prevent dangerous anthropogenic interference with the climate system.*

1.8 GREEN AND RENEWABLE ENERGY SOURCES

To meet carbon reduction targets, it is important that the world begin to use sources of energy that are renewable and sustainable. The need for environmentally friendly methods of transportation and stationary power are urgent. The need to replace traditional fossil-fuel-based systems by sustainable energy sources must be implemented to address the global warming.

1.8.1 Hydrogen

Besides renewable sources, such as wind and the sun, hydrogen (H) is an important source of clean, renewable energy. Hydrogen is abundantly available in the universe. Hydrogen is found in small quantities in the air. It is nontoxic. It is colorless and odorless.

Hydrogen can be used as an energy carrier, stored, and delivered to where it is needed. When hydrogen is used as a source of energy, it gives off only water and heat with no carbon emissions. Hydrogen has three times as much energy as the same quantity of oil.[14,15] A hydrogen fuel cell[14] is fundamentally different from a hydrogen combustion engine. In a hydrogen fuel cell, hydrogen atoms are divided into protons and electrons. The negatively charged electrons from hydrogen atoms create an electrical current with water as a byproduct (H_2O). Hydrogen fuel cells are used to generate electric energy at stationary electric power-generating stations for residential, commercial, and industrial loads. The fuel cell can also be used to provide electric energy for an automotive system, that is, a hydrogen combustion engine. Hydrogen-based energy has the potential to become a major energy source in the future, but there are many applied technical problems that must be solved to reduce the cost of hydrogen-based energy sources; a new infrastructure will also be needed for this technology to take hold.

1.8.2 Solar and Photovoltaic

Solar and photovoltaic (PV) energy are also important renewable energy sources. The sun, the earth's primary source of energy, emits electromagnetic waves. It has invisible infrared (heat) waves, as well as light waves. Infrared (IR) radiation has a wavelength between 0.7 and 300 micrometers (μm) or a frequency range between approximately 1 THz (terahertz; 10 to the power of 12) and 430 THz.[16] Sunlight is defined by irradiance, meaning radiant energy of light. We define one sun as the brightness to provide an irradiance of about one kilowatt (kW) per square meter (m^2) at sea level and 0.8 sun about 800 W/m^2. One sun's energy has 523 watts of IR light, 445 watts of visible light, and 32 watts of ultraviolet (UV) light.

Example 1.1 Compute the area in meter per square and square feet needed to generate 5000 kW of power. Assume the sun irradiant is equivalent to 0.8 sun of energy.

Solution
Power capacity of PV at 0.8 sun = 0.8 kW/m^2
Capacity in kW = (Sun irradiance in kW/m^2) $\cdot$ (Required area in m^2)
Required area in m^2 = 5000/0.8 = 6250 m^2
1 m^2 = 10.764 ft^2
Required area in ft^2 = (6250) $\cdot$ (10.764) = 67,275 ft^2

Plants, algae, and some species of bacteria capture light energy from the sun and through the process of photosynthesis they make food (sugar) from carbon dioxide and water. As the thermal IR radiation from the sun reaches the earth, some of the heat is absorbed by earth's surface and some heat is reflected back into space. Highly reflective mirrors can be used to direct thermal radiation from the sun to provide a source of heat energy. The heat energy from the sun — solar thermal energy — can be used to heat water to a high temperature and pressurized in a conventional manner to run a turbine generator.

Solar PV sources are arrays of cells of silicon materials that convert solar radiation into direct current electricity. The cost of a crystalline silicon wafer is very high, but new light-absorbent materials have significantly reduced the cost. The most common materials are amorphous silicon (a-Si), mainly O for p-type Si and C and the transition metals, mainly Fe. These materials are silicon put into different forms or polycrystalline materials, such as cadmium telluride (CdTe) and copper indium (gallium) (CIS and CIGS). The front of the PV module is designed to allow maximum light energy to be captured by the Si materials. Each cell generates approximately 0.5 V. Normally, 36 cells are connected together in series to provide a PV module producing 12 V.

Example 1.2   Compute the area in meter per square and square feet needed to generate 1000 kW of power. Assume the sun irradiant is equivalent to 0.4 sun of energy.

Solution

Power capacity of PV at 0.4 sun $= 0.4$ kW/m^2

Capacity in kW $=$ (Sun irradiance in kW/m^2) $\cdot$ (Required area in m^2)

Required area in m^2 $= 1000/0.4 = 2500$ m^2

1 m^2 $= 10.764$ ft^2

Required area in ft^2 $= (2500) \cdot (10.764) = 26{,}910$ ft^2

1.8.3 Wind Power

Wind is developed by uneven heating of water and land which causes the flow of air. Therefore wind is air in motion. As the sun rises, the air over the land heats up faster than the air over water. The heated air above the land swells and rises, and the denser, cooler air flow rapidly to take the place of heated air and in process generating wind. However, during the night, the process reverses. A wind turbine has blades and captures the wind's kinetic energy. The blades are connected to a drive shaft that turns an electric generator to produce electricity.

In 2012, wind power production in the United States generated about 3% of total US electricity generation. This is very small when compared with wind production around the world. However, electricity generation from wind in the United States increased from about 6 billion kilowatthours in 2000 to about 140 billion kilowatthours in 2012 (http://www.eia.gov and http://energy.gov/eere/wind/about-doe-wind-program).

1.8.4 Geothermal

Renewable geothermal energy refers to the heat produced deep under the earth's surface. It is found in hot springs and geysers that come to the earth's surface or in reservoirs deep beneath the ground. The earth's core is made of iron surrounded by a layer of molten rocks, or magma. Geothermal power plants are built on geothermal reservoirs and the energy is primarily used to heat homes and commercial industry in the area.[17]

1.8.5 Biomass

Biomass is a type of fuel that comes from organic matter like agricultural and forestry residue, municipal solid waste, or industrial waste. The organic matter used may be trees, animal fat, vegetable oil, rotting waste, and sewage. Biofuels, such as biodiesel fuel, are currently mixed with gasoline for fueling cars, or are used to produce heat or as fuel (wood and straw) in power stations to produce electric power. Rotting waste and sewage generate methane gas, which is also a biomass energy source.[18] However, there are a number of controversial issues surrounding the use of biofuel. Producing biofuel can involve cutting down forests, transforming the organic matter into energy can be expensive with higher carbon footprints, and agricultural products may be redirected instead of being used for food.

1.8.6 Ethanol

Another source of energy is ethanol, which is produced from corn and sugar as well as other means. However, the analysis of the carbon cycle and the use of fossil fuels in the production of "agricultural" energy leaves many open questions: per year and unit area solar panels produce 100 times more electricity than corn ethanol.[19]

As we conclude this section, we need always to remember the Royal Society of London's 1662 motto: "Nullius in verba" (Take Nobody's Word).

1.9 ENERGY UNITS AND CONVERSIONS

To estimate the carbon footprint of different classes of fossil fuels, we need to understand the energy conversion units.[17,20,21] Because fossil fuels are supplied from different sources, we need to convert to equivalent energy measuring units to evaluate the use of all sources. The energy content of different fuels is measured in terms of heat that can be generated. One British thermal unit (BTU) requires 252 calories (cal); it is equivalent to 1055 joules. The joule (J) is named after James Prescott Joule[18] (born December 24, 1818) an English physicist and brewer, who discovered the relationship between heat and mechanical work, which led to the fundamental theory of the conservation of energy. One BTU of heat raises one pound of water one degree Fahrenheit (°F). To measure the large amount of energy, the term "quad" is used. One quad is equivalent to 1015 BTU.[22]

From your first course in Physics, you may recall that one joule in the metric system is equal to the force of one Newton (N) acting through one meter (m). In terms of dimensions, one joule is equal to one Newton (N) times one meter (m) ($1 J = 1 N \times 1 m$); it is also equal to one watt times one second (s) ($1 J = 1 W \times 1 s$). Therefore, one joule is the amount of work required to produce one watt of power for one second. Therefore, 1 watt, normally shown as power "P" is 3.41 BTU per hour.[23–32]

Example 1.3   Compute the amount of energy in watts/s needed to bring 100 lb of water from 0°F to 212 °F.

Solution

Heat required = (100 lb) · (212°F) = 21,200 BTU

Energy in (Watt-hour) Wh = (1055) · (21200)/3600 = 6212.77 Wh.

One watt = 3.41 BTU/hour or simply 3.41 BTU

P = 1 BTU/hour = 0.293 watts

For direct current electricity (DC)

$P = V \cdot I$

where I represents the current through the load and V is the voltage across the load and unit of power, P is in watts if the current is in amperes (amp) and voltage in volts. Therefore, one kilowatt is a thousand watts. The energy use is expressed in kilowatt-hour (kWh), and one kWh is the energy used by a load for one hour. This can also be expressed in joules, and one kilowatt-hour (kWh) is equal to 3.6 million joules. Recall from your Introduction to Chemistry course that one calorie (cal) is equal to 4.184 J. Therefore, it follows that thousand BTU is equal to 0.293 kWh; one MWh to 3.41 million BTU. Because power grid generators are running on natural gas, oil, or coal, the energy from these types of fuel in terms of kilowatts per hour. For example, one thousand cubic feet of gas (Mcf) can produce 301 kWh and one hundred thousand BTU can produce 29.3 kWh of energy.[33]

Example 1.4   Compute the amount of heat in BTU needed to generate 10 kWh.

Solution

One watt = one joule/sec (J/s)

1000 W = 1000 J/s

1 kWh = 1 · 60 · 60 · 1000 = 3600 kJ/s

10 kWh = 36,000 kJ/s

One BTU = 1055.058 J/s

Heat in BTU needed for 10 kWh = 36,000,000/1055.058 = 34,121.3 BTU

The energy content of coal is measured in terms of BTU produced. For example, a ton of coal can generate 25 million BTU: equivalently, it can generate 7325 kWh. Furthermore, one barrel of oil (i.e., 42 gallons) can produce 1700 kWh. Other units of interest are: a barrel of liquid natural gas has 1030 BTU and one cubic foot of natural gas has 1030 BTU.

TABLE 1.2   Carbon Footprint of Various Fossil Fuels for Production of 1 kWh of Electric Energy[22,33]

Fuel Type	CO_2 Footprint (lb/kWh)
Wood	3.306
Coal-fired plant	2.117
Gas-fired plant	1.915
Oil-fired plant	1.314
Combined-cycle gas	0.992

Example 1.5   Compute how many kWh can be produced from 10 tons of coal.

Solution

One ton of coal = 25,000,000 BTU

10 ton of coal = 250,000,000 BTU

1 kWh = 3413 BTU

Energy used in kWh = (250,000,000)/3413 = 73,249.3 kWh

Table 1.2 presents carbon footprint of various fossil fuels for production of 1 kWh of electric energy. Table 1.3 presents carbon footprint of green and renewable sources for production of 1 kWh of electric energy.

Example 1.6   Compute the CO_2 footprint of a residential home using 100 kWh coal for 1 day.

Solution

1 kWh of electric energy using a coal fire plant has 2.117 lb.

Residential home carbon footprint for 100 kWh = (100) · (2.117) = 211.7 lb of CO_2

The carbon footprint can also be estimated in terms of carbon (C) rather CO_2. The molecular weight of C is 12 and CO_2 is 44. (Add the molecular weight of C, 12 to the molecular weight of O_2, 16 times 2 = 32, to get 44, the molecular weight of CO_2.) The emissions expressed in units of C can be converted to emissions in CO_2. The ratio of CO_2/C is equal to 44/12 = 3.67. Thus, CO_2 = 3.67 C. Conversely, C = 0.2724 CO_2.

TABLE 1.3   Carbon Footprint of Green and Renewable Sources for Production of 1 kWh of Electric Energy[22,33]

Fuel Type	CO_2 Footprint (lb/kWh)
Hydroelectric	0.0088
PV	0.2204
Wind	0.03306

TABLE 1.4   Fossil Fuel Emission Levels in Pounds per Billion BTU of Energy Input

Pollutant	Natural Gas	Oil	Coal
Carbon dioxide (CO_2)	117,000	164,000	208,000
Carbon monoxide (CO)	40	33	208
Nitrogen oxides	92	448	457
Sulfur dioxide	1	1122	2591
Particulates	7	84	2744
Mercury	0.000	0.007	0.016

Example 1.7   Compute the carbon footprint of 100 kWh of energy if coal is used to produce it.

Solution

$C = 0.2724\ CO_2$

$C = (0.2724) \cdot 211.7\ \text{lb} = \text{Antoine Becquerel (Bq) } 57.667\ \text{lb of C}$

The carbon footprints of coal are the highest among fossil fuels. Therefore, coal-fired plants produce the highest output rate of CO_2 per kilowatt-hour. The use of fossil fuels also adds other gasses to the atmosphere per unit of heat energy as shown in Table 1.4.

We can also estimate the carbon footprints for various electrical appliances corresponding to the method used to produce electrical energy. For example, one hour's use of a color television produces 0.64 pounds (lb) of CO_2 if coal is used to produce the electric power. For coal, this coefficient is approximated to be 2.3 lb CO_2 per kWh of electricity.

Example 1.8   A light bulb is rated 60 W. If the light bulb is on for 24 hours, how much electric energy is consumed?

Solution

The energy used is given as:

$$\text{Energy consumed} = (60\ \text{W}) \times (24\ \text{hours})/(1000) = 1.44\ \text{kWh}$$

Example 1.9   Estimate the CO_2 footprint of a 60 W bulb on for 24 hours.

Solution

Carbon footprint $= (1.44\ \text{kWh}) \times (2.3\ \text{lb } CO_2 \text{ per kWh}) = 3.3\ \text{lb } CO_2$

Large coal-fired power plants are highly economical if their carbon footprints and damage to the environment are overlooked. In general, a unit cost of electricity is an inverse function of the unit size. For example, for a 100 kW unit, the unit cost is $0.15 per kWh for a natural gas turbine and $0.30 per kWh for PV energy. Therefore, if the environmental degradation is ignored, the electric energy produced from fossil fuel is cheaper based on the present price of fossil fuel. For a large coal-fired power plant, the unit of electric energy is in the range $0.04–$0.08 per kWh. Green energy technology needs supporting governmental policies to promote electricity generation from green energy sources. Economic development in line with green energy policies is needed for lessening ecologic footprint of a developing world.

After thousands of years of burning wood and wood charcoal, CO_2 concentration was at 288 parts per million by volume (ppmv) in 1850 just on the dawn of Industrial Revolution. By the year of 2000, CO_2 had risen to 369.5 ppmv, an increase of 37.6% over 250 years. The exponential growth of CO_2 is closely related to the electric energy production (see Figures 1.4 and 1.6).

1.10 ESTIMATING THE COST OF ENERGY

As we discussed, the cost of electric energy is measured by the power used over time. The power demand of any electrical appliance is inscribed on the appliance and/or included in its documentation or in its nameplate. However, the power consumption of an appliance is also a function of the applied voltage and operating frequency. Therefore, the manufacturers provide on the nameplate of an appliance the voltage rating, the power rating, and the frequency. For a light bulb, which is purely resistive, the voltage rating and power rating are marked on the light bulb. A light bulb may be rated at 50 W and 120 V. This means that if we apply 120 volts to the light bulb, we will consume 50 watts.

Again, energy consumption for DC electricity and AC electricity with unity power factor can be expressed as follows:

$$P = V \cdot I \text{ W} \tag{1.2}$$

where the unit of power consumption, that is, P is in watts. The unit of V is in volts and unit I is in amperes.

The rate of energy consumption can be written as

$$P = \frac{dW}{dt} \tag{1.3}$$

We can then write the energy consumed by loads (i.e., electrical appliances) as

$$W = P \cdot t \tag{1.4}$$

In the above unit of W is in joules or watts-seconds. However, because the unit cost of electrical energy is expressed in dollars per kilowatts-hour, we express the electric power consumption, kilowatts-hour as

$$kWh = kW \times hour \tag{1.5}$$

Therefore, if we let λ represent the cost of electric energy in $ per kWh, the total cost can be expressed as

$$Energy\ cost\ (in\ dollars) = kWh \times \lambda \tag{1.6}$$

Example 1.10   Let us assume that you want to buy a computer. You see two computers: brand A is rated as 400 W and 120 V and costs $1000; brand B is rated as 100 W and 120 V and costs $1010. Your electric company charges $0.09 per kWh on your monthly bill. Compute the total cost of buying the computer and the operating cost if you use your computer for 3 years at the rate of 8 hours a day.

Solution

At 8 hours a day for 3 years, the total operating time is

Operating time $= 8 \times 365 \times 3 = 8760$ hours.

Brand A kWh energy consumption $=$ Operating time $\times$ kW of brand A

$$= 8760 \times 400 \times 10^{-3} = 3504\ kWh$$

Brand B kWh energy consumption $=$ Operating time $\times$ kW of brand B

$$= 8760 \times 100 \times 10^{-3} = 876\ kWh$$

Total cost for brand A $=$ Brand A kWh energy consumption $+$ Cost of brand A

$$= 3504 \times 0.09 + 1000 = \$1315.36$$

Total cost for brand B $=$ Brand B kWh energy consumption $+$ Cost of brand B

$$= 876 \times 0.09 + 1010 = \$1088.84$$

Therefore, the total cost of operation and price of brand B is much lower than brand A because the wattage of brand B is much less. Despite the fact that the price of brand A is lower, it is more economical to buy brand B because its operating cost is far lower than that of brand A.

Example 1.11   In Example 1.10, let us assume that the electric energy is produced using coal, what is the amount of CO_2 in pounds that is emitted over 3 years into the environment?

What is your carbon footprint?

Solution
From Table 1.4, the pounds of CO_2 emission per billion BTU of energy input for coal is 208.000.

1 kWh = 3.41 thousand BTU

Energy consumed for brand A over 3 years = $3504 \times 3.41 \times 10^3 = 11948640$ BTU

Therefore, for brand A, pounds of CO_2 emitted = $\dfrac{11948640 \times 208000}{10^9} = 2485.32$ lb

Energy consumed for brand B over 3 years = $876 \times 3.41 \times 10^3 = 2987160$ BTU

Therefore, for brand B, pounds of CO_2 emitted = $\dfrac{2987160 \times 208000}{10^9} = 621.33$ lb

Brand B has a much lower carbon footprint.

Example 1.12   Assume that you have purchased a new high-powered computer with a gaming card and an old CRT (cathode ray tube) monitor. Assume that the power consumption is 500 W and the fuel used to generate electricity is oil. Compute the following:

(i) Carbon footprints if you leave them on 24/7.
(ii) Carbon footprint if it is turned on 8 hours a day.

Solution
(i) Hours in one year = $24 \times 365 = 8760$ hours

Energy consumed in one year = $8760 \times 500 \times 10^{-3} = 4380$ kWh
$$= 4380 \times 3.41 \times 10^3 = 14935800 \text{ BTU}$$

From Table 1.4, pounds of CO_2 emission per billion The BTU of energy input for oil is 164000.

Therefore, the carbon footprint for one year = $\dfrac{14935800 \times 164000}{10^9}$
$$= 2449.47 \text{ lb}$$

(ii) Carbon footprint in the case of 8 hour/day use = $\dfrac{8}{24} \times$ footprint for use in 24 hours

$$= \frac{1}{3} \times 2449.47 = 816.49 \text{ lb}$$

1.11 CONCLUSION

In this chapter, we have studied a brief history of energy sources and their utilization. The development of human civilization is the direct consequence of
harnessing the earth's energy sources. We have used the power of the wind, the
sun, and wood for thousands of years. However, as new energy sources, such as
coal, oil, and gas have been discovered, we have continuously substituted a new
source of energy in place of an old source.

Global warming and environmental degradation have forced us to reexam-
ine our energy use and consequent carbon footprints, a topic we also have
addressed in this chapter. In the following chapters, we will study the basic
concept of power system operation, power system modeling, and the smart
power grid system as well as the design of smart microgrid rentable energy
systems.

PROBLEMS

1.1 Writing Assignment. Write a 3000 word report summarizing the Kyoto
Protocol. Compute the simple operating margin CO_2 factor for the year
2020 if the system load of 6000 MW is supplied by 50 coal units with a
capacity of 100 MW, 10 oil-fired generators with a capacity of 50 MW, and
10 gas-fired generators with a capacity of 50 MW.

   (i) The system load of 6000 MW supplied from only coal units

   (ii) The system load of 6000 MW supplied equally from gas units and
wind and solar power

1.2 Using the data given in Table 1.4, perform the following:

   (i) The carbon footprint of 500 W if coal is used to produce the electric
power

   (ii) The carbon footprint of a 500 W bulb if natural gas is used to produce the electric power

   (iii) The carbon footprint of a 500 W bulb if wind is used to produce the
electric power

   (iv) The carbon footprint of a 500 W bulb if PV energy is used to produce
the electric power

1.3 Compute the money saved in 1 month by using CFL (compact fluorescent
light) bulb (18 W) instead of using an incandescent lamp (60 W) if the
cost of electricity is $0.12/kWh. Assume the lights are used for 10 hours
a day.

1.4 Compute the carbon footprint of the lamps of Problem 1.3 if natural gas
is used as fuel to generate electricity. How much more will the carbon
footprint be increased if the fuel used is coal?

1.5 Will an electric oven rated at 240 V and 1200 W provide the same heat if connected to a voltage of 120 V? If not, how much power will it consume now?

1.6 If the emission factor of producing electric power by PV cells is 100 g of CO_2 per kWh, by wind power is 15 g of CO_2 per kWh, and by coal is 1000 g of CO_2 per kWh, then find the ratio of CO_2 emission when (a) 15% of power comes from wind farms, (b) 5% from a PV source, and (c) the rest from coal as opposed to when all power is supplied by coal-run power stations.

1.7 Compute the operating margin of the emission factor of a power plant with three units with the following specifications over 1 year:

Unit	Generation (MW)	Emission Factor (lb of CO_2 per MWh)
1	160	1000
2	200	950
3	210	920

1.8 If the initial cost to set up a thermal power plant of 100 MW is 2 million dollars and that of a PV farm of the same capacity is 300 million dollars, and the running cost of the thermal power plant is $90 per MWh and that of PV farm is $12 per MWh then find the time in years needed for the PV farm to become the most economical if 90% of the plant capacity is utilized in each case.

1.9 Consider a feeder that is rated 120 V and serving five light bulbs. Loads are rated 120 V and 120 W. All light loads are connected in parallel. If the feeder voltage is dropped by 20%, compute the following:

(i) The power consumption by the loads on the feeder in watts

(ii) The percentage of reduction in illumination by the feeders

(iii) The amount of carbon footprint if coal is used to produce the energy

1.10 The same as Problem 1.9, except a refrigerator rated 120 V and 120 W is also connected to the feeder and voltage is dropped by 30%.

(i) Compute the power consumption by the loads on the feeder in watts

(ii) Compute the percentage of reduction in illumination by the feeders

(iii) Do you expect any of the loads on the feeder to be damaged?

(iv) Compute the amount of carbon footprint if coal is used to produce the energy

(Hint: a 40 W incandescent light bulb produces approximately 500 lumens of light)

1.11   The same as Problem 1.9, except a refrigerator rated 120 W is also connected to the feeder and voltage is raised by 30%.

   (i) Compute the power consumption by the loads on the feeder in watts

   (ii) Compute the percentage of reduction in illumination by the feeders

   (iii) Do you expect any of the loads on the feeder to be damaged?

   (iv) Compute the amount of carbon footprint if coal is used to produce the energy

1.12   Compute the CO_2 emission factor in pounds of CO_2 per BTU for a unit in a plant that is fueled by coal, oil, and natural gas if 0.3 million tons of coal, 0.1 million barrels of oil, and 0.8 million cubic feet of gas have been consumed over 1 year. The average power produced over the period was 210 MW. Use the following data and the data in Table 1.4 for computation: a ton of coal has 25 million BTU; a barrel (i.e., 42 gallons) of oil has 5.6 million BTU; a cubic foot of natural gas has 1030 BTU.

1.13   Write a MATLAB program to compute CO_2 production for the year 2100.

REFERENCES

1. World Fossil Fuel Reserves and Projected Depletion White Paper 2002. Available at http://crc.nv.gov/docs/world%20fossil%20reserves.pdf. Accessed June 17, 2014.

2. International Energy Agency. Available at http://www.iea.org/publications/freepub lications/publication/gov_handbook.pdf. Accessed June 17, 2014.

3. Durant, W. The Story of Civilization (Ebook on the web). Available at http:// www.archive.org/details/storyofcivilizat035369mbp. Accessed November 9, 2010.

4. Brown, L.R. Mobilizing to Save Civilization. Available at http://www.earth-policy.org/index.php?/books/pb4/pb4_data (This is part of a supporting dataset for Lester R. Brown, Plan B 4.0: Mobilizing to Save Civilization, New York: W.W. Norton & Company, 2009). Accessed September 20, 2009.

5. BP Statistical Review of World Energy June 2010. Available at http://www.bp. com/liveassets/bp_internet/globalbp/globalbp_uk_english/reports_and_publicatio ns/statistical_energy_review_2008/STAGING/local_assets/2010_downloads/statist ical_review_of_world_energy_full_report_2010.pdf. Accessed September 20, 2009.

6. Clean Water Action. Available at http://www.cleanwateraction.org/page/fracking-dangers. Accessed December 17, 2013.

7. Table Elements, Los Alamos National Lab. Available at http://periodic.lanl.gov/ index.shtml. Accessed December 17, 2013.

8. British Petroleum (BP) Energy in 2012 – Adapting to a Changing World. Available at http://www.bp.com/content/dam/bp/pdf/statistical. Accessed December 11, 2013.

9. review/statistical_review_of_world_energy_2013.pdf. Accessed December 10, 2013.

10. Encyclopædia Britannica. Michael Faraday. (2010). Available at http://www.brit annica.com/EBchecked/topic/201705/Michael-Faraday. Accessed November 9, 2010.

11. Encyclopædia Britannica. Conte Alessandro Volta. Available at http://www.brit annica.com/EBchecked/topic/632433/Conte-Alessandro-Volta. Accessed November 9, 2010.

12. The Contribution of Francesco Zantedeschi at the Development of the Experimental Laboratory of Physics Faculty of the Padua University. Available at http://www.brera.unimi.it/sisfa/atti/1999/Tinazzi.pdf. Accessed November 9, 2010.

13. Arctic Surface-Based SeaIce Observations: Integrated Protocols and Coordinated Data Acquisition. Available at http://web.archive.org/web/20100724053830 or http://www.wmo.int/pages/prog/wcrp/documents/CliCASWSreportfinal.pdf. Accessed November 9, 2010.

14. Climate Change Observations. Available at http://www.climatechangeconnection. org/Science/Observations.htm. Accessed November 9, 2010.

15. The Potential of Fuel Cells to Reduce Energy Demands and Pollution From the UK Transport Sector. Available at http://oro.open.ac.uk/19846/1/pdf76.pdf. Accessed October 9, 2010.

16. Hydrogen FQA Sustainability. Available at http://www.formal.stanford.edu/jmc/pro gress/hydrogen.html. Accessed November 9, 2010.

17. Hockett, R.S. Analytical techniques for PV Si feedstock evaluation, in Proceeding of 18th Workshop on Crystalline Silicon Solar Cells &Modules: Material and Processes, edited by B. L. Sopori, Vail, CO, August 3–6, 2008, pp 48–59.

18. Patzek, T.W. (2004) Thermodynamics of the corn-ethanol biofuel cycle. *Critical Reviews in Plant Sciences,* 23(6), 519–567.

19. Carbon Footprints. Available at http://www.liloontheweb.org.uk/handbook/carbonf ootprint. Accessed November 9, 2010.

20. Managing the Nuclear Fuel Cycle: Policy Implications of Expanding Global Access to Nuclear Power Updated January 30, 2008.

21. Climate, Energy, and Transportation. Available at http://www.earth-policy.org/ data_center/C23. Accessed November 9, 2010.

22. Energy Quest, Chapter 8: Fossil Fuels – Coal, Oil and Natural Gas. Available at http://www.energyquest.ca.gov. Accessed September 26, 2010.

23. Encyclopædia Britannica. Eugène-Melchior Péligot. Available at http://www.brit annica.com/EBchecked/topic/449213/Eugene-Melchior-Peligot. Accessed November 9, 2010.

24. Encyclopædia Britannica. Antoine-César Becquerel. Available at http://www.brit annica.com/EBchecked/topic/58017/Antoine-Cesar-Becquerel. Accessed November 9, 2010.

25. Encyclopædia Britannica. Enrico Fermi. Available at http://www.britannica. com/EBchecked/topic/204747/Enrico-Fermi. Accessed November 9, 2010.

26. IPCC Third Annual Report (2006). Available at http://www.ipcc.ch/ and http:// www.grida.no/publications/other/ipcc_tar/?src=/climate/ipcc_tar/vol4/english/inde x.htm. Accessed September 27, 2010.

27. Encyclopædia Britannica. Hans Christian Ørsted. Available at http://www.brit annica.com/EBchecked/topic/433282/Hans-Christian-Orsted. Accessed November 9, 2010.

28. Encyclopædia Britannica. Joseph Henry. Available at http://www.britannica. com/EBchecked/topic/261387/Joseph-Henry. Accessed November 9, 2010.

29. Encyclopædia Britannica. Nikola Tesla. Available at http://www.britannica.com/EBchecked/topic/588597/Nikola-Tesla. Accessed November 9, 2010.
30. Energy Information Administration (2009). Official energy statistics from the US Government website. Available at www.eia.doe.gov. Accessed September 26, 2010.
31. World Population Data. Available at http://www.prb.org/pdf10/10wpds_eng.pdf. Accessed November 9, 2010.
32. Annual Energy Outlook and Projections to 2040. Available at http://www.eia.gov/forecasts/aeo/pdf/0383(2013).pdf. Accessed December 9, 2013.
33. Encyclopædia Britannica. Martin Heinrich Klaproth. Available at http://www.britannica.com/EBchecked/topic/319885/Martin-Heinrich-Klaproth.   Accessed November 9, 2010.

ADDITIONAL RESOURCES

http://assets.opencrs.com/rpts/R40797_20090908.pdf

Jacobs, J.A. Groundwater and Uranium: Chemical Behavior and Treatment, Water Encyclopedia, pp. 537–538. Available at www.onlinelibrary.wiley.com/doi/10.1002/047147844X.iw168/full. Accessed January 26, 2010.

Patzek, T.W. Thermodynamics of the corn-ethanol biofuel cycle. *Critical Reviews in Plant Sciences*, **23**(6), 519–567.

http://www cdiac.ornl.gov/pns/faq.html. Accessed January 20, 2009.

http://www.nrel.gov/docs/fy10osti/47523.pdf

Connecting Manitobans to climate change facts and solutions. Available at http://www.climatechangeconnection.org/science/Greenhouseeffect_diagram.htm. Accessed December 14, 2014.

Chen, Z., Marquis, M., Averyt, K.B., Tignor, M., and Miller, H.L. (eds.). Cambridge University Press, Cambridge, and New York. Available at http://www.ipcc.ch/pdf/assessment-report/ar4/wg1/ar4-wg1-chapter1.pdf.28. Accessed December 14, 2014.

Energy Information Administration, Official Energy Statistics from the US Government website: www.eia.doe.gov. Accessed September 26, 2010.

ENERGY QUEST

US Department of Energy. http://www1.eere.energy.gov/biomass/biomass_basics_faqs.html#ethanol. Accessed April 2010.

www.eia.doe.gov/electricity/page/co2_report/co2report.html.   Accessed   September 2010.

Emissions of Greenhouse Gases in the United States 1985–1990, DOE/EIA-0573. Available at http://www.eia.doe.gov/pub/oiaf/1605/cdrom/pdf/ggrpt/057306.pdf. Accessed December 14, 2014.

http://www.guardian.co.uk

Patzek, T.W. and Patzek, D.P. (2005) Thermodynamics of energy production from biomass. *Critical Reviews in Plant Sciences*, **24**, 327–364.

http://science.hq.nasa.gov/kids/imagers/ems/infrared.html. Accessed January 26, 2010.

Metz, B., Davidson, O.R., Bosch, P.R., Dave, R., and Meyer, L.A. (eds) Contribution of Working Group III to the Fourth Assessment Report of the Intergovernmental Panel on Climate Change, 2007. Cambridge University Press, New York/Cambridge.

Markvart, T. and Castaner, L. (2003) *Practical Handbook of Photovoltaics: Fundamentals and Applications*, Elsevier, Amsterdam.

CHAPTER 2

DESIGN OF PHOTOVOLTAIC MICROGRID GENERATING STATION

2.1 INTRODUCTION

The sun's energy is the primary source of energy for life in our planet. Solar energy is a readily available renewable energy; it reaches earth in the form of electromagnetic waves (radiation). Many factors affect the amount of radiation received at a given location on earth. These factors include location, season, humidity, temperature, air mass, and the hour of day. *Insolation* refers to exposure to the rays of the sun, that is, the word "insolation" has been used to denote the solar radiation energy received at a given location at a given time. The phrase *incident solar radiation* is also used; it expresses the average irradiance in watts per square meter (W/m^2) or kilowatt per square meter (kW/m^2).

The surface of the earth is coordinated with imaginary lines of latitude and longitude as shown in Figure 2.1a. Latitudes on the surface of the earth are imaginary parallel lines measured in degrees. The lines subtend to the plane of the equator. The latitudes vary from 90°S and 90°N. The longitudes are imaginary lines that vary from 180°E to 180°W. The longitudes converge at the poles (90°N and 90°S). The radiation of the sun on the earth varies with the location based on the latitudes. Approximately, the region between 30°S and 30°N has the highest irradiance as depicted in Figure 2.1. The latitude on which the sun shines directly overhead between these two latitudes depends on the time of the year. If the sun lies directly above the northern hemisphere, it is summer in the north and winter in the southern hemisphere. If it is above the southern hemisphere, it is summer in the southern hemisphere and winter in the north.

Figure 2.1b is a plot showing the irradiance at different locations on the earth marked by its latitude and longitude. The latitude varies from –90° to 90°,

Design of Smart Power Grid Renewable Energy Systems, Second Edition. Ali Keyhani.
© 2017 John Wiley & Sons, Inc. Published 2017 by John Wiley & Sons, Inc.
Companion website: www.wiley.com/go/smartpowergrid2e

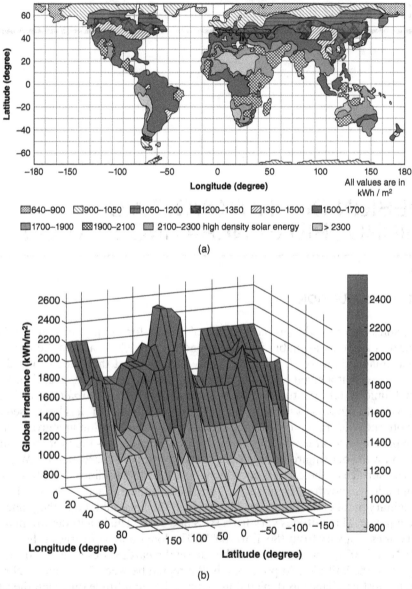

Figure 2.1   (a) The global irradiation values for the world (kWh/m^2).[1,2] Adapted from Elmhurst College Virtual Chembook. (b) The average northern hemisphere global irradiation source by longitude and latitude.[3,4] (c) The average southern hemisphere global irradiation source by longitude and latitude. (d) The average world global irradiation source by regions.[3,4] Region 1: Argentina, Chile; Region 2: Argentina, Chile; Region 3: Brazil, South Africa, Peru, Australia, Mozambique; Region 4: Indonesia, Brazil, Nigeria, Columbia, Kenya, Malaysia; Region 5: India, Pakistan, Bangladesh, Mexico, Egypt, Turkey, Iran, Algeria, Iraq, Saudi Arabia; Region 6: China, United States, Japan, Germany, France, United Kingdom, South Korea; Region 7: Russia, Canada, Sweden, Norway.

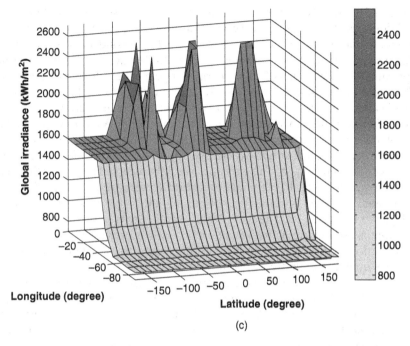

(c)

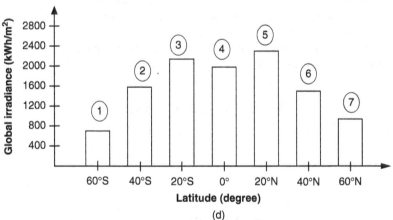

(d)

Figure 2.1   (*Continued*)

which is 90°N and 90°S, respectively. Similarly, the longitudes vary from –180° to 180°, which is 180°W and 180°E, respectively. The z-axis of the plot gives the irradiance in kWh/m². The points on the z-axis convey the amount of irradiance. The colors on the plot represent the intensity of the irradiance as given in Figure 2.1b.

The sun's position as seen from earth between latitudes 15°N and 35°N is the region with the most solar energy. This semiarid region, as shown in

Figures 2.1b and 2.1c, is mostly located in Africa, the Western United States, the Middle East, and India. These locations have over 3000 hours of intense sunlight radiation per year. The region with the second largest amount of solar energy radiation lies between 15°N latitude and the equator and has approximately 2500 hours of solar energy per year. The belt between latitudes 35°N and 45°N has limited solar energy. However, typical sunlight radiation is roughly about the same as the other two regions, although there are clear seasonal differences in both solar intensity and daily hours. As winter approaches, the solar radiation decreases; by midwinter, it is at its lowest level. The 45°N latitude and the region beyond experience approximately half of the solar radiation as diffused radiation. The energy of sunlight received by the earth can be approximated to equal 10,000 times the world's energy requirements.[2]

The sun's radiation is in the form of ultraviolet, visible, and infrared energy as depicted in Figure 2.2. The majority of the energy is in the form of a short wave that is used in the planet's heat cycle, weather cycle, wind, and waves. A small fraction of the energy is utilized for photosynthesis in plants and the rest of the solar energy is emitted back into space.[5]

The solar energy reaching the atmosphere is constant; hence the term *solar constant*. The solar constant is computed to be in the range of 1.4 kW/m^2, or 2.0 cal/cm^2/min. The sunlight's visible light has maximum solar energy.[1,6] The sunlight's shorter wavelengths scatter over a wider area than the longer wavelengths of light. The scattering may be due to gas molecules, pollution, and haze. The blue and violet lights have the maximum atmospheric scattering at sunrise and sunset without affecting the red rays of sunlight.[6]

Photovoltaic (PV) sources of energy trap the energy in the electromagnetic radiation in the form of light and convert that directly into electrical energy.

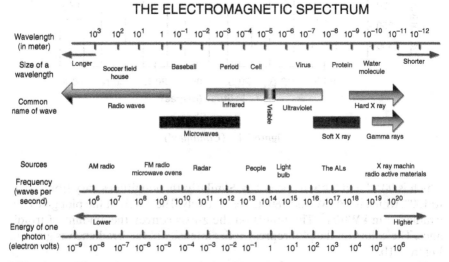

Figure 2.2   The electromagnetic spectrum. The visible light collects maximum solar energy.

When photovoltaic modules are installed at a location, it is desired that it is able to trap maximum energy available from the sunlight. As the time of the day and the day of the year changes, the position of the sun as seen in the sky also changes. The angle that the rays of the sun make with the PV, known as incident angle, changes as a result. The energy received by the PV from the rays is a function of this incident angle. The maximum energy is received when the rays from the sun are incident on the PV at normal angle. For that to take place, the PV modules need to be constantly oriented toward the sun. This requires a two-dimensional rotation of the PV modules. In order to achieve that, there should be mechanisms in place which will bring about this motion. Typically this is achieved by motors. However, this mechanism needs power which decreases the overall efficiency of the PV station. Also, it increases the cost of already expensive PV installation.

When a PV is installed without tracking mechanism, it is installed at an angle so that it receives maximum energy from the sun over a period of time. At a geographical location, PV is installed facing toward the equator and the angle which it makes with the horizontal plane is called the tilt angle. To determine the optimal tilt angle for the module to receive maximum energy from the sun, several information about the location are needed. The angle at which the rays of the sun reach the surface of the earth is determined from the latitude of the location. It also determines the amount of energy that location receives if there were no atmosphere. However, as the light travels through the atmosphere, some of it gets absorbed, some gets dispersed, some is reflected out and the remaining rays reach the surface. The amount that reaches the surface of the earth depends on the composition of the air of the location and the distance that the light is traveling though the atmosphere, defined by the air mass.

The light that reaches the module consists of three components. First, the direct or beam irradiation which reaches the module directly from the sun. Second, the light that gets dispersed in the atmosphere and reaches the module is called diffused irradiation. Third, the light that reaches the surface of the earth gets reflected and reaches the module and is known as reflection irradiation. The sum of the above three components gives the total irradiation received at the location.

Meteorological centers around the world keep record of the average irradiation that a particular location receives on a monthly basis for a surface tilted horizontally to the earth's surface. However, to determine what the irradiation will be for a module inclined at a given tilt angle, the components of the rays reaching the module has to be separated.

As the earth revolves around the sun, distance between them changes. Due to the acute angle which the earth makes with the plane of revolution, the angle at which a location receives the sun's rays also changes by the day. The extraterrestrial irradiation which is received outside the atmosphere is calculated from the day of the year. The global irradiation which is received by a horizontal surface is obtained from the meteorological charts.

The amount of diffusion irradiation depends on the clearness index which can be calculated from the ratio of the global irradiation recorded at the location and the extraterrestrial irradiation that the location receives at that particular day of the year. Diffusion irradiation is assumed to be isotropic: uniform at all angles.

The reflection irradiation of the location depends on the texture of the ground. The amount of reflection is defined by a factor called reflection coefficient having a value between zero and one. It is assumed that the ground refection is isotropic and maximum when the module is facing the ground and zero when it is facing the sky.

2.2 PHOTOVOLTAIC POWER CONVERSION

The PV photon cell charges with a high optical absorption offer a voltage of 1.1 electron volt (eV) to 1.75 electron volt (eV). Figure 2.3 depicts a solar cell structure.[1]

Figure 2.3 depicts a photovoltaic (PV) cell structure. A photovoltaic (PV) module connects a number of PV cells in series.[4] You may think of a PV cell as a number of capacitors that are charged by photon energy of light. Figure 2.4 depicts how the irradiance energy of the sun is converted to electric energy using PV cells. Solar cells are the building blocks of *photovoltaic modules*, otherwise known as solar panels. "The average solar panel output ranges from about 175 W to about 235 W, with an exceptionally powerful solar panel

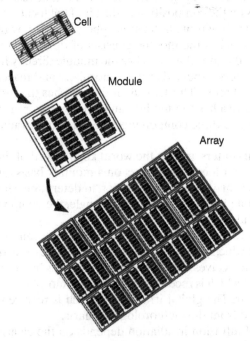

Figure 2.3 A photovoltaic cell structure.

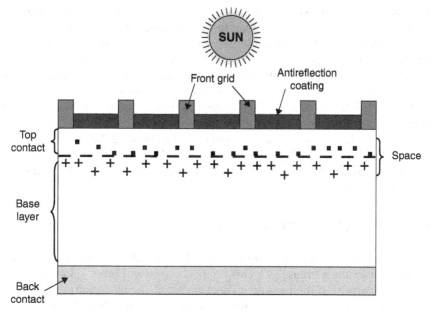

Figure 2.4   The structure of a photovoltaic cell.

measuring 315 W. Among the top ten manufacturers, the average wattage of a panel is about 200 W" (http://pureenergies.com/us/how-solar-works/solar-panel-output/. Accessed October 12, 2014).

2.3 PHOTOVOLTAIC MATERIALS

The manufacture of PV cells is based on two different types of material: (1) a semiconductor material that absorbs light and converts it into electron–hole pairs; and (2) a semiconductor material with junctions that separate photo-generated carriers into electrons and electron holes. The contacts on the front and back of the cells allow the current to the external circuit. Crystalline silicon cells (c-Si) are used for absorbing light energy in most semiconductors used in solar cells. Crystalline silicon cells are poor absorbers of light energy[1,7]; they have an efficiency in the range of 11–18% of that of solar cells. The most-efficient monocrystalline c-Si cell uses laser-grooved, buried grid contacts, which allow for maximum light absorption and current collection.[7] Each of the c-Si cells produces approximately 0.5 V. When 36 cells are connected in series, it creates an 18 V module. In the thin-film solar cell, the crystalline silicon wafer has a very high cost. Other common materials are amorphous silicon (a-Si), and cadmium telluride and gallium, which are another class of polycrystalline materials.[7] The thin-film solar cell technology uses a-Si and a p-i-n single-sequence layer, where "p" is for positive and "n" for negative, and "i" is the interface of a corresponding p- and n-type semiconductor.[8] Thin-film solar cells are constructed using lamination techniques, which promote their use under harsh weather conditions: they are environmentally robust modules. Due to the basic properties of c-Si devices, they may stay as the dominant PV technology for years to come. However, thin-film technologies are making rapid progress and a new material or process may replace the use of c-Si cells.[9]

Here we briefly introduce PV technology as it exists today. But as an evolving technology, students and engineers should recognize that these advances will come from basic research in material engineering and read the *IEEE Spectrum* to keep up with the developments in PV technology. We continue our discussion on how to develop models to study the integration of PV sources into the smart power grids.

2.4 PHOTOVOLTAIC CHARACTERISTICS

As sun irradiance energy is captured by a PV module, the open-circuit voltage of the module increases.[3–6] This point is shown in Figure 2.5 by V_{oc} with zero-input current. If the module is short circuited, the maximum short-circuit current can be measured. This point is shown in Figure 2.8 by I_{SC} with zero-output voltage. The point on the I versus V characteristic where maximum power (P_{MPP}) can be extracted lies at a current I_{MPP} and the corresponding voltage point, V_{MPP}. Typical data for a number of PV modules are given in Table 2.1. This information is used to design PV strings and PV generating power sources.

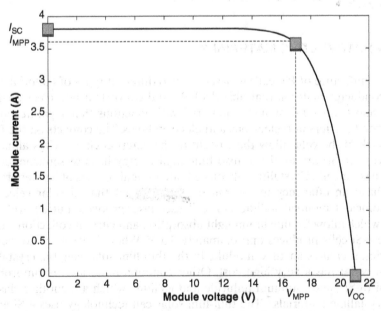

Figure 2.5   The operating characteristics of a photovoltaic module.

TABLE 2.1 Voltage and Current Characteristics of Typical Photovoltaic Modules

Module	Type 1	Type 2	Type 3	Type 4
Power (max), W	190	200	170	87
Voltage at maximum power point (MPP), V	54.8	26.3	28.7	17.4
Current at MPP, A	3.47	7.6	5.93	5.02
V_{OC} (open-circuit voltage), V	67.5	32.9	35.8	21.7
I_{SC} (short-circuit current), A	3.75	8.1	6.62	5.34
Efficiency, %	16.40	13.10	16.80	>16
Cost, $	870.00	695.00	550.00	397.00
Width, in.	34.6	38.6	38.3	25.7
Length, in.	51.9	58.5	63.8	39.6
Thickness, in.	1.8	1.4	1.56	2.3
Weight, lb	33.07	39	40.7	18.3

TABLE 2.2 Cell Temperature Characteristics of a Typical Photovoltaic Module

Typical Cell Temperature Coefficient		
Power	$T_k (P_p)$	–0.47% per °C
Open-circuit voltage	$T_k (V_{oc})$	–0.38% per °C
Short-circuit current	$T_k (I_{sc})$	0.1% per °C

PV module selection criteria are based on a number of factors[10]: (1) the performance warranty, (2) module replacement ease, and (3) compliance with natural electrical and building codes. A typical silicon module has a power of 300 W with 2.43 m² surface area; a typical thin film has a power of 69.3 W with an area of 0.72 m². Hence, the land required by a silicon module is almost 35% less. Typical electrical data apply to standard test considerations (STC). For example, under STC, the irradiance is defined for a module with a typical value such as 1000 W/m², spectrum air mass (AM) 1.5, and a cell temperature of 25°C.

Table 2.2 depicts cell temperature characteristics of a typical photovoltaic module.

Table 2.3 depicts the maximum operating characteristics of a typical photovoltaic module.

TABLE 2.3 Maximum Operating Characteristics of a Typical Photovoltaic Module

Limits	
Maximum system voltage	600 V DC
Operating module temperature	–40°C to 90°C
Equivalent wind resistance	Wind speed: 120 mph

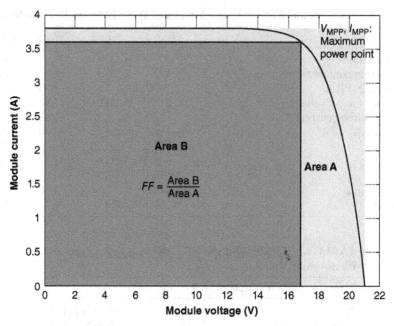

Figure 2.6 Photovoltaic module fill factor.

The PV fill factor (*FF*), as shown in Figure 2.6, is defined as a measure of how much solar energy is captured. This term is defined by PV module open-circuit voltage (V_{oc}), and PV module short-circuit current (I_{sc}).

$$FF = \frac{V_{MPP}I_{MPP}}{V_{OC}I_{SC}} \tag{2.1}$$

And

$$P_{max} = FF \cdot V_{OC}I_{SC} = V_{MPP}I_{MPP} \tag{2.2}$$

As seen in Figure 2.6, the maximum value for *FF* is unity. However, this value can never be attained. Some PV modules have a high fill factor. In the design of PV system a PV module with a high *FF* would be used. For high-quality PV modules, FFs can be over 0.85. For typical commercial PV modules, the value lies around 0.60. Figure 2.7 depicts a three-dimensional display of a typical PV module and the fixed irradiance energy received by the module. As shown, a typical PV module characteristic is not only a function of irradiance energy, but it is also a function of temperature.

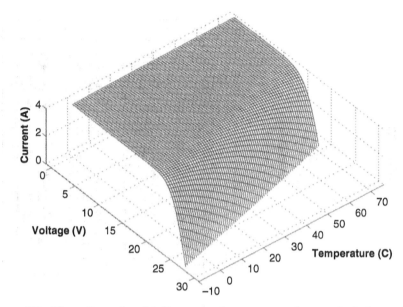

Figure 2.7 Three-dimensional *I–V* curve and temperature for a typical photovoltaic module.

2.5 PHOTOVOLTAIC EFFICIENCY

The PV module efficiency η is defined as

$$\eta = \frac{V_{\text{MPP}} I_{\text{MPP}}}{P_S} \tag{2.3}$$

where $V_{\text{MPP}} I_{\text{MPP}}$ is the maximum power output, P_{mpp} and P_S is the surface area of the module. The PV efficiency can be also defined as

$$\eta = FF \cdot \frac{V_{\text{OC}} I_{\text{SC}}}{\int\limits_0^\infty P(\lambda) \cdot d\lambda} \tag{2.4}$$

where $P(\lambda)$ is the solar power density at wavelength λ.

2.6 PV GENERATING STATION

Figure 2.8 depicts a PV module consisting of 36 PV cells. If each cell is rated at 1.5 V, the module rated voltage is 54 V.

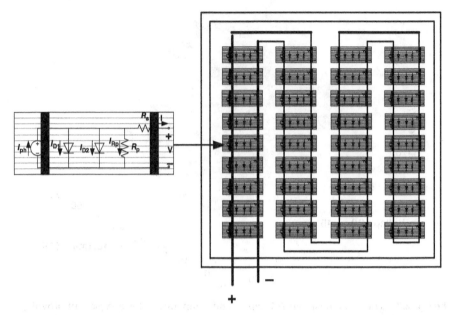

Figure 2.8   A photovoltaic module consisting of 36 photovoltaic cells.

A string is designed by connecting a number of PV modules in series. A number of strings connected in parallel make an array.[2] Two general designs of PV systems can be envisioned. Figure 2.9 presents the basic configuration of PV strings and arrays. Figure 2.10 depicts a PV design based on a central inverter. Figure 2.11 depicts the utilization of multiple inverters.

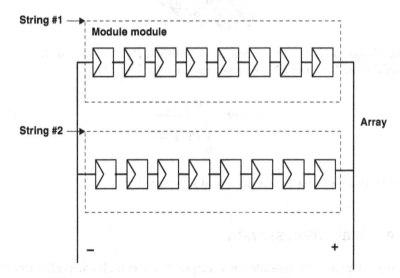

Figure 2.9   Basic configuration showing modules, strings, and an array.

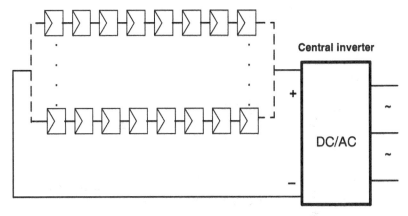

Figure 2.10   Central inverter for a large-scale photovoltaic power configuration.

Basically, to provide a higher DC operating voltage, modules are connected in series. To provide a higher operating current, the modules are connected in parallel.

$$V \text{ (series connected)} = \sum_{j=1}^{n} V_j \, ; n: \text{ number of series connected panels} \quad (2.5)$$

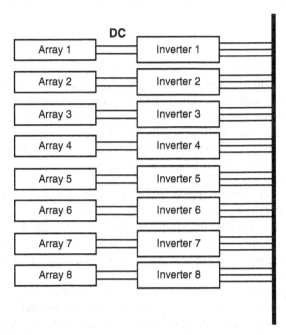

Figure 2.11   General structure of photovoltaic arrays with inverters.

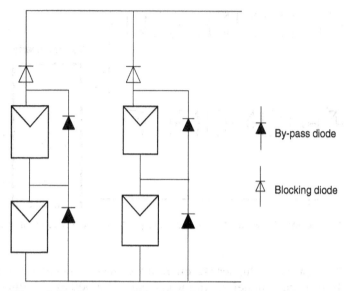

▲ By-pass diode

⍚ Blocking diode

Figure 2.12   Bypass and blocking diodes in a photovoltaic array.

For parallel connected panels,

$$I \text{ (parallel connected)} = \sum_{j=1}^{m} I_j \text{ ; } m: \text{ number of parallel connected panels}$$

(2.6)

In a PV system consisting of a number of arrays, all arrays must have equal exposure to sunlight: the design should place the modules of a PV system such that some of them will not be shaded. Otherwise, unequal voltages will result in some strings with unequal circulating current and internal heating producing power loss and lower efficiency. Bypass diodes are usually used between modules to avoid damage. Most new modules have bypass diodes in them as shown in Figure 2.12 to ensure longer life. However, it is very difficult to replace built-in diodes if a diode fails in a panel.

The photovoltaic (PV) industry, the International Society for Testing and Materials (ASTM),[9] and the U.S. Department of Energy have established a standard for terrestrial solar spectral irradiance distribution. The irradiance of a location is measured by an instrument called a pyranometer.[4] Figure 2.13 depicts irradiance in W/m^2 per nanometer (nm) as a function of wavelength in nanometers (nm).

The solar spectrum is the plot of the irradiance from the sun received on a particular location at a given temperature and air mass flow. The intensity (W/m^2/nm) of the radiation is plotted as a function of wavelength (nm). The photovoltaic modules convert the radiation energy to electrical energy. The

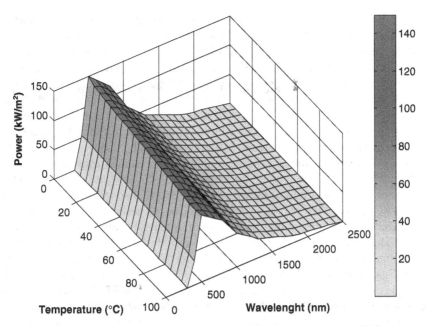

Figure 2.13 Spectra for photovoltaic performance evaluation.[11,12]

amount of energy produced by the PV module is directly proportional to the area of the module. The kW/m²/nm of PV module is plotted as a function of wavelength and estimated temperature from the solar spectrum. The above curve is plotted with an air mass of 1.5.

The PV modules are tested under a nominal temperature (*NT*). The *NT* is used to estimate the cell temperature based on the ambient temperature as shown below.

$$T_c = T_a + \left(\frac{NT - 20}{Kc}\right) \cdot S \qquad (2.7)$$

The cell temperature, T_c and the ambient temperature, T_a are in degrees centigrade. The operating nominal temperature, *NT* designates the cell temperature as tested by manufacturers; *S* is the solar insolation (kW/m²). *Kc* is a constant empirically computed from test data; it is in the range 0.7–0.8.

2.7 DESIGN OF PHOTOVOLTAIC GRIDS

The design of power grids is based on trial and error. Nevertheless, the design must demonstrate a clear understanding of the scientific and grids principles behind the proposed design. In this sense, "trial and error" is for fine-tuning the final design. However, all power grids must be based on the underlying physical process.

PV power plants use PV modules that have specific manufactured characteristics. However, in defining the overall objective of the PV plants some PV modules may satisfy the objectives of PV plants and some may not. Therefore, we need to test available PV modules against the PV plant design specifications.

For a PV power plant, the first specification is the power requirement in kW or MW that the plant is to produce. If the PV plant is to operate as an independent power generating station, the rated load voltage is specified including the storage systems. For example, for an independent residential PV plant, the PV electricity charges the storage system using DC–DC converters. Then the loads are supplied from the storage systems at nominal voltages 120 V and 207.8 V using a DC–AC inverter. If the residential PV plant is connected to the local utility, then the interconnection voltage must be specified for the design of utility connected PV plants. Consider the case of designing a residential PV plant. The PV modules are connected to generate a DC voltage source and DC power. The highest safe DC voltage to use in a design residential PV plant is determined by the electric codes of a particular locality. In principle, for safety, the design residential PV plants use a lower DC voltage. The higher PV DC voltage is used for commercial and industrial sites. Another design consideration for residential PV plants may be the weight and surface area restrictions. Finally, it is understood that the PV designer always seeks to design a PV plant to satisfy the constraints of sites and at the lowest installed and operating costs.

Table 2.4 defines nomenclature for design of PV plants.

The terms PV module and PV panel are used interchangeably. One PV module has a limited power rating; therefore, to design a higher power rating, we construct a string by connecting a number of PV modules in series.

$$SV = NM \times V_{OC} \tag{2.8}$$

where SV defines, the string voltage and V_{oc} is the open-circuit voltage of a module.

As an example, if the number of modules is five and the open-circuit voltage of the module is 50 V, we will have

$$SV = 5 \times 50 = 250 \, V \, DC$$

This may be a high voltage for a residential PV plant. A PV cell, a module, or an array acts as a charged capacitor. The amount of charge of a PV plant is a function of sun irradiance. At full sun, the highest amount of charge is stored that will generate the highest open-circuit voltage for a PV plant. In general, the open-circuit voltage for a residential PV plant might be set at a voltage lower than 250 V DC. The rated open-circuit voltage is governed by local electric codes. In general, the open-circuit string voltage is less than 600 V DC for a commercial and industrial PV plant at this time. However, higher DC voltage designs of PV systems are considered for higher power PV sites.

TABLE 2.4 Photovoltaic Design Nomenclature

Terms	Abbreviations	Descriptions
String voltage	SV	String voltage for series-connected modules
Power of a module	PM	Power produced by a module
String power	SP	Power that can be generated in one string
Number of strings	NS	Number of strings per array
Number of arrays	NA	Number of arrays in a design
Surface area of a module	SM	Surface area of a module
Total surface area	TS	Total surface area
Array power	AP	Array power is generated by connecting a number of strings in parallel
Number of modules	NM	Number of modules per string
Total number of modules	TNM	Total number of modules in all arrays put together
Array voltage for maximum power point tracking	V_{AMPP}	The operating voltage for maximum power point tracking of an array
Array current for maximum power point tracking	I_{AMPP}	The operating current for maximum power point tracking of an array
Array maximum power point	P_{AMPP}	The maximum operating power of an array
Number of converters	NC	Total number of DC–DC converters
Number of rectifiers	NR	Total number of AC–DC rectifiers
Number of inverters	NI	Total number of inverters

The string power, SP, is the power that can be produced by one string.

$$SP = NM \times PM \qquad (2.9)$$

where NM is the number of modules and PM is the power produced by a module. For example, if a design uses four PV modules, each rated 50 W, then the total power produced by the string is given as

$$SP = 4 \times 50 = 200\,W$$

For producing higher rated power from a PV generating station, the string voltage is increased. In addition, a number of strings are connected in parallel and create an array. Therefore, the array power, AP, is equal the number of string times the string power.

$$AP = NS \times SP \qquad (2.10)$$

If the number of strings is 10 and each string is producing 200 W, the array power is

$$AP = 10 \times 200 = 2000\,\text{W}$$

The DC power produced by an array is a function of the sun's position and irradiance energy received by the array. An array can be located on roofs or free-standing structures. The array power is processed by a converter to extract maximum power from the sun's irradiance energy. Because the transfer of DC power over a cable at low voltages results in high power losses, the array power is either converted to AC power by using a DC–AC inverter or in some applications, the array power is converted to a higher DC voltage power to produce higher AC voltage and to process higher rated power.

To obtain the maximum power out of an array, the maximum power point (MPP) tracking method is used. The MPP tracking method locates the point on the trajectory of power produced by an array where the array voltage and array current are at its maximum point and the maximum power output for the array.

The array maximum power point is defined as

$$P_{\text{MPP}} = V_{\text{AMPP}} \times I_{\text{AMPP}} \qquad (2.11)$$

where V_{AMPP} is the array voltage at MPP tracking and I_{AMPP} is the array current at its maximum power point tracking.

An array is connected to either an inverter or a boost converter and the control system operates the array at its MPP tracking.

The MPP operation of a PV array generating station is controlled by inverter. The DC–AC inverter control system is designed to locate the maximum operating point based on generated array voltage and array current, thus accommodating the changing irradiance energy received by an array as the sun's position changes during the day. The inverter output voltage is controlled by controlling the inverter amplitude modulation index. To process the maximum power by an inverter, the amplitude modulation index, M_a should be set at maximum value without producing the unwanted harmonic distortion. The value of M_a is set less than 1 and in the range of 0.95 to produce the highest AC output voltage.

2.7.1 Three-Phase Power

A PV generating station must be connected to local three-phase power for synchronous operation with power grid. The three-phase power grids consist of three single-phase grids. AC grid generators are designed to produce a three-phase alternating *current*. The three sinusoidal distributed windings or coils are designed to carry the same current. Figure 2.11 depicts the voltage of each phase of three-phase generator with respect to generator neutral.

Figure 2.14 represents the sinusoidal voltage (or current) as a function of time. In Figure 2.14, 0–360° (2π radians) is shown along the time axis.[13] Power

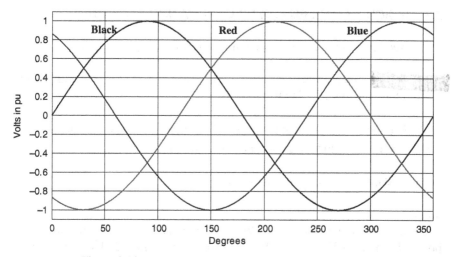

Figure 2.14 Three-phase generator voltage wave forms.

grids around the world each operate at a fixed frequency of 50 or 60 cycles per second. Based on a universal color-code convention,[7] black is used for one phase of the three-phase system; it denotes the ground as the reference phase with zero-degree angle. Red is used for the second phase, which is 120° out of phase with respect to the black phase. Blue is used for the third phase, which is also 120° out of phase with the black phase. For example for a three-phase system with line-to-line voltage of 460 V, the following expressions describe the three-phase generator voltage of Figure 2.14.

$$V_{ac} = \frac{460\sqrt{2}}{\sqrt{3}} \cdot \sin(2\pi 60 \cdot t) \qquad (2.12)$$

$$V_{ab} = \frac{460\sqrt{2}}{\sqrt{3}} \cdot \sin(2\pi 60 \cdot t + 90)$$

$$V_{bc} = \frac{460\sqrt{2}}{\sqrt{3}} \cdot \sin(2\pi 60 \cdot t - 90)$$

The three-phase AC system can be considered as three single-phase circuits. The first AC generators were single phase. However, it was recognized that the three-phase generators can produce three times as much power. However, the higher phase generators will not produce proportionally more power.[9]

The term V_{LL} is the line-to-line voltage and it is equal to $\sqrt{3} \cdot V_{L-N}$ where V_{L-N} is the line-to-neutral voltage. The detailed boost converter operation and three-phase power generators are presented in later chapters.

2.8 DESIGN EXAMPLES FOR PV GENERATING STATIONS

In this section, a number of designs of PV stations are presented.

Example 2.1   Design a PV system to process 10 kW of power at 230 V, 60 Hz single-phase AC. Determine the following:

(i) Number of modules in a string and number of strings in an array
(ii) Inverter specification and one-line diagram

The PV module data are given below.

Solution
The load voltage is specified as 230 V single-phase AC. To acquire maximum power from the PV array a DC/AC inverter is required. The operation of DC/AC inverters will be discussed in detail in a later chapter. For now, we need to know the fundamental relationships for DC power conversion of a single-phase inverter from DC voltage of PV station to inverter AC power. For single-phase inverter, the DC voltage of inverter is related to its AC voltage and is given by Equation (2.13).

$$V_{idc} = \frac{\sqrt{2}V_{ac}}{M_a} \tag{2.13}$$

The term V_{ac} is the root-mean square (RMS) of AC voltage as measured by a voltmeter. The M_a is amplitude modulation index of the inverter. The magnitude of the fundamental of AC output voltage is directly proportional to the ratio of the peaks of V_c (the desired control voltage) and V_T (the triangular voltage):

$$M_a = \frac{V_{C(max)}}{V_{T(max)}} \tag{2.14}$$

This ratio is defined as the amplitude modulation index, M_a and corresponding frequency modulation is given as

$$M_f = \frac{f_S}{f_e}$$

The term f_S is sampling frequency of desired AC voltage and the term f_e is the frequency of the output AC voltage of inverter. In United States the frequency is set at 60 hertz (cycles per second or cps)

The conversion of DC voltage (current and power) to AC power is based on pulse width modulation (PWM) method.

Let us select a modulation index of $M_a = 0.9$.

$$V_{idc} = \frac{\sqrt{2} \times 230}{0.9} = 361.4\,\text{V}$$

The inverter is designed to operate at the maximum power point tracking of PV array. Therefore, the number of modules to be connected in series in a string is given by

$$NM = \frac{V_{idc}}{V_{MPP}} \qquad (2.15)$$

where V_{MPP} is the voltage at maximum power point of PV of the module.

$$NM = \frac{361.4}{50.6} \approx 7$$

The string voltage is given as

$$SV = NM \times V_{MPP} \qquad (2.16)$$

Using this module, string voltage (see Table 2.5) for this design is

$$SV = 7 \times 50.6 = 354.2\,\text{V}$$

(i) The power generated by one string is given by

$$SV = NM \times V_{MPP}$$

where P_{MPP} is the nominal power generated at the maximum power point tracking.

The power generated by a string for this design is given as

$$\text{kW per string} = 7 \times 300 = 2100\,\text{W}$$

TABLE 2.5 The Voltage and Current Characteristics of a Typical Photovoltaic Module

Power (max) Maximum voltage, P_{MPP}	300 W
Voltage at maximum power point (MPP), V_{MPP}	50.6 V
Current at MPP, I_{MPP}	5.9 A
V_{oc} (open-circuit voltage)	63.2 V
I_{sc} (short-circuit current)	6.5 A

TABLE 2.6 Photovoltaic Specifications for 10 kW Generation

Modules per String	Strings per Array	Number of Arrays	String Voltage (V)
7	5	1	354.2

Table 2.6 depicts the photovoltaic specifications for 10 kW generation. To calculate the number of strings for 10 kW PV system, we divide the PV power rating by power per string

$$NS = \frac{AP}{SP} \tag{2.17}$$

where NS is the number of string and AP is the array power and SP is string power. For this design we have

$$NS = \frac{10 \times 10^3}{2100} = 5$$

Therefore, we have five strings and one array to generate 10 kW of power.

(ii) In the final design, the inverter should be rated such that it is able to process generation of 10 kW and supply the load at 230 V AC from its array at its maximum power point tracking. Based on the PV module of Table 2.5, the string voltage is specified as

$$V_{idc} = 354.2 \text{ V}$$

and the modulation index is given as follows:

$$M_a = \frac{\sqrt{2} V_{ac}}{V_{idc}}$$

$$M_a = \frac{\sqrt{2} \times 230}{354.2} = 0.92$$

Table 2.7 specifies data for a typical inverter. Let us select a sampling switching frequency of 6 kHz. Therefore, the frequency modulation index is given by

$$M_f = \frac{f_S}{f_e} = \frac{6000}{60} = 100$$

TABLE 2.7 Inverter Specifications

Input Voltage, V_{idc} (V)	Power Rating (kW)	Output Voltage, V_{AC} (V)	Amplitude Modulation Index, M_a	Frequency Modulation Index, M_f
354.2	10	230	0.92	100

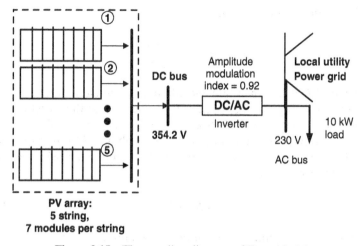

Figure 2.15 The one-line diagram of Example 2.1.

The one-line diagram is given in Figure 2.15.

The PV plant of Figure 2.15 cannot operate as a standalone PV plant. However, PV plant can operate in parallel with local utility grid in synchronous operation with power grid. For standalone operation, a storage system must be connected to DC bus of Figure 2.13. The frequency of operation is set by inverter, usually 60 Hz and the system acts as an uninterruptible power supply (UPS).

Example 2.2   Design a PV system to process 500 kW of power at 460 V, 60 Hz, three-phase AC, and using PV data of Example 2.1. Determine the following:

 (i) Number of modules in a string and number of strings in an array
 (ii) Inverter and boost specification
(iii) The output voltage as a function and total harmonic distortion
(iv) The one-line diagram of this system

Solution
Before we present the solution of this design, we need to give the basic operation of boost converter. The boost converter converts a low voltage DC power to a higher DC power and voltage at reduced current. For ideal boost converters, the input power and output power remain the same. The basic relationships of the input DC voltage, V_{in}, to output DC voltage, V_o, is given as

$$V_o = \frac{V_{in}}{1 - D} \qquad (2.17.1)$$

The duty ratio, D, is defined as

$$D = \frac{T_{on}}{T_S} = \frac{T_{on}}{T_{on} + T_{off}} \tag{2.17.2}$$

The value of D is less than one and where f_s is switching frequency and T_S is defined as $T_S = 1/f_S$ and $T_S = T_{on} + T_{off}$ is defined as the switching period. Therefore, the output voltage is always more than the input voltage. The relationship of the input current to output current is obtained from power balance. Because the input and output power must be balanced for a lossless system, the output current is given as

$$V_{in}.I_{in} = V_o.I_o \tag{2.17.3}$$

Continuing with solution presentation of Example 2.2, we recognize that the load is 500 kW rated at 460 V AC. Based on the voltage of the load and an amplitude modulation index of 0.9, the input DC voltage for a three-phase inverter:

$$V_{idc} = \frac{2\sqrt{2}V_{LL}}{\sqrt{3}M_a} \tag{2.17.4}$$

$$= \frac{2\sqrt{2} \times 460}{\sqrt{3} \times 0.9} = 835 \text{ V}$$

In Equation (2.17.4), V_{LL} is line–line RMS voltage value of load.

Let us limit the string voltage to 600 V DC. Therefore, a boost converter to boost the string voltage to 835 V is needed.

Let us select string approximate voltage of 550 V, and then the number of modules is

(i) The number of modules in a string is given by

$$\frac{V_{string}}{V_{MPP}} = \frac{550}{50.6} \approx 11$$

where V_{MPP} is the voltage of a module at maximum power point tracking.

The string power, SP can be computed as

$$SP = NM \times P_{MPP}$$

Using a module rated at 300 W, the string power is

$$SP = 11 \times 300 = 3300 \text{ W}$$

The string voltage is given as

$$SP = NM \times V_{MPP}$$

Therefore, the string voltage, SV, for this design is

$$SV = 11 \times 50.6 = 556.6\,V$$

If each array to generate a power of 20 kW, then the number of strings, NS, in an array is given by

$$NS = \frac{\text{Power of one array}}{\text{Power of one string}}$$
$$NS = \frac{20}{3.3} = 6$$

The number of array, NA, for total power generation is

$$NA = \frac{\text{PV generation}}{\text{Power of one array}} \tag{2.18}$$

Therefore, $NA = \frac{500 \cdot kW}{20 \cdot kW} = 25$.

Table 2.8 gives the data for design of the photovoltaic generating station.

The inverters should be rated to withstand the output voltage of boost converter and should be able to supply the required power. The inverter is rated at 100 kW with input voltage of 835 V DC and the amplitude modulation index of 0.9. The output voltage of inverter is 460 V AC.

The number of inverters, NI, needed to process a generation of 500 kW is given by

$$NI = \frac{\text{PV generation}}{\text{Power of one inverter}} \tag{2.19}$$

Therefore, $NI = \frac{500}{100} = 5$

Hence, we need to connect five inverters in parallel to supply the load of 500 kW. If a sampling switching frequency is set at 5.04 kHz.

Therefore, the frequency modulation index, M_f, is given by

$$M_f = \frac{f_s}{f_e} = \frac{5040}{60} = 84.$$

Table 2.9 gives the resulting data for the inverter specification.

TABLE 2.8   Photovoltaic Generating Station

Modules per String	Strings per Array	Number of Arrays	String Voltage(V)
11	6	25	556.6

TABLE 2.9   Inverter Specifications

Number of Inverters	Input Voltage, V_{idc} (V)	Power Rating (kW)	Output Voltage, V_{AC} (V)	Amplitude Modulation Index, M_a	Frequency Modulation Index, M_f
5	835	100	460	0.90	84

The number of boost converters needed is the same as the number of arrays, which is 25, and the power rating of each boost converter is 20 kW.

The boost converter input voltage is equal to the string voltage:

$$V_i = 556.6 \, \text{V}$$

The output voltage of the boost converter is equal to the inverter input voltage:

$$V_{idc} = V_o = 835\text{V}$$

Recall the basics operation of DC converter.

$$V_o = \frac{V_{in}}{1 - D}$$

The duty ratio of the boost converter is given by

$$D = 1 - \frac{V_i}{V_o}$$

$$D = 1 - \frac{556.6}{835} = 0.33$$

Table 2.10 depicts the boost converter design specification.

(ii) With a frequency modulation of 84, the harmonic content of the output voltage was computed from a simulation testbed using a fast Fourier method is given in Table 2.11.

TABLE 2.10   Boost Converter Specifications for Generation of 500 kW

Number of Boost Converters	Input Voltage, V_i (V)	Power Rating (kW)	Output Voltage, V_o (V)	Duty Ratio, D
25	556.6	20	835	0.33

TABLE 2.11   Harmonic Content of Line-to-Neutral Voltage Relative to the Fundamental

3rd Harmonic	5th Harmonic	7th Harmonic	9th Harmonic
0.01%	0.02%	0	0.03%

The output line-to-neutral voltage as a function of time series all frequencies present in PWM presentation of output voltage and currents. The voltage representation is given below.

$$
\begin{aligned}
V_{ac} &= \frac{460\sqrt{2}}{\sqrt{3}} \cdot \sin(2\pi 60 \cdot t) + \frac{0.01}{100} \times \frac{460\sqrt{2}}{\sqrt{3}} \cdot \sin(2\pi \times 3 \times 60.t) \\
&\quad + \frac{0.02}{100} \times \frac{460\sqrt{2}}{\sqrt{3}} \cdot \sin(2\pi \times 5 \times 60 \cdot t) \\
&\quad + \frac{0}{100} \times \frac{460\sqrt{2}}{\sqrt{3}} \cdot \sin(2\pi \times 7 \times 60 \cdot t) \\
&\quad + \frac{0.03}{100} \times \frac{460\sqrt{2}}{\sqrt{3}} \cdot \sin(2\pi \times 9 \times 60 \cdot t) \\
&= 376 \sin(2\pi 60 \cdot t) + 0.037 \sin(6\pi 60 \cdot t) + 0.075 \sin(10\pi 60 \cdot t) \\
&\quad + 0.113 \sin(18\pi 60 \cdot t)
\end{aligned}
$$

The total harmonic distortion is given by

$$
THD = \sqrt{\sum (\%\text{harmonic})^2} = \sqrt{0.01^2 + 0.02^2 + 0^2 + 0.03^2} = 0.04\%
$$

(iii) The one-line diagram is given in Figure 2.16.

The PV plant of Figure 2.16 cannot operate as a standalone PV plant. However, PV plant can operate in parallel with local utility grid in synchronous operation. For standalone operation, a storage system must be connected to DC bus of Figure 2.16. The frequency of operation is set by inverter, usually 60 or 50 Hz and the system acts as a UPS supply.

Example 2.3   Design a PV plant to process 1000 kW of power at 460 V, 60 Hz three-phase AC using the PV data given in Table 2.12. Determine the following:

(i) Number of modules in a string, number of strings in an array, number of arrays, surface area for PV, weight of PV, and cost

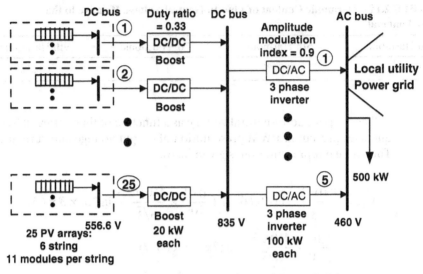

Figure 2.16   The one-line diagram of Example 2.2.

(ii) DC–AC inverter and boost converter specifications and the one-line diagram of the system

Solution
The load is 1000 kW rated at 460 V AC. Based on the voltage of the load and an amplitude modulation index of 0.85, the input DC voltage for a three-phase inverter is

$$V_{idc} = \frac{2\sqrt{2}V_{LL}}{\sqrt{3}M_a} = \frac{2\sqrt{2} \times 460}{\sqrt{3} \times 0.85} = 884\,V$$

TABLE 2.12   Photovoltaic Data for Example 2.3

Module	Type 1
Power (Max), W	190
Voltage at MPP, V	54.8
Current at MPP, A	3.47
V_{OC} (open-circuit voltage), V	67.5
I_{SC} (short-circuit current), A	3.75
Efficiency, %	16.40
Cost, $	870.00
Width, in.	34.6
Length, in.	51.9
Thickness, in.	1.8
Weight, lb	33.07

Limiting the maximum voltage that a string to have to 600 V, a boost converter is needed to boost the string voltage to 884 V.

(i) Let string approximate voltage of 550 V, then, the number of modules in a string, NM, is given as

$$NM = \frac{V_{string}}{V_{MPP}} = \frac{550}{54.8} \approx 10$$

where V_{MPP} is the voltage at MPP of the PV module.

$$SP = NM \times P_{MPP}$$

where P_{MPP} is the power generated by a PV module at MPP.

$$SP = 10 \times 190 = 1900\,W$$

And the string voltage, SV, is

$$SV = NM \times V_{MPP}$$

Therefore, the string voltage, SV, for this design is

$$SV = 10 \times 54.8 = 548\,V$$

If each array is to have a rating of 20 kW, the number of strings, NS, in an array is

$$NS = \frac{AP}{SP}$$
$$NS = \frac{20}{1.9} = 11$$

The number of arrays, NA, for this design is

$$NA = \frac{PV\ generation}{Power\ of\ one\ array}$$
$$NA = \frac{1000}{20} = 50$$

The total number of PV modules, TNM, is given by the product of the number of modules per string, the number of strings per array, and the number of arrays:

$$TNM = NM \times NS \times NA \tag{2.20}$$

$$TNM = 10 \times 11 \times 50 = 5500$$

The surface area of one module, SM, is given by the product of its length and width.

$$SM = \frac{34.6 \times 51.9}{144} = 12.5 \, \text{ft}^2$$

The total surface area, TS, is therefore given by the total number of modules and the surface area of each module.

$$TS = 5500 \times 12.5 = 68,750 \, \text{ft}^2 = \frac{68,750}{43,560} = 1.57 \, \text{acre}$$

The total cost of PV modules is given by the product of the number of PV modules and the cost of one module.

$$\text{The total cost} = 5500 \times 870 = \$4.78 \, \text{million}$$

The total weight of PV modules is given by the product of the number of PV modules and the weight of one module. The total weight $= 5500 \times 33.07 = 181,885 \, \text{lb}$.

The design data for the 1000 kW generation station are given in Table 2.13.

(ii) The inverters should be rated to withstand the output voltage of the boost converter and should be able to supply the required power. Selecting an inverter rated at 250 kW, we will have the following number of inverters:

The number of inverters, NI, needed to process the generation of 1000 kW is given by

$$NI = \frac{\text{PV generation}}{\text{Power of one inverter}}$$

$$NI = \frac{1000}{250} = 4$$

Hence, we need to connect four inverters in parallel to supply the load of 1000 kW.

TABLE 2.13   Photovoltaic Specifications for 1000 kW Generation

Modules per String	Strings per Array	Number of Arrays	String Voltage (V)	Total Area (ft^2)	Total Weight (lb)	Total Cost (million $)
10	11	50	548	68,750	181,885	4.78

TABLE 2.14   Inverter Specifications

Number of Inverters	Input Voltage, V_{idc} (V)	Power Rating (kW)	Output Voltage, V_{AC} (V)	Amplitude Modulation Index, M_a	Frequency Modulation Index, M_f
4	884	250	460	0.85	90

Selecting a switching frequency of 5.40 kHz, the frequency modulation index is given by

$$M_f = \frac{f_S}{f_e} = \frac{5400}{60} = 90.$$

The design data for the inverter specification are given in Table 2.14.

The number of boost converters needed is the same as the number of arrays, which are 50. Selecting a boost converter rating of 20 kW and the boost converter input voltage to be equal to the string voltage:

$$V_i = 548 \text{ V}$$

The output voltage of the boost converter is equal to the inverter input voltage:

$$V_{idc} = V_o = 884 \text{ V}$$

The duty ratio of the boost converter is given by

$$D = 1 - \frac{V_i}{V_o}$$
$$D = 1 - \frac{548}{884} = 0.38$$

The design data for the boost converter specification are given in Table 2.15. The one-line diagram is given in Figure 2.17.

The PV plant of Figure 2.17 cannot operate as a standalone PV plant. However, PV plant can operate in parallel with local utility grid in synchronous operation. For standalone, a storage system must be connected to DC bus.

TABLE 2.15   Boost Converter Specifications

Number of Boost Converters	Input Voltage, V_i (V)	Power Rating (kW)	Output Voltage, V_o (V)	Duty Ratio, D
50	548	20	884	0.38

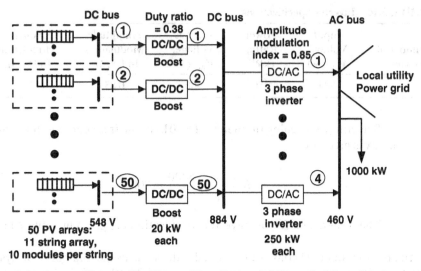

Figure 2.17 The one-line diagram of Example 2.3.

Example 2.4 Design a microgrid of PV system rated at 50 kW of power at 220 V, 60 Hz single-phase AC using a boost converter and single-phase DC–AC inverter. Use the data given in Table 2.16, Table 2.17, and Table 2.18 in design of microgrid of Example 2.4.

Determine the following:

(i) Number of modules in a string for each PV type, number of strings in an array for each PV type, number of arrays and surface area, weight, and cost for each PV plant

(ii) Boost converter and inverter specifications and the one-line diagram of this system

TABLE 2.16 Typical Photovoltaic Modules

Module	Type 1	Type 2	Type 3	Type 4
Power (Max), W	190	200	170	87
Voltage at max. power point (MPP), V	54.8	26.3	28.7	17.4
Current at MPP, A	3.47	7.6	5.93	5.02
V_{OC} (open-circuit voltage), V	67.5	32.9	35.8	21.7
I_{SC} (short-circuit current), A	3.75	8.1	6.62	5.34
Efficiency, %	16.40	13.10	16.80	>16
Cost, $	870.00	695.00	550.00	397.00
Width, in.	34.6	38.6	38.3	25.7
Length, in.	51.9	58.5	63.8	39.6
Thickness, in.	1.8	1.4	1.56	2.3
Weight, lb	33.07	39	40.7	18.4

TABLE 2.17 Single-Phase Inverter Data

Inverter	Type 1	Type 2	Type 3	Type 4
Power	500 W	5 kW	15 kW	4.7 kW
Input voltage DC	500 V	500 V max	500 V	500 V
Output voltage AC	230 V AC/60 Hz at 2.17 A	230 V AC/60 Hz at 27 A	220 V AC/60 Hz at 68 A	230 V AC/60 Hz at 17.4 A
Efficiency	Min 78% at full load	97.60%	>94%	96%
Length	15.5 in.	315 mm	625 mm	550 mm
Width	5 in.	540 mm	340 mm	300 mm
Height	5.3 in.	191 mm	720 mm	130 mm
Weight	9 lb	23 lb	170 kg	21 kg

Solution

The load is 50 kW rated at 220 V AC. Based on the voltage of the load and an amplitude modulation index of 0.9, the input DC voltage for the inverter is

$$V_{idc} = \frac{\sqrt{2}V_{ac}}{M_a} = \frac{\sqrt{2} \times 220}{0.9} = 345 \text{ V}$$

(i) If we select string voltage, *SV* of 250 V, the number of modules is

$$NM = \frac{\text{String voltage}}{V_{MPP}}$$

where V_{MPP} is the voltage at MPP of the PV module.

$$NM = \frac{250}{54.8} \approx 5 \text{ type 1}$$

TABLE 2.18 Typical Boost Converters

Input Voltage (V)	Output Voltage (V)	Power (kW)
24–46	26–48	9.2
24–61	26–63	12.2
24–78	26–80	11.23
24–78	26–80	13.1
24–98	26–100	12.5
80–158	82–160	15.2
80–198	82–200	14.2
80–298	82–300	9.5
200–600	700–1000	20.0

$$= \frac{250}{26.3} \approx 10 \text{ for type 2}$$

$$= \frac{250}{28.7} \approx 9 \text{ for type 3}$$

$$= \frac{250}{17.4} \approx 15 \text{ for type 4}$$

The string voltage SV is given as

$$SV = NM \times V_{\text{MPP}}$$

Therefore, the string voltage SV for this design is

$$\begin{aligned} SV &= 5 \times 54.8 = 274 \text{ V for type 1} \\ &= 10 \times 26.3 = 263 \text{ V for type 2} \\ &= 9 \times 28.7 = 258.3 \text{ V for type 3} \\ &= 15 \times 17.4 = 261 \text{ V for type 4} \end{aligned}$$

Selecting the 15.2 kW boost converter from Table 2.18, the number of boost converters, NC, is

$$NC = \frac{\text{PV generation}}{\text{Boost converter power rating}}$$
$$NC = \frac{50}{15.2} \approx 4$$

Therefore, the design should have four arrays: each with its boost converter. The array power, AP is

$$AP = \frac{\text{PV generation}}{\text{Number of arrays}}$$
$$AP = 50/4 = 12.5 \text{ kW}$$

String power, SP, is given as

$$SP = NM \times P_{\text{MPP}}$$

where P_{MPP} is the power generated by the PV module at MPP.

$$\begin{aligned} SP &= 5 \times 190 = 0.95 \text{ kW for type 1} \\ &= 10 \times 200 = 2.0 \text{ kW for type 2} \end{aligned}$$

$$= 9 \times 170 = 1.53 \, \text{kW for type 3}$$

$$= 15 \times 87 = 1.305 \, \text{kW for type 4}$$

The number of strings, *NS*, is given by

$$NS = \frac{\text{Power per array}}{\text{Power per string}}$$

$$NS = \frac{12.5}{0.95} = 14 \, \text{for type 1}$$

$$= \frac{12.5}{2} = 7 \, \text{for type 2}$$

$$= \frac{12.5}{1.53} = 9 \, \text{for type 3}$$

$$= \frac{12.5}{1.305} = 10 \, \text{for type 4}$$

The total number of modules, *TNM*, is given by

$$TNM = NM \times NS \times NA$$

$$TNM = 5 \times 14 \times 4 = 280 \, \text{for type 1}$$

$$= 10 \times 7 \times 4 = 280 \, \text{for type 2}$$

$$= 9 \times 9 \times 4 = 324 \, \text{for type 3}$$

$$= 15 \times 10 \times 4 = 600 \, \text{for type 4}$$

The surface area *TS* needed by each PV type is given by the product of the total number of modules, and the length and the width of one PV module:

$$TS = \frac{280 \times 34.6 \times 51.9}{144} = 3492 \, \text{ft}^2 \, \text{for type 1}$$

$$= \frac{280 \times 38.6 \times 58.5}{144} = 4391 \, \text{ft}^2 \, \text{for type 2}$$

$$= \frac{324 \times 38.3 \times 63.8}{144} = 5498 \, \text{ft}^2 \, \text{for type 3}$$

$$= \frac{600 \times 25.7 \times 39.6}{144} = 4241 \, \text{ft}^2 \, \text{for type 4}$$

The total weight needed for each PV type is the product of the number of modules and the weight of one module:

$$\text{The total weight} = 280 \times 33.07 = 9260 \text{ lb for type 1}$$

$$= 280 \times 39.00 = 10,920 \text{ lb for type 2}$$

$$= 324 \times 40.70 = 13,187 \text{ lb for type 3}$$

$$= 600 \times 18.40 = 11,040 \text{ lb for type 4}$$

The total cost for each PV type is the product of the number of modules and the cost of one module:

$$\text{The total cost} = 280 \times 870 = \$243,600 \text{ for type 1}$$

$$= 280 \times 695 = \$194,600 \text{ for type 2}$$

$$= 324 \times 550 = \$178,200 \text{ for type 3}$$

$$= 600 \times 397 = \$238,200 \text{ for type 4}$$

The design data for each photovoltaic type are given in Table 2.19.
(ii) The boost converter rating is

$$\text{The boost converter power rating} = \frac{\text{PV generation}}{\text{Number of converters}}$$

$$\text{The boost converter rating} = \frac{50}{4} = 12.5 \text{ kW}$$

Selecting the boost converter output voltage of $V_{idc} = V_o = 345$ V and input voltage equal to string voltage:
$V_i = 274$ V for type 1, $V_i = 263$ V for type 2, $V_i = 258.3$ V for type 3, and $V_i = 261$ V for type 4
The duty ratio of the boost converter is given by

$$D = 1 - \frac{V_i}{V_o}$$

$$D = 1 - \frac{274}{345} = 0.205 \text{ for type 1 PV}$$

$$= 1 - \frac{263}{345} = 0.237 \text{ for type 2 PV}$$

TABLE 2.19 The Photovoltaic Specifications for Each Photovoltaic Type

PV Type	Number of Modules per String	Number of Strings per Array	Number of Arrays	String Voltage (V)	Total Area of the PV (ft²)	Total Weight of the PV (lb)	Total Cost of the PV ($)
1	5	14	4	274	3492	9260	243,600
2	10	7	4	263	4391	10,920	194,600
3	9	9	4	258.3	5498	13,187	178,200
4	15	10	4	261	4241	11,040	238,200

TABLE 2.20   Boost Converter Specifications

PV Type	Number of Boost Converters	Input Voltage, V_i (V)	Power Rating (kW)	Output Voltage, V_o (V)	Duty Ratio, D
1	4	274	12.5	345	0.205
2	4	263	12.5	345	0.237
3	4	258.3	12.5	345	0.251
4	4	261	12.5	345	0.243

$$= 1 - \frac{258.3}{345} = 0.251 \text{ for type 3 PV}$$

$$= 1 - \frac{261}{345} = 0.243 \text{ for type 2 PV}$$

The design data for boost converter are presented in Table 2.20.

The inverters should be rated to withstand the output voltage of the boost converter and should be able to supply the required power. Let us design with each inverter having a rating of 10 kW.

The input voltage of the inverter is $V_{idc} = 345$ V with an amplitude modulation index of 0.90. The output voltage of the inverter is at 220 V AC.

The number of inverters, NI, to process a generation of 50 kW is given by

$$NI = \frac{\text{PV generation}}{\text{Power of one inverter}}$$
$$NI = \frac{50}{10} = 5$$

Hence, we need to connect five inverters in parallel to supply the load of 50 kW. Of course, we can also use one inverter with a higher rating to convert the DC power to AC. Naturally, many other designs are also possible.

Selecting a switching frequency of 5.1 kHz, the frequency modulation index will be given as

$$M_f = \frac{f_S}{f_e} = \frac{5100}{60} = 85$$

Students can compute the total harmonic distortion.

The design of inverter is presented in Table 2.21.

The one-line diagram of the system is shown in Figure 2.18.

The PV plant of Figure 2.18 cannot operate as a standalone PV plant. However, PV plant can operate in parallel with local utility grid in synchronous operation with power grid. For standalone operation, a storage system must be connected to DC bus. The frequency of operation is set by inverter, usually 60 Hz and the system acts as a UPS.

TABLE 2.21   Inverter Specifications

Number of Inverters	Input Voltage, V_{idc} (V)	Power Rating (kW)	Output Voltage, V_{AC} (V)	Amplitude Modulation Index, M_a	Frequency Modulation Index, M_f
5	345	10	220	0.90	85

The selection of the type of PV system may be based on the weight and cost of the system. For residential and commercial systems with existing roof structures, the PV modules that have minimum weight are normally selected.

The selection of the boost converter is based on the power rating of the boost converter and its output voltage. The boost converter must be rated at the minimum output voltage of the PV system and the required DC input voltage of the inverter. The amplitude modulation index is selected to be less than one, but close to one for processing the maximum power of the DC source to the AC power. The frequency modulation index selected is based on the highest sampling time recommended by the manufacturer of the inverter to limit the total harmonic distortion. The number of strings and the number of modules in the string is based on the rating of the input voltage of the boost converter. The number of boost converters and inverters is based on the required output power of the PV generating station.

Example 2.5   Design a microgrid of PV system using the data in Table 2.17, assume the total load is 500 kW at 460 V, 60 Hz three-phase AC.

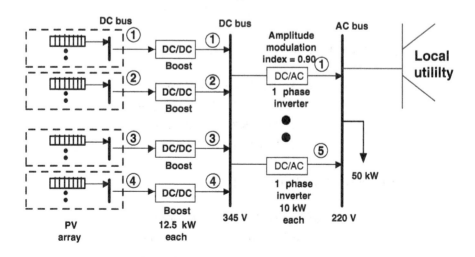

Figure 2.18   The one-line diagram of Example 2.4.

For each type of PV system, determine the following:

(i) The number of modules in a string, number of strings in an array, number of arrays, weight, and surface area

(ii) The boost converter and inverter specifications, and the one-line diagram

Solution
The load is 500 kW rated at 460 V AC. Based on the voltage of the load and an amplitude modulation index of 0.9, the input DC voltage for the three-phase inverter from Equation (2.17) is

$$\frac{2\sqrt{2} \times 460}{\sqrt{3} \times 0.9} = 835\,\text{V}$$

(i) Limiting the maximum voltage for a string to 600 V DC, a boost converter has to be used to boost the string voltage.

If we select string voltage, SV, to 550 V, the number of modules, NM, from Equation (2.13) is

$$NM = \frac{550}{54.8} \approx 10 \text{ for type 1}$$
$$= \frac{550}{26.3} \approx 21 \text{ for type 2}$$
$$= \frac{550}{28.7} \approx 20 \text{ for type 3}$$
$$= \frac{550}{17.4} \approx 32 \text{ for type 4}$$

The string voltage, SV, from Equation (2.16) is

$$SV = 10 \times 54.8 = 548\,\text{V for type 1}$$
$$= 21 \times 26.3 = 552\,\text{V for type 2}$$
$$= 20 \times 28.7 = 574\,\text{V for type 3}$$
$$= 32 \times 17.4 = 557\,\text{V for type 4}$$

String power from Equation (2.9) is

$$SP = 10 \times 190 = 1.9\,\text{kW for type 1}$$
$$= 21 \times 200 = 4.2\,\text{kW for type 2}$$
$$= 20 \times 170 = 3.4\,\text{kW for type 3}$$
$$= 32 \times 87 = 2.784\,\text{kW for type 4}$$

If we design each array to generate power of 20 kW, then the number of strings, *NS*, from Equation (2.17) is

$$NS = \frac{20}{1.9} = 11 \text{ for type 1}$$
$$= \frac{20}{4.2} = 5 \text{ for type 2}$$
$$= \frac{20}{3.4} = 6 \text{ for type 3}$$
$$= \frac{20}{2.784} = 8 \text{ for type 4}$$

The number of arrays, *NA*, is given by Equation (2.18) is

$$NA = \frac{500}{20} = 25$$

The total number of PV modules, *TNM*, as given by Equation (2.20) is

$$TNM = 10 \times 11 \times 25 = 2750 \text{ for type 1}$$
$$= 21 \times 5 \times 25 = 2625 \text{ for type 2}$$
$$= 20 \times 6 \times 25 = 3000 \text{ for type 3}$$
$$= 32 \times 8 \times 25 = 6400 \text{ for type 4}$$

The total surface area, *TS*, needed by each PV type can be computed from the number of modules and each module's area in square feet.

$$TS = \frac{2750 \times 34.6 \times 51.9}{144} = 34,294\,\text{ft}^2 = 0.787 \text{ acre for type 1}$$
$$= \frac{2625 \times 38.6 \times 58.5}{144} = 41,164\,\text{ft}^2 = 0.944 \text{ acre for type 2}$$
$$= \frac{3000 \times 38.3 \times 63.8}{144} = 50,907\,\text{ft}^2 = 1.169 \text{ acre for type 3}$$
$$= \frac{6400 \times 25.7 \times 39.6}{144} = 45,232\,\text{ft}^2 = 1.038 \text{ acre for type 4}$$

The total weight is

$$\text{Total weight} = 2750 \times 33.07 = 90,943\,\text{lb for type 1}$$
$$= 2625 \times 39.00 = 102,375\,\text{lb for type 2}$$
$$= 3000 \times 40.70 = 122,100\,\text{lb for type 3}$$
$$= 6400 \times 18.40 = 117,760\,\text{lb for type 4}$$

The total cost for each design is

Total cost = $2750 \times 870 = \$2.39$ million for type 1

$$\dot{=} 2625 \times 695 = \$1.82 \text{ million for type 2}$$

$$= 3000 \times 550 = \$1.65 \text{ million for type 3}$$

$$= 6400 \times 397 = \$2.54 \text{ million for type 4}$$

The design specification for each photovoltaic type is given in Table 2.22.

(ii) We need to use one boost converter for each array. The total number of converters is 25—each can be selected with 20 kW rating. The nominal output voltage of the boost converter is the same as the input of the inverter.

$$V_{idc} = V_o = 835 \text{ V}$$

The input voltage, V_i, of the boost converter is equal to the string voltage.

$$V_i = 548 \text{ V for type 1}$$

$$V_i = 552 \text{ V for type 2}$$

$$V_i = 574 \text{ V for type 3}$$

$$V_i = 557 \text{ V for type 4}$$

The duty ratio of the boost converter is given by

$$D = 1 - \frac{V_i}{V_o}$$

$$D = 1 - \frac{548}{835} = 0.34 \text{ for type 1 PV}$$

$$= 1 - \frac{552}{835} = 0.34 \text{ for type 2 PV}$$

$$= 1 - \frac{574}{835} = 0.31 \text{ for type 3 PV}$$

$$= 1 - \frac{557}{835} = 0.33 \text{ for type 2 PV}$$

The design data fir the boost converter are presented in Table 2.23.

The inverters should be rated to withstand the output voltage of boost converter and should be able to supply the required power. We can select each inverter having a rating of 100 kW, the input voltage of the inverter is to be $V_{idc} = 835$ V with amplitude modulation index of 0.9 and output voltage of the inverter at 460 V AC.

TABLE 2.22 The Photovoltaic Specifications for Each Photovoltaic Type

PV Type	Number of Modules per String	Number of Strings per Array	Number of Arrays	String Voltage (V)	Total Area of the PV (ft²)	Total Weight of the PV (lb)	Total Cost of the PV (Million $)
1	10	11	25	548	34,294	90,943	2.39
2	21	5	25	552	41,164	102,375	1.82
3	20	6	25	574	50,907	122,100	1.65
4	32	8	25	557	45,232	117,760	2.54

TABLE 2.23   Boost Converter Specifications

PV Type	Number of Boost Converters	Input Voltage V_i (V)	Power Rating (kW)	Output Voltage, V_o (V)	Duty Ratio, D
1	25	548	20	835	0.34
2	25	552	20	835	0.34
3	25	574	20	835	0.31
4	25	557	20	835	0.33

The number of inverters, NI, from Equation (2.19), needed to process a generation of 500 kW is calculated as

$$NI = \frac{500}{100} = 5$$

Hence, we need to connect five inverters in parallel to supply the load of 500 kW

Selecting a switching frequency of 10 kHz to limit the total harmonic distortion, the frequency modulation index is given by

$$M_f = \frac{f_S}{f_e} = \frac{10000}{60} = 166.67$$

The design data for inverter specification are presented in Table 2.24.

The one-line diagram of the system is shown in Figure 2.19.

The PV plant of Figure 2.19 cannot operate as a standalone PV plant. However, PV plant can operate in parallel with local utility grid in synchronous operation with power grid. For standalone operation, a storage system must be connected to DC bus of Figure 2.19. The frequency of operation is set by inverter, usually 60 Hz and the system acts as a UPS.

Example 2.6   Consider a microgrid with 2 MW. The system operates as part of a microgrid connected to the local utility at 13.2 kV.

The local utility uses the following design data:

(i) The transformer specifications are as follows: 13.2 kV/460 V, 2 MVA, negligible reactance, and 460 V/220 V, 20 kVA, with negligible reactance.

(ii) The data for the PV system are given in Table 2.17.

TABLE 2.24   Inverter Specifications

Number of Inverters	Input Voltage, V_{idc} (V)	Power Rating (kW)	Output Voltage, V_{AC} (V)	Amplitude Modulation Index, M_a	Frequency Modulation Index, M_f
5	835	100	460	0.90	166.67

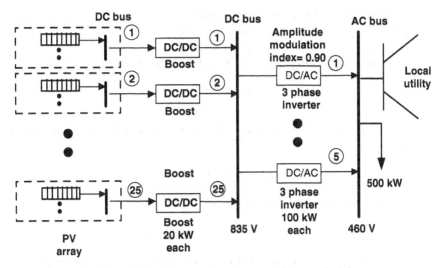

Figure 2.19   The one-line diagram of Example 2.5.

Determine the following:

  (i) The number of modules in a string, number of strings, number of arrays, surface area, weight, and cost
 (ii) The inverter specification and the one-line diagram

Solution
A three-phase inverter rated at AC voltage 220 V can process approximately 20 kW. Based on 20 kW three-phase inverter and 220 volts AC output and the modulation index of 0.9, the input DC voltage is as given by Equation (2.17).

$$V_{idc} = \frac{2\sqrt{2} \times 220}{\sqrt{3} \times 0.9} = 399 \text{ V}$$

  (i) Table 2.25, tabulates the data for each of the four types of PV modules.

TABLE 2.25   Weight per Unit Power

PV Type	Surface Area of One Module (ft^2)	Power Rating (W)	Weight per Unit Power (lb per W)
1	33.07	190	0.174
3	39.00	200	0.195
3	40.70	170	0.239
4	18.30	87	0.210

Table 2.25 shows that the PV module type 1 has the minimum weight per unit power; hence, our choice. The string voltage, SV of the PV system should be close to the rated inverter input voltage, V_{idc}.

Using string voltage, SV of 400, from Equation (2.15), we have

$$NM = \frac{399}{54.8} \approx 7$$

Using seven modules, string voltage, SV, as given by Equation (2.16) is

$$SV = 7 \times 54.8 = 384 \text{ V}$$

The string power, SP, from Equation (2.9) is

$$SP = 7 \times 190 = 1.33 \text{ kW}$$

Designing an array to generate 20 kW, the number of strings, NS, in an array is as given by Equation (2.17):

$$NS = \frac{20}{1.33} = 15$$

The number of arrays, NA, needed for this design is given by Equation (2.18):

$$NS = \frac{2000}{20} = 100$$

And, the total number of PV modules from Equation (2.20), $TNM = 7 \times 15 \times 100 = 10,500$

The total surface area, TS, needed by the type 1 PV module is computed from the area of one module and the total the number of modules.

$$TS = \frac{10,500 \times 34.6 \times 51.9}{144} = 130,940 \text{ ft}^2 = 3 \text{ acre}$$

The total weight, TW, for the type 1 PV module can be computed in a similar manner as

$$TW = 10,500 \times 33.07 = 347,235 \text{ lb}$$

The total cost for the type 1 PV module is the product of the number of modules and the cost of each module as given below.

Total cost of PV modules $= 10,500 \times 870 = \$9.14$ million.

The design data for the 2000 kW photovoltaic specification are given in Table 2.26.

TABLE 2.26 The Photovoltaic Specifications for 2000 kW

PV Type	Number of Modules per String	Number of Strings per Array	Number of Arrays	String Voltage (V)	Power per String (kW)	Total Area of the PV (ft^2)	Total Weight of the PV (lb)	Total Cost of the PV (million $)
1	7	15	100	384	1.33	130,940	347,235	9.14

(ii) We will use one inverter for each array. Hence, the rating of each inverter will also be 20 kW.

The number of inverters, NI, needed for a generation of 2000 kW as from Equation (2.19)

$$NI = \frac{2000}{20} = 100$$

For 2000 kW PV generating station, we need 100 inverters to operate in parallel. Other designs using higher rating inverters are also possible. For this design, the final input DC voltage of the inverter is specified by the string voltage as

$$V_{idc} = V_{string} = 384 \text{ V}$$

With nominal DC input, the voltage of the inverter is selected, the amplitude modulation index, M_a is given as

$$M_a = \frac{2\sqrt{2} \times 220}{\sqrt{3} \times 384} = 0.93 \text{ V}$$

If the switching frequency is selected at 10 kHz, the frequency modulation index, M_f is given as

$$M_f = \frac{f_S}{f_e} = \frac{10000}{60} = 166.67.$$

The design data for the inverter specification are given in Table 2.27.

The one-line diagram of the system is given in Figure 2.20. We will use 100 transformers of 20 kVA, which are connected to the inverters to step the voltage up from 220 V to 460 V. Finally, one transformer of 460/13.2 kV of 2 MVA is used to connect the system to the local power grid.

Again, students should recognize that many designs are possible. Each design specification has its own limitation that must be taken into account. The above analysis can be altered as needed to satisfy any design requirements.

TABLE 2.27   Inverter Specifications

Number of Inverters	Input Voltage, V_{idc} (V)	Power Rating (kW)	Output Voltage, V_{AC} (V)	Amplitude Modulation Index, M_a	Frequency Modulation Index, M_f
100	384	20	220	0.93	166.67

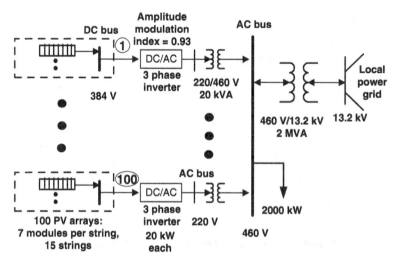

Figure 2.20   The one-line diagram of the system of Example 2.6.

2.9 MODELING OF A PHOTOVOLTAIC MODULE

As we discussed, a commercial PV module is constructed from a number of PV cells. A PV cell is constructed from a p–n homojunction material. When a p-type doped semiconductor is joined with an n-type doped semiconductor, and a p–n junction is formed. If the p- and the n-type semiconductors have the same band gap energy, a homojunction is formed. The homojunction is a phenomenon that takes place between layers of similar semiconductor materials. These types of semiconductors have equal band gaps and they normally have different doping. The absorption of photons of energy generates DC power.

A PV cell is shown in Figure 2.20. As irradiance energy of sun is received by the module, it is charged with electric energy. The model of a PV cell is similar to that of a diode and can be expressed by the well-known Shockley–Read equation.

The PV module can be modeled by a single exponential model. The model is presented in Figure 2.21. The current is expressed in terms of voltage, current, and temperature as shown in Equation (2.21).

In the above model in Figure 2.22, the PV module is represented by a current source I_{ph} in parallel with the shunt resistance R_{sh}. The current flowing through the shunt resistance is designated as I_{Rsh}. The output DC voltage, V is in series with the internal resistance R_S. The PV model of Figure 2.19 also depicts the power loss through current, I_{D1} circulating through the diode. The single exponential model is given by Equation (2.21).

$$I = I_{ph} - I_0 \left\{ \exp \left[\frac{q(V + IR_s)}{n_c A k T} \right] - 1 \right\} - \frac{V + IR_s}{R_{sh}} \qquad (2.21)$$

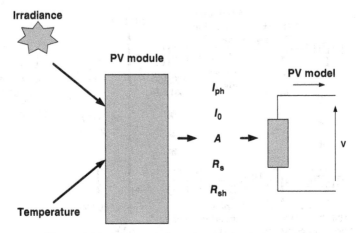

Figure 2.21   The modeling of a photovoltaic module.

Other parameters are the diode quality factor (A), n_c is the number of cells in the module, Boltzmann's constant (k), 1.38×10^{-23} J/K; the electronic charge (q), 1.6×10^{-19} C, and the ambient temperature (T) in Kelvin.

Equation (2.21) is nonlinear and its parameters I_{ph}, I_o, R_s, R_{sh}, and A are functions of temperature, irradiance, and manufacturing tolerance. We can use numerical methods and curve fitting to estimate the parameters from test data provided by manufacturers. The estimation of PV array model is quite involved. The formulation of this problem and solution is given in References 14–17.

2.10 MEASUREMENT OF PHOTOVOLTAIC PERFORMANCE

A PV module at a maximum constant level of irradiance can produce 1000 W/m². 1000 W/m² is also termed as one sun. The power output of the PV module is calibrated in relation to exposure to the sun. The sun energy in W/m² is given in Table 2.28.

The sun irradiance energy is calibrated for a PV array system based on the angle of incidence and depicted in Figure 2.23. These data are used to operate

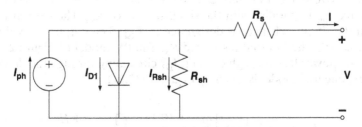

Figure 2.22   The single exponential model of a photovoltaic module.

TABLE 2.28   Sun Energy versus Incident Irradiance

Sun Energy	Incident Irradiance, W/m^2
1	1000
0.8	800
0.6	600
0.4	400
0.2	200

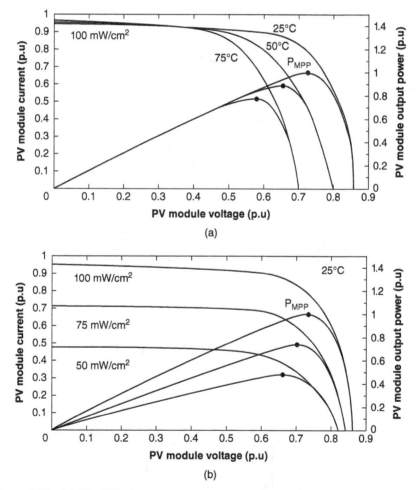

Figure 2.23   (a) The PV output current versus output voltage and output power as a function of temperature variation. (b) The output power in W/m^2 at various irradiances as a function of module current and output voltage.[8]

the PV array at its maximum power point. In the next section, we will discuss how a DC–AC inverter and digital controller operate a PV plant at its maximum power point.

2.11   MAXIMUM POWER POINT OF A PHOTOVOLTAIC PLANT

First, let us review the maximum power transfer in a resistive circuit. Consider the circuit of Figure 2.24. Assume a voltage source with an input resistance, R_{in}. This source is connected to a load resistance, R_L.

The current supplied to the load is

$$I = \frac{V}{R_{in} + R_L} \tag{2.22}$$

The power delivered to the load R_L is

$$P = I^2 R_L \tag{2.23}$$

$$P = V^2 \frac{R_L}{(R_{in} + R_L)^2} \tag{2.24}$$

Differentiating with respect to R_L:

$$\frac{dP}{dR_L} = V^2 \frac{(R_{in} + R_L)^2 \frac{dR_L}{dR_L} - R_L \frac{d(R_{in}+R_L)^2}{dR_L}}{(R_{in} + R_L)^4}$$

$$\frac{dP}{dR_L} = V^2 \frac{R_{in} - R_L}{(R_{in} + R_L)^3} \tag{2.25}$$

Setting the above to zero, we can calculate the operating point for the maximum power. The maximum power point can be delivered to the load when, $R_L = R_{in}$.

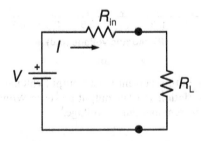

Figure 2.24   A DC source with a resistive load.

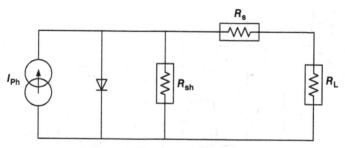

Figure 2.25   A photovoltaic plant model and its load.

A PV module output power is the function of irradiance solar energy. Figure 2.19 depicts the output power in W/m² at various irradiances as a function of module current and output voltage.

The PV plant should be operated to extract the maximum power from its PV station as the environmental conditions change in relation to the position of the sun, cloud cover, and daily temperature variations. The equivalent circuit model of a PV array depicted can be presented during its power transfer mode to a load R_L as shown in Figure 2.25.

Figure 2.25 presents the circuit model for a PV plant by a current source that has a shunt resistance, R_{sh} and series resistance, R_s.[18] The shunt resistance has a large value and series resistance is very small. The load resistance is represented by R_L. In Figure 2.25, R_L is the reflected load because in practice the load is connected to the inverter AC side. When the PV is connected to the power grid, the load is based on the injected power to the power grid. The equivalent voltage source circuit model of current source model is depicted in Figure 2.26.

Figure 2.25 indicates that the characteristics of a PV plant are highly nonlinear. The input impedance of a PV plant is affected by irradiance variation and temperature.

To achieve maximum power transfer from the PV array, the input impedance of the PV generator must match the load. The maximum power point tracking control algorithm seeks to operate at a point on the PV plant current and voltage characteristics where the maximum output power can be obtained. For a

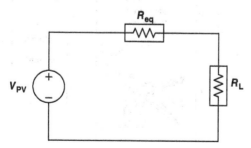

Figure 2.26   A simple voltage source equivalent circuit model of a photovoltaic plant.

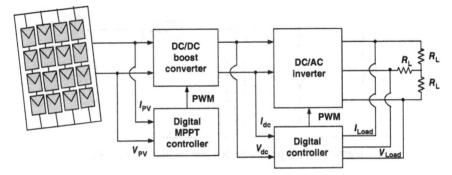

Figure 2.27 A photovoltaic plant using a boost converter to step up the voltage and an inverter.

PV plant, the control algorithm computes the $dP/dV > 0$ and $dP/dV < 0$ to identify if the pick power has been achieved. Figure 2.20 depicts the control algorithm. If the PV plant is to supply power to DC loads such as DC lighting, then a DC–AC inverter is not needed.

2.12 CONTROL OF MAXIMUM POWER POINT OF PHOTOVOLTAIC PLANTS

A number designs of power plants can be considered depending on the plant design A photovoltaic plant using a boost converter to step up the voltage and an inverter are employed. The design presented in Figure 2.27 has two control loops. The first control loop is designed to control DC–DC converter and the second control loop can control the total harmonic distortion and output voltage.

Figure 2.28 is used to detect the maximum power point and then invert DC voltage to AC voltage.

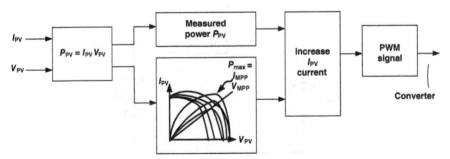

Figure 2.28 A maximum power point tracking control algorithm.

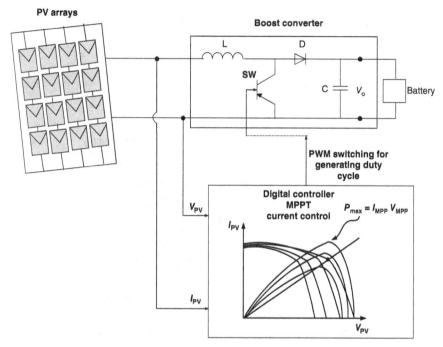

Figure 2.29   Maximum power point tracking using only a boost converter.

When the PV system is to charge a battery storage system, the PV plant is designed as depicted in Figure 2.29 using a boost converter or as in Figure 2.30 using a buck converter.

Figure 2.30 uses a buck converter to charge a battery.

Figure 2.31 depicts the design of a PV plant and MPP using an inverter when the PV generating station is connected to a local utility. Again, the digital controller tracks the PV station output voltage and current and computes the MPP point according to the control algorithm. The control algorithm issues the PWM switching policy to control inverter current such that the PV station operates at its maximum power point. However, the resulting control algorithm may not result in minimum total harmonic distortion.

When the MPP control is performed as part of the inverter as shown in Figure 2.31, the tracking of MPP may not be optimum. In this type of MPP, the current to the inverter flows though all modules in the string. However, the I–V curves may not be the same and some strings will not operate at their MPP. Therefore, the resulting energy capture may not be as high and some energy will be lost in such systems.

Figure 2.32 depicts a PV generating station with a battery storage system when the PV system is connected to the local utility. The DC–DC converter and its MPP are referred to a charger controller. The charger controllers have a number of functions. Some charger controllers are used to detect the variations

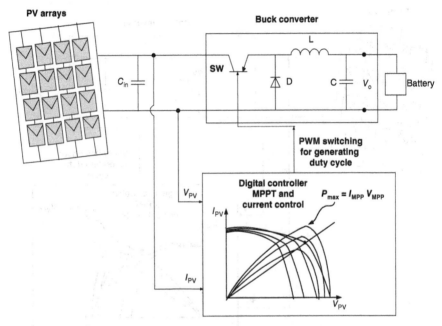

Figure 2.30 Maximum power point tracking using a buck converter.

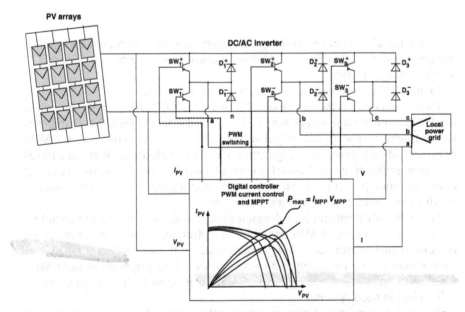

Figure 2.31 A photovoltaic generating station operating at maximum power point tracking when the photovoltaic system is connected to a local power grid.

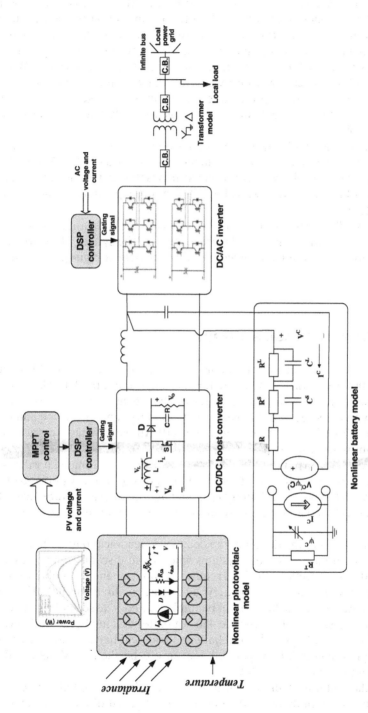

Figure 2.32 A photovoltaic generating station operating at maximum power point tracking with a battery storage system when the photovoltaic system is connected to a local power grid.

in the current–voltage characteristics of a PV array. MPP controllers are used for a PV plant to operate at voltage close to maximum power point to draw maximum available power. The charger controllers also perform battery power management. For normal operation, the controllers control the battery voltage, which varies between the acceptable maximum and minimum values. When the battery voltage reaches a critical value, the charge controller function is to charge the battery and protect the battery from an overcharge. This control is accomplished by two different voltage thresholds, namely, battery voltage and PV module voltage.

At lower voltage, typically 11.5 V, a controller switches the load off and charges the battery storage system. At higher voltage, usually 12.5 V for 12 V battery storage system charge, a controller switches the load to the battery. The control algorithm adjusts the two voltage thresholds depending on the battery storage system.

DC–DC MPP PV charger controllers facilitate standardization of integration of PV system for use in a local storage system. The system of Figure 2.32 also can be used as a standalone microgrid that can deliver high-quality power for a UPS.

2.13 BATTERY STORAGE SYSTEMS

The design of battery storage follows the same procedure as design of PV station except battery storage systems are rated in ampere-hours (Ah). The capacity of a battery is rated in ampere-hours (Ah). The Ah measures the capacity of a battery to hold energy. A 1 Ah means that a battery can deliver one ampere for one hour. Based on the same concept, a 110 Ah battery has a capacity to deliver 10 amps for 11 hours. However, after a battery is discharged for an hour, the battery will need to be charged longer than 1 hour. It is estimated it will take 1.25 Ah to restore the battery to the same state of charge. Battery performance also varies with temperature, battery type, and age. Lead-acid battery technology is well established and is a widely adopted energy source for various power industries.

Recent advances in the design of the deep-cycle lead-acid battery have promoted the use of battery storage systems when rapid discharge and charging are required. For example, if the load requires a 900 Ah bank, a number of battery storage systems can be designed. As a first design, three parallel strings of deep-cycle batteries rated 300 Ah can be implemented. The second design can be based on two strings of deep-cycle 450 Ah batteries. Finally, the design can be based on a single large industrial battery. Lead-acid batteries are designed to have approximately 2.14 V per cell. For an off-the-shelf 12 V battery the voltage rating is about 12.6–12.8 V.

The fundamental problem is that if a single cell in a string fails, the entire storage bank will rapidly discharge beyond the required discharge level; this will permanently destroy the bank.

It is industry practice not to install more than three parallel battery strings. Each string is monitored to ensure equal charging and discharging rates. If a storage bank loses its equalization, it will result in accelerated failure of any weak cells and the entire storage bank. The battery characteristics change with age and charging and discharging rates. It is industry practice not to enlarge an old battery bank by adding new battery strings. As the battery ages, the aging is not uniform for all cells. Some cells will establish current flow to the surrounding cells and this current will be difficult to detect. If one cell fails, changing the resistance in one battery string, the life of the entire string can be reduced substantially. Therefore, the storage system will fail. By paralleling several strings, the chances of unequal voltages across the strings increases; therefore, for optimized battery storage systems, the system should be designed using a single series of cells that are sized for their loads.

The battery capacity can be estimated for a given time duration by multiplying the rated load power consumption in watts by the number of hours that the load is scheduled to operate. This results in energy consumption in watt hours (or kilowatt hours [kWh]), stated as

$$kWh = kV \cdot Ah \tag{2.26}$$

For example, a 60 W light bulb operating for 1 hour uses 60 Wh. However, if the same light bulb is supplied by a 12 V battery, the light will consume 5 Ah. Therefore to compute Ah storage required for a given load, the average daily usage in W should be divided by the battery voltage. As another example, if a load consumes 5 kWh per day from a 48 V battery storage system, we can determine the required Ah by dividing the watt-hours by the battery voltage. For this example, we will need 105 Ah. However, because we do not want to discharge the battery more than 50%, the battery storage needed should be 210 Ah. If this load has to operate for 4 days, the required capacity is 840 Ah. If the battery cabling is not properly insulated from earth, the capacitive coupling from the DC system with earth can cause stray current flow from the DC system to underground metallic facilities, which will corrode the underground metallic structure.

The data for typical battery storage are given in Table 2.29.

TABLE 2.29 Typical Battery Storage Systems

Class 1	24–40 Ah	12 V
Class 2	70–85 Ah	12 V
Class 3	85–105 Ah	12 V
Class 4	95–125 Ah	12 V
Class 5	180–215 Ah	12 V
Class 6	225–255 Ah	12 V
Class 7	180–225 Ah	6 V
Class 8	240–415 Ah	6 V

Battery energy storage is still expensive for large-scale stationary power applications under the current electric energy rate. However, the battery storage system is an important technology for the efficient utilization of an intermittent renewable energy system such as wind or PV in the integration of the renewable energy in electric power grids. Utility companies are interested in the large-scale integration of a battery storage system in their substations as community storage to capture the high penetration of solar energy and wind in their distribution system. The community storage system with the ramping capability of at least an hour can be utilized in power grid control. This is an important consideration for utility companies because the installed energy storage system can be used as spinning reserve.

The storage system must be effectively and efficiently scheduled for the charging, discharging, and rest time of each string. Therefore, the battery storage system requires extensive monitoring and control. These issues are currently being studied.[19,20]

2.14   STORAGE SYSTEMS BASED ON A SINGLE-CELL BATTERY

The rapid electrification of the automotive industry has a large impact on storage technology and its use in stationary power grids. At present, nickel metal hydride (NiMH) batteries are used in most electric and hybrid electric vehicles available to the public. Lithium-ion batteries have the best performance of the available batteries. The cost of energy storage for grid-level renewable energy storage at present is approximately around $300–500 per kWh. The price is rapidly decreasing as more companies are developing new technologies.

The large battery storage system is constructed from single-cell batteries[12,19,20] and is considered a multicell storage system. The performance of multicell storage is a function of output voltage, internal resistance, cell connections, the discharge current rate, and cell aging.[12,19,20]

Single-cell battery technology is rapidly making new advances, for example, a new lithium-ion battery is being developed. The price of a single-cell battery has also been dropping dramatically. In comparison to the regular lead-acid battery where 12 cells are internally connected in a 12 V battery, the single-cell batteries can be individually connected and reconfigured. Also, because the individual single-cell's rate of charging and discharging can be monitored, the health and performance of all cells in a string can be evaluated. Figure 2.26 depicts three single cells in a string.

Figure 2.33 depicts a battery storage system consisting of two strings of three single cells that are connected in parallel making an array.

Table 2.30 presents the energy densities of two types of batteries. Figure 2.34 depicts two strings of three cells connected in parallel.

Table 2.31 presents the energy density and the cost of a storage system of typical batteries.

In all batteries, the performance will change with repeated charging at the discharge current rates. As expected, the higher the discharge current rate is

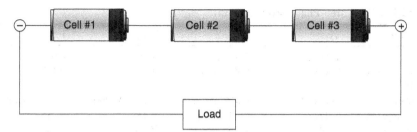

Figure 2.33   Three single cells in a string.

TABLE 2.30   Comparison of Battery Energy Density and Power Density[12]

Application/Battery Type	Energy Density (Wh/kg)	Energy Stored (kWh)	Fraction of Usable Energy (%)
NiMH	65	40–50	80
Lithium ion	130	40–50	80

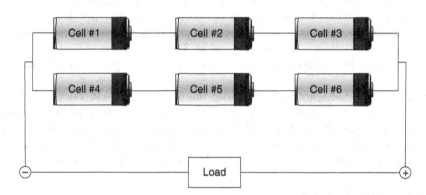

Figure 2.34   Two strings of three single cells connected in parallel.

the lower will be the remaining capacity and output voltage and the higher the internal resistance. However, the reduced capacity as the result of a higher discharge current rate will be recovered after the battery system is allowed to rest before the next discharge cycle.

TABLE 2.31   The Energy Density and the Cost of a Storage System

Type	Wh/kg	Wh	Weight kg (1)	$/kg	$/kWh	$/kW
Standard	25	1875	75	2.5	100	9.35
Thin film	20	1000	50	4.0	200	10.0
NiMH	45	1800	40	22.5	500	45.0
Lithium ion	65	1170	18	45	700	41.0

TABLE 2.32   Three-Phase Inverter Data

Inverter	Type 1	Type 2	Type 3	Type 4
Power	100 kW	250 kW	500 kW	1 MW
Input voltage DC	900 V	900 V max	900 V	900 V
Output voltage AC	660 V AC/60 Hz	660 V AC/60 Hz	480 V AC/60 Hz	480 V AC/60 Hz
Efficiency	Peak efficiency 96.7%	Peak efficiency 97.0%	Peak efficiency 97.6%	Peak efficiency 96.0%
Depth, in.	30.84	38.2	43.1	71.3
Width, in.	57	115.1	138.8	138.6
Height, in.	80	89.2	92.6	92.5
Weight, lb	2350	2350	5900	12,000

Therefore, the design and optimization of a multicell storage system require an understanding of the storage system's discharge performance under various operation conditions. Furthermore, if the battery storage system is to be used as a community storage system in a power grid distribution system, dynamic models of the battery systems are needed. The dynamic model of a storage system will facilitate dispatching power from intermittent green energy sources.

Example 2.7   Design a microgrid with the load of 1000 kW rated at 460 V AC and connected to the local power grid at 13.2 kV using an ideal transformer rated 2 MVA, 460 V/13.2 kV. To support the emergency loads, microgrid needs 200 kWh storage systems to be used for 8 hours a day. Data for a three-phase inverter are given in Table 2.32. The data for PV system are given in Example 2.4.

Determine the following:

(i) The ratings of PV plant, converters, inverters, storage systems and a single-line diagram of this design based on the minimum surface area. Also, compute the cost, weight, and square feet area of each PV type and give the results in a table.

(ii) Per unit model

Solution

(i) The load is 1000 kW rated at 460 V AC. Based on the voltage of the load and an amplitude modulation index of 0.9, we have the following input DC voltage for the inverter:

$$V_{idc} = \frac{2\sqrt{2}V_{LL}}{\sqrt{3}.M_a} = \frac{2\sqrt{2} \times 460}{\sqrt{3} \times 0.9} = 835 \text{ V}$$

TABLE 2.33 Inverter Specifications

Number of Inverters	Input Voltage, V_{idc} (V)	Power Rating (kW)	Output Voltage, V_{AC} (V)	Amplitude Modulation Index, M_a	Frequency Modulation Index, M_f
4	835	250	460	0.90	84

Selecting an inverter rated 250 kW, the total number of inverters, *NI*, for the processing of 1000 kW is given as

$$NI = \frac{\text{PV generation}}{\text{Rating of inverters}}$$
$$NI = \frac{1000}{250} = 4$$

For this design, four inverters should be connected in parallel. If we select a switching frequency of 5.04 kHz, the frequency modulation index is

$$M_f = \frac{f_S}{f_e} = \frac{5040}{60} = 84$$

Table 2.33 presents data for the inverter specification.

Students should recognize that other designs are also possible. The input DC voltage of PV specifies the output AC voltage of inverters. Table 2.34 gives the data for each PV type.

As noted in Table 2.34, the PV module of type 1 requires the minimum of surface area. Selecting PV type 1 and string open-circuit voltage of 550 V DC, the number of modules, *NM*, is

$$NM = \frac{\text{String voltage}}{V_{MPP}}$$

TABLE 2.34 A Photovoltaic System Design Based on Photovoltaic Type

PV Type	Surface Area of One Module (ft²)	Power Rating (W)	Area per Unit Power (ft²/W)
1	12.47	190	0.066
2	15.68	200	0.078
3	16.97	170	0.100
4	7.07	87	0.081

where V_{MPP} is the voltage at maximum power point of the PV module from the PV data.

$$NM = \frac{550}{54.8} \approx 10 \text{ for type 1 PV}$$

The string voltage, SV, under load is given as

$$SV = NM \times V_{MPP}$$
$$SV = 10 \times 54.8 = 548 \text{ V}$$

The string power, SP, is given as

$$SV = NM \times P_{MPP}$$

where P_{MPP} is the power generated by the PV module at maximum power point.

$$SP = 10 \times 190 = 1.9 \text{ kW for type 1}$$

If we design each array to generate a power of 20 kW, then the number of strings, NS, is given by

$$NS = \frac{\text{Power of one array}}{\text{Power of one string}}$$
$$NS = \frac{20}{1.9} = 11$$

The number of arrays, NA, is given by

$$NA = \frac{\text{PV generation}}{\text{Power of one array}}$$
$$NA = \frac{1000}{20} = 50$$

The total number of PV modules, TNM, in an array is given by

$$TNM = NM \times NS \times NA$$

where NM is number of modules in a string, NS, is the number of strings, and NA is the number of arrays in a PV station.

$$NA = 10 \times 11 \times 50 = 5500 \text{ for PV module of type 1}$$

The total surface area needed, TS, for type 1 PV module is

$$TS = \frac{5500 \times 34.6 \times 51.9}{144} = 68,586\,\text{ft}^2 = 1.57\,\text{acre}$$

The total weight, TW, needed for a type 1 PV module is the product of the number of modules and the weight of each module.

$$TW = 5500 \times 33.07 = 181,885\,\text{lb}$$

The total cost for a PV module is the product of the number of modules and the cost of each module.

$$\text{Total cost} = 5500 \times 870 = \$4.78\,\text{million for PV module type 1.}$$

The design data for the photovoltaic specification are presented in Table 2.35.

The output voltage of the boost converter, V_o, is the same as the input voltage, of the inverter, V_{idc}.

$$V_o = V_{idc} = 835\ \text{V}$$

The boost input voltage, V_i, is same as the string voltage, $SV = V_i = 548$ V. The duty ratio of the boost converter is given by

$$D = 1 - \frac{V_i}{V_o}$$

For this design, it is,

$$D = 1 - \frac{548}{835} = 0.34$$

We need one boost converter for each array. Therefore, the number of boost converters is 50 and each is rated 20 kW. The design data for the boost converter are given in Table 2.36.

In storage design, we need to limit the number of batteries in a string and limit the number of arrays to three. These limitations are imposed on lead-acid-type batteries to extend the life of the storage system. We select the Class 6 batteries that are rated at 255 Ah at 12 V. In this design, three batteries per string and three strings in each array are used. The string voltage, SV, of the storage system is

$$SV = 3 \times 12 = 36\ \text{V}$$

TABLE 2.35 The Photovoltaic Specifications

PV Type	Number of Modules per String	Number of Strings per Array	Number of Arrays	String Voltage (V)	Total Area of the PV (ft^2)	Total Weight of the PV (lb)	Total Cost of the the PV (million $)
1	10	11	50	548	68,586	181,885	4.78

TABLE 2.36 Boost Converter Specifications

Number of Boost Converters	Input Voltage, V_i (V)	Power Rating (kW)	Output Voltage, V_o (V)	Duty Ratio, D
50	548	20	835	0.34

The string energy stored, SES, in each battery is given by the product of the Ah and the battery voltage.

$$SES = 255 \times 12 = 3.06\,\text{kWh}$$

Each array has nine batteries. Therefore, the array energy stored, AES, is given as

$$AES = 9 \times 3.06 = 27.54\,\text{kWh}$$

The number of arrays, NA, needed to store 200 kWh is given by

$$NA = \frac{\text{Total energy}}{\text{Energy in each array}}$$
$$NA = \frac{200}{27.54} \approx 8$$

Table 2.37 presents a number of batteries for storing 200 kWh of energy.

Because we have eight storage arrays, we use one buck–boost converter for each array storage system. We need a total of eight buck–boost converters. The buck–boost converters are used to charge–discharge the battery storage system.

In this design, the buck–boost converter input is 835 V of the DC bus and its output must be 36 V DC to charge the battery storage system. If the storage systems are to be used for 8 hours, they can be discharged to 50% of their capacity. Therefore, they can be used to supply 100 kWh. The power, P, supplied by the storage system is given by

$$P = \frac{\text{kWh}}{\text{hour}}$$
$$P = \frac{100}{8} = 12.5\,\text{kW}$$

TABLE 2.37 Battery Array Specifications

Battery Class	Number of Batteries per String	Number of Strings per Array	Number of Arrays	String Voltage (V)	Energy Stored per Array (kWh)
6	3	3	8	36	27.54

TABLE 2.38 Buck–Boost Converter Specifications

Number of Buck–Boost Converters	Input Voltage, V_i (V)	Power Rating (kW)	Output Voltage, V_o (V)	Duty Ratio, D
8	835	1.56	36	0.04

The array power, AP, rating is given by

$$AP = \frac{\text{Power}}{\text{Number of arrays}}$$

$$AP = \frac{12.5}{8} = 1.56 \text{ kW}$$

Let us select a buck–boost converter rated at 1.56 kW. The duty ratio is given by

$$D = \frac{V_o}{V_i + V_o}$$

The detailed presentation of buck–boost converter is given in the next chapter.

$$D = \frac{36}{835 + 36} = 0.04$$

The design data for the buck-boost converter specification are given in Table 2.38.

Figure 2.35 depicts the one-line diagram of the PV system.

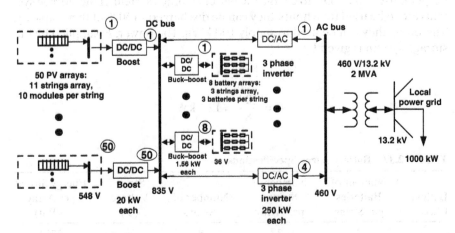

Figure 2.35 The one-line diagram of Example 2.6.

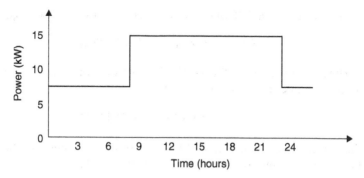

Figure 2.36   Plot of the daily load cycle for Example 2.7.

Example 2.8   Assume that a residential house total load is 7.5 kW from 11 P.M. to 8 A.M. and 15 kW for the remaining 15 hours. Determine the following:

(i) Plot load cycle for 24 hours
(ii) Total kWh energy consumption for 24 hours
(iii) If the sun irradiance is 0.5 sun for 8 hours daily, what is the roof space needed to generate adequate kWh for 24 hours operation?
(iv) Assume the maximum kWh to be used during the night is 40% of the total daily load. Search the Internet to select a battery storage system and give your design data.

Solution

(i) The load is 7.5 kW for 9 hours (11:00 P.M. to 8:00 A.M.) and 15 kW for 15 hours (8:00 A.M. to 11:00 P.M.). The load cycle is as given in Figure 2.36.
(ii) The total kWh energy consumption for 24 hours is the area under the curve of the daily load cycle and is given by kWh = kW × hours.
      Therefore, the energy consumption = 7.5 × 9 + 15 × 15 = 292.5 kWh
(iii) The type 1 PV is selected because it needs the minimum area per unit power produced.
      The amount of power produced by type 1 PV (see Table 2.17) is equal to 190 W per module for 1 sun. Therefore, the energy produced for 0.5 sun for 8 hours is given by 0.5 × 190 × 8 = 0.76 kWh.
      The number of modules, NM, needed is given by

$$NM = \frac{\text{Total energy demand}}{\text{Energy of one panel}}$$

$$NM = \frac{292.5}{0.76} \approx 385$$

The surface area, *SM*, of one module is given by *width* × *length*

$$SM = 34.6 \times 51.9/144 = 12.47 \text{ ft}^2$$

The total area, *TS*, for 292.5 kWh is given by the product of the number of modules and the area of one module:

$$TS = 385 \times 12.47 = 4801.11 \text{ ft}^2$$

(iv) The energy used during the night is 40% of the total energy. Therefore, the energy demand for one night = 0.4 × 292.5 = 117 kWh.

From Table 2.23, a Class 6 battery storage system is chosen to store the kWh needed for the night. For battery storage conservation, the batteries should not be discharged more than 50% of their capacity.

The energy stored per battery is given by amp-hours × voltage (Ah × V). Therefore, the energy stored in one battery = 255 × 12 = 3.06 kWh. The number of batteries, *NB*, needed is given by

$$NB = \frac{2 \times \text{Energy demand}}{\text{Energy stored per battery}}$$

$$NB = \frac{2 \times 117}{3.06} = 77$$

We can use three batteries in a string; the maximum number of strings in an array is equal to three. Therefore, the maximum number of batteries in the array is 3 × 3 = 9.

The number of arrays of battery is given by

$$NA = \frac{\text{Total number of batteries}}{\text{Number of batteries per array}}$$

$$NA = \frac{77}{9} = 8$$

2.15 THE ENERGY YIELD OF A PHOTOVOLTAIC MODULE AND THE ANGLE OF INCIDENCE

To estimate the energy yield of a photovoltaic module, the angle of inclination for a module with respect to the position of the sun must be determined. The angle of inclination is defined as the position that a magnetic needle makes with the horizontal plane at any specific location. The magnetic inclination is 0° at the magnetic equator and 90° at each of the magnetic poles. The irradiance is defined as the density of radiation incident on a given surface expressed in W/m² or W/ft². When the earth rotates, the angle at which the rays of the sun reach a PV module change. The PV energy yield at a location as a function the PV module inclination angle is given in Appendix C.

2.16 PHOTOVOLTAIC GENERATION TECHNOLOGY

In recent years, the shift toward the development and installation of PV sources of energy has resulted in an explosion of growth in the research, development, and manufacture of PV plants. The global installed PV generation plants were approximately 69 GW in 2011.[21] The yearly PV energy production was around 85 TWh of electricity. More than 75% of the world's solar PV energy is produced in Europe.[22] Italy is the top in market followed by Germany. In 2011, the cumulative installed capacity of Germany was 24 GW as the leading country in Europe, followed by Italy, with more than 12 GW.

According to solar industry data,[23] 832 MW of PV plants were installed in United States in 2013. Most of the installed capacity was commissioned by the US utility industry. As of now, there are over 9370 MW of cumulative PV plants operating in United States. Cost of PV plant is about $3.05/W in United States.[23]

Importantly, in recent years, this growth has been fueled by an increase in grid-connected systems. Today, a large share of grid-connected cumulative installed capacity is comprised of grid-PV-connected centralized applications. This further highlights the growing relevance of PV systems in fulfilling the ever-increasing energy demands of the twenty-first century.

The current PV modules have about 12–17% efficiency.[24] However, there are modules in production with 36% efficiency that would change the panels from 300 W to 900 W in the same footprint. Research into the development of more efficient PV modules is ongoing worldwide. The theoretical limit for a PV module constructed from multilayered cells is 60%. In the future, we can expect PV panes rated 1500 W; concentrated solar PV has the potential to go up to 200 suns today.

The technology of high power inverters is reaching into the 2 MW class. Solar panels are being designed at 600 V bus voltages. In Italy, the Rende installation uses one MW inverter and produces 1.4 GWh per year.[25] This design uses 180 W panels. In 5 or 10 years from now, we can envision a scalable design of a rooftop solar PV system that can produce 2 MW. It will not take much to get there from the PV side. Students are urged to search the Internet for up-to-date developments in PV systems.[1,2,3,4,5,7]

2.17 THE ESTIMATION OF PHOTOVOLTAIC MODULE MODEL PARAMETERS

Recall the equivalent circuit of a single-diode model of a PV cell.

The model of Figure 2.37 presents the current–voltage characteristics for a single cell of a PV module.[18,26,27,28] The model of a module consisting of a number of cells, n_c, can be presented as

$$I = I_{ph} - I_o \left(e^{\frac{V + IR_s}{n_c V_t}} - 1 \right) - \frac{V + IR_s}{R_{sh}} \tag{2.27}$$

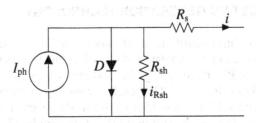

Figure 2.37   The single exponential model of a PV module.[26]

In Equation (2.27), V_t, the junction thermal voltage, is given as

$$V_t = \frac{AkT_{stc}}{q} \tag{2.28}$$

It is helpful to express the equations in terms of V_t rather than A. The value of A can be determined easily if V_t is found, by simply rearranging the terms of Equation (2.28), we will have

$$A = \frac{qV_t}{kT_{stc}} \tag{2.29}$$

Table 2.39 describes the variables used in modeling a module presented by Equation (2.27).

The term STC denotes the standard conditions used to measure the nominal output power of photovoltaic cells. The cell junction temperature at STC is 25°C, the irradiance level is 1000 W/m^2, and the reference air mass is 1.5 solar spectral irradiance distributions. In Equation (2.27), the term "–1" is much smaller than the exponential term and it is generally ignored.

The problem of estimating the model parameters is to determine the five parameters, I_{ph}, I_o, R_s, R_{sh}, and A from the data sheet provided by the manufacturer of the PV module measured under STC. Because A can be expressed

TABLE 2.39   The Photovoltaic Module Parameters of Single Diode Model

I_{ph}	Photo-generated current at STC
I_O	Dark saturation current at STC
R_s	Panel series resistance
R_{sh}	Panel parallel (shunt) resistance
n_c	Number of cells in the panel connected in series
V_t	Junction thermal voltage
A	Diode quality (ideality) factor
k	Boltzmann's constant
T_{stc}	Temperature at STC in Kelvin
q	Charge of the electron

TABLE 2.40   The Measured Data Used in Model Estimation

I_{SC}	Short-circuit current at STC
V_{OC}	Open-circuit voltage at STC
V_{mmp}	Voltage at the maximum power point (MPP) at STC
I_{mmp}	Current at the MPP at STC

easily in terms of V_t, q, k, and T_{stc}, of which only the former is unknown, the approach will be to first obtain V_t, and then solve Equation (2.28) for A.

The V–I characteristic will be employed to estimate the model parameters. These characteristics are the short-circuit current, the open-circuited voltage, and the maximum power point. Table 2.40 summarizes the measured data at STC used for model development.

The model of Equation (2.27) is evaluated at the measured data point of Table 2.40.

$$I_{sc} = I_{ph} - I_o e^{\frac{I_{sc}R_s}{n_c V_t}} - \frac{I_{sc}R_s}{R_{sh}} \tag{2.30}$$

$$I_{mmp} = I_{ph} - I_o e^{\frac{V_{MPP}+I_{MPP}R_s}{n_c V_t}} - \frac{V_{MPP} + I_{MPP}R_s}{R_{sh}} \tag{2.31}$$

$$I_{oc} = 0 = I_{ph} - I_o e^{\frac{V_{oc}}{n_c V_t}} - \frac{V_{oc}}{R_{sh}} \tag{2.32}$$

The maximum power point corresponds to the point where rate of change of power with respect to the voltage is zero on the V–I characteristic, then the derivative of power with respect to voltage is zero.

$$\left.\frac{dP}{dV}\right|_{\substack{V=V_{MPP} \\ I=I_{MPP}}} = 0 \tag{2.33}$$

For estimating five parameters, a fifth equation is still needed. The derivative of the current with the voltage at short circuit is given as the negative of the reciprocal of R_{sho},

$$\left.\frac{dI}{dV}\right|_{I=I_{sc}} = -\frac{1}{R_{sho}} \tag{2.34}$$

Hence, five equations with five variables have been established. The detailed derivation of five parameters from five equations are given in References 26–28.

2.18 CONCLUSION

In this chapter, the photovoltaic energy planets have been studied. Specifically the design of PV plants, how to estimate the energy yield of a photovoltaic module based on the angle of inclination for a PV string with respect to the

TABLE 2.41 Voltage and Current Characteristics of a Typical PV Module

Power (max)
Voltage at maximum power point (MPP)
Current at MPP
V_{oc} (open-circuit voltage)
I_{sc} (short-circuit current)
Efficiency
Cost
List five operating temperatures for V_{oc} vs. I_{sc}
Width
Length
Height
Weight

position of the sun, the irradiance as the density of radiation incident on a given surface in W/m^2 or W/ft^2 and an estimation method to construct a model for a PV plant were described.

PROBLEMS

2.1 Search the Internet and specify four PV modules. Give a table as shown in Table 2.41 and compare the rated voltage, cost, width, length, and weight.

2.2 Search the Internet to find the voltage–current characteristic of four PV modules. Make a table of input impedances as current varies for each operating temperature. Develop a plot of input impedance as a function of PV load current for each operating temperature.

2.3 Design a microgrid of PV plant rated at 100 kW of power at 230 V AC using a PV module with the following voltage and current characteristics shown in Table 2.42.

Determine the following:

(i) Number of modules in a string for each PV type

(ii) Number of strings in an array for each PV type

(iii) Number of arrays

TABLE 2.42 Photovoltaic Module Data for Problem 2.3

Power (max)	400 W
Voltage at Maximum Power Point (MPP)	52.6 V
Current at MPP	6.1 A
V_{oc} (open-circuit voltage)	63.2 V
I_{sc} (short-circuit current)	7.0 A

(iv) Inverter specifications

(v) One-line diagram of the PV plant

2.4 Design a microgrid of PV plant rated at 600 kW of power at 460 V AC using a PV module with the data given in Table 2.46.

Determine the following:

(i) Number of modules in a string for each PV type

(ii) Number of strings in an array for each PV type

(iii) Number of arrays

(iv) Inverter specifications

(v) One-line diagram of this system

2.5 Search the Internet for four single-phase inverters and summarize the operating conditions in a table and discuss the results.

2.6 Search the Internet for DC–DC boost converters and DC–AC inverters and create a table and summarize the operating conditions of four DC–DC boost converters and DC–AC inverters in a table and discuss the results and operations.

2.7 Design a PV plant of 50 kW, rated at 230 V AC. Use the PV module of Problem 2.3 and the converters of Problem 2.6. The design should use the least number of converters and inverters. Determine the following:

(i) Number of modules in a string for each PV type

(ii) Number of strings in an array for each PV type

(iii) Number of arrays

(iv) DC–DC converter and inverter specifications

(v) One-line diagram of PV plant

2.8 Design a PV plant of 600 kW of power rated at 230 V AC. Use the PV module of Problem 2.6. The design should use the least number of converters and inverters. Determine the following:

(i) Number of modules in a string for each PV type

(ii) Number of strings in an array for each PV type

(iii) Number of arrays

(iv) DC–DC convert and inverter specifications

(v) One-line diagram of this PV plant

2.9 Design a PV plant rated at 2 MW and connected through a smart net metering to the local utility at 13.2 kV. The local loads consists of 100 kW of lighting loads rated at 120 V and 500 kW of AC load rated at 220 V. The storage system has a 700 kWh capacity. Local transformer specifications are 13.2 kV/460 V, 2 MVA with negligible reactance, and each load transformer is rated 460 V/230 V, 250 kVA, with negligible

TABLE 2.43   Photovoltaic Module Data

Panel	Type 1	Type 2	Type 3	Type 4
Power (Max), W	190	200	170	87
Voltage at max. power point (MPP), V	54.8	26.3	28.7	17.4
Current at MPP, A	3.47	7.6	5.93	5.02
V_{OC} (open-circuit voltage), V	67.5	32.9	35.8	21.7
I_{SC} (short-circuit current, A	3.75	8.1	6.62	5.34
Efficiency, %	16.40	13.10	16.80	>16
Cost, $	870.00	695.00	550.00	397.00
Width, in.	34.6	38.6	38.3	25.7
Length, in.	51.9	58.5	63.8	39.6
Thickness, in.	1.8	1.4	1.56	2.3
Weight, lb	33.07	39	40.7	18.3

reactance; and 460 V/120 V, 150 kVA, with negligible reactance. The data for this problem are given in Table 2.43.

Search the Internet for four DC–DC boost converters, rectifier, and inverters and create a table. Summarize the operating conditions in a table and discuss the results and operations as relates to this design problem. Develop a MATLAB testbed to perform the following.

(i) Select boost converter, bidirectional rectifier, and inverters for the design of a PV plant from commercially available converters. If commercial converters are not available, specify the data for a new design of a boost, bidirectional rectifier, and inverters.

(ii) Give the one-line diagram of your design. Make tables and give the number of modules in a string for each PV type, number of strings in an array for each PV type, number of arrays, converters, weight and surface area required for each PV module type.

(iii) Design a 700 kWh storage system. Search online and select a deep-cycle battery storage system. Give the step in your design and including the dimension and weight of the storage system.

(iv) Develop the PV plant one-line diagram.

2.10 Design a PV plant operating at voltage of 400 V DC serving a load of 50 kW and at 220 V AC. Use the data sets given in Tables 2.44 through 2.47 as needed in your design. Perform the following:

(i) Select a deep-cycle battery to store 100 kWh.

(ii) Select a boost converter, bidirectional rectifier, and inverters for the design of a microgrid from commercially available converters (see data in Tables 2.44 and 2.45). If commercial converters are not

TABLE 2.44 Typical Deep-Cycle Battery Data

Types	Volts	Length (mm)	Width (mm)	Height (mm)	Unit Wt lb (kg)	Capacity Ampere-Hours							
						1-H Rate	2-H Rate	4-H Rate	8-H Rate	24-H Rate	48-H Rate	72-H Rate	120-H Rate
PVX-340T	12	7.71 (196)	5.18 (132)	6.89 (175)	25 (11.4)	21	27	28	30	34	36	37	38
PVX-420T	12	7.71 (196)	5.18 (132)	8.05 (204)	30 (13.6)	26	33	34	36	42	43	43	45
PVX-490T	12	8.99 (228)	5.45 (138)	8.82 (224)	36 (16.4)	31	39	40	43	49	52	53	55
PVX-560T	12	8.99 (228)	5.45 (138)	8.82 (224)	40 (18.2)	36	45	46	49	56	60	62	63
PVX-690T	12	10.22 (260)	6.60 (168)	8.93 (277)	51 (23.2)	42	53	55	60	69	73	76	79
PVX-840T	12	10.22 (260)	6.60 (168)	8.93 (277)	57 (25.9)	52	66	68	74	84	90	94	97
PVX-1080T	12	12.90 (328)	6.75 (172)	8.96 (228)	70 (31.8)	68	86	88	97	108	118	122	126
PVX-1040T	12	12.03 (306)	6.77 (172)	8.93 (227)	66 (30.0)	65	82	85	93	104	112	116	120
PVX-890T	12	12.90 (328)	6.75 (172)	8.96 (228)	62 (28.2)	55	70	72	79	89	95	98	102

TABLE 2.45   Boost Converters

Input Voltage (V)	Output Voltage (V)	Power (kW)
24–46	26–48	9.2
24–61	26–63	12.2
24–78	26–80	11.23
24–78	26–80	11.23
24–78	26–80	13.1
24–98	26–100	12.5
80–158	82–160	15.2
80–198	82–200	14.2
80–298	82–300	9.5

available, specify the data for the new design of a boost, bidirectional rectifier, and inverters.

(iii) Give the one-line diagram of your design. Make tables and give the number of modules in a string for each PV type; number of strings in an array for each PV type; number of arrays, converters, weight, and surface area required for each PV module type.

2.11 Write a MATLAB testbed for designing a PV system with minimum weight and minimum number of inverters using the data of Tables 2.47–2.48

Perform the following:

(i) PV system for 5000 kW at 3.2 kV AC: Specify the inverter operating condition.

(ii) PV system for 500 kW at 460 V AC: Specify the inverter operating condition.

(iii) PV system for 50 kW at 120 V AC: Specify the inverter operating condition.

TABLE 2.46   Single-Phase Inverter Data

Inverter	Type 1	Type 2	Type 3	Type 4
Power	500 W	5 kW	15 kW	4.7 kW
Input Voltage DC	500 V	500 V max	500 V	500 V
Output Voltage AC	230 V AC/60 Hz at 2.17 A	230 V AC/60 Hz at 27 A	220 V AC/60 Hz at 68 A	230 V AC/60 Hz at 17.4 A
Efficiency	Min 78% at full load	97.60%	>94%	96%
Length	15.5 in.	315 mm	625 mm	550 mm
Width	5 in.	540 mm	340 mm	300 mm
Height	5.3 in.	191 mm	720 mm	130 mm
Weight	9 lb	23 lb	170 kg	20 lb

TABLE 2.47   Three-Phase Inverter Data

Inverter	Type 1	Type 2	Type 3	Type 4
Power	100 kW	250 kW	500 kW	1 MW
Input voltage DC	900 V	900 V max	900 V	900 V
Output voltage AC	660 V AC/60 Hz	660 V AC/60 Hz	480 V AC/60 Hz	480 V AC/60 Hz
Efficiency	Peak efficiency 96.7%	Peak efficiency 97.0%	Peak efficiency 97.6%	Peak efficiency 96.0%
Depth, in.	30.84	38.2	43.1	71.3
Width, in.	57	115.1	138.8	138.6
Height, in.	80	89.2	92.6	92.5
Weight, lb	2350	2350	5900	12,000

TABLE 2.48   Data for Problem 2.21

a_1 (I_{sc})	3.87 A
a_2 (V_{oc})	42.1 V
$a_3(V_{MPP})$	33.7 V
a_4 (I_{MPP})	3.56 A
a_5 (n_c)	72

2.12   Consider the residential home in Figure 2.38. Perform the following:

   (i) Estimate the load consumption of the house.

   (ii) Plot the daily load cycle operation of the house's loads over 24 hours and calculate the total energy consumption.

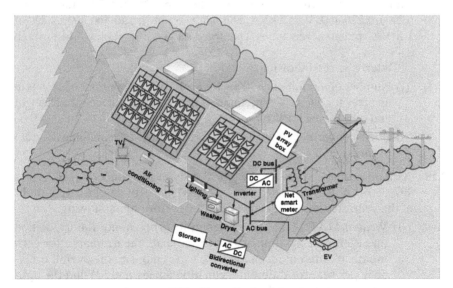

Figure 2.38   Figure for Problem 2.12.

(iii) Search the Internet and select a PV module and design the PV plant for the house. Compute cost, weight of PV array, and roof areas needed for the PV system. Search the Internet and select an inverter, battery storage, and bidirectional converter.

2.13 For Problem 2.11, if only 25% of the load is operated during the night, use the data of Problem 2.10 and specify a battery storage system to store the required energy for operating 25% of the load during the night.

2.14 If the price of kWh from a utility company is $0.3 for buying or selling energy, estimate the net operating cost or revenue for the house of Problem 2.12.

2.15 Design a PV system rated 50 kW using a boost converter and a DC–AC inverter. The system operates as a standalone and supports a water pumping system with a rated load voltage of 120 V AC. Use the data given in Problem 2.10.

2.16 Design a residential PV plant. The load cycle is 10 kW from 11 P.M. till 8 A.M. and 14 kW for the remaining 15 hours. Determine the following:
  (i) Total kWh energy consumption for 24 hours.
  (ii) What is the roof space needed to generate adequate kWh for 24 hours operation?
  (iii) Assume the maximum kWh to be used during the night is 40% of the total daily load. Search the Internet to select a battery storage system and compute the required energy for nightly operation. Give your design data.

2.17 Design a microgrid of PV system rated one MW of power at 220 V, 60 Hz with all the PV strings connected to the same DC bus. The transformer data are 220/460, V 250 kVA, and 5% reactance; and 460 V/13.2 kV of 1 MVA, and 10% reactance. Use the data given in Tables 2.44 through 2.48.
  Determine the following:
  (i) Number of modules in a string for each PV type, number of strings in an array for each PV type, number of arrays and surface area, weight and cost for each PV type.
  (ii) Boost converter and inverter specifications and the one-line diagram of this system.

2.18 Assume a sample value for the global daily irradiation, $G = [1900, 2690, 4070, 5050, 6240, 7040, 6840, 6040, 5270, 3730, 2410, 1800]$ for 12 months of the year. Assume a reflectivity of 0.25. Perform the following:
  (i) Write a MATLAB m-file program to (a) compute the irradiation on different inclination angles, (b) tabulate the irradiance for each month at different inclination angles, (c) tabulate the overall irradiance per year for different inclination angles, and (d) find the optimum inclination angle for each month and a year.

(ii) If the sun irradiance is 0.4 sun for 8 hour daily for this location what is the roof space needed to capture 20 kW at an optimal angle?

(iii) If the sun irradiance is 0.3 sun on the average over a year for 5 hours daily for this location what total kW can be captured over 1500 ft^2 at the optimum inclination angle?

2.19 Assume the global daily irradiation (G) for the city of Columbus on the horizontal surface is as follows:

G = [1800, 2500, 3500, 4600, 5500, 6000, 5900, 5300, 4300, 3100, 1900, 1500] for 12 months of the year. The latitudinal location of Columbus is 40 degrees. Assume a reflectivity of 0.25. Perform the following:

(i) Write a MATLAB m-file to (a) compute the irradiation on different inclination angles, (b) tabulate the irradiance for each month at different inclination angles, (c) tabulate the overall irradiance per year for different inclination angles, and (d) find the optimum inclination angle for each month and a year.

(ii) If the sun irradiance is 0.4 sun for 8 hours daily for this location what is the roof space needed to capture 50 kW at an optimum inclination angle?

(iii) If the sun irradiance is 0.3 sun on the average over a year for 5 hours daily for this location what is the total kWh that can be captured over 1500 ft^2 at the optimum inclination angle?

2.20 For your city, search the Internet for solar irradiation data, G, on the horizontal surface and its latitudinal location. Perform the following:

(i) Write a MATLAB m-file to (a) compute the irradiation on different inclination angle, (b) tabulate the irradiance for each month at different inclination angles, (c) tabulate the overall irradiance per year for different inclination angles, and (d) find the optimum inclination angle for each month and a year.

(ii) If the sun irradiance is 0.3 sun on the average over a year for 5 hours daily for this location what is the total kW that can be captured over 1500 ft^2 at the optimum inclination angle?

2.21 For a PV module given below (Table 2.48), write a MATLAB simulation testbed using Gauss–Seidel iterative approximation and estimate the module parameters (use Internet resources and learn about Gauss–Seidel iterative approximation).

REFERENCES

1. Elmhurst College. *Virtual Chembook*. Energy from the Sun. Available at http://www.elmhurst.edu/~chm/vchembook/320sunenergy.html. Accessed July 10, 2009.

2. Planning and installing photovoltaic systems: a guide for installers, architects and engineers.

3. British Petroleum (BP). Solar. Available at http://www.bp.com/genericarticle.do?categoryId=3050421&contentId=7028816. Accessed July 20, 2009.

4. Markvart, T. and Castaner, L. (2003) *Practical Handbook of Photovoltaics: Fundamentals and Applications*, Elsevier, Amsterdam.

5. Clevelan, C.J. (2006) The Encyclopedia of Earth. Mouchout, Auguste. Available at http://www.eoearth.org/article/Mouchout,_Auguste. Accessed November 9, 2010.

6. California Energy Commission. *Energy Quest, the Energy Story*. Chapter 15: Solar energy. Available at www.energyquest.ca.gov/story. Accessed June 10, 2009.

7. U.S. Department of Energy, Energy Information Administration. Official Energy Statistics from the US Government. Available at http://www.eia.doe.gov/. Accessed September 10, 2009.

8. Georgia State University. The doping of semiconductors. Available at http://hyperphysics.phy-astr.gsu.edu/hbase/solids/dope.html. Accessed November 26, 2010.

9. Carlson, D.E. and Wronski, C.R. (1976) Amorphous silicon solar cells. *Applied Physics Letters*, 28, 671–673.

10. EnergieSolar. Homepage. Available at http://www.energiesolar.com/energie/html/index.htm. Accessed November 26, 2010.

11. American Society for Testing and Materials (ASTM) Terrestrial. ASTM Standards and Digital Library. Available at http://www.astm.org/DIGITAL_LIBRARY/index.shtml. Accessed November 26, 2010.

12. U.S. Department of Energy National Renewable Energy Laboratory. Available at http://www.nrel.gov/. Accessed October 10, 2010.

13. Wikipedia. Augustin-Jean Fresnel. Available at http://en.wikipedia.org/. Accessed October 9, 2009.

14. Keyhani, A. (2011) *Design of Smart Power Grid Renewable Energy Systems*, John Wiley & Sons, Inc. and IEEE Publication.

15. Chatterjee, A. and Keyhani, A. (2012) Neural Network Estimation of Microgrid Maximum Solar Power. *IEEE Transactions on Energy Conversion*.

16. Chatterjee, A., Keyhani, A., and Kapoor, D., (2011) Identification of photovoltaic source models. *IEEE Transactions on Energy Conversion*, 26(3), 883–889.

17. Chatterjee, A., Keyhani, A., and Kapoor, D. (2011) Identification of photovoltaic source models. *IEEE Transaction on Energy Conversion*, 26(3), 883–889.

18. Quaschning, V. Understanding renewable energy systems. Available at http://theebooksbay.com/ebook/understanding-renewable-energy-systems/. Accessed December 20, 2009.

19. Nourai, A. (2002).Large-scale electricity storage technologies for energy management, in Proceedings of the Power Engineering Society Summer Meeting, Vol. 1, Piscataway, NJ: IEEE; pp. 310–315.

20. Song, C., Zhang, J., Sharif, H. and Alahmad, M. (2007). A novel design of adaptive reconfigurable multicell battery for poweraware embedded network sensing systems, in Proceedings of Globecom. Piscataway, NJ: IEEE, pp. 1043–1047.

21. http://en.wikipedia.org/wiki/Solar_energy_in_the_European_Union#Photovoltaic_solarpower. Accessed December 5, 2013.

22. http://www.epia.org/news/publications/. Accessed December 5, 2013.

23. http://www.seia.org/research-resources/solar-industry-data. Accessed December 5, 2013.
24. http://sroeco.com/solar/most-efficient-solar-panels. Accessed December 5, 2013.
25. Siemens. Photovoltaic power plants. Available at http://www.energy.siemens.com/hq/en/power-generation/renewables/solar-power/photovoltaic-power-plants.htm. Accessed October 10, 2010.
26. Gow, J.A. and Manning, C.M. (1999) Development of a photovoltaic array model for use in power-electronics simulation studies in electric power application, in IEEE Proceedings, Vol. 146, Piscataway, NJ: IEEE, pp. 193–200.
27. Esram, T. and Chapman, P.L. (2007) Comparison of photovoltaic array maximum power point tracking techniques. *IEEE Transactions on Energy Conversion*, **22**(2), 439–449.
28. Sera, D., Teodorescu, R., and Rodriguez, P. (2007). PV panel model based on datasheet values, in Proceedings of the IEEE International Symposium on Industrial Electronics. Piscataway, NJ: IEEE, pp. 2392–2396.

ADDITIONAL RESOURCES

Alahmad, M. and Hess, H.L. (2005) Reconfigurable topology for JPLs rechargeable micro-scale batteries. Paper presented at: 12th NASA Symposium on VLSI Design; October 4–5, 2005; Coeur dAlene, ID.

Alahmad, M.A. and Hess, H.L. (2008) Evaluation and analysis of a new solid-state rechargeable microscale lithium battery. *IEEE Transactions on Industrial Electronics*, **55**(9), 3391–3401.

ASTM International. Homepage. Available at http://www.astm.org/

Burke, A. (2005) Energy storage in advanced vehicle systems. Available at http://gcep.stanford.edu/pdfs/ChEHeXOTnf3dHH5qjYRXMA/14_Burke_10_12_trans.pdf. Accessed June 10, 2009.

Chan, D. and Phang, J. (1987) Analytical methods for the extraction of solar-cell single- and double-diode model parameters from I-V characteristics. *IEEE Transactions on Electronic Devices*, **34**(2), 286–293.

Davis, A., Salameh, Z.M., and Eaves, S.S. (1999) Evaluation of lithium-ion synergetic battery pack as battery charger. *IEEE Transactions on Energy Conversion*, **14**(3), 830–835.

Delta Energy Systems. ESI 48/120V Inverter specifications. Available at http://www.delta.com.tw/product/ps/tps/us/download/ESI%20120V%20Inverter.pdf. Accessed June 15, 2009.

Hahnsang, K. (2009) On dynamic reconfiguration of a large-scale battery system, in Proceedings of the 15th IEEE Real-Time and Embedded Technology and Applications Symposium, Piscataway, NJ, IEEE, 2009.

International Energy Agency. (2009). Trends in photovoltaic applications, survey report of selected IEA countries between 1992 and 2008. Report IEA-PVPS Task 1 IEA PVPS T1-18:2009.

Maui Solar Software. Homepage. Available at http://www.mauisolarsoftware.com.

National Renewable Energy Laboratory. Renewable Resource Data Center. Solar radiation for flat-rate collectors—Ohio. Available at http://rredc.nrel.gov/solar/pubs/redbook/PDFs/OH.PDF. Accessed June 15, 2009.

Rodriguez, P. (2007) PV panel model based on datasheet values, in IEEE International Symposium on Industrial Electronics. Piscataway, NJ, IEEE, 2007.

Sandia National Laboratories. Photovoltaic research and development. Available at http://photovoltaics.sandia.gov.

Siemens. Solar inverter systems. Available at http://www.automation.siemens.com/photovoltaik/sinvert/html_76/referenzen/. Accessed June 15, 2009.

U.S. Department of Energy. Solar Energies Technologies Program. Available at http://www1.eere.energy.gov/solar/. Accessed June 25, 2009.

U.S. General Services Administration. Homepage. Available at http://www.gsa.gov/.

CHAPTER 3

FUNDAMENTALS OF POWER CIRCUIT ANALYSIS

3.1 INTRODUCTION

To understand the basic concept of power grid and renewable energy from solar and wind energy, we must understand the basic concept of electric circuit. The electricity can be generated in the form of direct voltage and direct current (DC). The term direct current is designated as DC electricity and the term alternating current (AC) electricity is used when the generated current is described with the sinusoidal current waves.

Thomas Edison is recognized as a tireless inventor and the designer of the first DC generating power plant in 1883. However, the generated DC current could not be transmitted over long distances. With the invention of AC generating stations, DC generating systems were replaced by the AC generating systems.[1–3] Alternating current (AC) and voltage were first developed in France, Italy, and Germany. The AC voltage was stepped up by a transformer and the current was reduced. Therefore, the transmission losses were reduced and the AC power could be transmitted over a large distance. In the United States, it is Nicola Tesla to whom we owe credit for the invention and design of the power grid. Tesla developed a competing electrical grid to Edison's DC-based system. For AC systems,[2,3] a frequency of 60 cycles per second was adopted in the United States and 50 cycles per second was adopted in many parts of world (CPS or Hertz). Alternating voltage is designated as AC voltage or the AC generating system. The AC current is supplied to all energy users. A power grid provides electric energy to end users, who use electricity in their homes and businesses. In power grids, any device that consumes electric energy is referred to as a *load*. In the residential electrical grids, the loads are air

Design of Smart Power Grid Renewable Energy Systems, Second Edition. Ali Keyhani.
© 2017 John Wiley & Sons, Inc. Published 2017 by John Wiley & Sons, Inc.
Companion website: www.wiley.com/go/smartpowergrid2e

conditioning, lighting, television, refrigeration, washing machine, dishwasher, etc. Similarly, the industrial loads are composite loads with induction motors forming the bulk of these loads. Commercial loads consist largely of lighting, office computers, copy machines, laser printers, communication systems, etc. All electrical loads are served at rated nominal voltages. The nominal-rated voltage of each load is specified by the manufacturer. The rated voltage can be a maximum of 5% above or 5% below the rated nominal values. The residential voltages are rated at 120 V, 208 V, and 200 V. Residential electricity service is rated 150 amps or 300 amps. The service in the United States consists of two 208 V lines and neutral and ground conductors. The ground is created by connecting the neutral point of service pole mounted transformer to earth. The commercial and industrial services are provided by three wires and neutral and ground lines. The voltage rating may be 230 V or 415 V or higher.

3.2 BATTERIES

The potential difference between two terminals of a battery is measured in volt, which is named in honor of Alessandro Volta.[4,5] The volt expresses the potential energy between two points. Current flow in a circuit represents energy flow of electrons. As electrons impart energy to the next electrons, the current flow is established from higher potential to lower potential in an electric circuit. When a conductor is connected to the two terminals of a battery, the electron energy flows from negative terminal to positive terminal of the battery establishing the current flow. This is similar to water flow in a water distribution system. Water on the top of a dam has potential energy that can be used to generate electricity. When the water reaches the bottom of the dam, its potential energy is exhausted. In the city water distribution, the pressure causes the water to flow throughout the water piping system. If the pressure is low, at a given point in the water distribution, and if a faucet is then opened at that point, the water flow can be very slow. The electric circuits energized by energy sources can be compared to water-filled pipes. The voltage (difference in electric potential) can be compared to the difference in water pressure at two points of the water distribution system. The battery voltage provides DC. For an ideal battery, DC voltage and current do not change with time. However, a practical battery ages with time, the terminal voltage drops, and it discharges the internal energy over time. In a car that is not driven over a few months, the battery voltage drops substantially and the battery becomes empty of charge and the car will not start.

3.3 DC CIRCUITS AND OHMS LAW

The ratio of voltage over current defines the resistance of the line as given by Equation (3.1)[6]:

$$R = \frac{V}{I} \tag{3.1}$$

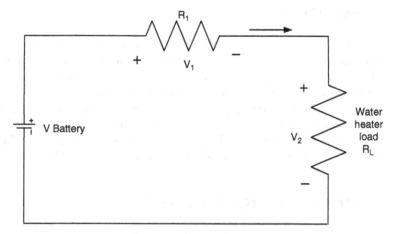

Figure 3.1   The battery voltage source with cable and water heater as load.

The above equation is named after German physicist Georg Simon Ohm and it is called Ohm's law. The ohm (symbol: Ω) describes the unit of electrical resistance in metric system.

Figure 3.1 depicts a battery voltage source and a cable with an electric water heater as the load. The battery source can be supplied from a solar photovoltaic voltage source.

Kirchhoff's circuit law states that the algebraic current from any point in a circuit is zero. That is, the summation of currents entering and leaving is equal at any point in the circuit.[7] The voltage law states that the algebraic sum of voltages around a loop in a circuit is equal to zero. The electrical circuit laws were first described in 1845 by Gustav Kirchhoff.[8]

For example, in Figure 3.1

$$V_2 + V_1 = V_{battery} \tag{3.2}$$

Or algebraically

$$V_2 + V_1 + V_{battery} = 0$$
$$V_{battery} = I \cdot (R_1 + R_{load}) \tag{3.3}$$

The current I in the circuit can be computed from Equation (3.3) as

$$I = V/(R_1 + R_{load}) \tag{3.4}$$

The voltage drop across the heating element such as electric stoves or water heaters are expressed as,

$$V_2 = I \cdot R_{load} \tag{3.5}$$
$$P = V_2 \cdot I = (V_2)^2/R_{load} = I^2 R_{load} \tag{3.6}$$

The unit of P is watts and current in amp and voltage in volts.

3.4 COMMON TERMS

First, let us define a few common terms. The highest level of current that a conductor can carry defines its capacity: this value is a function of the cross-section area of the conductor. The power capacity of an element of power grid is rated in volt amps (VA), and one thousand VA is one kilovolt amp (kVA). One thousand kVA is one megavolt amp (MVA). The energy is the use of electric power by loads over time; it is given in kilowatt-hours (kWh) and one thousand kWh is one megawatt-hours (MWh) and one thousand MWh is one gigawatt-hours (GWh). The reactive power is var and 1000 var is kvar and 1000 kvar is Mvar.[9]

3.5 ELEMENTS OF ELECTRICAL CIRCUITS

The power grid is an interconnected electrical circuit. The elements of electrical circuits are inductors, capacitors, and resistances. The resistance represents the power loss in the circuit or power used by loads. For example, incandescent light bulbs, electrical water heaters, and electrical stoves are constructed of resistances. An incandescent light bulb is a filament wire heated to a high temperature that will results in light. It was invented by Edison. In DC circuits, a load is specified by its power consumption in watts and voltage rating in volts. In AC circuits, a load is specified by its VA or kVA or MVA and voltage rating in volts. Equation (3.7) defines the power consumption of inductive or capacitive loads.

$$P = V \cdot I \cdot pf \tag{3.7}$$

where, the unit of P is watts and pf is power factor. The power factor is either lagging or leading. For inductive load current phase angel is lagging the voltage and for capacitive load current phase angel is leading where the voltage is the reference. The power generation or consumptions and power factor are discussed in detail later in this chapter.

If the load is purely resistive, power factor is unity, then the power consumption is represented by Equation (3.9)

$$P = V \cdot I \tag{3.8}$$

where, the unit of P is watts.

Consider a light bulb. For example, consider a 50 W light bulb. The power and voltage ratings are stamped on the light bulb as shown in Figure 3.2:

The manufacturer of the light bulb is telling us that if we apply 120 volts across the light bulb, the light bulb consumes 50 watts. From the above values, we can calculate the resistance of the light bulb.

$$R_{\text{Light\_bulb}} = \frac{V^2}{P} = \frac{(120)^2}{50} = \frac{14400}{50} = 288\,\Omega$$

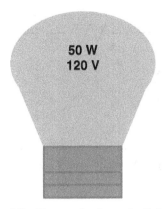

Figure 3.2   The rated values of a light bulb.

If 480 volts is applied to the light bulb, which is four times the rated value, then the light bulb will have a bright glow and possibly the light bulb will explode. Therefore, the rated values tell the safe operating condition of an electrical device.[10]

An electrical circuit is constructed from four elements: resistor (R), inductor (L), capacitor (C), and supply voltage. These concepts are introduced next.

3.5.1 Inductors

An inductor is a coil that is constructed from wires wound around a bobbin and supplied with electricity. If inside of the bobbin or spool has an iron core, then it is called an iron core inductor. The bobbin with wires wrapped on it represents a coil or a winding. The winding has two terminals. If the wire is wound around a piece of iron as shown in Figure 3.3 where the direction of turns is as shown, the direction of magnetic field can be specified using the right hand rule. For the right hand rule, place the fingers of right hand around the wire on the bobbin in the direction of current and the direction of winding enclosing the current, then the thumb points to the north pole. This is the basic

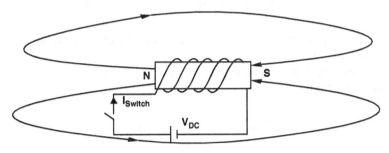

Figure 3.3   An iron core inductor.

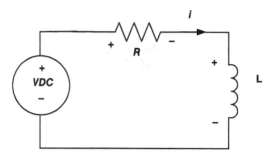

Figure 3.4   The equivalent circuit of Figure 3.3.

concept of an electromagnet. The circles around the coil indicate the magnetic field.

Figure 3.3 depicts an iron core inductor supplied by DC current from a DC voltage source such as a battery. When the switch is open no magnetic field is present. When the switch is closed, the current rises from zero. This is called transient current, or in power grid it is called inrush current or surge. This phenomenon is similar to rush of water in a pipe. When the water valve is opened, water rushes into the pipe (transient) and eventually the pipe is full of water and pressure stabilizes in the pipe. This is called steady-state condition of the water in the pipe. Similarly, the steady state flow of current is established in a circuit after closing of the switch as shown in Figure 3.3 and the transient flow of current gets reduced to a steady state flow.[11] The equivalent circuit of Figure 3.3 is shown in Figure 3.4.

The equivalent circuit in Figure 3.4 depicts the resistance of wire and inductance of iron core after the switch is closed. It is important to keep reference polarities of current sources and voltages sources. In Figure 3.4, a DC voltage source (such as a battery) is designated with the term VDC. The rise in voltage from the VDC source is equal to the sum of voltage drops across the resistor R and the inductor L. As soon as the switch is closed, the direction of current determines the direction of voltage drop across each element of the circuit.[12]

$$V = L \cdot \frac{\text{Change in current}}{\text{Change in time}} = L\frac{I(t + \delta t) - I(t)}{(t + \delta t) - (t)} \tag{3.9}$$

As current rushes through the circuit as time changes, the voltage across inductor changes accordingly.

The following condition defines the voltage rise and voltage drop across each element of the circuit.

$$\text{Voltage rise} = \text{Voltage drop}$$

$$V_{\text{battery}} = V_{\text{R}} + V_{\text{L}} \tag{3.10}$$

where

$$V_R = I \cdot R$$

To monitor energy efficiency of an electric load, the power in kWh, the energy usage must be measured in kW then multiplied by the number of hours the appliance is operating. Recall Equation (3.11) for resistive loads as restated below.

$$P = V \cdot I \tag{3.11}$$

According to above we must measure current injected into the appliance by the voltage across it. The measurement of voltage is not necessary because the voltage across the appliance must be maintained around the rated value plus or minus 5%. However, we need to measure the current. In general, current measurement is more difficult in terms of cost of measuring the current accurately. The least costly way is to measure the strength of magnetic field. The magnetic field around a conductor is a function of current through the conductor. The magnetic field measurement is a function of the current enclosed. Therefore, a meter called clamp ammeter can be used but must enclose one conductor. Otherwise, if both conductors supplying power to the load clamped around, the net enclosed current is zero and the clamp ammeter will read zero current as expected. Figure 3.5 depicts the magnetic field in parallel conductors. For example, suppose, it is desired to measure the energy used by a toaster, then one conductor from the socket supplies 120 V and the other conductor is the neutral conductor and it is at ground potential. The current supplied return to its source through the neutral conductor. To measure the energy used by a toaster, a splitter is used to separate the two conductors.

The electromagnetic energy can be stored in inductors. The energy stored is expressed as

$$W = \frac{1}{2}L \cdot i^2 \tag{3.12}$$

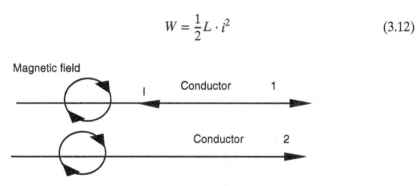

Figure 3.5   Magnetic field around current carrying conductors.

In Equation (3.12), the energy stored in an inductor is designated W and its unit is Joules. We should always remember that the unit of power is watts. One watt is equal to the energy used in one second. W = Joules/Second. One kilowatt hour is the energy used in one hour.

3.5.2 Capacitors

The function of capacitors in electric circuit is to store electric field charges. For example, when electric lighting takes place, there is a flash over between charged clouds and ground (earth). The flash over occurs because of buildup of charges in the clouds. Batteries are also another example of stored charges by electrochemical process. A capacitor is a two-terminal circuit element. The current flow in a capacitor is proportional to the change of voltage across its terminals. The coefficient of change in voltage, designated as capacitance, is proportional to the change in electric field over the change in time. Therefore, a capacitor is the element that is used to model the effect of the electric field in a circuit. The energy stored in electric fields effect the voltage and current in the circuit. There is a capacitance whenever there are electric fields, conversely there are electric fields whenever there is a voltage between conductors. In power grid, the charge conductors carrying electric power, there are capacitors distributed along the length as depicted in Figure 3.6.

However, if the line is short, the distributed capacitance is ignored. For long and medium transmission lines, the capacitance is represented by lump sum in middle of line or one half is placed at each end of the line.

The unit of capacitance is Farad (or F), and it is named after Michael Faraday[13] (1719–1867), an English scientist. Farad (F) is equal to an ampere-second/volt. Since an ampere is a Coulomb/second (C), F is equal to Coulomb per volt, that is, C/V.

The capacitance is presented by a schematic symbol shown in Figure 3.7.

An electric circuit with resistance R and capacitance C is presented in Figure 3.8.

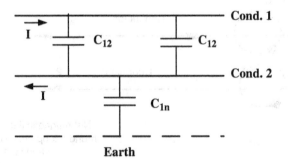

Figure 3.6   Distributed capacitances between two charged conductors and ground.

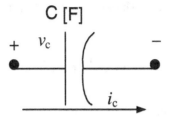

Figure 3.7   The schematic representation of a capacitor.

Before the switch is closed, the current flow in the circuit is zero and the capacitor can have charge on it. Assuming the capacitor has zero charge, the circuit can be represented by

$$\text{Voltage rise} = \text{Voltage drop}$$

$$V_{\text{battery}} = V_R + V_C \tag{3.13}$$

where

$$V_R = I \cdot R \tag{3.14}$$

And
Current flowing in the circuit can be represented by

$$I = C \cdot \frac{\text{Change in voltage}}{\text{Change in time}} = C \cdot \frac{V(t + \delta t) - V(t)}{(t + \delta t) - (t)} \tag{3.15}$$

Equation (3.15) represents the current change as a function of the voltage change in the circuit. Therefore, after the switch is closed and after a few seconds, the capacitor will be charged and reaches the battery voltage and current will not flow and will be zero and circuit reaches steady state. When a battery is being charged by battery charger, the circuit can be approximated by R and C circuit. Just replace the capacitor by another battery and the source voltage battery by a DC voltage source. A rectifier is used to convert AC power to DC

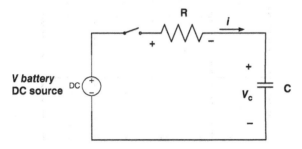

Figure 3.8   A resistor and capacitor circuit.

power for charging a battery. The energy stored in the capacitor is expressed by Equation (3.16).

$$W_C = \frac{1}{2} \cdot C \cdot V^2 \tag{3.16}$$

The unit of W_C is in Joules.

It is easier to understand energy stored in battery and electric field because of our experience with static charges. However, it is difficult to understand the inductors and electromagnetic field. To understand inductance, it is necessary to understand magnetic field. Recalling the permanent magnet, and asking why a magnet attracts metals such as nails? Figure 3.9 presents a piece of iron before being magnetized.

Figure 3.9 shows the random polarity of magnetic field in a nonmagnetized iron. However, when this piece of iron comes in contact with magnet, the random field of the iron will have polarized orientation as shown in Figure 3.10.

If the magnetized iron comes in contact with nails, the same process happens and nails are attracted to the iron, that is, north pole attracted to south pole. The same magnetization can be established if DC current is used as shown in Figure 3.11 using a winding.

The development of electromagnetic induction by Faraday laid the foundation for invention of AC generator, AC motors, by Tesla. Tesla laid the foundation for interconnected power grids and electrification of the world. The DC electricity is being generated by photovoltaic solar energy by converting light energy to DC electricity. The battery development is the last step in electrification of transportation and reduction of carbon footprints on modern civilization.

n	s	n←s	s→n
n ↑ s	s ↓ n	n ↑ s	s ↓ n
s→n	n←s	s	n
n	s	s ↓ n	n ↑ s

Figure 3.9 Domain polarity in a piece of iron.

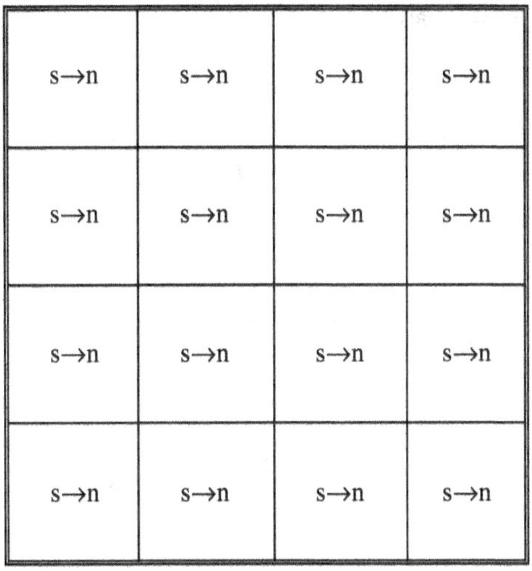

s→n	s→n	s→n	s→n
s→n	s→n	s→n	s→n
s→n	s→n	s→n	s→n
s→n	s→n	s→n	s→n

Figure 3.10   The magnetized iron magnetic field orientation.

3.6 CALCULATING POWER CONSUMPTION

The power consumption in a DC circuit can be computed as

$$P = V \cdot I = \frac{V^2}{R} = I^2 R \tag{3.17}$$

In an AC circuit currents and voltages are complex numbers. For example, the voltage must be represented in complex domain in polar notation and is expressed as $|V| \angle(\theta_V$. It can also be expressed in Cartesian domain. $V = |V| \ (\text{Cos}\theta_V + j\,\text{Sin}\theta_V)$.

If you are not familiar with complex numbers, you should study the summary presented in Section 3.6.1.

3.6.1 Complex Domain

Complex number operations are very important in the steady state of power systems and power flow computation and power factor correction.

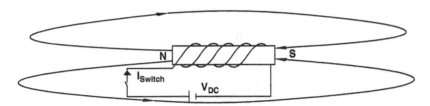

Figure 3.11   Electromagnetization by using DC current.

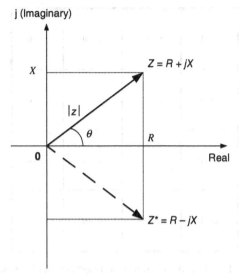

Figure 3.12   The complex number representation.

In AC circuit, the ratio of voltage over current is expressed as impedance.

$$Z = R + jX$$

where the unit of Z is ohms and R and X are real numbers, and $j = \sqrt{-1}$: $R =$ real part of Z; $X=$ imaginary part of Z.

Figure 3.12 depicts a complex number plane.

where $Z^* = R - jX$ is called the conjugate of Z.

A complex number can also be written in phasor form:

$$Z = Z(\cos\theta + j\sin\theta)$$
$$Z = |Z|\, e^{j\theta}$$

Which can also be represented as $Z = |Z| \angle\theta$ where $|Z|$ is the magnitude of impedance θ is angle of impedance in degrees.

Conversion between Cartesian and polar coordinates can be expressed as:

$$Z = R + jX \Rightarrow Z = Z\angle\theta$$
$$|Z| = \sqrt{(R^2 + X^2)}$$
$$\theta = \tan^{-1}(X/R)$$
$$Z = |Z| \angle\theta \Rightarrow Z = R + jX$$
$$R = |Z|\cos\theta$$
$$X = |Z|\sin\theta$$

We can add or subtract two complex numbers, Z_1 and Z_2, as shown:

$$Z_1 = R_1 + jX_1, Z_2 = R_2 + jX_2$$

Addition/subtraction

$$Z_1 + Z_2 = (R_1 + jX_1) + (R_2 + jX_2)$$
$$Z_1 - Z_2 = (R_1 - jX_1) - (R_2 - jX_2)$$

We can multiply two complex numbers, Z_1 and Z_2, in two ways.

$$Z_1 \times Z_2 = (R_1 R_2 - X_1 X_2) + j(X_1 R_2 + R_2 X_1)$$
$$Z_1 \cdot Z_2 = |Z_1| \angle\theta_1 \cdot |Z_2| \angle\theta_2 = |Z_1| + |Z_2| \angle\theta_1 + \theta_2$$

We can obtain the conjugate of Z_1 as:

$$Z_1 Z_2^* = (R_1 + jX_1)(R_2 - jX_2) = (R^2 + X^2)\angle 0 = |Z_1|^2 \angle 0$$

where

$$Z_1^* \text{ is the conjugate of } Z_1$$

We can divide two complex numbers in Cartesian form or polar form.

$$Z_1 = R_1 + jX_1, Z_2 = R_2 + jX_2$$
$$\frac{Z_1}{Z_2} = \frac{(R_1 + jX_1)}{(R_2 + jX_2)} = \frac{(R_1 + jX_1)(R_2 - jX_2)}{(R_2 + jX_2)(R_2 - jX_2)}$$

The results are given below.

$$\frac{Z_1}{Z_2} = \frac{R_1 R_2 + X_1 X_2}{R^2 + X^2} + j\frac{X_1 R_2 - R_1 X_2}{R^2 + X^2}$$

The addition and subtraction can be easily done in polar form, whereas multiplication and division will have more computation if they are done in Cartesian form.

Let us use the polar form and multiply Z_1 and Z_2,

$$Z_1 = |Z_1| e^{j\theta_1}, \ \ Z_2 = |Z_2| e^{j\theta_2}$$

To multiply, we simply multiply the magnitudes and add the phase angles.

$$Z_1 \times Z_2 = |Z_1| |Z_2| e^{j(\theta_1 + \theta_2)}$$

In general, we can take a complex number to the power of n as:

$$Z_1^n = |Z_1|^n e^{j(n\theta_1)} = |Z_1|^n (\cos n\theta_1 + j \sin n\theta_1)$$

To divide two complex numbers, we can simply divide the magnitudes and subtract the angles:

$$\frac{Z_1}{Z_2} = \frac{|Z_1|}{|Z_2|} e^{j(\theta_1 - \theta_2)}$$

Multiplication and division are easier to perform in phasor form.

To calculate the power consumption in a single-phase AC circuit, we need to use the complex conjugate of current and multiply it by voltage across the load.

$$S = V.I^* = |V| \ |I| \angle (\theta_V - \theta_I) \tag{3.18}$$

In Equation (3.19), V and I are the root mean square (RMS) values of voltage and currents. The power factor (pf) is computed based on the phase angle between the voltage and current where the voltage is the reference phase.

$$p.f. = \cos (\theta_V - \theta_I)$$
$$Z = \frac{|V| \angle \theta_V}{|I| \angle \theta_I} \tag{3.19}$$

In the above equation, $\theta = (\theta_V - \theta_I)$, is also the angle of impedance. The complex power has two components; active power consumption, P and reactive power consumption, Q as shown in Equation (3.20)

$$S = |V| \ |I| (\cos \theta + j \sin \theta) = P + jQ \tag{3.20}$$

For complex power, the voltage across the load can also be expressed as:
$V = I \cdot Z$

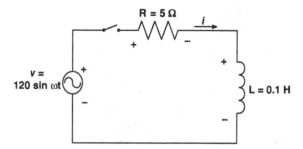

Figure 3.13   An R–L circuit.

Therefore, we also have the complex power expressed by Equations (3.21) and (3.22).

$$S = V \cdot I^* = I \cdot Z \cdot I^* = |I|^2 Z \qquad (3.21)$$

$$S = V \cdot I^* = V \left(\frac{V}{Z}\right)^* = \frac{|V|^2}{Z^*} \qquad (3.22)$$

Let us review the basic inductive circuit, R–L as given below in Figure 3.13.

In the circuit depicted in Figure 3.13, we are using the standard polarity notation. We mark the polarity of each element by a plus and a minus. These markings facilitate the application of Kirchhoff's law of voltage. The positive terminal indicates the direction of current flow in the circuit. For example, in this circuit, the source voltage rise—from the minus terminal to the positive terminal—is equal to the drop across the resistance and inductance in the circuit.

If, at $t = 0$, the switch is closed, the response current is given by the differential equation:

$$v = Ri + L\frac{di}{dt} \qquad (3.23)$$

In Equation (3.23), v and i are instantaneous values of voltage and current. After some time, the current reaches steady state. This condition is shown in Figure 3.14.

At the steady state, v and i can be represented by phasor forms as V and I. In this book, we are interested in the design of a power grid for steady-state operation. We can express the steady-state operation of an R–L circuit as:

$$V = V_{rms} \angle \theta_V, \quad V_{rms} = \frac{120}{\sqrt{2}}$$

$$I = I_{rms} \angle \theta_I, \quad \theta_I = \omega t + \theta_{I,0}$$

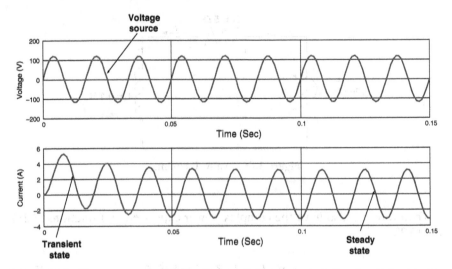

Figure 3.14 The responses of the voltage and the current of the circuit depicted in Figure 3.13.

where V is the RMS value and θ_V is the phase angle of V and θ_I is the phase angle of I (RMS) that is determined by θ_V and impedance angle $\theta(\theta_I = \theta_V - \theta)$, and $\theta_{I,0}$ is the initial angle of current.

$$I = I_{rms}e^{j(\omega t + \theta_{I,0})} \Rightarrow \frac{dI}{dt} = j\omega I_{rms}e^{j(\omega t + \theta_{I,0})} = j\omega I$$

The above differential equation can now be presented in its steady-state forms as:

$$V = (R + j\omega L)I \tag{3.24}$$

Let $X_L = \omega L$ represent the inductor reactance.

$$V = (R + jX_L)I \tag{3.25}$$

Let $Z = R + jX_L = |Z| \angle\theta$, $\theta = \tan^{-1}\left(\frac{X_L}{R}\right) > 0$

We will have $\theta_V - \theta_I = \theta$

Generally, we choose V as the reference voltage, then $\theta_V = 0$

$$V = |V| \angle 0, \quad I = \frac{V}{Z} = |I| \angle -\theta$$

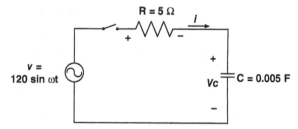

Figure 3.15 An R–C circuit supplied by an AC source.

Therefore, the power supplied by a power source to an inductive load can be expressed by Equation (3.26):

$$S = VI^* = |V|\,|I|\cos\theta + j\,|V|\,|I|\sin\theta = P + jQ \qquad (3.26)$$

As can be seen from the above definition, an inductive load will have a lagging pf Therefore, the power source supplies reactive power to the load. The reactive power supplied by the source is consumed by the inductive load. This can also be expressed as:

$$S = VI^* = (R + j\,X_L)I \cdot I^* = |I|^2\,R + j\,|I|^2\,X_L = P + jQ$$

Now let us look at an R–C circuit given by Figure 3.15.

If at $t = 0$ the switch is closed, the system differential equation can be presented as:

$$v = RC\frac{dv_c}{dt} + v_c \qquad (3.27)$$

In Equation (3.27) v and v_c are instantaneous values. After some time, the transient response will die out and the voltage across the capacitor reaches steady state. This is shown in Figure 3.16.

At steady state, the source voltage V and capacitor voltage V_C, can be represented by phasors V and V_C. By now, you should have noticed that we are using capital letters V and I to depict RMS values.

$$V = V_{rms}\angle\theta_V, \quad V_{rms} = \frac{120}{\sqrt{2}}$$

$$V_C = V_{C,rms}\angle\theta_{Vc}$$

$$\frac{dV_C}{dt} = j\omega V_C I = C\frac{dV_C}{dt} = j\omega C V_C$$

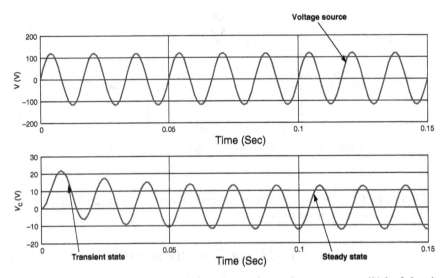

Figure 3.16 The voltage response (V) and capacitor voltage response (V_C) of the circuit depicted in Figure 3.15.

Therefore, we will have

$$V_C = I \cdot \left(\frac{1}{j\omega C} \right) = I \cdot \left(-j\frac{1}{\omega C} \right)$$

For the R–C circuit of Figure 3.15, we will have

$$V = \left(R - j\frac{1}{\omega C} \right) .I$$
$$V = (R - jX_C)I$$

where $X_C = \frac{1}{\omega C}$ and X_C is the capacitor reactance.

Let $Z = R - jX_C = |Z| \angle -\theta$

Then, $\theta = \tan^{-1}\left(\frac{X_C}{R} \right)$

Because we choose the load V as reference, then set θ_V to zero ($\theta_V = 0$) and we have $\theta_V - \theta_I = -\theta$

$$V = |V| \angle 0, \quad I = \frac{V}{Z} = |I| \angle \theta$$

Therefore, for a capacitor load, the current through the capacitor leads the voltage. The power supplied by the AC source can be expressed as:

$$S = VI^* = |V|\,|I|\cos\theta - j\,|V|\,|I|\sin\theta = P - jQ \tag{3.28}$$

The capacitive loads will supply reactive power to the source. We can also write the complex power as:

$$S = VI^* = (R - j\,X_C)I \cdot I^* = |I|^2\,R - j\,|I|^2\,X_C = P - jQ \tag{3.29}$$

The pf is computed based on the angle between the voltage and current where the voltage is reference phase, and p.f. $= \cos(\theta_V - \theta_I)$ with designation of leading or lagging. That is, for inductive loads, the load current lags the voltage and for the capacitive load, the load current will lead the voltage. Because the impedance is the ratio of voltage over the current flowing through impedance, the impedance of inductive loads has a positive-phase angle and the impedance of capacitive loads has a negative-phase angle.

3.6.2 Diodes

Diodes are two-terminal devices. An ideal diode has zero resistance to current flow in one direction called the diode's *forward* direction and high resistance, ideally infinite, resistance in the other direction (the reverse direction). Diodes are made from silicon, or selenium, or germanium. A **semiconductor diode** is constructed from a crystalline piece of semiconductor material with a p–n junction. Semiconductor diodes conduct electricity only if a certain threshold voltage or cut-in voltage in the forward direction called forward bias is established. When the voltage drops below a certain threshold voltage, the current flow is cut off and the diode is reverse biased. If a diode is placed in a battery-lamp circuit, the diode will either allow or prevent current through the lamp, depending on the polarity of the applied voltage. If forward biased, the lamp will light up. If reverse biased, the lamp will not light up since the current flow is cut off by the diode. The positive terminal is called anode and negative terminal cathode, as shown in Figure 3.17.

3.6.3 Controllable Switch

The controllable semiconductor switches are three-terminal devices.

Figure 3.17 A diode with direction of current flow specified.

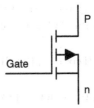

Figure 3.18   A controllable semiconductor switch.

The switch is controlled through its gate. When a pulse appears at the gate, the switch is forward biased and current can flow from "p" to "n." Otherwise, the switch is reverse biased and no current flows through the switch and acts as an open circuit Figure 3.18 depicts a controllable switch.

3.6.4 The DC–DC Converters in Green Energy Grids

The block diagram of a DC–DC converter is depicted in Figure 3.19. The DC–DC converter is a three-terminal device. The input voltage is converted to a higher or lower output voltage by controlling the switching frequency. The duty cycle D defines the relationships between the input voltage and output voltage. A switching signal provides the command to the switch of the converter, which can be used to vary the value of duty cycle D. Depending on whether the output voltage is lower or higher than the input value; the converter is called a *buck* or *boost* or *buck–boost* converter.

The components of DC–DC converters are an inductor L, a capacitor C, a controllable semiconductor switch S, a diode D, and load resistance R. The power electronic switches and diodes are the key element for stepping up the input DC voltage to higher level of DC voltage in the case of a boost converter and to step down the voltage for a buck converter. The exchange of energy between an inductor and a capacitor is used for designing DC–DC converters.[14] The current slew rate through a power switch is limited by an inductor. The inductor stores the magnetic energy for the next cycle of transferring the energy to the capacitor. The slew rate is defined to describe how quickly

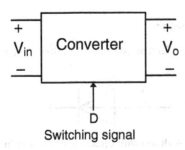

Figure 3.19   Block diagram of a DC–DC converter.

a circuit variable changes with respect to time. The high transient switching current is damped by the switching resistance due to switching losses. The stored energy is expressed in Joules as a function of the current by Equation (3.30).

Energy stored in inductor$= \frac{1}{2}L \cdot i_L^2$

$$\text{Energy stored in capacitor} == \frac{1}{2}C \cdot v_C^2 \tag{3.30}$$

where i_L is the inductor current and v_c is the voltage across the capacitor as a function of time.

The stored energy in an inductor is recovered by a capacitor. The capacitor provides a new level of controlled output DC voltage.

The fundamental voltage and current dynamic response are represented by Equations (3.31) and (3.32).

$$V_1 = L \cdot \frac{\text{Change in current}}{\text{Change in time}} = L \cdot \frac{i(t + \delta t) - (t)}{(t + \delta t) - (t)} = L\frac{di}{dt} \tag{3.31}$$

$$i = C \cdot \frac{\text{Change in voltage}}{\text{Change in time}} = C.\frac{V(t + \delta t) - V(t)}{(t + \delta t) - (t)} = C\frac{dv_c}{dt} \tag{3.32}$$

The power switch is used to charge the current through inductor upon closing the switch. However, because the current and voltage are related by Equations (3.31) and (3.32), the energy stored in the inductor is transferred to the capacitor in the discharge phase of the switching cycle. The basic elements of the circuit topology can be reconfigured to reduce the output voltage or to increase the output voltage. In Section 3.6.5, the step-up converter and the step-down converter in Section 3.6.6 are presented.

3.6.5 The Step-Up Converter

A PV module is a variable DC power source. Its output increases as the sun rises and it has its maximum output at noon when the maximum solar energy can be captured by the solar module. The same is true for a variable-speed wind energy source. When the wind speed is low, a limited mechanical energy is converted to variable frequency electrical power; in turn, the rectified variable AC power provides a low-voltage DC power source. A DC–DC boost converter allows capturing a wider range of DC power by boosting the DC voltage. The higher DC voltage is controlled to be in a range that can be converted to AC power at a system operating frequency by DC/AC inverters.

A step-up converter is called a boost converter; it consists of an inductor L, capacitor C, and controllable semiconductor switch S, diode D, and load resistance R as depicted in Figure 3.20.

When the switch S is on, the inductor draws energy from the source and stores it in the inductor as magnetic field energy. When the switch is turned off that energy is transferred to the capacitor. When steady state is reached,

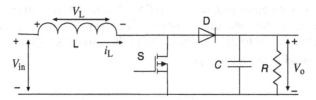

Figure 3.20   A boost converter circuit.

the output voltage is controlled to be higher than the input voltage and the magnitude depends on the duty ratio of the switch.

To explain how the boost converter operates, let us assume that the inductor is charged in the previous cycle of operation and the converter is at steady-state operation. Let us start the cycle of operation with the power switch S open. This condition is depicted in Figure 3.21a. Since the inductor is fully charged in the previous cycle, it continues to force its current through the diode D to the output circuit and charge the capacitor. In the next cycle, the power switch, S is closed. Now, the diode will be reverse biased by the capacitor voltage; hence, it will act as open. The equivalent circuit is shown in Figure 3.21b. Now the voltage across the inductor is equal to the input voltage, $+V_{in}$. Therefore, by Equation (3.31), the current in the inductor starts to rise from its initial value with a slope V_{in}/L. The output voltage now is supplied by the charged capacitor alone. The inductor current i_L, is shown in Figure 3.22.

Figure 3.23 shows the steady-state current and voltage waveforms of the inductor for a few cycles of operation. In steady state, the average inductor

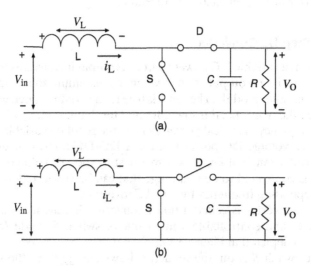

Figure 3.21   The equivalent circuit of boost converter when the switch S is (a) open and (b) closed.

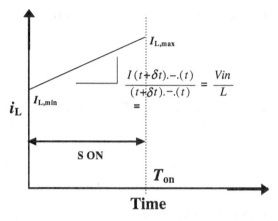

Figure 3.22 Charging phase: when the switch is closed, the current ramps up through the inductor.

voltage is zero. Therefore, in the steady-state operation, as expected, the inductor current is constant. If the average voltage of the inductor had not been zero, the average value of the inductor current would have continued to rise or fall depending on the polarity of the inductor voltage. For this case, the inductor current would not have returned to the value it started from, when the cycle began, at the end of the switching period according to inductor rate of current change.

$$V_1 = L \cdot \frac{\text{Change in current}}{\text{Change in time}} = L \cdot \frac{i(t + \delta t) - (t)}{(t + \delta t) - (t)}$$

The waveforms of a boost converter in steady state are shown in Figure 3.23. If the initial voltage of the capacitor is zero, the inductor current slowly charges up the capacitor. The output voltage across the capacitor rises over

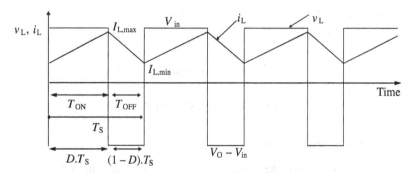

Figure 3.23 The steady-state voltage and current waveform of a boost converter.

each cycle until a steady state is reached are presented by Equations (3.33) through (3.37).

$$V_{\text{o}} = \frac{V_{\text{in}}}{1-D} \tag{3.33}$$

Where f_s is switching frequency and T_S is defined as $T_S = 1/f_S$ and $T_S = T_{\text{on}} + T_{\text{off}}$ (see Figure 3.23) is defined as the switching period. The duty ratio is defined as

$$D = \frac{T_{\text{on}}}{T_S} = \frac{T_{\text{on}}}{T_{\text{on}} + T_{\text{off}}} \tag{3.34}$$

The value of D is less than one. Therefore, the output voltage is always more than the input voltage. The relationship of the input current to output current is obtained from power balance. Because the input and output power must be balanced for a lossless system, the output current is given as

$$V_{\text{in}} \cdot I_{\text{in}} = V_{\text{O}} \cdot I_{\text{O}} \tag{3.35}$$

Where I_{in} and I_{O} are the average input and output currents, respectively, therefore the input current is given as:

$$I_{\text{in}} = I_{\text{L}} = \frac{V_{\text{O}}}{V_{\text{in}}} \cdot I_{\text{O}} \tag{3.36}$$

$$I_{\text{in}} = \frac{I_{\text{O}}}{1-D} \tag{3.37}$$

Here, I_{L} is the average value of inductor current. The ripple in the inductor current, ΔI_{L}, is the difference in its maximum and minimum values in steady state.

$$\Delta I_{\text{L}} = I_{\text{L,Max}} - I_{\text{L,Min}} = \frac{V_{\text{in}}}{L} \cdot D T_S \tag{3.38}$$

The maximum and minimum values of the inductor current can be determined from its average values.

$$I_{\text{L,Max}} = I_{\text{L}} + \frac{\Delta I_{\text{L}}}{2}, \quad I_{\text{L,Min}} = I_{\text{L}} - \frac{\Delta I_{\text{L}}}{2} \tag{3.39}$$

Figure 3.24 depicts the no-load voltage condition in a boost converter.

3.6.6 The Step-Down Converter

The buck and boost converters are used in a green energy microgrid to lower DC bus voltage or step up the DC bus voltage. To charge a battery, the DC bus

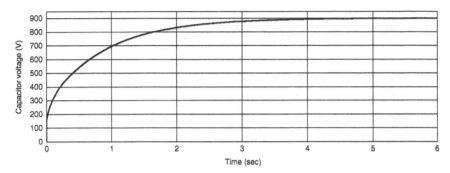

Figure 3.24 No-load output voltage build-up of a boost converter with a supply voltage of 40 V and a duty ratio of 50%.

voltage is stepped down to battery bus voltage. To use the DC power stored in storage system, the storage system DC bus voltage is stepped up using a boost DC–DC converter.

At present, the storage battery in residential and commercial systems is rated at 6 V and 12 V DC. Three 12 V DC batteries can be used in series to create a 36 V DC bus voltage for a community storage system. To obtain higher kilowatts of storage, a number of 36 V DC bus systems can be connected in parallel. The storage systems with high kilowatts and DC voltage are used as community storage systems that are placed in substations of utilities.

A step-down converter is a basic converter and is known as a buck converter. As inferred from its name, the main function of this converter is to convert the input DC voltage level to another and lower voltage level. The main components in a buck converter are a semiconductor switch S, diode D, inductor filter L, and capacitor filter C as shown in Figure 3.25.

The operating cycle of the converter can be explained starting from the closing of the switch S. During this time, the diode D is reverse biased by the input voltage and is open. The equivalent circuit is shown in Figure 3.26.

As can be seen in Figure 3.26, the voltage across the inductor is $V_{in} - V_o$. Because the input voltage is more than the output voltage, the voltage across the inductor is positive and the inductor is being charged. The current in the

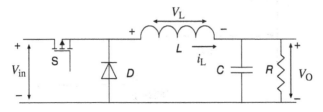

Figure 3.25 A buck converter circuit.

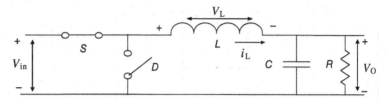

Figure 3.26 The equivalent circuit of a step-down converter when the switch "S" is closed.

inductor can determined from Equation (3.40)

$$V_1 = L \cdot \frac{\text{Change in current}}{\text{Change in time}} = L \cdot \frac{i(t + \delta t) - (t)}{(t + \delta t) - (t)}$$

$$di_L/dt = \frac{I(t + \delta t) - (t)}{(t + \delta t) - (t)} \tag{3.40}$$

Because v_L is positive, change in di_L/dt is positive and the current rises from its initial value. This phase of the operation takes place as long as the switch S is on. This time is the on time, T_{on}, as shown in Figure 3.27. Here, the inductor continues to be charged and stores energy in it as magnetic field energy, which is released in the next phase of operation to the capacitor.

At the end of time T_{on}, the switch S is opened and the next phase of operation starts. The inductor current needs to maintain itself. The only path for that occurs when the switch is off and the inductor forces the diode into conduction. In the next phase of operation, the switch S is opened and the diode D starts to conduct. The equivalent circuit is shown in Figure 3.28.

The inductor plays the role of energy transfer from the input to the output circuit. The steady-state operation of a buck converter is shown in Figure 3.28.

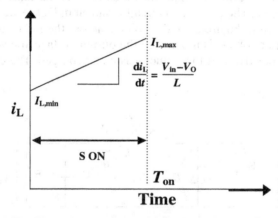

Figure 3.27 The charging phase of the inductor for a buck converter.

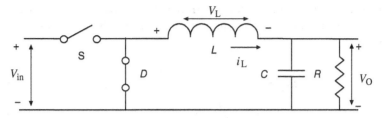

Figure 3.28   The equivalent circuit when S acts as open and D acts as a closed switch.

Figure 3.29 presents the voltage and current waveforms of a buck converter. To derive the input–output voltage relationship, a continuous conduction mode is assumed. The average voltage across the inductor must be zero over a switching period at steady state.

$$V_o = D \cdot V_{in} \tag{3.41}$$

where D is the duty ratio, f_s is the switching frequency, and T_S is defined as $T_S = 1/f_S$, the switching period. The duty ratio is defined as

$$D = \frac{T_{on}}{T_S} \quad \text{and} \quad 1 - D = \frac{T_{off}}{T_S} \tag{3.42}$$

Because the input and output power must be balanced for a lossless system, the input current can be computed as.

$$V_{in} \cdot I_{in} = V_O \cdot I_O$$

where, I_{in} and I_O are the average input and output currents, respectively.

$$I_{in} = \frac{V_o}{V_{in}} \cdot I_o = DI_o \tag{3.43}$$

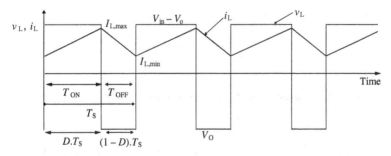

Figure 3.29   The voltage and current waveforms of a buck converter.

Similar to the boost converter, the ripple in the inductor current ΔI_L is the difference in its maximum and minimum values in steady state. From Equation (3.43)

$$\Delta I_L = I_{L,Max} - I_{L,Min} = \frac{V_o}{L} \cdot (1 - D)\, T_S \qquad (3.44)$$

$$I_{L,Max} = I_L + \frac{\Delta I_L}{2}, \; I_{L,Min} = I_L - \frac{\Delta I_L}{2} \qquad (3.45)$$

In the above equation, I_L is the average value of inductor current. The average capacitor current is zero at steady state; therefore, the inductor current is equal to the output current.

$$I_L = I_O \qquad (3.46)$$

Example 3.1   Assume a PV system with DC bus voltage of 120 V DC. A DC–DC buck converter, which has a duty ratio of the switch of 0.75 with a switching frequency of 5 kHz, is supplied by the PV system DC bus. If the value of the inductor L is 1 mH, the capacitor C is 100 μF, and the battery system is assumed to act as a load resistance of 4 Ω, compute the following:

(i) The output voltage
(ii) The minimum and maximum inductor current
(iii) The rating of the switch "S" and diode "D"

Solution

(i) The output voltage is given as

$$V_o = D \cdot V_{in}$$
$$V_o = 0.75 \times 120 = 90 \text{ V}$$

(ii) The output current is

$$I_o = \frac{V_o}{R} = \frac{90}{4} = 22.5 \text{ A}$$

The average inductor current can be computed as

$$I_L = I_o = 22.5 = 22.5 \text{ A}$$
$$\Delta I_L = \frac{V_o}{L} \cdot (1 - D)\, T_S = \frac{V_o}{L} \cdot (1 - D)\frac{1}{f} = \frac{90}{0.001} \times (1 - 0.75) \times \frac{1}{5000} = 4.5 \text{ A}$$

The maximum inductor current is

$$I_{\text{L,Max}} = I_{\text{L}} + \frac{\Delta I_{\text{L}}}{2} = 22.5 + \frac{4.5}{2} = 24.75 \text{ A}$$

The minimum inductor current is

$$I_{\text{L,Min}} = I_{\text{L}} - \frac{\Delta I_{\text{L}}}{2} = 22.5 - \frac{4.5}{2} = 20.25 \text{ A}$$

(iii) The switch S is being subjected to the input voltage when it is off. Therefore, the voltage rating of the switch should be more than 120 V. The maximum current flowing through the switch is $I_{\text{L,Max}} = 24.75$. Hence, the current rating should be at least 25 A.

Similarly, the diode is also subjected to a maximum voltage equal to input voltage and maximum current of inductor maximum current. Hence, the diode should also have at least a rating of 120 V and 25 A. Figure 3.30 depicts the inductor voltage and current of Example 3.1.

Just like a boost converter, the buck converter will also take several cycles to reach the steady state. The capacitor voltage is initially zero and as the inductor

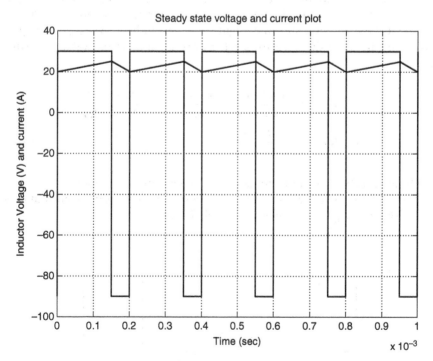

Figure 3.30   The inductor voltage and current for Example 3.1.

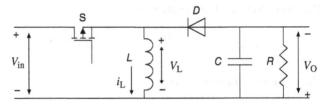

Figure 3.31   A buck–boost converter circuit.

transfers its energy to the capacitor over several cycles, the capacitor voltage picks up and attains a steady value depending on the duty ratio D. If the converter is not loaded—in other words, the output terminal is open instead of having a resistance connected—the capacitor voltage stays at its state value determined by the duty ratio.

3.6.7 The Buck–Boost Converter

A buck–boost converter consists of input voltage V_{in}, inductor L, capacitor C, load R, and controllable switch S, The circuit diagram is shown in Figure 3.31.

The operation of the buck–boost converter is explained starting from the closing of the switch S. When the switch S is closed, the diode D is reverse biased both by the input and output voltages (the voltage at the upper output DC bus is negative), and the input voltage is impressed across the inductor. The equivalent circuit is shown in Figure 3.32. The inductor voltage being positive, the current in the inductor starts to rise from its initial value following equation.

$$V_L = L \cdot \frac{\text{Change in current}}{\text{Change in time}} = L.\frac{i(t + \delta t) - (t)}{(t + \delta t) - (t)}$$

$$di_L/dt = \frac{i(t + \delta t) - (t)}{(t + \delta t) - (t)} \tag{3.47}$$

During this time, the inductor starts to increase its stored energy.

The inductor current wave is shown in Figure 3.33. The inductor current rises with a slope of V_{in}/L as long as the switch S is closed.

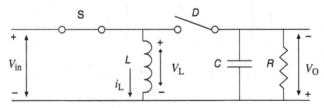

Figure 3.32   An equivalent circuit of a buck–boost converter when switch S is closed.

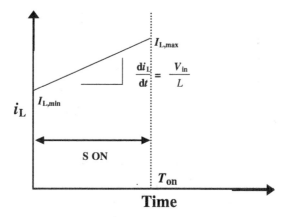

Figure 3.33 The inductor current when the switch is closed.

At the end of the duty period, the switch is turned off; the diode is forced to conduct the inductor current and it turns on. The equivalent circuit is shown in Figure 3.34.

Now, the stored energy in the inductor is transferred to the capacitor C. The current waveform of the inductor is shown in Figure 3.35.

At steady state, the inductor current must return to the value it started from at the end of the cycle. The steady-state waveforms are shown in Figure 3.36.

Like other converters, we equate the average value of inductor voltage to zero:

$$\frac{V_{in} \cdot DT_S - V_O \cdot (1-D) \cdot T_S}{T_S} = 0 \tag{3.48}$$

From this equation, it is found that the transfer ratio for the input–output relation will be

$$V_o = \frac{D \cdot V_{in}}{1-D} \tag{3.49}$$

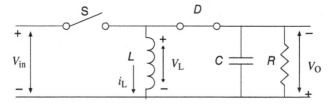

Figure 3.34 An equivalent circuit of a buck–boost converter when switch S is open.

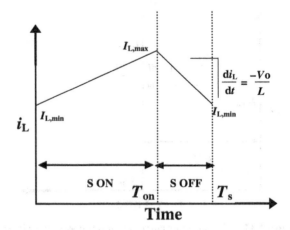

Figure 3.35   The inductor current when the switch is open.

where f_s is switching frequency and T_S is defined as $T_S = 1/f_S$, the switching period.

The duty ratio is defined as

$$D = \frac{T_{on}}{T_S} = \frac{T_{on}}{T_{on} + T_{off}} \qquad (3.50)$$

The value of the output voltage depends on the duty ratio of the switch. If D is less than half, the converter steps the voltage down, and if it is more than half, it boosts the voltage.

From an input–output energy balance,

$$V_{in} \cdot I_{in} = V_O \cdot I_O \qquad (3.51)$$

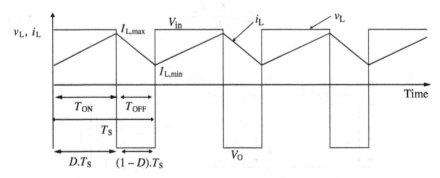

Figure 3.36   The voltage and current waveform of the buck–boost converter.

where I_{in} and I_O are the average input and output currents, respectively. We can compute:

$$I_{in} = \frac{V_O}{V_{in}} \cdot I_O = \frac{D \cdot I_O}{1 - D} \tag{3.52}$$

Now, from Figure 3.36, assuming that the average input current is (area of the trapezium divided by the time):

$$I_{in} = \frac{1}{T_S} \cdot \left(\frac{I_{L,Max} + I_{L,Min}}{2} \right) \cdot D \cdot T_S = D \cdot \frac{I_{L,Max} + I_{L,Min}}{2} \tag{3.53}$$

Again, from Figure 3.36, the average inductor current is

$$I_L = \frac{1}{T_S} \left[\left(\frac{I_{L,Max} + I_{L,Min}}{2} \right) \cdot DT_S + \left(\frac{I_{L,Max} + I_{L,Min}}{2} \right) \cdot (1 - D) T_S \right] \tag{3.54}$$

$$= \frac{I_{L,Max} + I_{L,Min}}{2}$$

From Equations (3.52) and (3.53),

$$I_{in} = D \cdot I_L \tag{3.55}$$

Figure 3.37 depicts the input current waveform of a buck-boost converter. The ripple in the inductor current, ΔI_L, is the difference in its maximum and minimum values in steady state.

$$\Delta I_L = I_{L,Max} - I_{L,Min} = \frac{V_{in}}{L} \cdot DT \tag{3.56}$$

$$I_{L,Max} = I_L + \frac{\Delta I_L}{2}, I_{L,Min} = I_L - \frac{\Delta I_L}{2} \tag{3.57}$$

For all three converters, we have assumed that the output voltage is constant and stiff. However, in practice, the capacitor voltage has a small ripple on top of the average value. The quality of the output voltage is determined by the amount of ripple in it. The smaller the ripples, the better the quality of the output voltage. The ripples in the output voltage can be reduced by increasing the capacitance C or by increasing the switching frequency f_s. But increasing the capacitance increases the size of the capacitor, which increases the cost and size of the converter. Again, by increasing the switching frequency, the switching losses is increased, decreasing the efficiency of the converter. Hence, a tradeoff is reached between the quality of output voltage and the size of capacitor and switching frequency. The inductor current should be maintained at a positive value. A minimum value of inductance is needed for this purpose. Moreover,

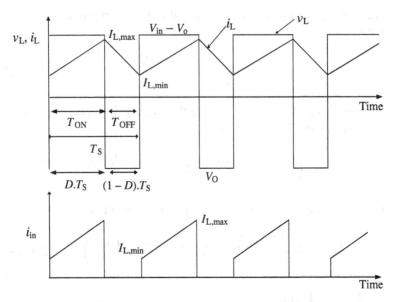

Figure 3.37 The input current waveform of a buck–boost converter.

if the ripple in inductor current is high, the maximum current in the switches will increase, increasing their rating and cost. However, to limit the ripple in the inductor current, the inductance should be as high as possible; however, the size and cost of the converter increases with higher size inductor. So a tradeoff must be reached between the size of the inductor, capacitor, switching frequency, and the cost and size of the converter, and its efficiency.

Example 3.2   Design a buck–boost converter that is supplied from a PV source whose voltage varies from 80 V to 140 V—depending on the available sun irradiant energy—and supplies a load of 10 kW at fixed 120 V DC. Give the range of duty ratio needed to supply the load at the rated voltage.

Solution
The output voltage is given as

$$V_o = 120\,\text{V}$$

The power rating is specified as

$$P = 10\,\text{kW}$$

Therefore, the output current rating can be computed as

$$I_o = \frac{10 \times 10^3}{120} = 83.33\,\text{A}$$

The maximum average input current rating is

$$I_{ave} = I_{in} = \frac{10 \times 10^3}{80} = 125 \text{ A}$$

The duty ratio for an input voltage of 80 V can be computed as

$$D = \frac{V_o}{V_{in} + V_o} = \frac{120}{80 + 120} = 0.6$$

The duty ratio for an input voltage 140 V can be computed as

$$D = \frac{V_o}{V_{in} + V_o} = \frac{120}{140 + 120} = 0.46$$

Therefore, $0.46 \leq D \leq 0.6$.

3.7 SOLAR AND WIND POWER GRIDS

A DC/AC inverter converts direct current (DC) power generated by a DC power source to sinusoidal alternating currents (AC). The photovoltaic cells (PV) are sources of DC power. The high-speed microturbine generators are sources of high-frequency AC power. Because these generators are designed for high-speed operation, they are lightweight and low volume. Since battery storage systems are expensive, they are used in combination with solar PV and load control to operate as standalone smart grids. Microturbines use natural gas and have less carbon footprints. Fuel cells are also green energy source, however; they are expensive. Variable wind-speed generators have the same operating principle as microturbine generators, except they run at variable speed and generate variable AC power. To utilize the variable frequency AC power sources, the generated power is rectified to DC power using an AC/DC rectifier. AC/DC rectifies are presented later. DC/AC inverters are used to convert the generator DC power output to AC power at the system operating frequency. Variable-speed wind power source is depicted in Figure 3.38. Figure 3.39 presents a radial PV grid with a local storage.

In Figures 3.38 and 3.39, transformers (see Figure 3.40) are used to step up the voltage or step down the voltage. When the voltage is stepped up, as a result, the current is stepped down. Therefore, the same power inputted are outputted. If side 1 is low-voltage side, side 2 is high-voltage side, then, when the voltage is stepped up on one side, the current is stepped down on the same side. Therefore, the high-voltage side has lower current then the low-voltage side. Conversely,

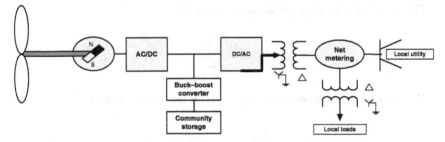

Figure 3.38   A variable-speed permanent magnet wind generator system.[1]

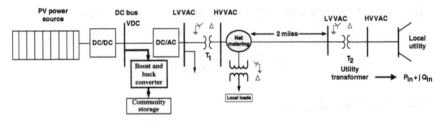

Figure 3.39   A radial photovoltaic microgrid distributed generation system.[1]

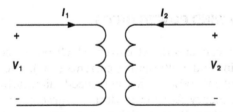

Figure 3.40   A power transformer.

the low-voltage side has a higher current. The ratio of step up and step down is determined by voltage ratio.

$$\frac{Vhv}{Vlv} = \frac{Nhv}{Nlv} = \text{Turn ratio} \tag{3.58}$$

Transformers are important elements of AC grids and they are presented in detail in Chapter 7.

3.8 SINGLE-PHASE DC–AC INVERTERS WITH TWO SWITCHES

Figure 3.41 depicts a single-phase converter with two switches. The function of inverters is to convert power from DC to AC at the system frequency of

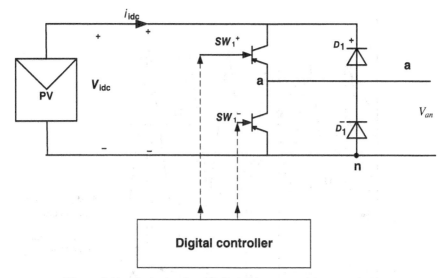

Figure 3.41   A single-phase DC–AC inverter with two switches.

operation. The two-switch single-phase inverter is used for low-power applications and is also referred to as a single-phase half-bridge inverter. The pulse width modulation (PWM) technique is used to achieve an AC voltage with a fundamental frequency of 60 or 50 Hz of the power grid. The two power switches of the single-phase inverter are sequentially turned on and off with antiparallel diodes connected between the DC link. By turning on and off the power switches, a time-varying voltage is developed. If the switch SW_1^+ is on, the potential at point "a" is the same as that of the positive DC bus (see Figure 3.41). If the switch SW_1^- is on, the potential of node "a" is that of the negative DC bus. The voltage between point "a" and "point "n" is depicted in Figure 3.42. The potential of the positive DC bus with respect to the point "n" is $+V_{idc}$. Therefore, if the switch SW_1^+ is on, the voltage V_{an} is $+V_{idc}$ and when SW_1^- is on, V_{an} is zero. To generate a time-varying voltage at point "a," a sine wave with the desired frequency is compared with a triangular wave to determine the switching policy. The objective is to sample the DC voltage such that the sample pulse voltage will have the same fundamental frequency of the sine wave. The turn on and off sequence of the two switches is decided by the relative value of the sine wave with respect to the triangular wave.

Figure 3.42 shows the waveforms of sine-PWM: two waves are being compared. One is a sine wave, V_c, that is designated as controlled voltage. The second wave is a triangular wave, V_T, with amplitude and frequency higher than the sine wave. The sampling time of time is defined as $T_s = \frac{1}{f_s}$ where f_s is the frequency of the triangular wave. To sample the DC bus voltage such that a pulse-width-generated voltage will have the same frequency as the control reference sine wave, the switching policy is based on comparing V_T with V_C. Therefore,

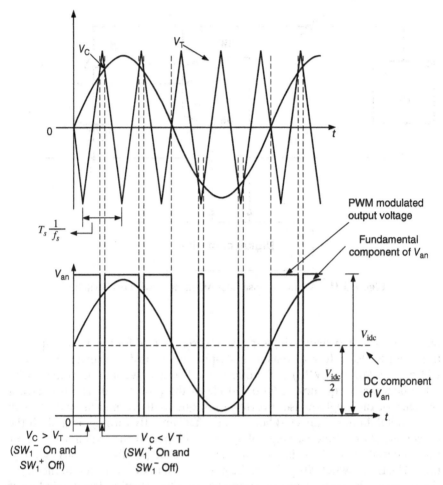

Figure 3.42 A pulse width modulation (PWM) voltage waveform of a single-phase DC–AC inverter with two switches.

by turning on the switch $SW_1{}^+$ when V_c is greater than V_T with V_{an} equal to V_{idc} and by turning on the switch $SW_1{}^-$ when V_c is smaller than V_T with V_{an} is equal to zero. The voltage V_{an} is not a pure sinusoid, but it has a DC component, a fundamental AC voltage at the same frequency as the control reference voltage, and the harmonic voltages.

Therefore, the switching policy for two power switches is:

(i) If $V_c > V_T$ $SW_1{}^+$ is *on*; $SW_1{}^-$ is *off*, and $V_{an} = V_{idc}$ and
(ii) If $V_c < V_T$ $SW_1{}^-$ is *on*, $SW_1{}^+$ is *off* and $V_{an} = 0$.

As the magnitude of the sine wave is varied relative to the triangular wave, the fundamental component of V_{an} varies in direct proportion with the relative

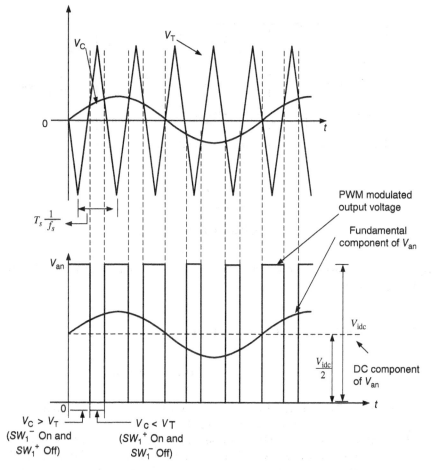

Figure 3.43 A pulse width modulation (PWM) voltage waveform of a single-phase DC–AC inverter with two switches when the sine wave has reduced amplitude.

amplitude of the triangular wave and sine wave. Figure 3.43 shows the waveform when the amplitude of the reference sine wave is reduced. The fundamental component of V_{an} has the same frequency as that of a triangular wave.

The magnitude of the fundamental of AC output voltage is directly proportional to the ratio of the peaks of V_c and V_T. This ratio is defined as the amplitude modulation index, M_a:

$$M_a = \frac{V_{C(max)}}{V_{T(max)}}$$

The peak of the fundamental component of the output voltage is

$$V_{an,1} = \frac{V_{idc}}{2} M_a \qquad (3.59)$$

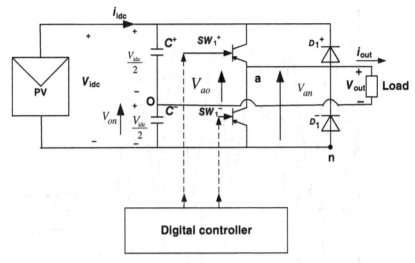

Figure 3.44 A single-phase inverter with two switches and load connected to the center-tap position.

The instantaneous value of the output voltage will be

$$V_{an} = \frac{V_{idc}}{2} + \frac{V_{idc}}{2} M_a \cdot \sin \omega_e t + \text{harmonics} \qquad (3.60)$$

where the $\omega_e = 2\pi \cdot f_e$ is the frequency of the sine wave in radian per second and f_e is the frequency of the sine wave in Hz.

The output voltage has a DC component of value $V_{idc}/2$ and a fundamental with an amplitude of $M_a \cdot V_{idc}/2$ and harmonics. M_a varies from 0 to 1. The presence of DC components would cause any inductive load to get magnetically saturated. Hence, an inductive load cannot be connected between nodes "a" and "n." However, we can eliminate the DC voltage in the output by connecting loads across the terminal V_{ao}.

Figure 3.44 shows the topology of the inverter having two capacitors with available center tap. The load is connected between node "a" and the center-tap point "o." The two capacitors C^+ and C^- are of equal capacitance. Therefore, each has a voltage of $V_{idc}/2$ across it. The potential of the point "o" is $+V_{idc}/2$ with respect to point "n."

Two capacitors used in this topology with the center-tap point "o" make both the positive and negative voltages available. When SW_1^+ is on, SW_1^- is off, the voltage V_{an} is equal to V_{idc}. Because V_{on} is $+V_{idc}/2$; V_{ao} $(= V_{an} - V_{on})$ is $+V_{idc}/3$. Similarly, when SW_1^- is on and SW_1^+ is off, then V_{an} is zero and V_{ao} is equal to $-V_{idc}/2$. If the capacitors are sufficiently large, they maintain a constant voltage of $V_{idc}/2$ across them irrespective of the load.

With the center-tap point available, the voltage across the load is between positive and negative values of $V_{idc}/2$. The DC component of the output voltage

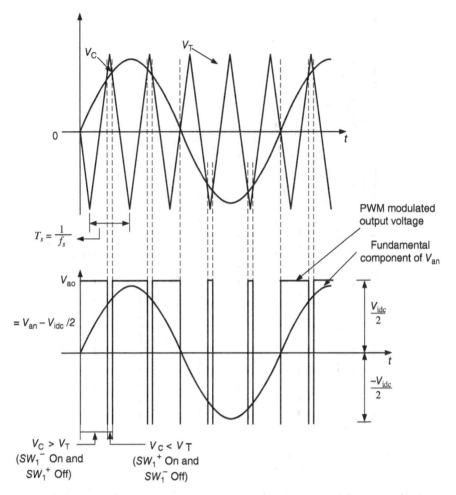

Figure 3.45   The sine pulse width modulation (PWM) of a single-phase inverter with two switches for load connected to the center tap of the capacitor.

is zero and output voltage V_{ao} is AC. The logic involved in switching is the same as in the previous case. The resulting output voltage is shown in Figure 3.45.

Figure 3.45 depicts the control voltage (V_c) and the triangular wave, (V_T) that samples the control voltage V_c. The sampling of V_c at the sampling frequency of V_T produces the waveform of Figure 3.45 where V_{ao} is output PWM wave voltage and the fundamental of the PWM voltage is shown in Figure 3.45.

The switching policy for two power switches is

(i) If $V_c > V_T$ $SW_1{}^+$ is on; $SW_1{}^-$ is off, and $V_{an} = V_{idc}/2$
(ii) If $V_c < V_T$ $SW_1{}^-$ is on; $SW_1{}^+$ is off, and $V_{an} = -V_{idc}/2$

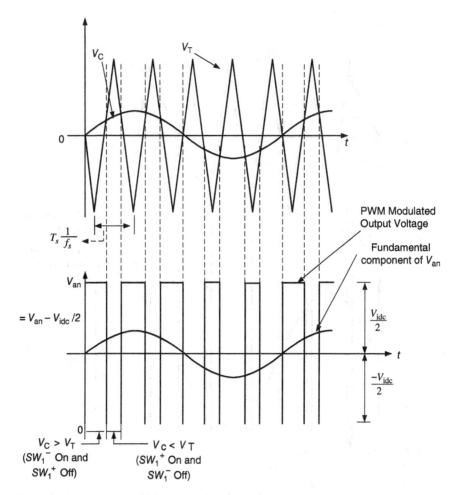

Figure 3.46   The pulse width modulation (PWM) voltage waveform of a single-phase DC–AC inverter with two switches when the sine wave has reduced amplitude.

Again, the fundamental component of the output voltage can be varied by controlling the amplitude of the sine wave. Figure 3.46 gives the voltage waveforms for a reduced peak value of a sine wave.

Figure 3.46 (top) depicts the control voltage (V_c) and the triangular wave, (V_T) that samples the control voltage V_c. However, the amplitude of control voltage is half of the control voltage of Figure 3.45. The sampling of V_c at the sampling frequency of V_T produces the waveform of Figure 3.46 (bottom) where V_{ao} is output PWM wave voltage and the fundamental of the PWM voltage is shown in Figure 3.46 (bottom). The output PWM voltage is also half of Figure 3.45 (bottom).

The output voltage in this case has no DC component. The output voltage is PWM voltage with the fundamental frequency of the reference control voltage and harmonics. The output voltage is given by

$$V_{ao} = \frac{V_{idc}}{2} M_a \cdot \sin \omega_e t + \text{harmonics} \qquad (3.61)$$

where the $\omega_e = 2\pi \cdot f_e$ is the frequency of the sine wave in radians per second and f_e is the frequency of the sine wave in Hz. When $0 \le M_a \le 1$, the amplitude of the fundamental varies linearly with the amplitude modulation index. When M_a is greater that one, it enters the nonlinear region and as it is increased further, the fundamental output voltage saturates at $\frac{4}{\pi} \cdot \frac{V_{idc}}{2}$ and does not increase with M_a.

The fundamental frequency of the output voltage is the same as the frequency of the sine voltage. Thus, by adjusting the peak of the sine wave, the amplitude of the output voltage can be varied. Similarly, by changing the frequency of the sine wave, the output frequency is varied.

From the point of view of the voltage quality, the output of the inverter should be as close to the sine wave as possible. The harmonic contents in the voltage should be minimized. To achieve low harmonic distortion the frequency of the triangular wave is increased as much as possible relative to the frequency of the sine wave. To measure the effect of higher switching frequency, the frequency modulation index is increased. The frequency of modulation index, M_f is defined as

$$M_f = \frac{f_s}{f_e} \qquad (3.62)$$

where f_s is the frequency of the triangular wave and f_e is the frequency of the sine wave.

The amount of harmonics in the output voltage is determined by the frequency modulation index. The amount of harmonics relative to the fundamental for amplitude modulation index of 0.6 and for different frequency modulation indices is tabulated in Table 3.1.

The first column of Table 3.1 lists the order of the harmonics. The power frequency in the United States is 60 Hz. Therefore, the fundamental has a frequency of 60 Hz. The third harmonic will have three times the frequency of the fundamental, which is 180 Hz. Similarly, the fifth, seventh, and ninth harmonics will have a frequency of 300 Hz, 420 Hz, and 540 Hz, respectively. The last four columns of Table 3.1 give the harmonic content as a percentage of the fundamental. The order of the harmonic present in the output voltage is a direct consequence of M_f. The third harmonic content when M_f is 3 is 163% of the fundamental. Again, the fifth, seventh, and ninth harmonics are also more than 100% when M_f is 5, 7, and 9, respectively. When M_f is 9, the third and

TABLE 3.1   The Harmonic Content of the Output Voltage for a Different M_f with M_a Fixed at 0.6

Order of Harmonic	$M_f = 3$ (%)	$M_f = 5$ (%)	$M_f = 7$ (%)	$M_f = 9$ (%)
1	100	100	100	100
3	163	22	0.42	0.05
5	61	168	22	0.38
7	73	25	168	22
9	37	62	22	168

the fifth harmonics are practically absent (0.05% and 0.38%), while the seventh and ninth harmonics are considerable. As M_f is increased, the order of the harmonics, which are a high percentage of the fundamental, also increases. The load on an inverter is usually inductive, which acts like a low-pass filter. Therefore, the higher-order harmonics are easily filtered out. But the low-order harmonics are not. So, if the M_f is high, the low-order harmonic content of the voltage will be very small and the high-order harmonics will be filtered out giving near sinusoidal currents. M_f should be made as high as possible to reduce the harmonic content of the load current. Another consideration is the high switching frequency that increases the switching loss. Thus, a tradeoff has to be reached between the switching loss, audible noise, and the harmonic distortion.[15]

Example 3.3   A half-bridge single-phase inverter with load connected to the center-tap point of the capacitors, is to provide 60 Hz at its AC output terminal and it is supplied from a PV source with 380 VDC. Assume the switching frequency of 420 Hz (M_f is 7) and amplitude modulation, M_a of 0.9. Write a MATLAB m-file code to present the waveforms of the inverter. Include the sine wave, triangular wave and the output wave V_{ao}.

Solution
Steps followed to write the program are

1. Input DC value (V_{dc}) and the peak values of $V_T(t)$ and V_C (V_{TMax} and V_{Cmax}, respectively) are assigned.
2. The frequencies of $V_T(t)$ and V_C are assigned (f_T and f_C, respectively).
3. T_s, the period of V_T, is calculated as the inverse of f_T.
4. "For loop" is now used to plot V_T, V_C, and V_o by varying k from 0 to 2/60 (two power cycles) in steps of $T_s/500$.
5. V_T is defined by equations of line.
6. Here, after plotting the wave for one period, it is shifted through integral periods to plot it for the entire time range.
7. V_T is plotted in black color: plot (k, V_T, "k"), where "k" indicates black.:

8. V_C is plotted in red color: plot (k, V_T, "r"), where "r" indicates red (V_T and V_C are plotted on the same axis).

9. Value of V_o is decided by considering whether $V_C > V_T$ or $V_C \leq V_T$.

$$\begin{array}{ll} \text{If} & V_C > V_T \\ \text{then} & V_{an} = V_{dc}/2 \\ \text{else} & V_{an} = -V_{dc}/2 \end{array}$$

10. V_o is plotted with offset –500 and in blue color indicated by "b":
plot (k, –500 + V_{an}_new, "b")

11. The points where V_T and V_C intersect are the points of discontinuity for V_O. Here the solid lines and dotted lines are plotted using for loops keeping x-axis constant and varying the y-axis and using appropriate colors and limits.

The MATLAB code provides the following result.

Figure 3.47 depicts PWM waveforms of a single phase with voltage switching for a half-bridge inverter.

With single-phase inverters with two switches, two capacitors are needed to make the center-tap point available. This makes the inverter bulky. Moreover, the switches are subjected to a voltage equal to the DC link voltage, but the output of the load can be a maximum of half that value. Therefore, the switches are underutilized. We can use four switches and increase the voltage across the load. This topology will be presented next.[1–4,14]

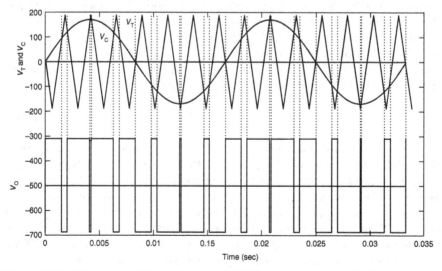

Figure 3.47 The pulse width modulation (PWM) waveforms of a single phase with voltage switching for a half-bridge inverter from MATLAB coding.

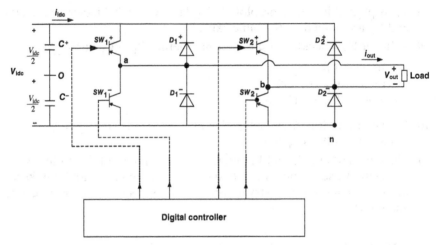

Figure 3.48 A single-phase DC–AC converter with four switches.

3.9 SINGLE-PHASE DC–AC INVERTERS WITH A FOUR-SWITCH BIPOLAR SWITCHING METHOD

This DC/AC converter topology has two legs. Each leg has two controllable switches and each switch can be turned on by sending a pulse to its base. The operation of each leg is as described before for the half-bridge single-phase converters. For the converter of Figure 3.48, the SW_1^+ and SW_2^- and SW_1^- and SW_2^+ are switched as pairs. This topology allows the load voltage to vary between $+V_{idc}$ and $-V_{idc}$, which is double the variation available in inverters with two switches. This results in higher voltage and allows more power handling capability of these inverters. When SW_1^+ and SW_2^- are on, the effective voltage across the load terminals is the difference in the voltages at nodes "a" and "b" and in this case the voltage is $+V_{idc}$. Again, when SW_2^+ and SW_1^- are on, load voltage is $-V_{idc}$. When sine PWM as shown in Figure 3.48 is applied, the peak of the fundamental output voltage is given by

$$V_{o,1} = V_{idc}M_a \tag{3.63}$$

Similar logic is followed for an inverter with two switches. If $V_c > V_T$ then the switches SW_1^+ and SW_2^- are on and other switches are off, results V_{an} at V_{idc} and V_{bn} at zero resulting in an output voltage V_{ab} (= $V_{an} - V_{bn}$) of $+V_{idc}$. When $V_c < V_T$ and the switches SW_1^- and SW_2^+ are on, we will have V_{an} at zero voltage and V_{bn} at V_{idc} voltage giving an output voltage V_{ab} (= $V_{an} - V_{bn}$) of $-V_{idc}$. Therefore, the output voltage varies between the $+V_{idc}$ and $-V_{idc}$ as shown in Figure 3.49. This modulation technique is known as bipolar sine PWM as the output voltage jumps between positive and negative values.

Therefore, for bipolar output voltage waveform, we have the following switching policy.

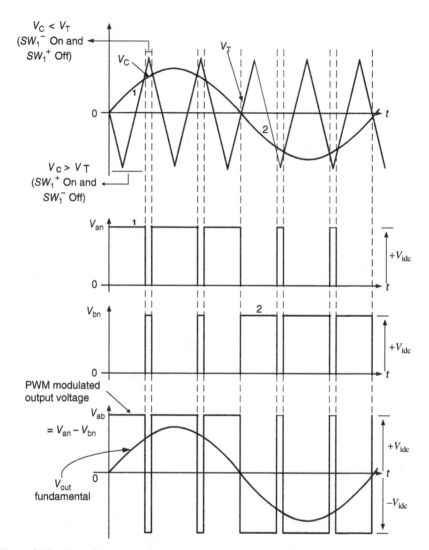

Figure 3.49 Waveforms showing sine pulse width modulation (PWM) for single-phase inverter bipolar switching.

If $V_c > V_T$ SW_1^+ and SW_2^- are on, other switches are off and $V_{ab} = V_{idc}$.
If $V_c < V_T$ SW_1^- and SW_2^+ are on, other switches are off and $V_{ab} = -V_{idc}$.

The sequences of switching operations are shown in Figure 3.49 with the details of the direction of flow of current and the output voltage when different combinations of switches are on.

SW_1^+ is on , D_1^+ is on and conducting ($V_{an} = V_{idc}$); SW_2^- is on, D_2^- is on and conducting

($V_{bn} = 0$), $V_{out} = V_{an} - V_{bn} = V_{idc}, i_{out} < 0$.

$SW_1{}^+$ is on and conducting ($V_{an} = V_{idc}$); $SW_2{}^-$ is on and conducting ($V_{bn} = 0$),

$$V_{out} = V_{an} - V_{bn} = V_{idc}, i_{out} > 0$$

$SW_1{}^-$ is on, $D_1{}^-$ is on and conducting ($V_{an} = 0$); $SW_2{}^+$ is on, $D_2{}^+$ is on and conducting ($V_{bn} = V_{idc}$), $V_{out} = V_{an} - V_{bn} = -V_{idc}, i_{out} > 0$.

$SW_1{}^-$ is on and conducting ($V_{an} = 0$); $SW_2{}^+$ is on and conducting ($V_{bn} = V_{idc}$),

$$V_{out} = V_{an} - V_{bn} = V_{idc}, i_{out} < 0$$

$SW_1{}^+$ is on , $D_1{}^+$ is on and conducting ($V_{an} = V_{idc}$); $SW_2{}^-$ is on, $D_2{}^-$ is on and conducting

$(V_{bn} = 0)$, $V_{out} = V_{an} - V_{bn} = V_{idc}, i_{out} < 0$.

$SW_1{}^+$ is on and conducting ($V_{an} = V_{idc}$); $SW_2{}^-$ is on and conducting $(V_{bn} = 0)$, $V_{out} = V_{an} - V_{bn} = V_{idc}, i_{out} > 0$.

$SW_1{}^-$ is on, $D_1{}^-$ is on and conducting ($V_{an} = 0$); $SW_2{}^+$ is on, $D_2{}^+$ is on and conducting ($V_{bn} = V_{idc}$), $V_{out} = V_{an} - V_{bn} = -V_{idc}, i_{out} > 0$.

$SW_1{}^-$ is on and conducting ($V_{an} = 0$); $SW_2{}^+$ is on and conducting ($V_{bn} = V_{idc}$), $V_{out} = V_{an} - V_{bn} = -V_{idc}, i_{out} < 0$.

Figure 3.50 depicts schematic representation of status of power switches in a single-phase full-bridge inverter. The output voltage in this case is given by

$$V_{ao} = V_{idc}M_a \cdot \sin \omega_e t + \text{harmonics} \qquad (3.64)$$

The $\omega_e = 2\pi \cdot f_e$ is the frequency of the sine wave in radians per second and f_e is the frequency of the sine wave in Hz.

The peak of the fundamental component is $V_{idc}M_a$ with $0 \leq M_a \leq 1$. When $0 \leq M_a \leq 1$, the amplitude of the fundamental varies linearly with the amplitude modulation index. When M_a is greater that one, it enters the nonlinear region and as it is increased further, the fundamental output voltage saturates at $\frac{4}{\pi} \cdot V_{idc}$ and does not increase with M_a.

The switching scheme given above results in an output voltage that jumps back and forth between positive and negative DC link voltage because the transition always takes place from positive value to negative value directly, this scheme is called bipolar sine PWM.

The harmonic content of the output voltage in this scheme is the same as that of inverter with two switches (see Table 3.1). The only difference is that the DC offset voltage in this case is zero.

Like the inverter with two switches, the amplitude of the fundamental component is decided by the sine wave by varying the amplitude modulation index ($M_a = V_{C(Max)}/V_{T(Max)}$). The harmonic content can be reduced by increasing the frequency modulation index ($M_f = f_S/f_e$).[1–4,14]

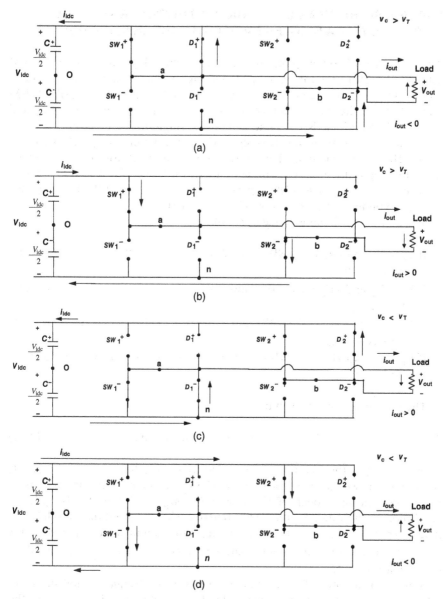

Figure 3.50 (a–d) Schematic representation of status of power switches in a single-phase full-bridge inverter.

3.10   PULSE WIDTH MODULATION WITH UNIPOLAR VOLTAGE SWITCHING FOR A SINGLE-PHASE FULL-BRIDGE INVERTER

The output PWM voltage jumps between $+V_{idc}$ and $-V_{idc}$ for the bipolar PWM scheme. Therefore, the load is subjected to high-voltage fluctuations. The insulations on the load side are also subjected to high stress. The unipolar PWM method allows the output voltage to jump between $+V_{idc}$ and 0 or $-V_{idc}$ and 0. The switching logic is the same as that for single-phase inverters with two switches, but here there are two legs switched independently with two sine waves. The upper switch of a leg is turned on when the V_c for that leg is having a greater magnitude than V_T. The same logic is followed for the other leg with its own sine wave, which is 180° apart from the sine wave of the first leg. This results in voltage waveforms shown in Figure 3.51.

The unipolar switching policy is as follows:

If $V_c > V_T$ and $-V_c < V_T$: $SW_1{}^+$ and $SW_2{}^-$ are "on," other switches are "off" and $V_{ab} = V_{idc}$.

If $V_c < V_T$ and $-V_c < V_T$: $SW_1{}^-$ and $SW_2{}^-$ are "on," other switches are "off" and $V_{ab} = 0$.

If $V_c < V_T$ and $-V_c > V_T$: $SW_1{}^-$ and $SW_2{}^+$ are "on," other switches are "off" and $V_{ab} = -V_{idc}$.

If $V_c > V_T$ and $-V_c > V_T$: $SW_1{}^+$ and $SW_2{}^+$ are "on," other switches are "off" and $V_{ab} = 0$.

The output voltage is given by

$$V_{ao} = V_{idc} M_a \cdot \sin \omega_e t + \text{harmonics} \tag{3.65}$$

$\omega_e = 2\pi \cdot f_e$ is the frequency of the sine wave in radians per second and f_e is the frequency of the sine wave in Hz. Therefore, the peak of the fundamental component has a peak of $V_{idc} M_a$ with $0 \leq M_a \leq 1$. When $0 \leq M_a \leq 1$, the amplitude of the fundamental varies linearly with the amplitude modulation index. When M_a is greater that one, it enters the nonlinear region and as it is increased further, the fundamental output voltage saturates at $\frac{4}{\pi} \cdot V_{idc}$ and does not increase with M_a.

Similar to the other PWM schemes, here too the frequency and the magnitude of the output voltage is controlled by the reference sine wave and the harmonic content of the output voltage is determined by the frequency modulation index.

Example 3.4   Consider a PV source of 60 V. A single-phase inverter with four switches is used to convert DC to 50 Hz AC using a unipolar scheme. Assume M_a of 0.5 and M_f of 7. Develop a MATLAB program to generate the waveforms of the inverter showing the control voltage and the output voltage.

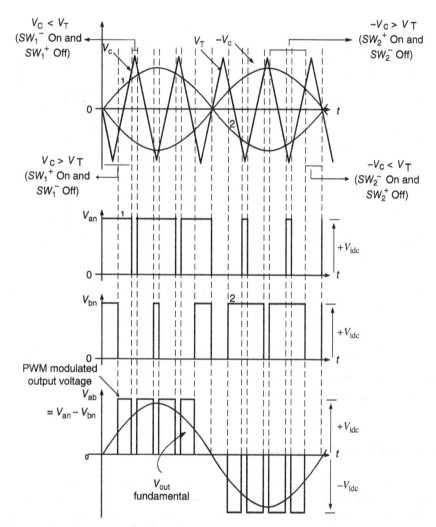

Figure 3.51 Waveforms for the unipolar switching scheme of a single-phase inverter with four switches and a frequency modulation index of 5.

Solution

The following plots were obtained

Figure 3.52 depicts the waveforms of a unipolar switching scheme for a single-phase inverter with four switches.

Similar to Table 3.1, Table 3.2 tabulates the harmonic content of the output voltage. Here, it tabulates the output voltage of the unipolar switching scheme. It can be seen from the table that for a frequency modulation index higher than 5, the lower-order harmonics are practically absent. The higher-order harmonics in the voltage are not usually a matter of concern because they are filtered

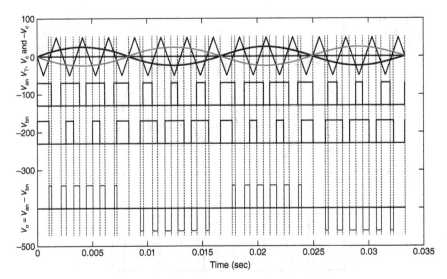

Figure 3.52   A MATLAB waveforms plot of a unipolar switching scheme for a single-phase inverter with four switches.

out by the low pass characteristics of the inductive loads. The unipolar switching, therefore, has the advantage of reduced harmonics and reduced voltage fluctuations in the load. The number of times a particular switch turns on and off is the same as bipolar switching.[1–4,14]

3.11   THREE-PHASE DC–AC INVERTERS

A three-phase inverter has three legs, one for each phase, as shown in Figure 3.53. Each inverter leg operates as a single-phase inverter.[2,3] The output voltage of each leg V_{an}, V_{bn}, V_{cn}, where "n" refers to negative DC bus voltage is computed from the input voltage, V_{idc}, and the switch positions. The inverter has three terminals, two inputs, and one output. The inverter input is DC power that can be supplied from a storage battery system, a PV power source, a fuel

TABLE 3.2   The Harmonic Content of the Output Voltage for a Different M_f with M_a Fixed at 0.6 for a Unipolar Switching Scheme

Order of Harmonic	$M_f = 3$ (%)	$M_f = 5$ (%)	$M_f = 7$ (%)	$M_f = 9$ (%)
1	100	100	100	100
3	12	0.01	0.01	0.01
5	61	0.55	0.01	0.03
7	67	12	0.03	0.02
9	33	62	0.58	0.01

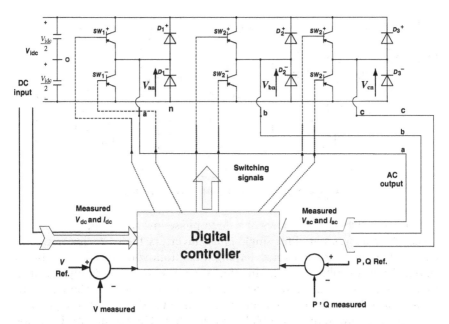

Figure 3.53   The operation of an inverter as a three-terminal device.

cell, or a green energy DC power source such as high-speed generators or variable-speed wind generators. The sine wave signal is supplied to a digital signal processor (DSP) controller to control the output AC voltage, power, and frequency. The DSP controller sends a sequence of switching signals to control the six power switches to produce the desired output AC power.[1–4,14]

The three-phase inverters have six switches and six diodes as shown in Figure 3.53. A switch is formed by the pair SW_i and D_i ($i = a, b, c$), which can conduct current in both directions. The three-phase inverter consists of three limbs that lie between the DC links, with each limb having two switches. By turning on the upper switch, the output node (a, b, or c) acquires a voltage of the upper DC line. Conversely, when the lower switch of a limb is on, the output node of that limb attains a voltage of the lower DC line. By alternately turning on the upper and the lower switch, the node voltage oscillates between the upper and lower DC line voltages.

$$V_{L-L,rms} = M_a \cdot \sqrt{\frac{3}{2}} \cdot \frac{V_{idc}}{2} \tag{3.66}$$

The above can be approximated as

$$V_{L-L,rms} = 0.612 V_{idc} M_a \tag{3.67}$$

where $V_{L\text{-}L,rms}$ denotes the RMS value of the fundamental of output line voltage.

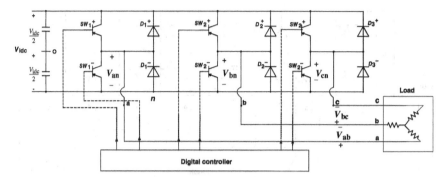

Figure 3.54 Three-phase inverter topology.

Figure 3.54 depicts schematics of a three-phase inverter topology.

Therefore, as was stated for single-phase inverters, the switching frequency is decided by the triangular wave, V_T and the output voltage is decided by the reference control voltage V_C, modulating wave. The modulated output voltage of each leg has the same waveform of a single-phase inverter. The frequency modulation index determines the harmonic content in the output voltage waveform. The standard frequency is 60 Hz for the United States and 50 Hz in most other countries. It is generally desirable that the frequency of the triangular wave be very high to make a high-frequency modulation index that will result in low harmonics in output voltage and the load currents. However, with an increase of the triangular frequency, the switching frequency increases and the high switching frequency results in high switching losses.

Several other factors affect the selection of switching frequency; it is also desirable to use high switching frequency due to the relative ease of filtering harmonic voltages. In addition, if the switching frequency is too high, the switches may fail to turn on and off properly; this may result in a short circuit of the DC link buses and damage the switches. In a residential and commercial grid, the switching frequency selected has to be out of the audible frequency range to reduce the high-pitch audible noise. Therefore, for most applications the switching frequency is selected to be below 6 kHz or above 20 kHz. For most applications in a residential system, the switching frequency is selected below 6 kHz.

To develop a simulation model for the above inverter operation, the above waveforms are expressed by mathematical expressions.

Let V_C (Phase a) be expressed as

$$V_{C(a)} = M_a \sin(\omega_e t) \qquad (3.68)$$

where $\omega_e = \frac{2\pi}{T_e}t$, $T_e = \frac{1}{f_e}$ is sampling time and f_e is the frequency of desired controlled voltage and

$$M_a = \frac{V_{C(max)}}{V_{T(max)}}$$

The expression for the triangular waveform, V_T is given by

$$x_1 = -1 + \frac{T_s}{2}t \quad \text{for} \quad 0 \leq t < \frac{T_s}{2}$$

Expressing the unit of time t in radian and recalling $f_s = M_a f_e$. The above is rewritten as

$$x_1 = -1 + \frac{2N}{\pi} \cdot \omega_e t \quad \text{for} \quad 0 \leq \omega_e t < \frac{\omega_e T_s}{2} \tag{3.69}$$

where $N = 1, 2, 3$, and so on. In this formulation, M_f is selected as a variable of the simulation.

Rewriting the above as

$$x_1(t) = -1 + \frac{2N}{\pi} \cdot \omega_e t \quad \text{for} \quad 0 \leq \omega_e < \frac{\pi}{M_f} \tag{3.70}$$

Similarly,

$$x_2(t) = 3 - \frac{2M_f}{\pi}\omega_e t \quad \text{for} \quad \frac{\pi}{N} \leq \omega_e t < \frac{2\pi}{M_f} \tag{3.71}$$

The triangular waveform $V_T(t)$ can be modeled by Equations (3.70) and (3.71).

To develop a simulation test bed, let us express V_C (for Phase a), V_C (for Phase b), and V_C (for Phase c) as

$$V_C(a) = M_a \sin \omega_e t \tag{3.72}$$

$$V_C(b) = M_a \sin\left(\omega_e t - \frac{2\pi}{3}\right) \tag{3.73}$$

$$V_C(c) = M_a \sin\left(\omega_e t - \frac{4\pi}{3}\right) \tag{3.74}$$

And the triangular wave of $V_T(t)$ as

$$x_1(t) = -1 + 2N\frac{\omega_e t}{\pi} \quad \text{for} \quad 0 \leq \omega_e t < \frac{\pi}{N} \tag{3.75}$$

$$x_3(t) = 3 - 2N\frac{\omega_e t}{\pi} \quad \text{for} \quad \frac{\pi}{N} \leq \omega_e t < \frac{2\pi}{N} \tag{3.76}$$

Therefore, the algorithm PWM voltage generation steps are as follows.

$$\text{If } V_c(a) \geq x_1(t) \quad \text{or} \quad x_2(t), \quad \text{then} \quad V_{an} = V_{idc} \tag{3.77}$$

Similarly:

$$\text{If } V_c(b) \geq x_1(t) \quad \text{or} \quad x_2(t), \quad \text{then} \quad V_{bn} = V_{idc} \tag{3.78}$$
$$\text{If } V_c(c) \geq x_1(t) \text{ or } x_2(t), \quad \text{then} \quad V_{cn} = V_{idc} \tag{3.79}$$
$$\text{Otherwise, } V_{an} = 0; V_{bn} = 0 \quad \text{and} \quad V_{cn} = 0 \tag{3.80}$$

The i_{idc} is computed as:

$$i_{idc} = i_a \, SW_1^+ + i_b SW_2^+ + i_c \, SW_3^+ \tag{3.81}$$

Figure 3.55 depicts the pulse width modulation (PWM) operation of a three-phase converter.

As discussed in the operation of a single-phase converter, one switch is turned on in each leg of the three-phase inverter. A sequence of switching operations for one cycle of inverter operation is shown in Figure 3.56. For example, in phase a, V_{an} with respect to the negative DC bus depends on V_{idc} and the switch status SW_1^+ and SW_1^-. This is depicted by Figure 3.56a.

$$V_{an}, V_{bn} = V_{idc}, V_{cn} = 0. \; I_a, I_b > 0, I_c < 0, SW_1^+ \text{ on}, SW_2^+ \text{ on}, SW_3^- \text{ on}$$
$$V_{an}, V_{bn} = V_{idc}, V_{cn} = 0. \; I_a > 0, I_b, I_c < 0, SW_1^+ \text{ on}, D_2^+ \text{ on}, SW_3^- \text{ on}$$
$$V_{an} = 0, V_{bn} = V_{cn} = V_{idc}, I_a < 0, I_b, I_c > 0, SW_1^- \text{ on}, SW_2^+ \text{ on } SW_3^+ \text{ on}$$
$$V_{an} = 0, V_{bn} = V_{cn} = V_{idc}, I_a, I_c < 0, I_b > 0, SW_1^- \text{ on}, SW_2^+ \text{ on}, D_3^+ \text{ on}$$
$$V_{an} = V_{cn} = V_{idc}, V_{bn} = 0, I_a, I_c > 0, I_b < 0, SW_1^+ \text{ on}, SW_2^- \text{ on}, SW_3^+ \text{ on}$$
$$V_{an} = V_{cn} = V_{idc}, V_{bn} = 0, I_a, I_b < 0, I_c > 0, D_1^+ \text{ on}, SW_2^- \text{ on}, SW_3^+ \text{ on}$$

The control objective is the same as discussed for a single-phase converter, that is the PWM seeks to control the modulated output voltage of each phase such that the magnitude and frequency of the fundamental inverter output voltage is the same as the control voltage. The PWM samples the DC bus voltage and the sampled voltage is the inverter output voltage.

Example 3.5   Compute the minimum DC input voltage if the switching frequency is set at 5 kHz. Assume the inverter is rated 207.6 V AC, 60 Hz, 100 kVA.

Solution

$$V_{rated} = 207.6 \, \text{V}, \text{kVA} = 100 \, \text{kVA}, f = 60 \, \text{Hz}$$

The AC-side voltage of the inverter is given as

$$V_{peak} \sin(2 \cdot \pi \cdot f \cdot t) = 207.6\sqrt{2} \cdot \sin(2 \cdot \pi \cdot 60 \cdot t) = 293.59 \sin(2 \cdot \pi \cdot 60 \cdot t)$$

Here, $V_{peak} = 293.59 \, \text{V}$

For the three-phase inverter using sine PWM, the line-line peak voltage is given as $V_{\text{L-L,peak}} = M_a \cdot \sqrt{3} \cdot \dfrac{V_{dc}}{2}$

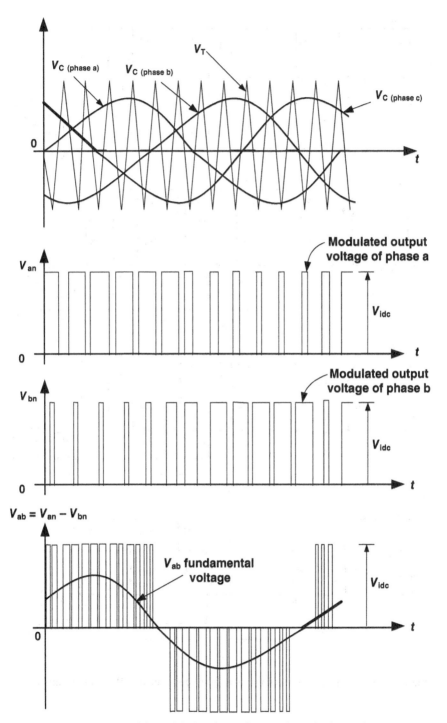

Figure 3.55 The pulse width modulation (PWM) operation of a three-phase converter.

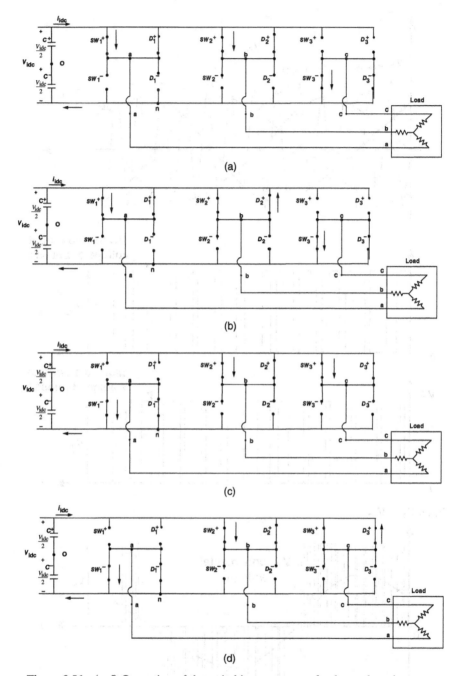

Figure 3.56   (a–f) Operation of the switching sequence of a three-phase inverter.

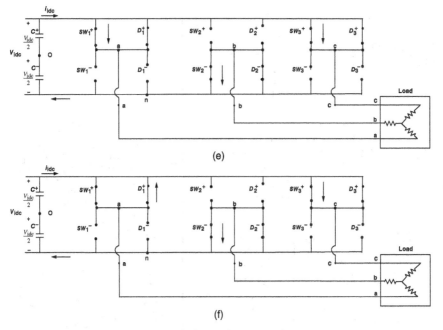

Figure 3.56   (*Continued*)

Therefore,

$$V_{dc} = \frac{2}{\sqrt{3}} \cdot \frac{V_{L\text{-}L,peak}}{M_a}$$

For the inverter to operate at maximum value of M_a is 1, the minimum DC voltage is given as

$$V_{dc,min} = \frac{2}{\sqrt{3}} \cdot \frac{V_{L\text{-}L,peak}}{M_{a,max}} = \frac{2}{\sqrt{3}} \cdot \frac{293.59}{1} = 339.01 \text{ V}$$

Therefore, the minimum $V_{dc} = 339.01$ V

3.12 MICROGRID OF RENEWABLE ENERGY

Figure 3.57 depicts a community microgrid operating in parallel with the local utility grid. The transformer T_1 voltage is in the range of 240 V AC/ 600 V AC and the transformer T_2 that connects the PV system to the utility bus system is rated to step up the voltage to the utility local AC bus substation.

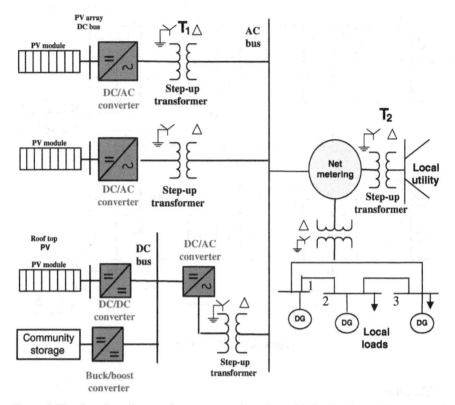

Figure 3.57 One-line diagram for a community microgrid distribution grid connected to a local utility power grid.

Example 3.6 Consider a microgrid that is fed from a PV generating station with AC bus voltage of 120 V as shown in Figure 3.58.
The modulation index for the inverter is 0.9.

(i) Compute the DC side voltage
(ii) Can the microgrid operate as standalone?

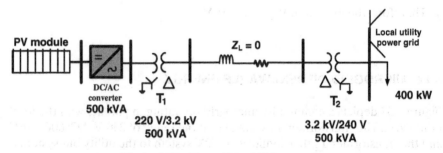

Figure 3.58 The microgrid of example.

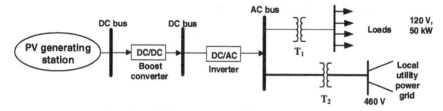

Figure 3.59   One-line diagram of Example 3.8.

Solution

(i) For the inverter:

$$V_{L,rms} = \frac{\sqrt{3}}{\sqrt{2}} \cdot M_a \frac{V_{dc}}{2}$$

$$V_{dc} = \frac{\sqrt{2}}{\sqrt{3}} \cdot \frac{2.V_{L,rms}}{M_a} = \frac{\sqrt{2}}{\sqrt{3}} \cdot \frac{2 \times 120}{0.9} = 217.7\,V$$

(ii) The microgrid cannot operate alone. The inverter of microgrid control system must be synchronized with the local utility power grid and operate as a generator of the local utility power grid.

Example 3.7   Consider the microgrid in Figure 3.59. Assume the PV generating station can produce 175 kW of power, transformer T_1 is an ideal with rated values of 240 V/120 V and 75 kVA, and transformer T_2 is also ideal and rated at 240 V/460 V, and 150 kVA

(i) Compute the DC bus voltage of PV station.
(ii) If the local power grid has outage and the breaker connecting microgrid to local utility opened, can PV microgrid operates as standalone.

Solution

(i) $V_{L,rms} = \frac{\sqrt{3}}{\sqrt{2}} \cdot M_a \frac{V_{dc}}{2}$

$$V_{dc} = \frac{\sqrt{2}}{\sqrt{3}} \cdot \frac{2.V_{L,rms}}{M_a} = \frac{\sqrt{2}}{\sqrt{3}} \cdot \frac{2 \times 240}{0.8} = 489.9\,V$$

Now, for the DC/DC converter:

$$V_o = \frac{V_{in}}{1 - \text{Duty ratio}}$$

$$V_{in} = (1 - \text{Duty ratio}) \times V_o = (1 - 0.6) \times 489.9 = 195.96\,\text{V}$$

(ii) If the local PV power grid has outage and the breaker connecting microgrid to local utility opened, the PV microgrid cannot operates as standalone.

3.13   THE SIZING OF AN INVERTER FOR MICROGRID OPERATION

As stated before for sinusoidal-triangle PWM, we can control the magnitude of the output voltage by controlling the amplitude modulation index.

In a microgrid system, three-phase inverters are used. The input of the inverter is a DC source that can be obtained from a PV source, a fuel cell, or variable frequency AC voltage from a variable-speed wind generator, which is rectified into DC. The DC voltage is typically converted into AC by employing sine PWM. The output AC voltage is obtained by comparing a triangular wave and a sine wave. The magnitude of the fundamental component of the output AC voltage depends on the amplitude modulation index, M_a that is defined as the ratio of the amplitude of the reference sine wave and the triangular wave. The frequency of the output voltage is the same as the reference sine wave. The harmonic content is decided by the frequency modulation index M_f that is defined as the ratio of the frequencies of the triangular wave and the reference sine wave. The harmonic content of the voltage should be as low as possible. To reduce the harmonics, the frequency modulation index should be high. The value of the fundamental component of the output AC line-to-line voltage is given by:

$$V_{\text{PWM,ab}} = \frac{\sqrt{3}}{\sqrt{2}} \cdot \frac{V_{dc}}{2} M_a \qquad (3.82)$$

Table 3.3 illustrates the variation of amplitude modulation index required to generate a fixed-amplitude AC voltage from the DC bus voltage. The switches of the inverter should be rated at the voltage of the DC link and should be able to carry current for the rated condition. The frequency modulation index should be chosen from the commutation characteristics of the switches used.

Although the amplitude modulation index decides the magnitude of the output voltage, the frequency modulation index decides the harmonic content. The amplitude modulation index plays a minor role in the harmonic content.

TABLE 3.3   The Values of Modulation Index Required to Keep the AC Voltage (Line-to-Line) Fixed for a Three-Phase Inverter

$V_{ac} = 120$ V	
V_{DC} (V)	M_a
200	0.98
250	0.78
300	0.65

Table 3.3 gives the harmonic content in terms of per unit of the fundamental. The second column of the table shows that if the fundamental of the voltage wave has a magnitude of 1, for $M_a = 1$, then the magnitude of the harmonic of order $1 \cdot M_f \pm 2$ is 0.318 or 31.8% of the fundamental (M_f is the frequency modulation index). For example, if the frequency modulation index is chosen as 50 (switching frequency of 3 kHz with a power frequency of 60 Hz), then from $1 \cdot M_f \pm 2$, the order of the harmonics that will be significant are 48 and 52. In other words, harmonics of $48 \times 60 = 2.88$ kHz and $52 \times 60 = 3.12$ kHz will have magnitudes comparable to the fundamental. Similarly, the harmonic of order $2 \cdot M_f \pm 1$ is 18.1% of the fundamental. Which means the 99th and 101st harmonics will have significant amplitudes. These harmonics will have frequencies 99 and 101 times the fundamental frequency, respectively (5.94 kHz and 6.06 kHz). Next, the third column gives the harmonic contents as a fraction of the fundamental if the amplitude modulation index is 0.9. This means that if M_a is 0.9, then the 48th and 52nd harmonics will be 29.8% of the fundamental. The subsequent columns show the harmonic content for different amplitude modulation indices.

Table 3.4 depicts the per unit harmonic content of the output phase voltage. In general, the orders of the harmonics that are significant or comparable to the fundamental are $N \cdot M_f \pm M$ where N and M are integers such that their sum results in an odd integer. Most electrical circuits consist of R–L structures. The impedance of the inductance is directly proportional to the frequency ($X_{L,1} = L \cdot 2 \cdot \pi \cdot f$). As the order of the harmonic voltage increases, its frequency also increases. The harmonic impedance offered by the inductor becomes $X_{L,n} = L \cdot 2 \cdot \pi \cdot n \cdot f = n \cdot X_{L,1}$. In other words, the harmonic impedance becomes the order of the harmonic times the fundamental impedance. Therefore, the higher

TABLE 3.4   The Per Unit Harmonic Content of the Output-Phase Voltage

Order of Harmonic	$M_a = 1$	$M_a = 0.9$	$M_a = 0.8$	$M_a = 0.7$	$M_a = 0.6$
1	1.000	1.000	1.000	1.000	1.000
$1 \cdot M_f \pm 2$	0.318	0.298	0.275	0.248	0.218
$2 \cdot M_f \pm 1$	0.181	0.283	0.393	0.506	0.617

the order of the harmonic the more is the harmonic impedance and less is the current. Therefore, with a higher M_f the more is the order of harmonic and less is the harmonic current. If the harmonic content of the current is less, the closer is the current to the desired sinusoidal wave. Hence, it is always desired to have a higher-frequency modulation index. However, there is some practical constraint to it. If the sampling frequency is high, then the number of turn on and offs of the switch per second increases and the switching losses increase. Moreover, if the sampling frequency is too high, the switches may fail to turn on and off properly. Therefore, a tradeoff is reached between harmonic content and the switching loss to determine the sampling frequency. Furthermore, the sampling frequency is selected out of audio range.

PROBLEMS

3.1 Consider the microgrid of Figure 3.60. A three-phase transformer, T1, is rated 500 kVA, 220 V (Y-grounded)/440 V (delta) transformer with the reactance of 3.5%. The microgrid is supplied from an AC bus of a PV generating station with its DC bus rated at 540 V. The distribution line is 10 miles long and has a series impedance of $0.1 + j\,1.0\ \Omega$ per mile and local load of 100 kVA at 440 V. The microgrid is connected to the local power grid using a three-phase transformer T2, rated at 440 V (Y-grounded)/13.2 kV (delta) with the reactance of 8%. Compute the per unit impedance diagram of the microgrid system. Assume the voltage base of 13.2 kV on the local power grid side and kVA base of 500.

3.2 Consider the microgrid of Figure 3.61. Assume the inverter AC bus voltage of 240 V and transformer T1 is rated 5% impedance, 240 V (delta)/120 V (Y-grounded), and 150 kVA. The transformer T2 is rated at 10% impedance, 240 V (delta)/3.2 kV (Y-grounded), and 500 kVA. The local loads are rated at 100 kVA and a power factor of 0.9 lagging. The inverter modulation index is 0.9.

   Compute the following:
   (i) The DC bus voltage and inverter rating
   (ii) The boost converter PV bus input voltage and input current ratings for the required DC bus voltage of the inverter

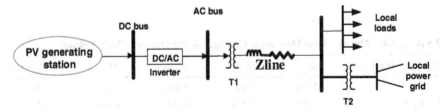

Figure 3.60 A single-line diagram for Problem 3.1.

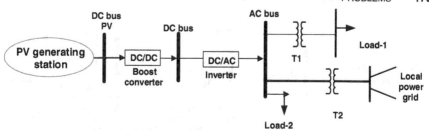

Figure 3.61   A one-line diagram of Problem 3.2.

(iii) The size of the microgrid PV station

(iv) The per unit model of the microgrid

3.3   For Problem 3.2 depicted in Figure 3.61, assume the DC bus voltage of inverter is equal to 800 V DC. The transformer T1 is rated 9% impedance, 400 V (delta)/220 V (Y-grounded), and 250 kVA. The transformer T2 is rated at 10% impedance, 400 V (Y-grounded)/13.2 kV (delta), and 500 kVA. The load number 1 is rated at 150 kVA with a power factor of 0.85 lagging. The load number 2 is rated at 270 kVA with a power factor of 0.95 leading.

Compute the following:

(i) The modulation index and inverter rating

(ii) The boost converter PV bus voltage and input current for the required DC bus voltage of the inverter

(iii) The minimum size of the microgrid PV station

3.4   For Problem 3.3 depicted in Figure 3.61, assume transformer T1 is rated at 5% impedance, 440 V (Y-grounded)/120 V (delta), and 150 kVA. The transformer T2 is rated at 10% impedance, 440 V (Y-grounded)/3.2 kV (delta), and 500 kVA. The inverter modulation index is 0.9. Select the power base of 500 kVA and the voltage base of 600 V. The local power grid internal reactance is 0.2 per unit based on 10 MVA and 3.2 kV. Assume the microgrid is not loaded.

Compute the following.

(i) The per unit equivalent impedance model

(ii) The fault current if a balance three-phase fault occurs on the inverter bus

3.5   Consider the microgrid in Figure 3.62. A three-phase 500 kVA, 440 V (Y-grounded)/3.2 kV (delta) transformer with the per unit reactance of 3.5% feeds from an AC source of a PV generating station. The distribution line is 10 miles long and has a series impedance of $0.01 + j\,0.09\ \Omega$ per mile. The local load is 250 kVA. Balance of power can be injected into the local utility using a 3.2/13.2 kV transformer and with the per unit reactance of 8%. Assume the voltage base of 13.8 kV on the local power grid side, kVA base of 500, and the DC bus voltage of 800 V.

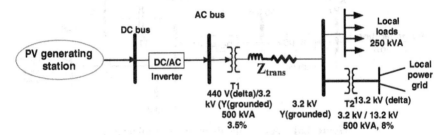

Figure 3.62   A one-line diagram of Problem 3.5.

Compute the following:

(i) The inverter and the PV generating station ratings

(ii) The per unit impedance diagram of the microgrid

(iii) If the local utility power grid has an outage due to a fault on distribution grid feeding the PV microgrid and the breaker connecting the PV microgrid to the local utility is opened, can PV microgrid operate as standalone?

3.6   Consider the microgrid in Figure 3.62 and the data of Problem 3.5. The local power grid net input reactance can be approximated as 9% based on the voltage base of 13.2 kV and 500 kVA base. Assume DC bus voltage of 880 V.

Compute the following.

(i) The inverter modulation index and inverter rating

(ii) The short circuit current of AC bus if a three-phase short circuit occurs on the AC bus of the inverter; ignore the PV station fault current contribution

(iii) The same as part (ii), except assume that the internal reactance of PV generating station and its inverter can be approximated as 1 per unit based on inverter rating

3.7   Consider a three-phase DC/AC inverter with a DC bus voltage rate at 600 V. Assume the DC bus voltage can drop during a discharge cycle to 350 V DC. Determine the AC bus voltage range for a modulation index of 0.9.

3.8   Consider a rectifier mode of operation for the following system data of a three-phase rectifier. Assume the data given in Table 3.5.

TABLE 3.5   Data for Problem 3.8

V_{AN}	f_1	P	X	pf	M_a
120 V RMS	60 Hz	1500 W	15 Ω	1	0.85

Perform the following:

(i) Find the angle δ and the DC link voltage. Ignore the resulting harmonics.

(ii) Write a MATLAB program to simulate the rectifier operation to plot the current I_a and the voltage V_{AN} considering V_{PWMa} and δ are known and have values the same as the result of part (i). Assume the PWM voltage is calculated from the incoming AC power. Ignore the resulting harmonics.

3.9 A three-phase rectifier with the supply-phase voltage of 120 V and the reactance between the AC source and the rectifier is 11 Ω. Write a MATLAB program to plot the modulation index necessary to keep the V_{DC} constant at 1200 V when the input power factor angle is varied from – 60° to 60°. Assume that the active power supplied by the system remains fixed at 5.5 kW.

3.10 Develop MATLAB testbed and compute the triangular wave V_T using the Fourier series. Assume the following specifications: peak 48 V, frequency 1 kHz, and order of harmonics ≤ 30. Plot the waveform.

Hint: $V_T(t)$ can be constructed using the Fourier series of a triangle wave. An advantage to this method is that you can create the waveform over a long period. The equation and solution of a Fourier series can be found in many sites online. It should be quickly noticed that the triangle waveform has an odd symmetry, thus $a_0 = 0$ and $a_k = 0$.

The Fourier equation becomes

$$F(t) = a_0 + \sum_{k=1}^{\infty} a_k \cdot \cos(k \cdot \omega t) + b_k \cdot \sin(k \cdot \omega t)$$

$$F(t) = \sum_{k=1}^{\infty} b_k \cdot \sin(k \cdot \omega t)$$

b_k is all that needs to be solved for. The triangle function can then be split into sections, somewhat like the first method, to find b_k.

The equation to form a triangular wave has been discussed in Section 3.13

$$b_k = \frac{-8 \cdot (-1)^k}{k^2 \cdot \pi^2} \sin\left(k \cdot \frac{\pi}{2}\right)$$

Now, we will substitute b_k and we will obtain the equations for the triangle wave as shown by equation:

$$F(t) = A \cdot \sum_{k=1,3,5}^{\infty} \frac{-8 \cdot (-1)^k}{k^2 \cdot \pi^2} \sin\left(k \cdot \frac{\pi}{2}\right) \cdot \sin(k \cdot \omega t)$$

where A is amplitude. This equation can then be put into MATLAB simulation.

We can use this equation to describe the operation of a converter in a MATLAB simulation.

For example, let us define the frequency, sampling time, and amplitude as

$$f_s = 5000; \ T_s = 1/f_s; VT_{Max} = 48$$

To see the full waveform, the step size should be at least half of T_s and a power of 10 smaller to create enough points.

$$t = 0 : T_s/20 : 10$$

The above MATLAB codes will create a waveform 10 seconds long. We can use a loop to iterate the equation constantly adding it to itself as needed by a series $\sum\limits_{k=1,3,5}^{\infty}$.

3.11 Develop MATLAB testbed and compute the triangular wave for PWM using the identity mapping method. Assume a peak value of 48 and a frequency of 1 kHz. Plot the wave form.

3.12 Develop MATLAB testbed for the DC/AC inverter of Figure 3.63 given below.

Assume V_{idc} is equal to 560 V (DC), $V_{C \ (max)}$ is equal to 220 V, f_e is equal to 60 Hz and modulation index of $0.5 \le M_a \le 1.0$, and change M_f from 2 to 20. Perform the following:

(i) The ratio of the fundamental frequency of the line–line (RMS) to the input DC voltage, V_{idc}, ($V_{(line–line \ (RMS))}/V_{idc}$).

(ii) Plot the $V_{(line–line \ (RMS))}/V_{idc}$ as a function of M_a for $M_f = 2$ and $M_f = 20$.

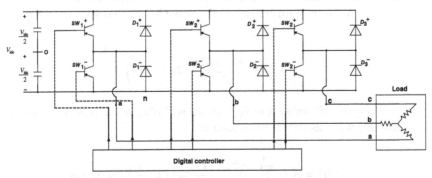

Figure 3.63 The linear and overmodulated operation of a three-phase converter.

3.13 Consider a PV source of 60 V. A single-phase inverter with four switches is used to convert DC to 50 Hz AC using a unipolar scheme. Select the following modulation indices:

   a. $M_a = 0.5$ and $M_f = 7$
   b. $M_a = 0.5$ and $M_f = 10$
   c. $M_a = 0.9$ and $M_f = 4$

   Write a MATLAB program to generate the waveforms of the inverter showing V_{an} voltage. Make tables and discuss your results.

REFERENCES

1. http://edisontechcenter.org/HistElectPowTrans.html. Accessed October 25, 2013

2. http://www.greatachievements.org. Accessed September 18, 2009.

3. http://en.wikipedia.org/. Accessed September 28, 2009.

4. http://web.mit.edu/newsoffice/2013/rechargeable-flow-battery-enables-cheaper-large-scale-energy-storage-0816.htm. Accessed October 20, 2013.

5. http://en.wikipedia.org/wiki/Volt. Accessed October 18, 2013.

6. http://en.wikipedia.org/wiki/Ohm.

7. Keyhani, A. (2011) *Design of Smart Power Grid Renewable Energy Systems,* John Wiley & Sons, Inc. and IEEE Publication.

8. Clayton, P.R. (2001). *Fundamentals of Electric Circuit Analysis,* John Wiley & Sons, ISBN 0-471-37195-5.

9. Mohan, N., Undeland, T., and Robbins, W. (1995) *Power Electronics*, John Wiley & Sons, Inc., New York.

10. Rashid, M.H. (2003) *Power Electronics, Circuits, Devices and Applications,* 3rd edn. Pearson Prentice Hall, Englewood Cliffs, NJ.

11. Enjeti, P. An advanced PWM strategy to improve efficiency and voltage transfer ratio of three-phase isolated boost dc/dc converter. Paper presented at: 2008 Twenty-Third Annual IEEE Applied Power Electronics, Austin, TX, February 24–28, 2010.

12. Dowell, L.J., Drozda, M., Henderson, D.B., Loose,V.W., Marathe, M.V., and Roberts, D.J. ELISIMS: comprehensive detailed simulation of the electric power industry, Technical Rep. No. LA-UR-98-1739, Los Alamos National Laboratory, Los Alamos, NM.

13. https://en.wikipedia.org/wiki/Michael_Faraday. Accessed October 8, 2015.

14. http://www.weatherwizkids.com/weather-lightning.htm. Accessed October 15, 2013.

15. Keyhani, A., Marwali, M.N., and Dai, M. (2010) *Integration of Green and Renewable Energy in Electric Power Systems*, John Wiley & Sons, Inc., Hoboken, NJ.

CHAPTER 4

SMART DEVICES AND ENERGY EFFICIENCY MONITORING SYSTEMS

4.1 INTRODUCTION

The energy conversion process from one state to another state is accompanied by energy loss. The loss is usually in the form of heat. The conversation of energy is the universal law of physical world. The input energy must be equal to the output energy plus energy loss because of the conversion. The local electric energy providers install metering system to measure the electric energy used by the energy users for the billings and record keeping. In the smart distributed energy distribution, when the end-energy users are also energy producers, the smart net metering is used. The metering records provide to the growth of energy usage and or energy production. The recoded data provide the data base for predicting the trend in penetration of electric power in the local power grid distribution systems. The energy records used by end users are also used for optimum market planned operation and participation in selling or buying electric energy in the energy market.

The key elements of the power grid are loads, generators, and transmission and distribution systems. The focus of this section is on power grid loads. The loads are the lighting systems, air handling motors, refrigerators; electric stoves, computers; televisions and many types of industrials and commercial loads. The efficiency of electric loads is increasing yearly and new electric loads are supplied from electric grids. The energy efficiency measurement and monitoring systems are important facilities to evaluate the performance of the electric loads. The capabilities of energy and power monitoring devices are improving with the development of smart grid technologies. More advanced meters

Design of Smart Power Grid Renewable Energy Systems, Second Edition. Ali Keyhani.
© 2017 John Wiley & Sons, Inc. Published 2017 by John Wiley & Sons, Inc.
Companion website: www.wiley.com/go/smartpowergrid2e

and recording items can display the time-varying energy usage of single and multiple loads. These meters can record important energy related data, such as voltage and current waveforms. For detailed analyses of power consumption in residential, commercial, industrial premises active, reactive power, power factor, and time of the day usage are recorded. The data capacity, measurement methods, and cost of smart meters and other energy efficiency element must be considered and compared when designing distributed smart microgrid systems.

4.2 MEASUREMENT METHODS

The local power provider meters take a variety of measurements. A standard power meter records the energy consumption of the end-energy users for billing. Advanced smart meters provide net metering of energy delivered and energy supplied to the grid and can record the voltages and currents, active and reactive power (load power factor), and frequency and harmonic distortion by power electronic loads that are injected to microgrids. Advanced smart meters are equipped with local storage and are able to take 50 of 60 such data points every fraction of a second.[1] The various measurements are used by end users and local power providers to analyze the efficiency degradation of loads (i.e., appliances) and quality of power of a local power grid.

4.2.1 Kilowatt-Hours Measurements

The standard unit for meter energy measurements is the kilowatt-hour (kWh). This unit refers to the average amount of power drawn from local power provider or injected to the local grids over one hour. As the amount of power drawn from a feeder line varies with time, the kilowatt-hour measurement will in turn rise and fall. A kilowatt-hour measurement is made by recording the real power (kW) drawn or supplied by a device under testing (DUT) over a period of time. The same term is also used as equipment under test (EUT) or unit under test (UUT). All these terms are used to test a manufactured product for performance evaluation. The monitoring meters measure current and voltage and computes and records kW, kVAr, voltage (V), power factor using an input pulse triggering and a stop pulse triggering specifying the integration period usually over an hour. In a laboratory setting, to measure the power of a DUT element the instantaneous voltage (V) across and current (A) through the DUT device must be recorded. These two variables can be measured by placing an ideal voltmeter in parallel with the DUT and an ideal ammeter in series with the DUT. The ideal voltmeters and ammeters, and how accurate voltage and current measurements are captured are presented. The instantaneous real power consumption of the DUT (kW) is determined by multiplying the instantaneous voltage by the instantaneous current and integrating the product and computing the average value of the product of the voltage times the current.

4.2.2 Current and Voltage Measurements

The voltmeters and ammeters are needed to measure the instantaneous power flowing through a device under test. The voltmeters measure the potential difference (voltage) across a device, while ammeters measure the rate of charge motion (current) through a device. An ideal voltmeter has infinite input resistance. When the resistance is infinite, that is, when the voltmeter has a very high input resistance, the voltmeter draws no current. Such a voltmeter measures accurate readings of voltage. An ideal ammeter has zero resistance and acts as a short circuit because it has zero resistance and measures the actual current passing through a device. Early voltmeters and ammeters were analog moving coil meters based on the galvanometer. A galvanometer is constructed using a needle connected to a winding coil. This structure is placed inside a permanent magnetic field. A DUT element is placed in series with the circuit. As current moves through the winding, a magnetic force is generated and acts on the rotating winding and the needle shows the strength of the generating force that is proportional to the current though the DUT element.[2] The basic design of a galvanometer schematic is given in Reference[2].

Modern voltmeters use solid state electronics for measurement. The current measurements are made by measuring a voltage signal through a current viewing resistor. Operational amplifiers are used to convert a voltage or current signal to a DC voltage signal for measurement. Solid state meters are more accurate since these meters use direct sampling of the signals. In high voltage transmission and distribution systems, currents and voltages cannot be directly measured by solid state devices.

A voltage transformer is used for high voltage measurement which is called the "instrument potential transformer." It is also called "potential transformer," or simply designated as PT. A PT is a conventional transformer with two coupled windings. One winding is connected to the high voltage circuit. The second winding is used to measure the proportional voltage of the high voltage winding. The low side measurement can then be recorded. A current transformer, also called CT, is used to measure current in high voltage lines. When measuring large alternating voltages or currents, potential and current transformers (PTs and CTs) are used to step down a large-valued signal using the mutual flux linkage between the high and low sides. The low-side measurement can then be used by a sensor or voltmeter or ammeter for display or to be recorded. The true voltage or current can then be found using the turns ratio N of the PT or CT, and the ideal transformer relationships as presented by Equation (4.1).

$$N = V_{\text{primary}}/V_{\text{secondary}} = I_{\text{secondary}}/I_{\text{primary}} \tag{4.1}$$

4.2.3 Power Measurements at 60 or 50 Hz

To measure the efficiency of conversion of a load in consumption of electric energy in producing useful work, the power factor of a DUT can be measured as well. The power factor of a load is the ratio of the real power delivered to a load (kW) to the total apparent power given to the load (kVA). The apparent power is the product of voltage (V) times current (I) or in high power consumption the kV times current (I). The power factor ranges from 0 to 1. Loads with a power factor of 1 are purely resistive loads. Examples of resistive loads are space heaters and electric stoves. The power factor of a load is the cosine of the phase shift between the voltage and current wave signals. The reactive power of a load (kVAr) is the square root of the load real power (kW) squared subtracted from the apparent power (kVA) squared. Therefore, real, reactive, and apparent power and power factor angle can be derived if two of the values are known.

One method for determining the power factor of a DUT is to measure the phase shift between the zero crossings of the device AC voltage and current. This test can be performed by using an electronic oscilloscope or digitally sensed by a sensor or can be recorded. In the first step, the positive-to-negative voltage zero crossing is recorded. In the second step, the positive-to-negative current zero crossing is recorded. The time delay between these two events can determine the power factor angle of the load. If the device is operating at 60 Hz, the power factor angle could be found by Equation (4.2).

$$\text{Zero-crossing time}/60 \times 360 \text{ Degrees} = \text{Power factor (PF)} \qquad (4.2)$$

A simpler method for finding a device power factor is to measure the real (kW) and apparent power (kVA) delivered to the load. The real power consumed by a device is the instantaneous voltage multiplied by the instantaneous current, which can be digitally sampled by sensors or measured with standard meters. The apparent power delivered to a device can be digitally sampled by sensors or measured by using AC voltmeters or AC ammeters. The AC meters measures the root mean square (RMS) voltage and RMS current which is the peak voltage and current values divided by $\sqrt{2}$. The power factor can be determined by dividing the device instantaneous power consumption (kW) by its apparent power (kVA). If the data are digitally sampled in a substation, the substation computer can process the data and display or send the data to real-time power management control center.

Proper data sampling must be used to ensure no information is lost during measurements. If a waveform is sampled infrequently, signal aliasing can occur, providing an inaccurate picture of the load energy consumption. Using high sampling sensors to sample a power frequency is unnecessary and adds to the cost of power meters. According to well-established methods of signal processing, aliasing is avoided if a signal is sampled at twice the frequency of the highest-frequency component of the waveform. For 60 Hz power signals, the sampling signal must have at least a rate of 120 Hz or greater (at least 120 samples taken per second). The normal sampling rate is in the range of 1–5 kHz.[3]

4.2.4 Analog-to-Digital Conversions

Power monitoring elements converts analog measurements to discrete digital values for data recording and storage. Digital data are interfaced with smart meters for net metering and real time pricing. Normally, up to 60 values are recorded during each measurement cycle. A smart meter makes use of digital data and its software for control and reporting. Standard analog-to-digital converters (ADCs) are used to record the sampled analog data. The resolution of the ADC depends on the number of bits used in the conversion process. Normally 12 bit ADC converters are used by smart meters. The voltage and current measurements can take on 2^{12}, or 4096, discrete values. For sinusoidal AC waveforms, half of these digital values are negative to account for the alternating waves. In this case, the sensor can output 2048 discrete magnitudes for measured data. The exact value of these discrete readings is dependent on the converter's range. For example, if the maximum voltage magnitude that can be read by the meter is 2000 V, the 12-bit converter has a resolution of 2000 V/2048 = 0.9766 V.

4.2.5 Root Mean Square (RMS) Measurement Devices

The RMS measurements are key values when studying efficiency of a load. RMS values of voltage, current, and power through a load device can be analyzed and used to determine the efficiency, stress, and power factor of a load. Many commercial smart meters are able to record RMS measurements for regular sinusoidal power waveforms.

P3 International makes a smart meter that records the voltage, current, real power, and apparent power of loads that is placed in series (plugged) into the meter. This meter has a small LCD interface and five buttons that switch the displayed measurement.[4] These meters are practical for local power monitoring for older appliances that plug into wall outlets. As of 2014, these meters cost $20 each.[5] The EPM6000 smart meter offered by General Electric can record more data, such as power factor and load reactive power. These rugged meters can be used in a variety of locations and can record load data of a feeder or an outlet. The EPM6000 meters can be used to monitor the power flow and efficiency power consumption for a home or commercial or industrial premises or local microgrid. As of 2014, the GE meters are sold for $1037 each.[6] Siemens also offers meters that can record various RMS data. The Siemens SICAM P50 meter can measure true RMS values of voltage, current, real, reactive, and apparent power, as well as the frequency of the load signal. As of 2014, these devices are available for $1200 each.[7]

4.3 ENERGY MONITORING SYSTEMS

A power monitoring system is comprised of data recording and analyzing software that collects and organizes the measurements collected by smart meters. Local power providers use monitoring system to track the time of day pricing and net metering if the end-energy users have local generation and inject power into the grid. The residential, commercial, and industrial systems also use the monitoring systems to track the efficient operation of loads. The smart meters and supporting software provide an end-to-end efficiency evaluation in all types of premises and microgrids. General Electric (GE)[3,8] offers a smart grid operations suite that can collect and organize data from multiple meters. This system is supported by an advanced metering infrastructure (AMI) that uses point-to-multipoint (P2MP) communication to connect numerous data nodes in a microgrid. The collected data are transmitted continuously between load devices and the operations software. The GE software is equipped with full energy monitoring system that includes smart meters, communication devices, data collectors, and computational engine to calculate energy efficiency and to track fatigue and thermal stress and degradation of the performance of loads. Siemens offers comparable measurement systems as well. The Siemens EnergyIP solution also uses AMI-based meter data management system that gathers and organizes various data from a distributed monitoring network.[6,9]

4.4 SMART METERS: CONCEPTS, FEATURES, AND ROLES IN SMART GRID

The major challenge of smart grid development is creating new metering systems that can become a point of contact between end-energy users and energy providers. These smart meters are to control efficient use of energy through load control and provide market information to end-energy users. The energy providers can use the data collected in the new paradigm of market-controlled grid operation and provide efficient production of energy. The smart efficient grid system can be accomplished through effective two-way communication between load and generation scheduling, and between suppliers and energy users. Over the past few decades, suppliers of equipment to energy providers have developed new meters that perform electrical billing, grid monitoring, security, and grid automation for residential, commercial, and industrial premises. The penetration of smart meters are rapidly increasing in the power grid and they offer the ability to all stock holders to effectively manage power consumption. Residences with local power generation or storage devices can use smart meters to track the market real-time price to efficiently operate and control the net power flow through the grid connection. Smart meters use several types of communication and data sharing protocols that allow system management. The effectiveness, security, and cost of different smart meter designs can establish a new market-controlled operation of distribution systems. The continued development of smart meters is critical for the establishment of an effective microgrid of distributed energy systems.[10]

4.4.1 Power Monitoring and Scheduling

Power meters have a primary responsibility to track the power consumption of end-energy users. This consumption recording is used by power providers for billing purposes. The energy flow through a meter, measured in kilowatt-hours (kWh), is recorded to determine electricity bills. Smart meters also include this basic feature. The smart meters, however, also contain the ability to record the energy consumption of individual loads. An electrically efficient building or industrial complex is designed to track energy consumption of loads for optimal operation and to reduce overall energy consumption and waste. The industrial grade smart meters are equipped to measure power quality. Low power quality has excessive harmonics which causes voltage fluctuation. The power harmonics are measured and the total harmonic distortion (THD) is calculated.[11] Excessive THD causes torsional stress to machines. Smart meters can track efficient operation of industrial equipment and give warning signals on the health of equipment.

For smart meters to measure data on energy use of load, the meters must have a connection to the power outlets of the load. A monitoring device is placed on the outlet of the load and data are collected and send to smart meter for recoding and analysis. Since the use and operation of all loads are not necessary, optimum operation of building loads is to divide the building loads into separate categories based on importance of operational needs. By dividing electrical loads into essential, periodically used, aesthetic, and nonessential groups, the net power consumptions of critical and noncritical loads can be more easily studied. The different load groups are each fed to different feeders with circuit breakers. Choosing the size and type of monitoring equipment depends on the aggregate power rating and current capacity of the loads. The design of smart distribution panels facilitate monitoring of energy use of buildings.[12]

To monitor and schedule loads of buildings, both the smart meter and loads must have communication capability. Two-way communication between the meter and load monitoring devices are used for local measurement of data and sending the data to smart meters. The smart meters control local monitoring devices and issue scheduling commands to local device for operation of loads. Current communication methods include radio frequency (RF) broadcasting, cellular, wired Ethernet, and wireless Wi-Fi communications. The broadcasting strategies are discussed in Section 4.4.2. New load equipment is equipped with built-in monitoring capabilities.[1,13] The local monitoring devices are under control of smart meters and are operating for optimum efficiency of premises. Smart load interfacing is performed by software control residing in smart meters or mobile phone applications.[1,14]

Older equipment can also be controlled with smart meters and new grid technology. Outlet local monitors that are placed between equipment and a wall outlet record the power consumption of the devices plugged into the monitor.[2,15] The data collected by the local monitor are send via a communication channel to smart meters.[16] The equipment connected to a local monitoring device can be remotely turned on and off, placed on a timer, or controlled by a proximity switch. The control of monitoring device by smart meters facilitates scheduling of power and reduces parasitic power drainage when a load is not in use. Figure 4.1 depicts a general monitoring and control structure of a grid load.

The local monitoring devices can be incorporated with modern smart meter systems to allow complete control of both new and old home appliances. The

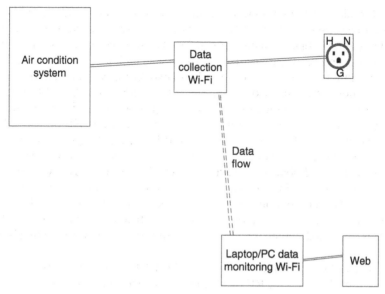

Figure 4.1   Monitoring device outlet of smart meter plug.[2]

numerous options for efficient and market-based control through smart grid connectivity will make the smart meter system a versatile solution for power monitoring and optimum operation of loads.

4.4.2 Communication Systems

The smart meters are equipped with control and data storage and communicate between end-energy users, local power providers, and building load devices. The communication system facilitates the efficient and real-time pricing control of power grids.[15,16] Both consumers and suppliers are able to monitor the power flow through a smart meter. Both parties are also able to remotely send commands to these devices to regulate loads in a residence or building per contract agreements. For example, per agreement between power provider and end-energy users for lower prices, load control can be achieved by power providers during system emergency of order of 1–5 minutes. This load control creates virtual generation during high power demands and by rotating on–off periods, the power provider can efficiently sell and buy power in real time and offer lower prices to end-energy users. Smart grids allow real-time equipment outages detection and can be pinpointed in a short amount of time due to emergency communications between individual smart meters and local distribution grid control center. There are currently many types of smart meter designs that use various communication methods to share data between the different ends of the smart grid.

The most common transceiver systems used in multidevice networks is radio frequency (RF) meshes. This communication system is based on an RF network with a broadcasting antenna to both send and receive information. Radio frequency range used is from 3 kHz to 300 GHz.[15] This RF network is able to broadcast at the same frequency for proper two-way communication. A number of manufactures are specialized in smart grid technologies and have developed smart meter systems that use radio networks for communication.

Smart meter systems also comply with the ZigBee IEEE 802.15 protocol for wireless radio communication. This IEEE protocol is used to regulate communication in short-range and low-power networks. The IEEE protocol determines the rate of data transfer for different broadcast frequencies. The local RF networks normally operate on a 2.4 GHz frequency band. The data transfer rate is at 250 kilobytes per second. Each channel is 5 MHz wide.[3,16] This common operating point is chosen for its high transfer rate and safe operating frequency. The main public concerns over the development of wireless smart meters are the possible exposure to harmful electromagnetic waves. The IEEE 802.15 protocol is designed to avoid the risk of traveling waves affecting human health.

The Pacific Gas and Electric Company (PGE) developed a smart meter electrical system that uses an RF mesh to communicate between different smart devices. This RF mesh connects each antenna-equipped device to all other nearby transceivers. This communication system is equipped to send data to a local PGE station at prespecified period according to IEEE protocol. The RF-equipped system can likewise communicate with nearby smart meters.[5,7] By distributing the communications system, the smart meter network can keep running even when an individual device is damaged. This decentralization is critical to smart grid design.

Smart meters use a number of other communication systems to connect end-energy users and local power providers. Some manufacturers offer smart meters that use IEEE standard radio communication, as well as Ethernet and Wi-Fi capabilities, to communicate across the smart grid.[4] The Internet protocol options provide capabilities for distributed communications. Ethernet-capable systems provide a physical and more rugged connection between a local sensor in the grid and the smart meter network. Wi-Fi systems offer wireless communication portals that do not involve the health concerns. However, bandwidth limitations, and short broadcast ranges are used in smaller RF meshes.

General Electric provides smart meters that can use power line communication (PLC) to transmit data between local sensors and local network of power providers.[5,6] The PLC communication system involves sending high-frequency data transmission across a line that is also transmitting power at 50 or 60 Hz. The PLC systems are placed on both ends of the transmission line, which can filter and translate the high-frequency data being sent along the line. Siemens offers smart grid designs that use fiber optic cables to transmit data collected by sensors[5,7] as well. These fiber optic cables operate between smart meters and local sensors that are acquiring data from the grid network or end-energy users. The fiber optics cables are placed underground or alongside power lines. However PLC overhead communication systems are at risk to weather conditions and environmental damages which could stop data transfers. Underground system of fiber optics communication systems can be more secure.

Many smart devices are Bluetooth equipped for easy power monitoring on a mobile phone.[2,3] General Electric[5,6] offers 3G and 4G network solutions for cellular information transmitting. Using mobile devices and cellular networks to send smart meter data is a more established method to keep data secure through third-party networks. Mobile device connectivity and cellular network transmission provide a new method of data sharing. The communication method of a smart meter system is a key design component for creating a secure and reliable smart grid network. Efficient smart renewable power microgrids must have effective communication to reduce cost and participate in market controlled smart grid system.

4.4.3 Network Security and Software

Smart meter communication system not only needs state of the art hardware and storage to transfer data between various stakeholders, it also needs a secure network and processing software. Malicious attack in smart grid can start at any point in the communication system of smart. A secure network must be designed such that information is only transferred between smart meter of local microgrid and end-energy users and local power providers.

Local area networks (LANs) are networking solutions that involve interconnecting various nodes of a microgrid or several microgrids connected to a local power provider feeder line from distribution or subtransmission depending on the size of loads and size of distributed generation of microgrids.[8] A smart meter is connected to a microgrid of PV residential user, LAN solutions can be established using fiber optic or communication cabling between the various parts of residential microgrid and local power providers. The physical connection ensures that all data are kept within secure network. However, during installation, care must be taken to ensure malicious silent hardware with splitters is not installed for later attack in grids. A LAN should include multiple pathways between each smart meter in order to distribute data flow and prevent complete cutoff from the local power provider monitor. The requirement of physical connections can limit the placement of smart meters in the smart grid. The risk of cable damage also requires proper protection and maintenance of the LAN pathways. A LAN system may work well in individual residential microgrids, but may not be the most practical solution for larger microgrids. The Ethernet communications offered by Smart Utility Systems[4,13,14] and the fiber optic options offered by Siemens are both examples of LANs.

Home area networks (HANs) are small, IP-based area networks that allow communication between nearby devices.[8] While HANs can include wired pathways, using broadband Internet connections or other wireless communication methods to interconnect devices separates HANs from LANs. The cellular broadcasting, RF meshing, and Bluetooth linking solutions offered by General Electric and Pacific Gas and Electric are examples of HAN networking. A localized wireless network can interconnect various smart devices and meters for data transfer without physical attachments or cabling. HANs also decrease the risk of damaging the smart grid network. These networks make it possible to create a fully distributed system without a large amount of wiring. One of the drawbacks to wireless HANs, however, is the increased risk of security breaching and hacking. Like traditional wireless networks, HANs are at risk to remote breaches if the system does not include proper firewalls and other safety measures. A malicious user can access the smart grid system and penetrate and gain control of the smart grid. Designers must be cautious when creating secure HANs.

Wide area networks (WANs) are made of interconnected LANs and HANs. These networks are generally seen in large metropolitan areas. While a WAN would not be seen in a residential smart meter system, a WAN would be established in order to connect multiple microgrids across a city or large urban center. Like the distributed meter-to-meter connections of a HAN, the interconnections in a WAN system would improve the redundancy of the overall smart grid. A WAN is achieved though TCP/IP protocols that use firewalls, modems, and switches to pass data between nodes through third-party networks.[6,8] These devices improve the overall security of the network.

In order to interact with smart meters, many companies offer custom-built software solutions for local microgrids. These programs can be installed along with a smart meter system, or customized by the consumer to fit their needs. General Electric offers the Smart Meter Operations Suite (SMOS), which acts as an intermediary between metering devices and the local power provider.[6] This software is easily utilized by consumers to track the usage of multiple devices over the course of a day, as well as schedule and control when certain appliances are used. Figure 4.2 shows the SMOS's place within the microgrid. This software and others like it can be purchased by smart meter owners to easily track and organize data transferred across the network. Such programs allow users to easily interface with and control the various meters and devices within a localized network. General Electric smart metering operations flow chart is given in Reference[6]

There are many smart grid solutions available for residential, commercial, and industrial microgrids. A smart microgrid is connected to the local power provider grid through the smart meters and sensors by the communication system and control software. Smart meters are available from a variety of large suppliers, including Siemens and General Electric. These meters can be installed in place of standard utility meters, and many devices feature "plug-n'-play" capabilities that allow easy integration with the grid system already in place. GE sells numerous types of smart meters that can fit in a variety of systems. Available designs include single-phase and three-phase meters, Ethernet and Wi-Fi equipped models, and meters ready for HAN installation. As of 2014, these meters cost around $150 each.[6]

Siemens is another large energy company that specializes in smart grid solutions for various premises. The smart grid products offered by Siemens range

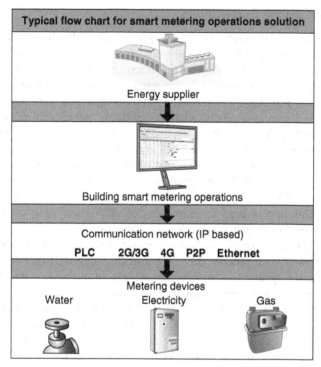

Figure 4.2   A smart metering operations suite flowchart[6] adapted from General Electric company, smart metering operations suite (SMOS), 2014.

from individual smart meters, to data acquisition hubs, to software control, and to end-to-end microgrid solutions. The smart meters offered by Siemens can use fiber optic cabling or wireless cellular networks for easy data transmission. Siemens currently provides only three-phase meters, making these meters suitable for commercial and industrial premises. As of 2014, these meters cost around $200 each.[7] Other suppliers also offer smart metering for a microgrid. Meters from Smart Utility Systems[4] are designed for efficiency in size and functionality, with AMI-interfacing capabilities, as well as optical ports for LAN or HAN network. More advanced models of these meters include internal backup batteries and supercapacitors that power the meter locally for a brief time during a power outage. These backup energy sources allow the meters to send a few last communications before power is lost. These messages will alert the utility company to the presence of a power loss at the meter's location, thus reducing communication and repair times. As of 2014, smart meters from Smart Utility Systems[4] cost $140–$160. Echelon[9] is another supplier that produces smart meters for many premises. Echelon smart meters are capable of connecting with HANs, and can be used in WANs[5] in medium and low voltage scenarios. As of 2014, Echelon[9] meters cost several hundred dollars each, and are suitable for large systems or interconnected microgrids.

4.4.4 Smart Phone Applications

Advanced smart monitoring systems offer a highly detailed and organized way of evaluating efficiency of local microgrids; these solutions may not be practical for all microgrids. Residential microgrids with a few loads that are to be monitored and a smart meter with PV generating plant may benefit from simple monitoring approach. Many independent software developers are designing smart phone applications that can track and control loads in a residential microgrid. These applications are cost effective and can connect wirelessly with microgrid devices for control and efficiency monitoring. Pacific Gas and Electric presents in an article[5] smart phone software application.

SunPower[15] offers an iPhone application (app) that is used in PV power plants to track PV plant energy output, as well as daily energy production and usage. This app is used to help schedule PV plant, load control in PV microgrids. The SunPower smart phone application is presented. http://www. weisersecurity.com/ provides a home control app that uses a 3G network. This app is used to remotely control connected microgrid devices, such as lights, and home appliances for efficient energy monument. The MeterRead (http://www.acrsystems.com/products/smartreader/) app is also used in place of smart meter to monitor the kilowatt-hour consumption of a microgrid load. This app can also be used to compare the benefits of adding more efficient devices, such as LED lights for energy efficiency monitoring and cost benefit analysis. The Control mobile (http://www.control4.com/solutions/mobile-app) app is also used for IPhone, Blackberry, and Android to remotely control appliances at home as well as any location with the app's wireless network. The IPhone Energy UFO (http://www.energyufo.com/) app is equipped with four electrical surge protectors to monitor the hour-by-hour energy consumption. This app also displays real time of electricity costs for efficient power usage. The five smart phone apps presented in this section are just some of the many independent pieces of software that are available for residential, commercial, and industrial microgrids for monitoring energy efficiency systems.[15]

4.5 SUMMARY

Power monitoring and metering has been a critical component of the electrical grid since its inception. Power providers use power meters for customer billing, data collection, forecasting and power grid efficiency evaluation. Smart power meters are used to provide information to end-energy users to monitor usage of electric energy and reduce cost of electric energy through efficient use. Smart meters record energy consumption of microgrid, each feeder to loads, monitor and control loads for cost effective microgrid operation by recoding voltage, current waveforms, real, reactive, and apparent power flow, and power factor. Multiple smart meters can be used in commercial, industrial, and condominiums together as a local microgrid with area network to record such data across microgrid. Currently, numerous companies provide both consulting and simple hardware choices for smart grid design. These smart meters are software controlled and provide end-to-end solutions and are utilized by residential, commercial, and industrial renewable power microgrids. Smart meters serve as an intermediary between the end-energy users and local power providers and are used to collect data and control individual load and buy and sell energy to grid. Some modern appliances, such as HVAC and refrigerators are equipped with communication capabilities to connect to local smart meters. Older appliances can be integrated into a microgrid by connecting these devices to a smart outlet meter with communication capabilities. The hardware and software manufactured facilitates development of smart renewable power microgrids as part of interconnected grid.

Smart meters use a variety of communication and data acquisition protocols to transfer information. Smaller metering networks can be built around wired LANs, while more distributed systems can use radio frequency or Internet protocols to establish HANs. Such networks can in turn be connected through secure firewalls and relays to create a WAN, which would bolster the overall smart grid. Private companies provide different designs and solutions for smart metering in a residence. Individuals and businesses must make proper choices when selecting the size, capacity, and type of smart meter to use in a system. The proper selection of smart meters is a necessary and critical point in microgrid and smart grid design.

PROBLEMS

4.1 Estimate the energy consumption of a 1500 ft^2 apartment consisting of three bedrooms, kitchen, washer, and dryer for a winter day operation and plot the hourly energy consumption. Compute the cost effective operation of appliances' energy usage if kWh metered by power provider's smart meter is 4 cent during the night and 8 cent during day time.

4.2 Specify a home consisting of two levels, 5000 ft^2 with four bedrooms, kitchen, dining room, family room, kitchen, washer, and dryer for a summer day operation and plot the hourly energy consumption. Compute the cost effective operation of appliances' energy usage if kWh metered by power provider's smart meter is 4 cent during the night and 10 cent during day time. For the assumed daily load cycle, compute monthly cost of energy.

REFERENCES

1. Whirlpool. Available at http://www.whirlpool.com/smart-appliances./. Accessed March 16, 2014.
2. Meter Plug. Available at http://meterplug.com/. Accessed March 16, 2014.
3. Wikipedia. Available at http://en.wikipedia.org/wiki/ZigBee. Accessed March 17, 2014.

4. Smart Utility Systems. Available at http://smartusys.com/hardware/. Accessed March 16, 2014.

5. Pacific Gas and Electric. Available at http://www.pge.com/en/myhome/customerservice/smartmeter/index.page. Accessed March 16, 2014.

6. General Electric. Available at http://www.gedigitalenergy.com/SmartMetering/catalog/p2mp.htm#p2mp2. Accessed February 13, 2015.

7. Siemens. Available at http://w3.siemens.com/smartgrid/global/en/Pages/Default.aspx. Accessed March 16, 2014.

8. Techopedia. Available at www.techopedia.com. Accessed March 17, 2014.

9. Echelon. Available at http://www.echelon.com/applications/smart-metering./. Accessed March 18, 2014.

10. General Electric. Available at http://www.gedigitalenergy.com/multilin/catalog/epmfamily.htm. Accessed March 21, 2014.

11. Hyperphysics. Available at http://hyperphysics.phy-astr.gsu.edu/hbase/magnetic/galvan.html. Accessed March 21, 2014.

12. P3 International. Available at http://www.p3international.com/manuals/p4400_manual.pdf. Accessed March 22, 2014.

13. Siemens. Available at http://www.energy.siemens.com/hq/en/automation/power-transmission-distribution/power-quality/power-monitoring-devices.htm. Accessed March 21, 2014.

14. General Electric. Available at http://www.gedigitalenergy.com/smartmetering/catalog/p2mp.htm. Accessed March 23, 2014.

15. Inhabitat. Available at http://inhabitat.com/5-smartphone-apps-that-will-help-you-save-energy/. Accessed March 24, 2014.

16. Siemens. Available at http://w3.siemens.com/smartgrid/global/en/products-systems-solutions/software-solutions/emeter/Pages/EnergyIP.aspx#. Accessed March 23, 2014.

CHAPTER 5

LOAD ESTIMATION AND CLASSIFICATION

5.1 INTRODUCTION

First step in the design of smart PV grid microgrid is load evaluation.[1-6] The load can be estimated through recording the load or load estimation using an Excel sheet. The load estimation can be expanded to daily usage through the knowledge of daily activities on the building. The yearly and monthly kWh consumption can be used as a source of kWh consumption for winter, spring, summer, and fall. The kWh consumption gives a realistic evaluation of expected peak load monthly load estimation. Through the peak load estimation, the size of the PV generating station can be determined.

Today, the cost of photovoltaic cells (PV) is rapidly dropping and becoming commercially attractive as the price of electric energy is increasing. Real-time pricing and time of day pricing are being implemented by energy providers. To install PV power generating plant, and evaluate the cost and payout of PV, it is desirable to install energy monitoring system to measure the residential, commercial, and industrial premises load consumption and energy efficiency. The load estimation and load classification facilitate pay out of PV installation and lighting system using light emitting diodes (LED).

5.2 LOAD ESTIMATION OF A RESIDENTIAL LOAD

As an example, the design steps are outlined to estimate the daily power consumption in a standard residence, a 2600 ft^2 building. The loads are estimated

Design of Smart Power Grid Renewable Energy Systems, Second Edition. Ali Keyhani.
© 2017 John Wiley & Sons, Inc. Published 2017 by John Wiley & Sons, Inc.
Companion website: www.wiley.com/go/smartpowergrid2e

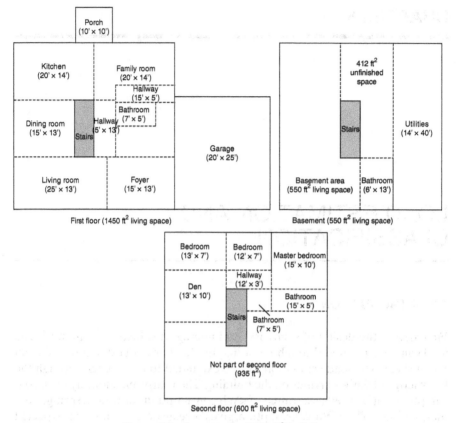

Figure 5.1   Layout of a 2600 ft^2 building.

based on expected load ratings. Figure 5.1 presents the dimensions and layout of the two-story building. The rooms and building facilities are labeled for estimating electric loads. The rooms with electrical elements were a foyer, a living room, a dining room, a kitchen, a family room, three hallways, a laundry room, a garage, a porch, a basement, a den, two bedrooms, a master bedroom, and four bathrooms.

As second step in design, an Excel sheet is used to identify all electric energy consumed by loads in each room and duration of operation and estimating the total peak loads if all loads are operating at the same time. The tables of load consumption for each room are presented in Tables 5.1–5.30.

Tables 5.1 through 5.16 present expected loads. The various devices and electric equipment of the building are classified as essential and nonessential loads for sizing the PV station. Essential loads are highlighted in Tables 5.1 through 5.16. The power ratings of each appliance are presented on these Excel tables as well, as the wattage of each device indicates the power consumption. If all loads are used at their full ratings at the same time, the total peak load for the building is 15,380 W, or about 15.4 kW. Using standard 120 V power, this peak

TABLE 5.1 Foyer Load Profile (Winter/Summer)

Load Element/Usage	12am	1am	2am	3am	4am	5am	6am	7am	8am	9am	10am	11am	12pm	1pm	2pm	3pm	4pm	5pm	6pm	7pm	8pm	9pm	10pm	11pm
Lighting (3 bulbs)	0W/ 0W	0W/ 0W	0W/ 0W	0W/ 0W	0W/ 0W	0W/ 0W	0W/ 0W	60W/ 0W	60W/ 60W	0W/ 0W	0W/ 0W	0W/ 0W	0W/ 0W	0W/ 0W	0W/ 0W	0W/ 0W	0W/ 0W	60W/ 0W	60W/ 60W	180W/ 180W	60W/ 60W	60W/ 60W	60W/ 60W	0W/ 0W

TABLE 5.2 Living Room Load Profile (Winter/Summer)

Load Element/Usage	12am	1am	2am	3am	4am	5am	6am	7am	8am	9am	10am	11am	12pm	1pm	2pm	3pm	4pm	5pm	6pm	7pm	8pm	9pm	10pm	11pm
Lighting (5 bulbs)	0W/ 0W	0W/ 0W	0W/ 0W	0W/ 0W	0W/ 0W	0W/ 0W	0W/ 0W	0W/ 0W	0W/ 0W	0W/ 0W	0W/ 0W	0W/ 0W	120W/ 120W	120W/ 120W	0W/ 0W	0W/ 0W	0W/ 0W	60W/ 0W	240W/ 240W	240W/ 240W	300W/ 300W	60W/ 60W	0W/ 0W	0W/ 0W

TABLE 5.3 Dining Room Load Profile (Winter/Summer)

Load Element/ Usage	12am	1am	2am	3am	4am	5am	6am	7am	8am	9am	10am	11am	12pm	1pm	2pm	3pm	4pm	5pm	6pm	7pm	8pm	9pm	10pm	11pm
Lighting (5 bulbs)	0W/ 0W	0W/ 0W	0W/ 0W	0W/ 0W	0W/ 0W	0W/ 0W	0W/ 0W	120W/ 120W	60W/ 60W	0W/ 0W	0W/ 0W	0W/ 0W	120W/ 120W	60W/ 60W	0W/ 0W	0W/ 0W	0W/ 0W	0W/ 60W	300W/ 300W	120W/ 120W	60W/ 60W	0W/ 0W	0W/ 0W	0W/ 0W

TABLE 5.4 Kitchen Load Profile (Winter/Summer)

Load Element/ Usage	12am	1am	2am	3am	4am	5am	6am	7am	8am	9am	10am	11am	12pm	1pm	2pm	3pm	4pm	5pm	6pm	7pm	8pm	9pm	10pm	11pm
Lighting (10 bulbs)	0W/ 0W	0W/ 0W	0W/ 0W	0W/ 0W	0W/ 0W	0W/ 0W	0W/ 0W	360W/ 360W	120W/ 120W	0W/ 0W	0W/ 0W	180W/ 180W	180W/ 180W	180W/ 0W	0W/ 0W	0W/ 0W	0W/ 0W	0W/ 0W	600W/ 600W	360W/ 360W	120W/ 120W	120W/ 120W	0W/ 0W	0W/ 0W
Refrigerator/ freezer	725W/ 725W	725W/ 725W	725W/ 725W	725W/ 725W	725W/ 725W	725W/ 725W	725W/ 725W	725W/ 725W	725W/ 725W	725W/ 725W	725W/ 725W	725W/ 725W	725W/ 725W	725W/ 725W	725W/ 725W	725W/ 725W	725W/ 725W	725W/ 725W	725W/ 725W	725W/ 725W	725W/ 725W	725W/ 725W	725W/ 725W	725W/ 725W
Dishwasher	0W/ 0W	0W/ 0W	0W/ 0W	0W/ 0W	0W/ 0W	0W/ 0W	0W/ 0W	0W/ 0W	0W/ 0W	0W/ 0W	0W/ 0W	0W/ 0W	0W/ 0W	0W/ 0W	0W/ 0W	0W/ 0W	0W/ 0W	0W/ 0W	0W/ 0W	2kW/ 2kW	2kW/ 2kW	0W/ 0W	0W/ 0W	0W/ 0W
Stove/oven	0W/ 0W	0W/ 0W	0W/ 0W	0W/ 0W	0W/ 0W	0W/ 0W	0W/ 0W	0W/ 0W	0W/ 0W	0W/ 0W	0W/ 0W	0W/ 0W	0W/ 0W	0W/ 0W	0W/ 0W	0W/ 0W	0W/ 0W	3kW/ 3kW	3kW/ 3kW	0W/ 0W	0W/ 0W	0W/ 0W	0W/ 0W	0W/ 0W
Microwave	0W/ 0W	0W/ 0W	0W/ 0W	0W/ 0W	0W/ 0W	0W/ 0W	0W/ 0W	0W/ 0W	0W/ 0W	0W/ 0W	0W/ 0W	0W/ 0W	1kW/ 1kW (15 min)	0W/ 0W	0W/ 0W	0W/ 0W	0W/ 0W	0W/ 0W	0W/ 0W	0W/ 0W	0W/ 0W	0W/ 0W	0W/ 0W	0W/ 0W
Toaster	0W/ 0W	0W/ 0W	0W/ 0W	0W/ 0W	0W/ 0W	0W/ 0W	0W/ 0W	800W/ 800W (15 min)	0W/ 0W	0W/ 0W	0W/ 0W	0W/ 0W	0W/ 0W	0W/ 0W	0W/ 0W	0W/ 0W	0W/ 0W	0W/ 0W	0W/ 0W	0W/ 0W	0W/ 0W	0W/ 0W	0W/ 0W	0W/ 0W

TABLE 5.5 Family Room Load Profile (Winter/Summer)

Load Element/ Usage	12am	1am	2am	3am	4am	5am	6am	7am	8am	9am	10am	11am	12pm	1pm	2pm	3pm	4pm	5pm	6pm	7pm	8pm	9pm	10pm	11pm
Lighting (6 bulbs)	0W/ 0W	0W/ 0W	0W/ 0W	0W/ 0W	0W/ 0W	0W/ 0W	0W/ 0W	120W/ 120W	60W/ 60W	60W/ 60W	60W/ 60W	60W/ 60W	60W/ 60W	60W/ 60W	60W/ 60W	60W/ 60W	120W/ 120W	120W/ 120W	240W/ 240W	240W/ 240W	240W/ 240W	120W/ 120W	120W/ 120W	0W/ 0W
Television	0W/ 0W	0W/ 0W	0W/ 0W	0W/ 0W	0W/ 0W	0W/ 0W	0W/ 0W	150W/ 150W	0W/ 0W	0W/ 0W	0W/ 0W	0W/ 0W	150W/ 150W	0W/ 0W	0W/ 0W	0W/ 0W	0W/ 0W	0W/ 0W	0W/ 0W	0W/ 0W	150W/ 150W	150W/ 150W	150W/ 150W	0W/ 0W
DVD/DVR	0W/ 0W	0W/ 0W	0W/ 0W	0W/ 0W	0W/ 0W	0W/ 0W	0W/ 0W	25W/ 25W	0W/ 0W	0W/ 0W	0W/ 0W	0W/ 0W	25W/ 25W	0W/ 0W	0W/ 0W	0W/ 0W	0W/ 0W	0W/ 0W	0W/ 0W	0W/ 0W	25W/ 25W	25W/ 25W	25W/ 25W	0W/ 0W
Stereo	0W/ 0W	0W/ 0W	0W/ 0W	0W/ 0W	0W/ 0W	0W/ 0W	0W/ 0W	60W/ 60W	0W/ 0W	0W/ 0W	0W/ 0W	0W/ 0W	60W/ 60W	0W/ 0W	0W/ 0W	0W/ 0W	0W/ 0W	0W/ 0W	0W/ 0W	0W/ 0W	60W/ 60W	60W/ 60W	60W/ 60W	0W/ 0W
Phone	15W/ 15W	15W/ 15W	15W/ 15W	15W/ 15W	15W/ 15W	15W/ 15W	15W/ 15W	15W/ 15W	15W/ 15W	15W/ 15W	15W/ 15W	15W/ 15W	15W/ 15W	15W/ 15W	15W/ 15W	15W/ 15W	15W/ 15W	15W/ 15W	15W/ 15W	15W/ 15W	15W/ 15W	15W/ 15W	15W/ 15W	15W/ 15W

TABLE 5.6 Hallway Load Profile (Winter/Summer)

Load Element/ Usage	12am	1am	2am	3am	4am	5am	6am	7am	8am	9am	10am	11am	12pm	1p	2p	3p	4p	5pm	6pm	7pm	8pm	9pm	10pm	11pm
Lighting (2 bulbs)	0W/ 0W	0W/ 0W	0W/ 0W	0W/ 0W	0W/ 0W	0W/ 0W	0W/ 0W	120W/ 60W	120W/ 60W	60W/ 0W	60W/ 0W	60W/ 0W	60W/ 0W	60W/ 0W	60W/ 0W	60W/ 0W	60W/ 0W	60W/ 0W	120W/ 60W	120W/ 60W	120W/ 60W	120W/ 60W	0W/ 0W	0W/ 0W

TABLE 5.7 Laundry Room Load Profile (Winter/Summer)

Load Element/ Usage	12 am	1 am	2 am	3 am	4 am	5 am	6 am	7 am	8 am	9 am	10 am	11 am	12 pm	1 pm	2 pm	3 pm	4 pm	5 pm	6 pm	7 pm	8 pm	9 pm	10 pm	11 pm
Lighting (3 bulbs)	0 W/ 0 W	0 W/ 0 W	0 W/ 0 W	0 W/ 0 W	0 W/ 0 W	0 W/ 0 W	0 W/ 0 W	0 W/ 0 W	0 W/ 0 W	0 W/ 0 W	0 W/ 0 W	0 W/ 0 W	0 W/ 0 W	180 W/ 180 W	180 W/ 180 W	0 W/ 0 W	0 W/ 0 W	0 W/ 0 W	0 W/ 0 W	0 W/ 0 W	0 W/ 0 W	0 W/ 0 W	0 W/ 0 W	0 W/ 0 W
Washer	0 W/ 0 W	0 W/ 0 W	0 W/ 0 W	0 W/ 0 W	0 W/ 0 W	0 W/ 0 W	0 W/ 0 W	0 W/ 0 W	0 W/ 0 W	0 W/ 0 W	0 W/ 0 W	0 W/ 0 W	0 W/ 0 W	400 W/ 400 W	0 W/ 0 W	0 W/ 0 W	0 W/ 0 W	0 W/ 0 W	0 W/ 0 W	0 W/ 0 W	0 W/ 0 W	0 W/ 0 W	0 W/ 0 W	0 W/ 0 W
Dryer	0 W/ 0 W	0 W/ 0 W	0 W/ 0 W	0 W/ 0 W	0 W/ 0 W	0 W/ 0 W	0 W/ 0 W	0 W/ 0 W	0 W/ 0 W	0 W/ 0 W	0 W/ 0 W	0 W/ 0 W	0 W/ 0 W	0 W/ 0 W	3 kW/ 3 kW	0 W/ 0 W	0 W/ 0 W	0 W/ 0 W	0 W/ 0 W	0 W/ 0 W	0 W/ 0 W	0 W/ 0 W	0 W/ 0 W	0 W/ 0 W

TABLE 5.8 Garage Load Profile (Winter/Summer)

Load Element/ Usage	12 am	1 am	2 am	3 am	4 am	5 am	6 am	7 am	8 am	9 am	10 am	11 am	12 pm	1 pm	2 pm	3 pm	4 pm	5 pm	6 pm	7 pm	8 pm	9 pm	10 pm	11 pm
Lighting (3 bulbs)	120 W/ 120 W	120 W/ 120 W	120 W/ 120 W	120 W/ 120 W	120 W/ 120 W	120 W/ 120 W	120 W/ 120 W	180 W/ 180 W	120 W/ 0 W	0 W/ 0 W	0 W/ 0 W	0 W/ 0 W	0 W/ 0 W	0 W/ 0 W	0 W/ 0 W	0 W/ 0 W	0 W/ 0 W	0 W/ 0 W	120 W/ 0 W	120 W/ 120 W	120 W/ 120 W	120 W/ 120 W	120 W/ 120 W	120 W/ 120 W
Garage door opener	0 W/ 0 W	0 W/ 0 W	0 W/ 0 W	0 W/ 0 W	0 W/ 0 W	0 W/ 0 W	0 W/ 0 W	0 W/ 0 W	350 W/ 350 W (5 minute total.)	0 W/ 0 W	0 W/ 0 W	0 W/ 0 W	0 W/ 0 W	0 W/ 0 W	0 W/ 0 W	0 W/ 0 W	0 W/ 0 W	0 W/ 0 W	0 W/ 0 W	0 W/ 0 W	0 W/ 0 W	0 W/ 0 W	0 W/ 0 W	0 W/ 0 W

TABLE 5.9 Porch Load Profile (Winter/Summer)

Load Element/ Usage	12 am	1 am	2 am	3 am	4 am	5 am	6 am	7 am	8 am	9 am	10 am	11 am	12 pm	1 pm	2 pm	3 pm	4 pm	5 pm	6 pm	7 pm	8 pm	9 pm	10 pm	11 pm
Lighting (4 bulbs)	0 W/ 0 W	0 W/ 0 W	0 W/ 0 W	0 W/ 0 W	0 W/ 0 W	0 W/ 0 W	0 W/ 0 W	0 W/ 0 W	0 W/ 0 W	0 W/ 0 W	0 W/ 0 W	0 W/ 0 W	0 W/ 0 W	0 W/ 0 W	0 W/ 0 W	0 W/ 0 W	120 W/ 0 W	240 W/ 0 W	240 W/ 120 W	240 W/ 240 W	240 W/ 240 W	240 W/ 240 W	0 W/ 0 W	0 W/ 0 W

TABLE 5.10 Basement Load Profile (Winter/Summer)

Load Element/ Usage	12 am	1 am	2 am	3 am	4 am	5 am	6 am	7 am	8 am	9 am	10 am	11 am	12 pm	1 pm	2 pm	3 pm	4 pm	5 pm	6 pm	7 pm	8 pm	9 pm	10 pm	11 pm
Lighting (15 bulbs)	0 W/ 0 W	0 W/ 0 W	0 W/ 0 W	0 W/ 0 W	0 W/ 0 W	0 W/ 0 W	0 W/ 0 W	0 W/ 0 W	300 W/ 300 W	300 W/ 300 W	300 W/ 300 W	300 W/ 300 W	300 W/ 300 W	300 W/ 300 W	300 W/ 300 W	300 W/ 300 W	300 W/ 300 W	600 W/ 600 W	600 W/ 600 W	600 W/ 600 W	600 W/ 600 W	120 W/ 120 W	120 W/ 120 W	0 W/ 0 W
Computer/ printer	210 W/ 210 W	210 W/ 210 W	210 W/ 210 W	210 W/ 210 W	210 W/ 210 W	210 W/ 210 W	210 W/ 210 W	210 W/ 210 W	430 W/ 430 W	430 W/ 430 W	430 W/ 430 W	430 W/ 430 W	580 W/ 580 W	430 W/ 430 W	430 W/ 430 W	430 W/ 430 W	430 W/ 430 W	430 W/ 430 W	430 W/ 430 W	210 W/ 210 W	210 W/ 210 W	210 W/ 210 W	210 W/ 210 W	210 W/ 210 W
Phone	15 W/ 15 W	15 W/ 15 W	15 W/ 15 W	15 W/ 15 W	15 W/ 15 W	15 W/ 15 W	15 W/ 15 W	15 W/ 15 W	15 W/ 15 W	15 W/ 15 W	15 W/ 15 W	15 W/ 15 W	15 W/ 15 W	15 W/ 15 W	15 W/ 15 W	15 W/ 15 W	15 W/ 15 W	15 W/ 15 W	15 W/ 15 W	15 W/ 15 W	15 W/ 15 W	15 W/ 15 W	15 W/ 15 W	15 W/ 15 W
Television	0 W/ 0 W	0 W/ 0 W	0 W/ 0 W	0 W/ 0 W	0 W/ 0 W	0 W/ 0 W	0 W/ 0 W	150 W/ 150 W	150 W/ 0 W	0 W/ 0 W	0 W/ 0 W	0 W/ 0 W	150 W/ 150 W	0 W/ 0 W	0 W/ 0 W	0 W/ 0 W	0 W/ 0 W	0 W/ 0 W	0 W/ 0 W	0 W/ 0 W	150 W/ 150 W	150 W/ 150 W	150 W/ 150 W	0 W/ 0 W
Video game console	0 W/ 0 W	0 W/ 0 W	0 W/ 0 W	0 W/ 0 W	0 W/ 0 W	0 W/ 0 W	0 W/ 0 W	0 W/ 0 W	0 W/ 0 W	0 W/ 0 W	0 W/ 0 W	0 W/ 0 W	0 W/ 0 W	0 W/ 0 W	0 W/ 0 W	0 W/ 0 W	0 W/ 0 W	0 W/ 0 W	0 W/ 0 W	0 W/ 0 W	195 W/ 195 W	195 W/ 195 W	0 W/ 0 W	0 W/ 0 W
DVD/DVR	0 W/ 0 W	0 W/ 0 W	0 W/ 0 W	0 W/ 0 W	0 W/ 0 W	0 W/ 0 W	0 W/ 0 W	25 W/ 25 W	0 W/ 0 W	0 W/ 0 W	0 W/ 0 W	0 W/ 0 W	25 W/ 25 W	0 W/ 0 W	0 W/ 0 W	0 W/ 0 W	0 W/ 0 W	0 W/ 0 W	0 W/ 0 W	0 W/ 0 W	0 W/ 0 W	0 W/ 0 W	25 W/ 25 W	25 W/ 25 W

TABLE 5.11 Den Load Profile (Winter/Summer)

Load Element/ Usage	12am	1am	2am	3am	4am	5am	6am	7am	8am	9am	10am	11am	12pm	1pm	2pm	3pm	4pm	5pm	6pm	7pm	8pm	9pm	10pm	11pm
Lighting (3 bulbs)	0W/ 0W	0W/ 0W	0W/ 0W	0W/ 0W	0W/ 0W	0W/ 0W	0W/ 0W	120W/ 0W	120W/ 120W	120W/ 120W	120W/ 120W	120W/ 120W	120W/ 120W	120W/ 120W	120W/ 120W	120W/ 120W	120W/ 120W	120W/ 120W	120W/ 120W	120W/ 120W	120W/ 120W	120W/ 0W	0W/ 0W	0W/ 0W
Computer/ printer	210W/ 210W	210W/ 210W	210W/ 210W	210W/ 210W	210W/ 210W	210W/ 210W	210W/ 210W	210W/ 210W	430W/ 430W	430W/ 430W	430W/ 430W	430W/ 430W	580W/ 580W	430W/ 430W	430W/ 430W	430W/ 430W	430W/ 430W	430W/ 430W	430W/ 430W	430W/ 430W	210W/ 210W	210W/ 210W	210W/ 210W	210W/ 210W
Phone	15W/ 15W	15W/ 15W	15W/ 15W	15W/ 15W	15W/ 15W	15W/ 15W	15W/ 15W	15W/ 15W	15W/ 15W	15W/ 15W	15W/ 15W	15W/ 15W	15W/ 15W	15W/ 15W	15W/ 15W	15W/ 15W	15W/ 15W	15W/ 15W	15W/ 15W	15W/ 15W	15W/ 15W	15W/ 15W	15W/ 15W	15W/ 15W

TABLE 5.12 Bedroom Load Profile (Winter/Summer)

Load Element/ Usage	12am	1am	2am	3am	4am	5am	6am	7am	8am	9am	10am	11am	12pm	1pm	2pm	3pm	4pm	5pm	6pm	7pm	8pm	9pm	10pm	11pm
Lighting (2 bulbs)	0W/ 0W	0W/ 0W	0W/ 0W	0W/ 0W	0W/ 0W	0W/ 0W	0W/ 0W	120W/ 120W	60W/ 60W	60W/ 60W	60W/ 60W	60W/ 60W	60W/ 60W	60W/ 60W	60W/ 60W	60W/ 60W	60W/ 60W	120W/ 60W	120W/ 60W	120W/ 60W	60W/ 60W	60W/ 60W	60W/ 60W	0W/ 0W
Laptop charger	50W/ 50W	50W/ 50W	50W/ 50W	50W/ 50W	50W/ 50W	50W/ 50W	50W/ 50W	50W/ 50W	0W/ 0W	0W/ 0W	0W/ 0W	0W/ 0W	0W/ 0W	0W/ 0W	0W/ 0W	0W/ 0W	0W/ 0W	0W/ 0W	0W/ 0W	0W/ 0W	0W/ 0W	0W/ 0W	0W/ 0W	50W/ 50W

TABLE 5.13 Master Bedroom Load Profile (Winter/Summer)

Load Element/ Usage	12am	1am	2am	3am	4am	5am	6am	7am	8am	9am	10am	11am	12pm	1pm	2pm	3pm	4pm	5pm	6pm	7pm	8pm	9pm	10pm	11pm
Lighting (4 bulbs)	0W/ 0W	0W/ 0W	0W/ 0W	0W/ 0W	0W/ 0W	0W/ 0W	0W/ 0W	240W/ 240W	60W/ 60W	60W/ 60W	60W/ 60W	60W/ 60W	60W/ 60W	60W/ 60W	60W/ 60W	60W/ 60W	60W/ 60W	240W/ 120W	240W/ 120W	240W/ 120W	120W/ 120W	120W/ 120W	120W/ 120W	0W/ 0W
Television	0W/ 0W	0W/ 0W	0W/ 0W	0W/ 0W	0W/ 0W	0W/ 0W	0W/ 0W	0W/ 0W	0W/ 0W	0W/ 0W	0W/ 0W	0W/ 0W	0W/ 0W	0W/ 0W	0W/ 0W	0W/ 0W	0W/ 0W	0W/ 0W	0W/ 0W	0W/ 0W	150W/ 150W	150W/ 150W	150W/ 150W	0W/ 0W
Laptop charger	50W/ 50W	50W/ 50W	50W/ 50W	50W/ 50W	50W/ 50W	50W/ 50W	50W/ 50W	50W/ 50W	0W/ 0W	0W/ 0W	0W/ 0W	0W/ 0W	0W/ 0W	0W/ 0W	0W/ 0W	0W/ 0W	0W/ 0W	0W/ 0W	0W/ 0W	0W/ 0W	0W/ 0W	0W/ 0W	0W/ 0W	50W/ 50W

TABLE 5.14 Bathroom Load Profile (Winter/Summer)

Load Element/ Usage	12am	1am	2am	3am	4am	5am	6am	7am	8am	9am	10am	11am	12pm	1pm	2pm	3pm	4pm	5pm	6pm	7pm	8pm	9pm	10pm	11pm
Lighting (4 bulbs)	0W/ 0W	0W/ 0W	0W/ 0W	0W/ 0W	0W/ 0W	0W/ 0W	0W/ 0W	240W/ 240W	0W/ 0W	0W/ 0W	0W/ 0W	0W/ 0W	0W/ 0W	240W/ 240W	0W/ 0W	0W/ 0W	0W/ 0W	0W/ 0W	0W/ 0W	0W/ 0W	0W/ 0W	240W/ 240W	0W/ 0W	0W/ 0W
Fan	0W/ 0W	0W/ 0W	0W/ 0W	0W/ 0W	0W/ 0W	0W/ 0W	0W/ 0W	65W/ 65W	0W/ 0W	0W/ 0W	0W/ 0W	0W/ 0W	0W/ 0W	0W/ 0W	0W/ 0W	0W/ 0W	0W/ 0W	0W/ 0W	0W/ 0W	0W/ 0W	0W/ 0W	0W/ 0W	0W/ 0W	0W/ 0W

TABLE 5.15 Utilities Load Profile (Winter/Summer)

Load Element/Usage	12am	1am	2am	3am	4am	5am	6am	7am	8am	9am	10am	11am	12pm	1pm	2pm	3pm	4pm	5pm	6pm	7pm	8pm	9pm	10pm	11pm
Freezer	1.2kW/ 1.2kW	1.2kW/ 1.2kW	1.2kW/ 1.2kW	1.2kW/ 1.2kW	1.2kW/ 1.2kW	1.2kW/ 1.2kW	1.2kW/ 1.2kW	1.2kW/ 1.2kW	1.2kW/ 1.2kW	1.2kW/ 1.2kW	1.2kW/ 1.2kW	1.2kW/ 1.2kW	1.2kW/ 1.2kW	1.2kW/ 1.2kW	1.2kW/ 1.2kW	1.2kW/ 1.2kW	1.2kW/ 1.2kW	1.2kW/ 1.2kW	1.2kW/ 1.2kW	1.2kW/ 1.2kW	1.2kW/ 1.2kW	1.2kW/ 1.2kW	1.2kW/ 1.2kW	1.2kW/ 1.2kW
Water heater	0W/ 0W	0W/ 0W	0W/ 0W	0W/ 0W	0W/ 0W	0W/ 0W	4kW/ 4kW	4kW/ 4kW	4kW/ 4kW	4kW/ 4kW	4kW/ 4kW	4kW/ 0W	0W/ 0W	0W/ 0W	0W/ 0W	0W/ 0W	4kW/ 4kW	0W/ 0W	4kW/ 4kW	4kW/ 4kW	4kW/ 4kW	4kW/ 4kW	4kW/ 4kW	4kW/ 4kW
Furnace fan	750 W/ 0W	750 W/ 0W	750 W/ 0W	750 W/ 0W	750 W/ 0W	750 W/ 0W	750 W/ 0W	750 W/ 0W	750 W/ 0W	750 W/ 0W	400 W/ 0W	400 W/ 0W	400 W/ 0W	400 W/ 0W	400 W/ 0W	400 W/ 0W	400 W/ 0W	400 W/ 0W	400 W/ 0W	400 W/ 0W	750 W/ 0W	750 W/ 0W	750 W/ 0W	750 W/ 0W
Air conditioner	0W/ 3kW	0W/ 3kW	0W/ 3kW	0W/ 3kW	0W/ 3kW	0W/ 3kW	0W/ 3kW	0W/ 3kW	0W/ 3kW	0W/ 5kW	0W/ 5kW	0W/ 5kW	0W/ 5kW	0W/ 5kW	0W/ 5kW	0W/ 5kW	0W/ 5kW	0W/ 5kW	0W/ 3kW	0W/ 3kW	0W/ 3kW	0W/ 3kW	0W/ 3kW	0W/ 3kW

TABLE 5.16 Foyer Load Profile (Winter/Summer Essential Items are Highlighted)

Load Element/Usage	12am	1am	2am	3am	4am	5am	6am	7am	8am	9am	10am	11am	12pm	1pm	2pm	3pm	4pm	5pm	6pm	7pm	8pm	9pm	10pm	11pm
Lighting (3 bulbs)	0W/ 0W	0W/ 0W	0W/ 0W	0W/ 0W	0W/ 0W	0W/ 0W	0W/ 0W	60W/ 0W	60W/ 0W	0W/ 0W	0W/ 0W	0W/ 0W	0W/ 0W	0W/ 0W	0W/ 0W	0W/ 0W	0W/ 0W	0W/ 0W	60W/ 0W	60W/ 0W	0W/ 0W	0W/ 0W	0W/ 0W	0W/ 0W

TABLE 5.17 Living Room Load Profile (Winter/Summer)

Load Element/ Usage	12am	1am	2am	3am	4am	5am	6am	7am	8am	9am	10am	11am	12pm	1pm	2pm	3pm	4pm	5pm	6pm	7pm	8pm	9pm	10pm	11pm
Lighting (5 bulbs)	0W/ 0W	0W/ 0W	0W/ 0W	0W/ 0W	0W/ 0W	0W/ 0W	0W/ 0W	0W/ 0W	0W/ 0W	0W/ 0W	0W/ 0W	0W/ 0W	60W/ 60W	60W/ 60W	0W/ 0W	0W/ 0W	0W/ 0W	60W/ 0W	60W/ 60W	60W/ 60W	0W/ 0W	0W/ 0W	0W/ 0W	0W/ 0W

TABLE 5.18 Dining Room Load Profile (Winter/Summer)

Load Element/ Usage	12am	1am	2am	3am	4am	5am	6am	7am	8am	9am	10am	11am	12pm	1pm	2pm	3pm	4pm	5pm	6pm	7pm	8pm	9pm	10pm	11pm
Lighting (5 bulbs)	0W/ 0W	0W/ 0W	0W/ 0W	0W/ 0W	0W/ 0W	0W/ 0W	0W/ 0W	60W/ 60W	0W/ 0W	0W/ 0W	0W/ 0W	0W/ 0W	60W/ 60W	0W/ 0W	0W/ 0W	0W/ 0W	0W/ 0W	60W/ 60W	180W/ 180W	60W/ 60W	0W/ 0W	0W/ 0W	0W/ 0W	0W/ 0W

TABLE 5.19 Kitchen Load Profile (Winter/Summer)

Load Element/ Usage	12am	1am	2am	3am	4am	5am	6am	7am	8am	9am	10am	11am	12pm	1pm	2pm	3pm	4pm	5pm	6pm	7pm	8pm	9pm	10pm	11pm
Lighting (10 bulbs)	0W/ 0W	0W/ 0W	0W/ 0W	0W/ 0W	0W/ 0W	0W/ 0W	0W/ 0W	120W/ 120W	0W/ 0W	0W/ 0W	0W/ 0W	0W/ 0W	60W/ 60W	60W/ 60W	0W/ 0W	0W/ 0W	0W/ 0W	120W/ 120W	120W/ 120W	60W/ 60W	0W/ 0W	0W/ 0W	0W/ 0W	0W/ 0W
Refrigerator/ freezer	725W/ 725W	725W/ 725W	725W/ 725W	725W/ 725W	725W/ 725W	725W/ 725W	725W/ 725W	725W/ 725W	725W/ 725W	725W/ 725W	725W/ 725W	725W/ 725W	725W/ 725W	725W/ 725W	725W/ 725W	725W/ 725W	725W/ 725W	725W/ 725W	725W/ 725W	725W/ 725W	725W/ 725W	725W/ 725W	725W/ 725W	725W/ 725W
Dishwasher	0W/ 0W	0W/ 0W	0W/ 0W	0W/ 0W	0W/ 0W	0W/ 0W	0W/ 0W	0W/ 0W	0W/ 0W	0W/ 0W	0W/ 0W	0W/ 0W	0W/ 0W	0W/ 0W	0W/ 0W	0W/ 0W	0W/ 0W	0W/ 0W	0W/ 0W	0W/ 0W	2kW/ 2kW	2kW/ 2kW	0W/ 0W	0W/ 0W
Microwave	0W/ 0W	0W/ 0W	0W/ 0W	0W/ 0W	0W/ 0W	0W/ 0W	0W/ 0W	0W/ 0W	0W/ 0W	0W/ 0W	0W/ 0W	0W/ 0W	200W/ 200W	0W/ 0W	0W/ 0W	0W/ 0W	0W/ 0W	0W/ 0W	0W/ 0W	0W/ 0W	0W/ 0W	0W/ 0W	0W/ 0W	0W/ 0W
Toaster	0W/ 0W	0W/ 0W	0W/ 0W	0W/ 0W	0W/ 0W	0W/ 0W	0W/ 0W	100W/ 100W	0W/ 0W	0W/ 0W	0W/ 0W	0W/ 0W	0W/ 0W	0W/ 0W	0W/ 0W	0W/ 0W	0W/ 0W	0W/ 0W	0W/ 0W	0W/ 0W	0W/ 0W	0W/ 0W	0W/ 0W	0W/ 0W

TABLE 5.20 Family Room Load Profile (Winter/Summer)

Load Element/ Usage	12am	1am	2am	3am	4am	5am	6am	7am	8am	9am	10am	11am	12pm	1pm	2pm	3pm	4pm	5pm	6pm	7pm	8pm	9pm	10pm	11pm
Lighting (6 bulbs)	0W/ 0W	0W/ 0W	0W/ 0W	0W/ 0W	0W/ 0W	0W/ 0W	0W/ 0W	0W/ 0W	0W/ 0W	0W/ 0W	0W/ 0W	0W/ 0W	0W/ 0W	0W/ 0W	0W/ 0W	0W/ 0W	60W/ 60W	120W/ 120W	120W/ 120W	120W/ 120W	120W/ 120W	60W/ 60W	60W/ 60W	0W/ 0W
Television	0W/ 0W	0W/ 0W	0W/ 0W	0W/ 0W	0W/ 0W	0W/ 0W	0W/ 0W	0W/ 0W	0W/ 0W	0W/ 0W	0W/ 0W	0W/ 0W	0W/ 0W	0W/ 0W	0W/ 0W	0W/ 0W	0W/ 0W	0W/ 0W	0W/ 0W	0W/ 0W	150W/ 150W	150W/ 150W	150W/ 150W	0W/ 0W
DVD/DVR	0W/ 0W	0W/ 0W	0W/ 0W	0W/ 0W	0W/ 0W	0W/ 0W	0W/ 0W	0W/ 0W	0W/ 0W	0W/ 0W	0W/ 0W	0W/ 0W	0W/ 0W	0W/ 0W	0W/ 0W	0W/ 0W	0W/ 0W	0W/ 0W	0W/ 0W	0W/ 0W	25W/ 25W	25W/ 25W	25W/ 25W	0W/ 0W
Stereo	0W/ 0W	0W/ 0W	0W/ 0W	0W/ 0W	0W/ 0W	0W/ 0W	0W/ 0W	0W/ 0W	0W/ 0W	0W/ 0W	0W/ 0W	0W/ 0W	0W/ 0W	0W/ 0W	0W/ 0W	0W/ 0W	0W/ 0W	0W/ 0W	0W/ 0W	0W/ 0W	60W/ 60W	60W/ 60W	60W/ 60W	0W/ 0W
Phone	0W/ 0W	0W/ 0W	0W/ 0W	0W/ 0W	0W/ 0W	0W/ 0W	0W/ 0W	0W/ 0W	0W/ 0W	0W/ 0W	15W/ 15W	0W/ 0W	0W/ 0W	0W/ 0W	0W/ 0W	0W/ 0W	0W/ 0W	15W/ 15W	0W/ 0W	0W/ 0W	0W/ 0W	0W/ 0W	0W/ 0W	0W/ 0W

TABLE 5.21 Hallway Load Profile (Winter/Summer)

Load Element/ Usage	12 am	1 am	2 am	3 am	4 am	5 am	6 am	7 am	8 am	9 am	10 am	11 am	12 pm	1 pm	2 pm	3 pm	4 pm	5 pm	6 pm	7 pm	8 pm	9 pm	10 pm	11 pm
Lighting (2 bulbs)	0 W/ 0 W	0 W/ 0 W	0 W/ 0 W	0 W/ 0 W	0 W/ 0 W	0 W/ 0 W	0 W/ 0 W	0 W/ 0 W	0 W/ 0 W	0 W/ 0 W	0 W/ 0 W	0 W/ 0 W	0 W/ 0 W	0 W/ 0 W	0 W/ 0 W	0 W/ 0 W	0 W/ 0 W	0 W/ 0 W	60 W/ 60 W	60 W/ 60 W	60 W/ 60 W	0 W/ 0 W	0 W/ 0 W	0 W/ 0 W

TABLE 5.22 Laundry Room Load Profile (Winter/Summer)

Load Element/ Usage	12 am	1 am	2 am	3 am	4 am	5 am	6 am	7 am	8 am	9 am	10 am	11 am	12 pm	1 pm	2 pm	3 pm	4 pm	5 pm	6 pm	7 pm	8 pm	9 pm	10 pm	11 pm
Lighting (3 bulbs)	0 W/ 0 W	0 W/ 0 W	0 W/ 0 W	0 W/ 0 W	0 W/ 0 W	0 W/ 0 W	0 W/ 0 W	0 W/ 0 W	0 W/ 0 W	0 W/ 0 W	0 W/ 0 W	0 W/ 0 W	0 W/ 0 W	0 W/ 0 W	0 W/ 0 W	0 W/ 0 W	0 W/ 0 W	0 W/ 0 W	0 W/ 0 W	0 W/ 0 W	60 W/ 60 W	0 W/ 0 W	0 W/ 0 W	0 W/ 0 W
Washer	0 W/ 0 W	0 W/ 0 W	0 W/ 0 W	0 W/ 0 W	0 W/ 0 W	0 W/ 0 W	0 W/ 0 W	0 W/ 0 W	0 W/ 0 W	0 W/ 0 W	0 W/ 0 W	0 W/ 0 W	0 W/ 0 W	0 W/ 0 W	0 W/ 0 W	0 W/ 0 W	0 W/ 0 W	0 W/ 0 W	0 W/ 0 W	0 W/ 0 W	400 W/ 400 W	0 W/ 0 W	0 W/ 0 W	0 W/ 0 W
Dryer	0 W/ 0 W	0 W/ 0 W	0 W/ 0 W	0 W/ 0 W	0 W/ 0 W	0 W/ 0 W	0 W/ 0 W	0 W/ 0 W	0 W/ 0 W	0 W/ 0 W	0 W/ 0 W	0 W/ 0 W	0 W/ 0 W	0 W/ 0 W	0 W/ 0 W	0 W/ 0 W	0 W/ 0 W	0 W/ 0 W	0 W/ 0 W	0 W/ 0 W	0 W/ 0 W	1.5 kW/ 1.5 kW	0 W/ 0 W	0 W/ 0 W

TABLE 5.23   Garage Load Profile (Winter/Summer)

Load Element/ Usage	12am	1am	2am	3am	4am	5am	6am	7am	8am	9am	10am	11am	12pm	1pm	2pm	3pm	4pm	5pm	6pm	7pm	8pm	9pm	10pm	11pm
Lighting (3 bulbs)	0W/ 0W	0W/ 0W	0W/ 0W	0W/ 0W	0W/ 0W	0W/ 0W	0W/ 0W	60W/ 60W	0W/ 0W	0W/ 0W	0W/ 0W	0W/ 0W	0W/ 0W	0W/ 0W	0W/ 0W	0W/ 0W	0W/ 0W	0W/ 0W	0W/ 0W	0W/ 0W	0W/ 0W	0W/ 0W	0W/ 0W	0W/ 0W
Garage door opener	0W/ 0W	0W/ 0W	0W/ 0W	0W/ 0W	0W/ 0W	0W/ 0W	0W/ 0W	0W/ 0W	30W/ 30W	0W/ 0W	0W/ 0W	0W/ 0W	0W/ 0W	0W/ 0W	0W/ 0W	0W/ 0W	0W/ 0W	0W/ 0W	0W/ 0W	0W/ 0W	0W/ 0W	0W/ 0W	0W/ 0W	0W/ 0W

TABLE 5.24   Porch Load Profile (Winter/Summer)

Load Element/ Usage	12am	1am	2am	3am	4am	5am	6am	7am	8am	9am	10am	11am	12pm	1pm	2pm	3pm	4pm	5pm	6pm	7pm	8pm	9pm	10pm	11pm
Lighting (4 bulbs)	0W/ 0W	0W/ 0W	0W/ 0W	0W/ 0W	0W/ 0W	0W/ 0W	0W/ 0W	0W/ 0W	0W/ 0W	0W/ 0W	0W/ 0W	0W/ 0W	0W/ 0W	0W/ 0W	0W/ 0W	0W/ 0W	0W/ 0W	0W/ 0W	0W/ 0W	60W/ 60W	60W/ 60W	0W/ 0W	0W/ 0W	0W/ 0W

TABLE 5.25 Basement Load Profile (Winter/Summer)

Load Element/ Usage	12 am	1 am	2 am	3 am	4 am	5 am	6 am	7 am	8 am	9 am	10 am	11 am	12 pm	1 pm	2 pm	3 pm	4 pm	5 pm	6 pm	7 pm	8 pm	9 pm	10 pm	11 pm
Lighting (15 bulbs)	0 W/ 0 W	0 W/ 0 W	0 W/ 0 W	0 W/ 0 W	0 W/ 0 W	0 W/ 0 W	0 W/ 0 W	0 W/ 0 W	120 W/ 120 W	120 W/ 120 W	120 W/ 120 W	120 W/ 120 W	120 W/ 120 W	120 W/ 120 W	120 W/ 120 W	120 W/ 120 W	120 W/ 120 W	120 W/ 120 W	120 W/ 120 W	60 W/ 60 W	60 W/ 60 W	60 W/ 60 W	60 W/ 60 W	0 W/ 0 W
Computer/ printer	0 W/ 0 W	0 W/ 0 W	0 W/ 0 W	0 W/ 0 W	0 W/ 0 W	0 W/ 0 W	0 W/ 0 W	0 W/ 0 W	280 W/ 280 W	280 W/ 280 W	280 W/ 280 W	280 W/ 280 W	580 W/ 580 W	280 W/ 280 W	280 W/ 280 W	280 W/ 280 W	280 W/ 280 W	280 W/ 280 W	0 W/ 0 W	0 W/ 0 W	0 W/ 0 W	0 W/ 0 W	0 W/ 0 W	0 W/ 0 W
Phone	0 W/ 0 W	0 W/ 0 W	0 W/ 0 W	0 W/ 0 W	0 W/ 0 W	0 W/ 0 W	0 W/ 0 W	0 W/ 0 W	0 W/ 0 W	15 W/ 15 W	0 W/ 0 W	0 W/ 0 W	0 W/ 0 W	0 W/ 0 W	0 W/ 0 W	0 W/ 0 W	0 W/ 0 W	0 W/ 0 W	0 W/ 0 W	0 W/ 0 W	0 W/ 0 W	0 W/ 0 W	0 W/ 0 W	0 W/ 0 W
Television	0 W/ 0 W	0 W/ 0 W	0 W/ 0 W	0 W/ 0 W	0 W/ 0 W	0 W/ 0 W	0 W/ 0 W	0 W/ 0 W	0 W/ 0 W	0 W/ 0 W	0 W/ 0 W	0 W/ 0 W	150 W/ 150 W	0 W/ 0 W	0 W/ 0 W	0 W/ 0 W	0 W/ 0 W	0 W/ 0 W	0 W/ 0 W	0 W/ 0 W	0 W/ 0 W	0 W/ 0 W	0 W/ 0 W	0 W/ 0 W
Video game console	0 W/ 0 W	0 W/ 0 W	0 W/ 0 W	0 W/ 0 W	0 W/ 0 W	0 W/ 0 W	0 W/ 0 W	0 W/ 0 W	0 W/ 0 W	0 W/ 0 W	0 W/ 0 W	0 W/ 0 W	195 W/ 195 W	0 W/ 0 W	0 W/ 0 W	0 W/ 0 W	0 W/ 0 W	0 W/ 0 W	0 W/ 0 W	0 W/ 0 W	0 W/ 0 W	0 W/ 0 W	0 W/ 0 W	0 W/ 0 W
DVD/DVR	0 W/ 0 W	0 W/ 0 W	0 W/ 0 W	0 W/ 0 W	0 W/ 0 W	0 W/ 0 W	0 W/ 0 W	0 W/ 0 W	0 W/ 0 W	0 W/ 0 W	0 W/ 0 W	0 W/ 0 W	25 W/ 25 W	0 W/ 0 W	0 W/ 0 W	0 W/ 0 W	0 W/ 0 W	0 W/ 0 W	0 W/ 0 W	0 W/ 0 W	0 W/ 0 W	0 W/ 0 W	0 W/ 0 W	0 W/ 0 W

TABLE 5.26 Den Load Profile (Winter/Summer)

Load Element/ Usage	12 am	1 am	2 am	3 am	4 am	5 am	6 am	7 am	8 am	9 am	10 am	11 am	12 pm	1 pm	2 pm	3 pm	4 pm	5 pm	6 pm	7 pm	8 pm	9 pm	10 pm	11 pm
Lighting (3 bulbs)	0 W/ 0 W	0 W/ 0 W	0 W/ 0 W	0 W/ 0 W	0 W/ 0 W	0 W/ 0 W	0 W/ 0 W	60 W/ 0 W	120 W/ 120 W	120 W/ 120 W	120 W/ 120 W	120 W/ 120 W	120 W/ 120 W	120 W/ 120 W	120 W/ 120 W	120 W/ 120 W	120 W/ 120 W	120 W/ 120 W	60 W/ 60 W	60 W/ 60 W	60 W/ 60 W	0 W/ 0 W	0 W/ 0 W	0 W/ 0 W
Computer/ printer	0 W/ 0 W	0 W/ 0 W	0 W/ 0 W	0 W/ 0 W	0 W/ 0 W	0 W/ 0 W	0 W/ 0 W	0 W/ 0 W	280 W/ 280 W	280 W/ 280 W	280 W/ 280 W	280 W/ 280 W	580 W/ 580 W	280 W/ 280 W	280 W/ 280 W	280 W/ 280 W	280 W/ 280 W	280 W/ 280 W	0 W/ 0 W	0 W/ 0 W	0 W/ 0 W	0 W/ 0 W	0 W/ 0 W	0 W/ 0 W
Phone	0 W/ 0 W	0 W/ 0 W	0 W/ 0 W	0 W/ 0 W	0 W/ 0 W	0 W/ 0 W	0 W/ 0 W	0 W/ 0 W	0 W/ 0 W	15 W/ 15 W	0 W/ 0 W	0 W/ 0 W	0 W/ 0 W	0 W/ 0 W	0 W/ 0 W	0 W/ 0 W	0 W/ 0 W	0 W/ 0 W	0 W/ 0 W	0 W/ 0 W	0 W/ 0 W	0 W/ 0 W	0 W/ 0 W	0 W/ 0 W

TABLE 5.27 Bedroom Load Profile (Winter/Summer)

Load Element/ Usage	12am	1am	2am	3am	4am	5am	6am	7am	8am	9am	10am	11am	12pm	1pm	2pm	3pm	4pm	5pm	6pm	7pm	8pm	9pm	10pm	11pm
Lighting (2 bulbs)	0W/ 0W	0W/ 0W	0W/ 0W	0W/ 0W	0W/ 0W	0W/ 0W	0W/ 0W	60W/ 60W	0W/ 0W	0W/ 0W	0W/ 0W	0W/ 0W	0W/ 0W	0W/ 0W	0W/ 0W	0W/ 0W	0W/ 0W	0W/ 0W	0W/ 0W	60W/ 60W	60W/ 60W	60W/ 60W	60W/ 60W	0W/ 0W
Laptop charger	50W/ 50W	50W/ 50W	50W/ 50W	50W/ 50W	50W/ 50W	50W/ 50W	50W/ 50W	0W/ 0W	0W/ 0W	0W/ 0W	0W/ 0W	0W/ 0W	0W/ 0W	0W/ 0W	0W/ 0W	0W/ 0W	0W/ 0W	0W/ 0W	0W/ 0W	0W/ 0W	0W/ 0W	0W/ 0W	0W/ 0W	50W/ 50W

TABLE 5.28 Master Bedroom Load Profile (Winter/Summer)

Load Element/ Usage	12am	1am	2am	3am	4am	5am	6am	7am	8am	9am	10am	11am	12pm	1pm	2pm	3pm	4pm	5pm	6pm	7pm	8pm	9pm	10pm	11pm
Lighting (4 bulbs)	0W/ 0W	0W/ 0W	0W/ 0W	0W/ 0W	0W/ 0W	0W/ 0W	0W/ 0W	120W/ 120W	0W/ 0W	0W/ 0W	0W/ 0W	0W/ 0W	0W/ 0W	0W/ 0W	0W/ 0W	0W/ 0W	0W/ 0W	0W/ 0W	0W/ 0W	120W/ 120W	120W/ 120W	60W/ 60W	120W/ 120W	0W/ 0W
Television	0W/ 0W	0W/ 0W	0W/ 0W	0W/ 0W	0W/ 0W	0W/ 0W	0W/ 0W	0W/ 0W	0W/ 0W	0W/ 0W	0W/ 0W	0W/ 0W	0W/ 0W	0W/ 0W	0W/ 0W	0W/ 0W	0W/ 0W	0W/ 0W	0W/ 0W	0W/ 0W	0W/ 0W	150W/ 150W	150W/ 150W	0W/ 0W
Laptop charger	50W/ 50W	50W/ 50W	50W/ 50W	50W/ 50W	50W/ 50W	50W/ 50W	50W/ 50W	0W/ 0W	0W/ 0W	0W/ 0W	0W/ 0W	0W/ 0W	0W/ 0W	0W/ 0W	0W/ 0W	0W/ 0W	0W/ 0W	0W/ 0W	0W/ 0W	0W/ 0W	0W/ 0W	0W/ 0W	0W/ 0W	50W/ 50W

TABLE 5.29 Bathroom Load Profile (Winter/Summer)

Load Element/ Usage	12 am	1 am	2 am	3 am	4 am	5 am	6 am	7 am	8 am	9 am	10 am	11 am	12 pm	1 pm	2 pm	3 pm	4 pm	5 pm	6 pm	7 pm	8 pm	9 pm	10 pm	11 pm
Lighting (4 bulbs)	0 W/ 0 W	0 W/ 0 W	0 W/ 0 W	0 W/ 0 W	0 W/ 0 W	0 W/ 0 W	0 W/ 0 W	120 W/ 120 W	0 W/ 0 W	0 W/ 0 W	0 W/ 0 W	0 W/ 0 W	0 W/ 0 W	60 W/ 60 W	0 W/ 0 W	0 W/ 0 W	0 W/ 0 W	0 W/ 0 W	0 W/ 0 W	0 W/ 0 W	0 W/ 0 W	120 W/ 120 W	0 W/ 0 W	0 W/ 0 W
Fan	0 W/ 0 W	0 W/ 0 W	0 W/ 0 W	0 W/ 0 W	0 W/ 0 W	0 W/ 0 W	0 W/ 0 W	65 W/ 65 W	0 W/ 0 W	0 W/ 0 W	0 W/ 0 W	0 W/ 0 W	0 W/ 0 W	0 W/ 0 W	0 W/ 0 W	0 W/ 0 W	0 W/ 0 W	0 W/ 0 W	0 W/ 0 W	0 W/ 0 W	0 W/ 0 W	0 W/ 0 W	0 W/ 0 W	0 W/ 0 W

TABLE 5.30 Utilities Load Profile (Winter/Summer)

Load Element/ Usage	12 am	1 am	2 am	3 am	4 am	5 am	6 am	7 am	8 am	9 am	10 am	11 am	12 pm	1 pm	2 pm	3 pm	4 pm	5 pm	6 pm	7 pm	8 pm	9 pm	10 pm	11 pm
Freezer	0.6 kW/ 0.6 kW	0.6 kW/ 0.6 kW	0.6 kW/ 0.6 kW	0.6 kW/ 0.6 kW	0.6 kW/ 0.6 kW	0.6 kW/ 0.6 kW	0.6 kW/ 0.6 kW	0.6 kW/ 0.6 kW	0.6 kW/ 0.6 kW	0.6 kW/ 0.6 kW	0.6 kW/ 0.6 kW	0.6 kW/ 0.6 kW	0.6 kW/ 0.6 kW	0.6 kW/ 0.6 kW	0.6 kW/ 0.6 kW	0.6 kW/ 0.6 kW	0.6 kW/ 0.6 kW	0.6 kW/ 0.6 kW	0.6 kW/ 0.6 kW	0.6 kW/ 0.6 kW	0.6 kW/ 0.6 kW	0.6 kW/ 0.6 kW	0.6 kW/ 0.6 kW	0.6 kW/ 0.6 kW
Water heater	0 W/ 0 W	0 W/ 0 W	0 W/ 0 W	0 W/ 0 W	0 W/ 0 W	0 W/ 0 W	0 W/ 0 W	0 W/ 0 W	0 W/ 0 W	0 W/ 0 W	0 W/ 0 W	0 W/ 0 W	0 W/ 0 W	0 W/ 0 W	0 W/ 0 W	0 W/ 0 W	0 W/ 0 W	0 W/ 0 W	3 kW/ 3 kW	3 kW/ 3 kW	3 kW/ 3 kW	3 kW/ 3 kW	3 kW/ 3 kW	0 W/ 0 W
Furnace fan	600 W/ 0 W	600 W/ 0 W	600 W/ 0 W	600 W/ 0 W	600 W/ 0 W	600 W/ 0 W	600 W/ 0 W	600 W/ 0 W	600 W/ 0 W	600 W/ 0 W	600 W/ 0 W	600 W/ 0 W	600 W/ 0 W	600 W/ 0 W	600 W/ 0 W	600 W/ 0 W	600 W/ 0 W	600 W/ 0 W	600 W/ 0 W	600 W/ 0 W	600 W/ 0 W	600 W/ 0 W	600 W/ 0 W	600 W/ 0 W
Air conditioner	0 W/ 3 kW	0 W/ 3 kW	0 W/ 3 kW	0 W/ 3 kW	0 W/ 3 kW	0 W/ 3 kW	0 W/ 2 kW	0 W/ 2 kW	0 W/ 2 kW	0 W/ 2 kW	0 W/ 2 kW	0 W/ 2 kW	0 W/ 2 kW	0 W/ 2 kW	0 W/ 2 kW	0 W/ 2 kW	0 W/ 2 kW	0 W/ 2 kW	0 W/ 3 kW	0 W/ 3 kW	0 W/ 3 kW	0 W/ 3 kW	0 W/ 3 kW	0 W/ 3 kW

loading requires a current of almost 130 amp. This building should therefore use a main circuit breaker of at least 150 amp for continuous load operation. However, due to inrush current of motors in heating, ventilating, and air condition systems (HVAC) and refrigerator, the main circuit breaker should be rated 300 amp.

5.3 SERVICE FEEDER AND METERING

The location of incoming service entrance and service panel and smart net metering including number feeders and rating of each breaker must be specified. The building uses 300 amp service with two service lines of 120 V and a neutral and a ground wire. The neutral conductor carries current, and is connected to ground (earth) at the main electrical distribution panel. The building PV station is connected at the main feeder after the main service breaker below the smart net metering serving the building.

The main electrical grid power for the building is supplied through underground lines that enter the residence through the basement of the building. The electrical grid service consists of two 120 V AC service lines and a neutral and a ground line. This line is fed to a metering system that is rated for 150 A for continuous maximum load current. The main circuit breaker for the building is rated for 300 A current. Between the metering station and breaker, a separate power line connects the proposed building PV plant system to a 120 V AC feeder that connects to both the breaker and the metering system. In this manner, PV-generated power can be either used by the building loads or distributed back to the main electric grid. To distribute power to the building loads, ten 15 A circuit breakers are used between the main breaker and various points of electricity use. Figure 5.2 shows a diagram of this metering and feeder design.

5.3.1 Assumed Wattages

Light Bulb—60 W, 820 lumens (lm)[1]

Refrigerator/Freezer—725 W (http://energy.gov/energysaver/articles/estima ting-appliance-and-home-electronic-energy-use)

Dishwasher—2000 W (http://energy.gov/energysaver/articles/estimating-appliance-and-home-electronic-energy-use)

Stove/Oven—3000 W (http://www.wholesalesolar.com/StartHere/HowtoSa veEnergy/PowerTable.html)

Microwave —1000 W (based on own microwave)

Toaster—800 W (http://energy.gov/energysaver/articles/estimating-applian ce-and-home-electronic-energy-use)

Television—150 W (http://energy.gov/energysaver/articles/estimating-appli ance-and-home-electronic-energy-use)

DVD/DVR—25 W (http://energy.gov/energysaver/articles/estimating-appli ance-and-home-electronic-energy-use)

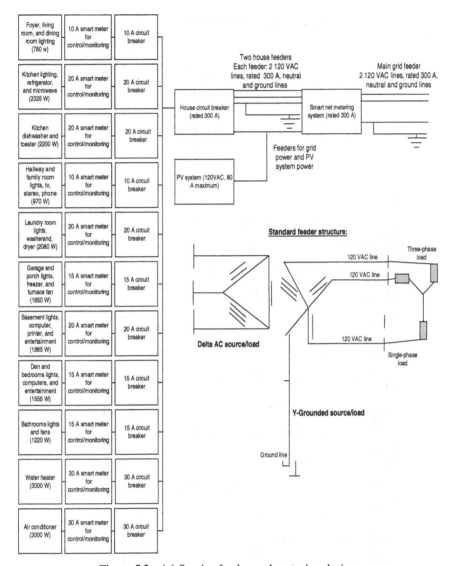

Figure 5.2   (a) Service feeder and metering design.

Stereo—60 W (http://www.wholesalesolar.com/StartHere/HowtoSaveEner gy/PowerTable.html)

Telephone—15 W (http://www.wholesalesolar.com/StartHere/HowtoSave Energy/PowerTable.html)

Washer—400 W (http://energy.gov/energysaver/articles/estimating-applia nce-and-home-electronic-energy-use)

Dryer—3000 W (http://energy.gov/energysaver/articles/estimating-applia nce-and-home-electronic-energy-use)

Physical metering layout:

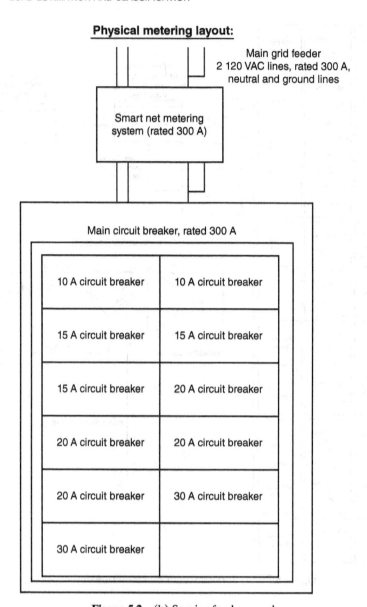

Figure 5.2 (b) Service feeder panel.

Garage Door Opener—350 W (http://www.absak.com/library/power-consu mption-table)

Video Game Console—195 W (http://www.wholesalesolar.com/StartHere/ HowtoSaveEnergy/PowerTable.html)

Desktop Computer—280 W awake/60 W asleep (http://energy.gov/energy saver/articles/estimating-appliance-and-home-electronic-energy-use)

Printer—300 W active/150 W asleep

Laptop Charger—50 W (http://energy.gov/energysaver/articles/estimating-appliance-and-home-electronic-energy-use)

Bathroom Fan—65 W (http://energy.gov/energysaver/articles/estimating-appliance-and-home-electronic-energy-use)

Standalone Freezer—1200 W (http://www.wholesalesolar.com/StartHere/HowtoSaveEnergy/PowerTable.html)

Water Heater—4000 W (http://energy.gov/energysaver/articles/estimating-appliance-and-home-electronic-energy-use)

Furnace Fan—750 W (http://energy.gov/energysaver/articles/estimating-appliance-and-home-electronic-energy-use)

Air Conditioner—5000 W (http://www.wholesalesolar.com/StartHere/HowtoSaveEnergy/PowerTable.html)

Based on the assumed power consumption by all appliances, utilities, and lighting elements of this building, the total number of kilowatt-hours used on a winter day is 166.5 kWh. The energy used on a summer day is 239.3 kWh. With these values, the average input wattage to the building on both days can be calculated by dividing the daily energy demand by 24 hours. The required average, hourly input power for the building is 6.94 kW on a winter day and 9.97 kW on a summer day. The unit power density of the building for both days can be found by dividing this average power requirement by the square footage of the building (2600 ft^2). The building power density is therefore 2.67 W/ft^2 on a winter day and 3.83 W/ft^2 on a summer day. Considering only the lighting elements of the building, the lighting power density is 0.41 W/ft^2 on a winter day and 0.34 W/ft^2 on a summer day. The building lighting makes up 15.4% of the building energy consumption on a winter day and 8.9% on a summer day. All lighting elements in this analysis are assumed to be 60 W bulbs with a rated luminous output of 820 lumens each. Thus, no matter how much energy the building used for lighting over 24 hours, either in summer or in winter, the lighting system efficiency of the building would always be 820 lumens per 60 W, or 13.67 LPW. Furthermore, incandescent lights are inefficient lighting devices, and are responsible for more residential energy consumption than is required for their desired output.

Figures 5.3 through 5.5 depict the daily energy consumption.

PROBLEMS

5.1 Estimate the energy consumption of the residential building given in Figure 5.6 for a summer day. This home has several rooms including three bedrooms, family room, dining room, kitchen, two bathrooms, and a basement. This home has three floors, the upper containing the bedrooms, the middle containing the living area, and the basement containing storage space. Plot the hourly energy consumption.

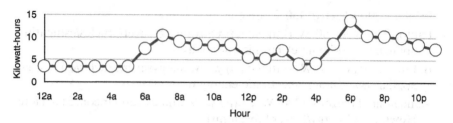

Figure 5.3 Residential loads consumed in kWh (winter day).

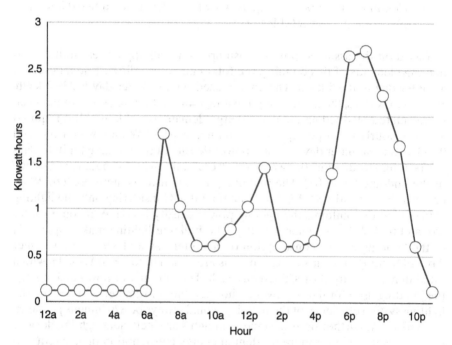

Figure 5.4 Residential lighting loads consumed in kWh (summer day).

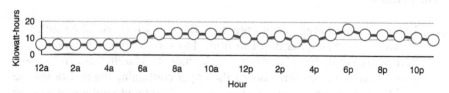

Figure 5.5 Residential loads energy consumed in kWh (summer day).

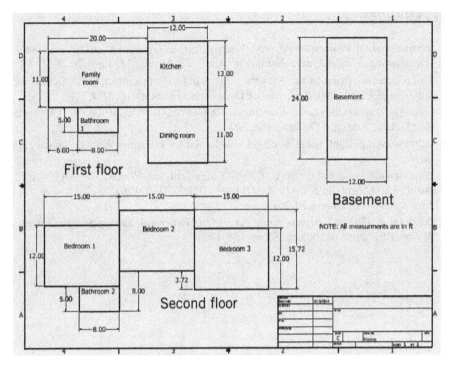

Figure 5.6 Schematic of house.

The exact dimensions of the rooms in the home can be found in Table 5.31 as well as the total area of the home. The thickness of walls in Table 5.31 are ignored dimensions.

5.2 Estimate the energy consumption of the residential building given in Figure 5.6 for a winter day and plot the hourly energy consumption.

5.3 Estimate the lighting load of residential building given in Figure 5.6 for a summer day and plot the hourly energy consumption.

TABLE 5.31 Room Dimensions and Area

List of Rooms			
Name	Width (ft)	Length (ft)	Area (ft^2)
Bedroom (×3)	12	15	540
Family room	11	20	220
Dining room	11	12	132
Kitchen	13	12	156
Bathroom (×2)	5	8	80
Basement	24	12	288
Total			1416

REFERENCES

1. Environmental Protection Agency, Energy Star. *Learn about LEDs*. http://www .energystar.gov/index.cfm?c=lighting.pr_what_are. Accessed January 20, 2014.
2. Encyclopædia Britannica. Available at http://www.britannica.com/EBchecked/ topic/340594/light-emitting-diode-LED. Accessed February 16, 2014.
3. http://www.amazon.com/GE-Lighting-41028-60-Watt-4-Pack/dp/B000BPILBY#pro ductDetails. Accessed December 18, 2014.
4. http://www.rapidtables.com/calc/light/how-lumen-to-watt.htm. Accessed January 20, 2014.
5. http://www.homedepot.com/p/Cree-60W-Equivalent-Soft-White-2700K-A19-Dim mable-LED-Light-Bulb-BA19-08027OMF-12DE26-2U100/204592770?N=5yc1vZ bm79. Accessed January 20, 2014.
6. Keyhani, A. (2011). *Design of Smart Power Grid Renewable Energy Systems*, 1st edn, Wiley-IEEE Press, Hoboken, NJ, pp. 138–140.

CHAPTER 6

ENERGY SAVING AND COST ESTIMATION OF INCANDESCENT AND LIGHT EMITTING DIODES

6.1 LIGHTING

The light bulb was first invented in 1806 by Humphrey Davy, an English scientist (http://www.unmuseum.org/lightbulb.htm). He invented a powerful electric arc lamp. The arc lamp was used for many years and it is similar to the light from a welding torch and it uses a lot power. Today, the arc lamps are used as search lights. Davy used electric power from fully charged batteries to light his arc lamps.

After the invention of practical direct current generators and motors, first mass produced incandescent light bulb was manufactured by Edison (February 11, 1847–October 18, 1931). Since early 1900, the standard incandescent light bulbs have been light sources for residences, commercial businesses, and industrial systems. The incandescent light bulbs are still in use today. An incandescent visible light is generated by passing current through a coiled filament with high resistance. A coiled wire makes is more efficient than a straight and parallel wire since more wire can be placed in a bulb and it produces more light. However, it uses more energy. Visible light is produced from heating up wire and glows at a high temperature.

Today, the coiled wire is tungsten and it is used in light bulbs. Incandescent light bulbs are cheap to manufacture, but inefficient since most of the power consumed by the bulbs are dissipated as heat (http://energy.gov/articles/history-light-bulb).

Edmund Germer (1901–1987) invented a high pressure vapor lamp and improved fluorescent lamp. In 1927, Edmund Germer co-patented an

Design of Smart Power Grid Renewable Energy Systems, Second Edition. Ali Keyhani.
© 2017 John Wiley & Sons, Inc. Published 2017 by John Wiley & Sons, Inc.
Companion website: www.wiley.com/go/smartpowergrid2e

experimental fluorescent lamp with Friedrich Meyer and Hans Spanner (http://inventors.about.com/od/filipinoscientists/a/Agapito_Flores.htm).

In 1974, Sylvania and General Electric companies developed the first compact fluorescent light (CFL). In the 1990s, substantial improvements were made in CFL performance, price, and efficiency and started to replace incandescent light bulbs. CFL bulbs consume 70% less energy than incandescent bulbs and lifetime of use is 10 times longer than incandescent bulbs. Florescent bulbs generate light by passing a current through a closed tube to excite a gas. The gas ionization produces ultraviolet light that is converted to visible light through a layer of phosphor powder coating the bulb tube.

The light-emitting diode (or LED) is the most efficient light bulb today. LED was invented by three physicists: Akasaki, Hiroshi Amano, and Shuji Nakamura. LEDs use a semiconductor to convert electricity into light. LED takes small area, approximately, less than 1 mm^2. LED emits light in a specific direction to desired space.

The LED diodes are semiconductor devices that are used in the LED bulb function. The diode is constructed by placing two different semiconductor materials in conjunction that easily allows current flow in only one direction. A diode can pass a high amount of current with an activation voltage of only 1–2 V.[1] The LED projects photons as current is passed across the device. Different types of semiconductor material combinations in the diode junction will produce different wavelengths of light. The majority of LEDs are produced with different types of Gallium Arsenide (GaAs).[1] Today, larger LED structures are being made to replace common incandescent and florescent light bulbs. Each LED produces light of a certain wavelength. White light is achieved by covering an array of LEDs in a phosphor powder in a bulb housing.[2] LED lights do not produce heat and they do not burn out. LEDs do not fail due to tube puncture like florescent lights. LEDs do not experience a gradual failure and decrease in luminous output.

Efficiency of the light bulb is measured in luminous efficacy. It is the ratio of luminous flux to power (Watts). "The luminous efficacy of a source is a measure of the efficiency with which the source provides visible light from electricity."[3] "A bulb that is 100 percent efficient at converting energy into light would have an efficacy of 683 lm/W. To put this in context, a 60- to 100-watt incandescent bulb has an efficacy of 15 lm/W, an equivalent CFL has an efficacy of 73 lm/W, and current LED-based replacement bulbs on the market range from 70–120 lm/W with an average efficacy of 85 lm/W."[4]

6.2 COMPARATIVE PERFORMANCE OF LED, INCANDESCENT, AND LFC LIGHTING

The industry comparative performance estimate of LED, incandescent and LFC lighting[5] is performed. The results are presented in Table 6.1.

The lists of the rooms in a 2600 ft^2 building are presented in Figure 6.1. Figure 6.1 shows a basic, dimensioned layout of the two-story building. The power

TABLE 6.1   Incandescent and LED Lighting Residential Power Consumptions

Life Span (average)	Incandescent 50,000 hours	Florescent 1200 hours	LED 8000 hours
Watts of electricity used (equivalent to 60 W bulb) LEDs use less power (watts) per unit of light generated (lumens). LEDs help reduce green building gas emissions from power plants and lower electric bills	**60 W**	**13–15 W**	6–8 W
Kilo-watts of electricity used (30 incandescent bulbs per year equivalent)	3285 kWh/yr	**767 kWh/yr**	**329 kWh/yr**
Annual operating cost (30 incandescent bulbs per year equivalent)	**$328.59/yr**	**$76.65/year**	$32.85 /yr

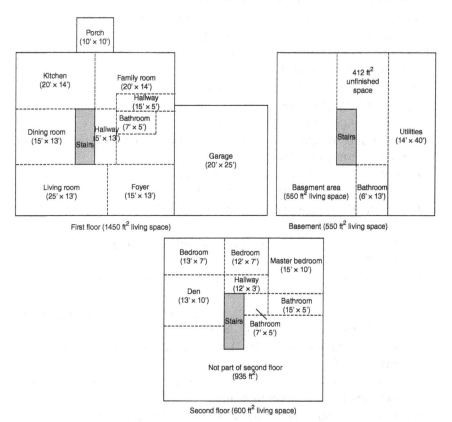

Figure 6.1   The layout of 2600 ft² building.

consumption of each load is estimated based on the rated wattages. The estimated power consumptions are used, along with estimated times of operation to calculate the kilowatt-hours of energy used by each load. All lighting elements in the building are assumed to be 60 W incandescent light bulbs with a rated luminous output of 820 lumens. The daily energy consumption of the building is estimated for both a summer and a winter day. The total energy consumption of the building and its lighting energy consumption are calculated as presented by the 24-hour load plots in Figures 6.4 through 6.6. The calculated energy values are then used to find the efficiency of the building electrical loads in terms of lumens provided per watt, and watts per square foot of building area.

The loads of foyer, living room, dining room, kitchen, family room, three hallways, laundry room, garage, porch, basement, den, two bedrooms, master bedroom, four bathrooms, and utility closet are estimated by using data provided from the following websites.

Light Bulb—60 W, 820 Lumen (lm) (http://www.amazon.com/Ella-Fashion-Lumen-Recessed-Dimmable/dp/B00RE1TKJA)

Refrigerator/Freezer—725 W (http://energy.gov/energysaver/articles/estima ting-appliance-and-home-electronic-energy-use)

Dishwasher—2000 W (http://energy.gov/energysaver/articles/estimating-appliance-and-home-electronic-energy-use)

Stove/Oven—3000 W (http://www.wholesalesolar.com/StartHere/HowtoSa veEnergy/PowerTable.html)

Microwave—1000 W (based on own microwave)

Toaster—800 W (http://energy.gov/energysaver/articles/estimating-applian ce-and-home-electronic-energy-use)

Television–150 W (http://energy.gov/energysaver/articles/estimating-appli ance-and-home-electronic-energy-use)

DVD/DVR–25 W (http://energy.gov/energysaver/articles/estimating-appli ance-and-home-electronic-energy-use)

Stereo—60 W (http://www.wholesalesolar.com/StartHere/HowtoSaveEner gy/PowerTable.html)

Telephone—15 W (http://www.wholesalesolar.com/StartHere/HowtoSave Energy/PowerTable.html)

Washer—400 W (http://energy.gov/energysaver/articles/estimating-applian ce-and-home-electronic-energy-use)

Dryer—3000 W (http://energy.gov/energysaver/articles/estimating-applian ce-and-home-electronic-energy-use)

Garage Door Opener—350 W (http://www.absak.com/library/power-con sumption-table)

Video Game Console—195 W (http://www.wholesalesolar.com/StartHere/ HowtoSaveEnergy/PowerTable.html)

Desktop Computer—280 W awake/60 W asleep (http://energy.gov/energ ysaver/articles/estimating-appliance-and-home-electronic-energy-use)

Printer—300 W active/150 W asleep

Laptop Charger—50 W (http://energy.gov/energysaver/articles/estimating-appliance-and-home-electronic-energy-use)

Bathroom Fan—65 W (http://energy.gov/energysaver/articles/estimating-appliance-and-home-electronic-energy-use)

Standalone Freezer—1200 W (http://www.wholesalesolar.com/StartHere/HowtoSaveEnergy/PowerTable.html)

Water Heater—4000 W (http://energy.gov/energysaver/articles/estimating-appliance-and-home-electronic-energy-use)

Furnace Fan—750 W (http://energy.gov/energysaver/articles/estimating-appliance-and-home-electronic-energy-use)

Air Conditioner—5000 W (http://www.wholesalesolar.com/StartHere/HowtoSaveEnergy/PowerTable.html)

An Excel sheet is used to estimate loads of each space. A normal daily energy usage is assumed. On the Excel sheet each appliance is listed. The energy consumption of each load is estimated based on hours of operation and rated power. Also, a data recorder can be used (see Chapter 4) for a typical daily operation to obtain accurate energy consumption. For commercial and industrial sites, the daily load operation can be obtained by conducting detailed discussions with the users of facilities Based on the daily operation of the building, the energy consumption of lighting loads can be calculated for a typical summer day and a winter day. These energy consumption estimates can be used to estimate the saving of energy by replacing incandescent light bulbs with LED bulbs.

The estimated peak load of the building can be used to the size of PV plant generation requirements of the building.[6] However, for PV generating station, the irradiance energy that can be captured by the sunny side of the building should also be taken into account.

For the building of Figure 6.1, the total number of kilowatt-hours used on a winter day is 166.5 kWh and the energy used on a summer day is 239.3 kWh. Figure 6.2 presents a summer day kilowatt-hours energy consumptions.

The average input wattage to the building on a summer day and a winter day can be calculated by dividing the daily energy demand by 24 hours. The required average, hourly input power for the building is 6.94 kW on a

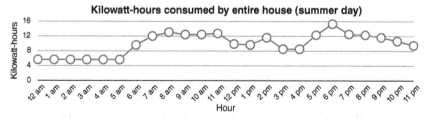

Figure 6.2   Kilowatt-hours consumed by the building in Figure 6.1 on a summer day.

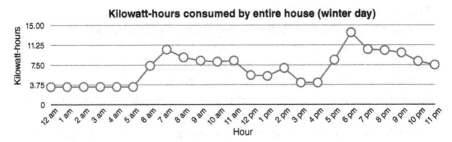

Figure 6.3   Kilowatt-hours consumed by the building in Figure 6.1 on a winter day.

winter day and 9.97 kW on a summer day. The unit power density of the building for both days can be found by dividing this average power requirement by the square footage of the building (2600 ft²). The building power density is therefore 2.67 W/ft² on a winter day and 6.83 W/ft² on a summer day. Considering only the lighting elements of the building, the lighting power density is 0.41 W/ft² on a winter day and 0.34 W/ft² on a summer day. The building lighting makes up 15.4% of the building energy consumption on a winter day and 8.9% on a summer day. All lighting elements in this analysis were assumed to be 60 W bulbs with a rated luminous output of 820 lumens each. Thus, no matter how much energy the building used for lighting over 24 hours, either in summer or in winter, the lighting system efficiency of the building would always be in lumen per watt (LPW) 820 lumens per 60 W, or 13.67 LPW.

Figure 6.3 presents the kilowatt-hour consumed by the lighting loads of the building in Figure 6.1 on a winter day.

Although these power and energy calculations are based on many simplifying assumptions about the electrical load of a building, these estimations show that there is a higher lighting energy demand during the wintertime, when there is less sunlight over a 24-hour period.

Figure 6.4, presents an estimate of daily kilowatt-hour consumed by incandescent lighting in the building on a summer day and Figure 6.5 on a winter day.

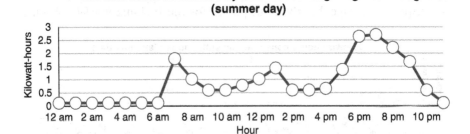

Figure 6.4   An estimate of daily kilowatt-hour consumed by incandescent lighting in the building on a summer day.

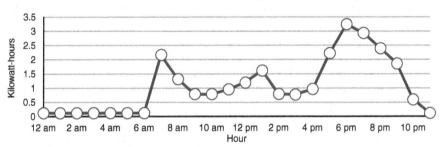

Figure 6.5 An estimate of daily kilowatt-hour consumed by incandescent lighting in the building on a winter day.

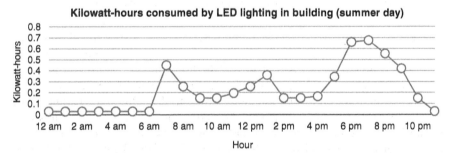

Figure 6.6 An estimate of daily kilowatt-hour consumed by LED lighting in the building on a summer day.

Incandescent lights are inefficient lighting devices, and are responsible for more residential energy consumption than is required for their desired output.

Figure 6.6 presents an estimate of daily kilowatt-hour consumed by LED lighting in the building on a summer day and Figure 6.7 on a winter day.

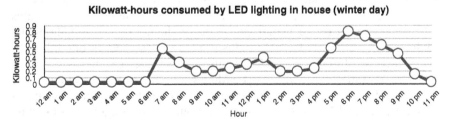

Figure 6.7 An estimate of daily kilowatt-hour consumed by LED lighting in the building on a winter day.

6.3 LED ENERGY SAVING

To estimate energy saving and cost analysis, the following assumptions are made:

  (i) Incandescent light bulb cost: about $0.50 each (http://www.gordonele
      ctricsupply.com/index~text~17233~path~product~part~17233~ds~
      dept~process~search?gclid=CPWi8pu2yLwCFQtgMgoduHcAPA)
 (ii) LED light bulb cost: about $6.00 each (https://www.google.com/sho
      pping/product/7760842826329356957?client=firefox-a&hs=4eB&rls=
      org.mozilla:en-US:official&sclient=psy-ab&q=LED+bulb&oq=LED
      +bulb&pbx=1&bav=on.2,or.r_qf.&bvm=bv.61190604,d.aWc,pv.xjs.s.
      en_US.OD_8LLFKFf4.O&biw=1230&bih=582&tch=1&ech=1ψ=
      W1_8UpWPM8aiyAHyjIDYCg.1392271196562.5&ei=aF_8UtK9Bsbly
      QGspoHIBA&ved=0CJUBEKYrMAA)
(iii) One kilowatt-hour of energy: $0.112 (http://www.npr.org/blogs/money/
      2011/10/27/141766341/the-price-of-electricity-in-your-state)

The LED energy saving can be computed based on estimated lighting loads. The price of one standard incandescent bulb and one LED bulb is estimated at 50 cents and 6 dollars, respectively.[7,8] The home lighting loads consists of 69 individual light bulbs. The light bulbs usage is estimated for winter and summer days. The light bulbs are assumed to operate for 426 hours in winter, and 355 hours during summer day. The average amount of time light bulbs is lit in the building over a year is estimated by the average of the summer day and winter day for total of 391 hours, times 365 days. Thus, the total hours that the light bulbs are lit in the building over a year is equal to 391 times 365 days that is equal to 142,715 hours. Assuming incandescent light bulbs have a lifetime of 1200 hours, and an LED light bulb has a lifetime of 50,000 hours, the building needs 119 new incandescent light bulbs a year and 3 new LED light bulbs a year. This analysis should be looked at as a comparison and quite a variation exists in actual usage. The average hourly price to cover these bulb purchases are the number of bulbs, times the bulb price, divided by 8760 hours in one year. The cost to buy new incandescent light bulbs is $59.50 per year. The cost to buy new LED bulbs is $18 per year. Assuming that all 69 light bulbs must be bought new at the start of a year, the hourly cost to cover this purchase is $0.0039 per hour for incandescent light bulbs and $0.047 per hour for LED bulbs. To calculate the hourly cost of lighting the building, the hourly cost for light bulbs is added to the cost of one kilowatt-hour, which is estimated to be 11.2 cents. These values can be used to find the hourly cost in cents to light the building with both incandescent and LED light bulbs. Figure 6.8 presents the hourly cost of lighting in building (incandescent bulbs) and Figure 6.9 presents the hourly cost of lighting in building (LED bulbs).

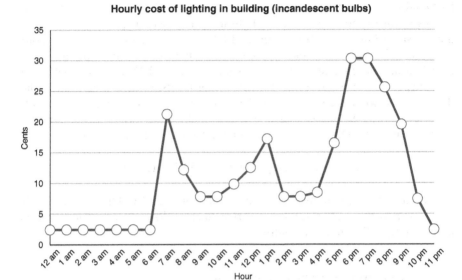

Figure 6.8   Hourly cost of lighting in building (incandescent bulbs).

6.4   RETURN ON INVESTMENT ON LED LIGHTING

Assume that the building lighting is used for 50,000 hours. The simple return on investment for both incandescent lights and LED lights can be calculated as

$$ROI = (\text{Annual savings}/\text{Installation cost}) \times 100\% \qquad (6.1)$$

The incandescent light bulbs have an estimated lifetime of 1200 hours, and LED light bulbs are expected to have a lifetime of 50,000 hours. The building contains 69 bulbs and the installation cost of incandescent bulbs would be $34.50, and $414.00 for LED bulbs. Assume the building lighting is used for 50,000 hours. Since incandescent light bulbs have lower life expectancy, 42 incandescent light bulbs will need to be replaced for a price of $21.00, and one

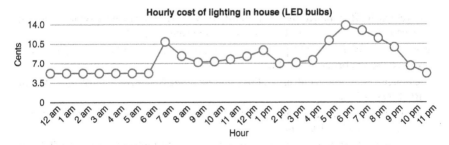

Figure 6.9   Hourly cost of lighting in building (LED bulbs).

LED bulb will need to be replaced for a price of $6.00. The average number of kilowatt-hours used by the building lighting system over a year is 8552 kWh for incandescent bulbs, and 2138 kWh for LED bulbs. Assuming a kilowatt-hour price of 11.2 cents, and including the cost of replacement bulbs, the incandescent lighting would cost $978.82 per year, and the LED lighting would cost $245.46 per year. Therefore, the annual savings for using LED light bulbs is the difference between the cost of incandescent lighting and LED lighting. The difference between the two cost estimates is $733.36. The simple return on investment for using LED lights can be calculated from Equation (6.6.1):

$$ROI = (\$733.36/\$414.00) \times 100\% = 177.14\%$$

The simple return on investment indicates that using LED lights can quickly break even from a higher installation price due to its lower energy consumption and the longer lifetime of its bulbs.

6.5 THE ANNUAL CARBON EMISSIONS

The annual carbon emissions of both lighting systems can be calculated in a similar way. Assuming that one 60 W incandescent bulb will produce 4500 pounds of CO_2 in one year, and 50 W of LED power will produce 450 pounds per year, the carbon emissions of both lighting systems can be estimated. These values assume the bulbs are left on continuously for one year, or 8760 hours. The filament bulbs thus produce 4500 pounds for every 525,600 kWh, and the LED bulbs produce 451 pounds of carbon dioxide for every 438,000 kWh. According to the residential lighting model used in this analysis, the building would use 8552 kWh per year of energy using incandescent bulbs and 2138 kWh per year with LED bulbs. These values indicate that the residential lighting system would annually produce 73.22 pounds of CO_2 with incandescent bulbs and only 2.2 pounds of carbon dioxide with LEDs.

REFERENCES

1. Encyclopædia Britannica. Available at http://www.britannica.com/EBchecked/topic/340594/light-emitting-diode-LED. Accessed February 16, 2014.
2. Energy Star. Available at http://www.energystar.gov/index.cfm?c=lighting.pr_what_are. Accessed February 16, 2014.
3. Messenger, R.A. and Ventre, J. (2004). *Photovoltaic Systems Engineering*, 2nd edn, CRC Press. p. 123.
4. Luminous efficacy. Available at: http://en.wikipedia.org/wiki/Luminous_efficacy. Accessed November 10, 2014.
5. Comparison chart: LED lights vs. incandescent light bulbs vs. CFLs. Available at: http://www.designrecycleinc.com/led%20comp%20chart.html. Accessed November 10, 2014.

6. National Public Radio. Available at http://www.npr.org/blogs/money/2011/10/27/141 766341/the-price-of-electricity-in-your-state. Accessed February 16, 2014.

7. Gordon Electric Supply. Available at http://www.gordonelectricsupply.com/index~te xt~17233~path~product~part~17233~ds~dept~process~search?gclid=CPWi8pu2 yLwCFQtgMgoduHcAPA. Accessed February 16, 2014.

8. Philips. Available at http://www.1000bulbs.com/product/94243/PHILIPS-420562. html?utm_source=SmartFeedGoogleBase&utm_medium=Shopping&utm_term= PHILIPS-420562&utm_content=LED+-+R40+-+2700K+- +Warm+White&utm_campaign=SmartFeedGoogleBaseShopping&gclid=CIPD1v z837wCFa9FMgodlyAAJg. Accessed February 22, 2014.

CHAPTER 7

THREE-PHASE POWER
AND MICROGRIDS

7.1 INTRODUCTION

In 1791, Faraday, an English scientist discovered the basic concept of induction. The Faraday's law of induction states that changing the magnetic field in coils or winding of wire creates a voltage in the coils. In 1832, Hippolyte Pixii of France built the first AC generator. This development expanded further in 1867 by Werner Von Siemens of Germany and Charles Wheatstone of England. The work of Siemens was further refined in 1886 in the United States, by William Stanley, George Westinghouse, Nikola Tesla, and Elihu Thomson. In 1890s, generator design has resulted in widely used commercial generators by Westinghouse, Siemens, Oerlikon, and General Electric. The Polyphase AC generators (1890s) were developed by C.S. Bradly (United States), August Haselwander (Germany), Mikhail Dolivo-Dobrovsky (Germany/Russia), and Galileo Ferraris (Italy).

7.2 THE BASIC CONCEPT OF AC GENERATOR

To generate alternating current (AC), an external source of power is needed.[1] There are two ways to generate electricity: (a) the mechanical power (b) solar and wind power. The mechanical power can be generated by water power through directing falling water to turbines to provide mechanical power to the shaft of electric generators. The mechanical power can be provided by conditioning steam at right temperature and pressure by burning coal or gas or nuclear reaction. Faraday's law of induction states that changing the magnetic

Design of Smart Power Grid Renewable Energy Systems, Second Edition. Ali Keyhani.
© 2017 John Wiley & Sons, Inc. Published 2017 by John Wiley & Sons, Inc.
Companion website: www.wiley.com/go/smartpowergrid2e

field in coils or winding of wire creates a voltage in the coils. Faraday's law of induction states that when windings are subjected to time-varying magnetic field, electromotive force (emf) is induced in the windings. The magnetic field forces the electrons in the copper metal to be pushed to the next atom. For example, the copper has 27 electrons. However, the last two electrons in their orbit can be forced on to the next atom by emf force. The flow of electrons is electrical current.

7.3 THREE-PHASE AC GENERATOR

A three-phase synchronous generator depicted in Figure 7.1 is a three-terminal device.[2,3] The mechanical power is input from a prime energy source such as hydropower or mechanical power from a thermal unit that drives the shaft of the generator. The field winding of the generator is located on the rotor of the machine. The field winding is depicted by North (N) and South (S) in Figure 7.1 and below it shown by a winding supplied by DC current. The field winding is energized through slip rings of segmented copper and brushes[3-5] The three-phase balanced windings are located on the stator of the machine. As the names of stator and rotor implies, the stator is stationary and rotor is the rotating element of the machine. The space between stator and rotor is referred to as the air gap of the machine. The air gap length is in millimeters and the rotor is supported by a well-designed bearing system to keep the air gap as small as possible yet keep the rotor from having contact with the stator (i.e., rubbing). Because the air gap length is designed to be as small as possible, the maximum magnetic fields cross the air gap and link to the stator of the machine.[1] By rotating the field winding in the air gap of the machine, the time-varying magnetic field is generated. The time-changing magnetic fields link the

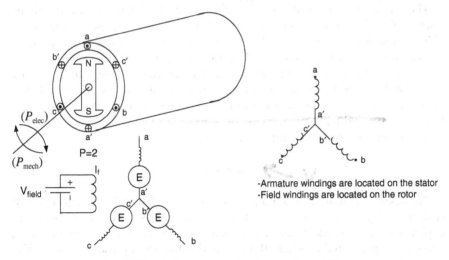

Figure 7.1   A three-phase synchronous generator.

three-phase winding located on the stator of the machine.[6,7] Two types of time-changing flux are produced. One is the linking magnetic flux that crosses the air gap of the machine and the second field is the leakage flux field that leaks through the air. The linkage field gives rise to the magnetizing inductance and to leakage inductance. Because the induced voltage and current is produced on the stator windings, the stator windings are referred to as armature windings.

The operating frequency of the generator is a function of shaft speed and number of magnetic poles of field winding.

$$\omega_{elec} = \frac{P}{2}\omega_{mech} \tag{7.1}$$

The unit of Equation (7.1) is radians per second.

$$2\pi f_{elec} = \frac{P}{2}\omega_{mech} \tag{7.2}$$

$$\omega_{mech} = n\frac{2\pi}{60} \tag{7.3}$$

In (7.2), the unit of n is converted from radians per second to revolutions per minute

$$f = \frac{P}{2}\frac{n}{2\pi}\frac{2\pi}{60} = \frac{Pn_{syn}}{120} \tag{7.4}$$

where $n_{syn} = \dfrac{120f}{P}$ Hz.

where P is the number of poles, and ω_{elec}, ω_{mech} represents electrical and mechanical speed in radians per second. Finally, n is revolutions per minute (rpm), f is cycles per second, and n_{syn} is called the synchronized speed in rpm.

The infinite bus concept is used in a power grid to define an ideal voltage source.[8] The ideal voltage source remains at constant voltage under any amount of loads. This is indeed the case when a generator is connected to large interconnected power grids. For example, when an independent power producer (IPP) connects 10 MW of wind generators to a large power grid with total capacity of 10,000 MW, the bus voltages of power grids do not change more that plus or minus 5% when the wind generators are connected to the power grid. The concept of synchronization is discussed in Section 7.4.

The voltage of the generator is expressed as

$$E = K.I_f.\omega \tag{7.5}$$

The open-circuit induced voltage, E is a function of machine dimensions that is depicted by constant K and field current, I_f and shaft speed, ω. Figure 7.2 depicts the circuit model of one phase of a three-phase generator. The induced voltage is represented by E and the terminal voltage is represented by V_T. The magnetizing inductance of the machine is not shown in Figure 7.2 circuit model.

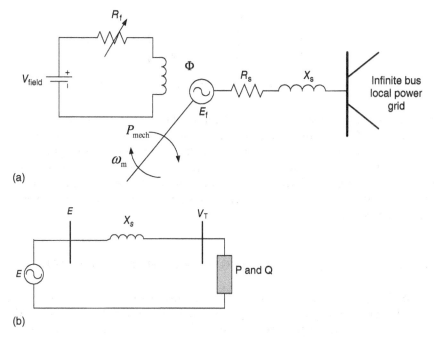

(a)

(b)

Figure 7.2 (a) One-machine generator connected to the local power grid bus. (b) A one-phase equivalent circuit model of three-phase generator.

The generator is designed to draw a small amount of current to be magnetized. The leakage inductance shown by X_s is called synchronized reactance because it is computed using the steady-state synchronized operation.

$$X_S = \omega e.Ll \tag{7.6}$$

where

$$\omega e = 2\pi f_{\text{elec}} = \frac{P}{2}\omega_{\text{mech}} \tag{7.7}$$

The power supplied by the generator is shown as active and reactive power P and Q. Active and reactive power load consumption are discussed later in Section 7.5.

There are a number of commonly used exciters. The DC field current is supplied from a DC connected to the positive terminal and the negative terminal of the DC source. This system is called exciter.[9] The exciter includes a DC generator located on rotor and two brush structures located on stator. A segmented copper structure located on rotor is called commutator and it rotates. The DC current is injected into the DC field winding. This exciter may be driven by a motor, prime mover, or the shaft of the synchronous machine.

Another type of exciter is called rectifier exciter based exciter. The rectifier converts AC current to DC current. This exciter derives its energy from an

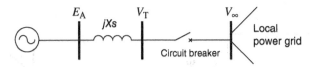

Figure 7.3   The generator operation before synchronization.

AC generator and is converted to DC by a rectifier system. This type of exciter includes an AC generator and power to rectifier that may be either noncontrolled or controlled. The generator may be driven by a motor, prime mover, or the shaft of the synchronous machine. The rectifier may be stationary or rotating with the generator shaft. A compound rectifier exciter derives its electric power from the terminals of the synchronous machine and then it is converted to DC by rectifier systems.[10]

7.4   THE SYNCHRONIZATION OF GENERATOR TO POWER GRIDS

Figure 7.3 depicts parallel operation of a generator with a local power grid. The parallel operation of generators requires that all generators be operated at the same electrical frequency or the same electrical speed which is called synchronized operation.

For synchronous operation, the following conditions are necessary:

(i) The electrical speed of the synchronous generator should be equal to the electrical speed of the power system generators.

(ii) The terminal voltage of the generator should be equal to the V_∞ bus and they must be in the same a-b-c sequence.

When the above conditions are satisfied, the circuit breaker can be closed. Synchronous operation means the parallel operation of synchronous generators.[11] The notation, V_∞ bus, has a special meaning. We call a bus voltage as an infinite bus to signify that its voltage is constant and cannot be changed by external events. This means that an infinite bus acts as an ideal voltage source. Figures 7.4a, 7.4b, and 7.5 depict the one machine connected to an infinite bus.

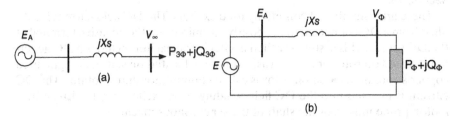

Figure 7.4   The operation of the generator in parallel with the power system. (a) One-line diagram. (b) One-phase equivalent circuit.

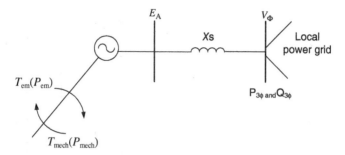

Figure 7.5   The operation of a generator injecting power into a local power grid.

Consider a synchronous generator connected to a local power grid as depicted in Figure 7.5.

If mechanical power supplied to the generator shaft is constant and losses are ignored, then mechanical power supplied, P_{mech} is equal to the electrical power injected into the local power grid.

7.5 POWER FACTOR AND ACTIVE AND REACTIVE POWER CONCEPTS

If the field current is adjusted to a value as $I_f = I_{f0}$, then the excitation voltage is equal to E_A.

If the field current is reduced to a new value of I'_f, then E_A reduces to a new value of E'. However, if P_{mech} is kept constant and continue to reduce I_f the same amount of electrical power equal to mechanical power is injected to the power grid. However, E' decreases to a new value of E''. The change of field current results in change of reactive power and change in power factor. For example, a load with resistive load of R and reactance of X, the active power consumption is given by

$$P = I^2.R \tag{7.8}$$

The reactive power consumed by the load is given by

$$Q = I^2X \tag{7.9}$$

Consider a generator supplying power at a lagging power factor to a local power grid as depicted in Figure 7.6. The power grid must be balanced in generation and loads to remain stable. This condition is shown in Figure 7.7.

$$\sum_{i=1}^{6} P_{Gi} = \sum_{i=1}^{6} P_{Li} + P_{losses} \tag{7.10}$$

$$\sum_{i=1}^{6} Q_{Gi} = \sum_{i=1}^{6} Q_{Li} + Q_{losses} \tag{7.11}$$

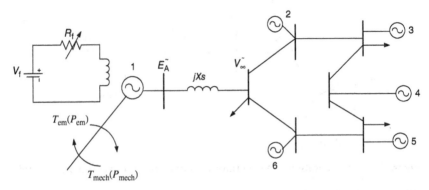

Figure 7.6   The operation of a generator injecting power into a local power grid.

Figure 7.7 depicts a generator connected to a power grid injecting power. If the generator is operating with lagging power factor, then the magnitude of excitation voltage is greater than the magnitude of terminal voltage (i.e., $|E_A| > V_\infty$). This condition is defined as an overexcited generator operation and terminal current lags the terminal voltage.[10–12]

To change the power factor of the generator, the supplied mechanical power constant must remain constant: P_{mech} = const and therefore, the shaft speed is also constant, that is, ω_m = const. By reducing the field current from a value of I_f to a new lower value of I'_f, the excitation voltage is reduced to a new lower value of E'_A. However, because the supplied mechanical power remains constant, the electric power generated by the generator must also remain constant. By reducing the field current I_f, the reactive power generated by the synchronous generator is reduced. However, the active power generated by the generator remains the same and a new operating condition is established where, $Q_{3\phi} < Q'_{3\phi}$. However, the balance of generations and load is maintained by power grid control system to maintain the stability of power grid.

$$\sum_{i=1}^{6} P_{Gi} = \sum_{i=1}^{6} P_{Li} + P_{losses} \tag{7.12}$$

$$\sum_{i=1}^{6} Q_{Gi} = \sum_{i=1}^{6} Q_{Li} + Q_{losses} \tag{7.13}$$

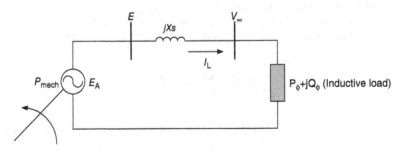

Figure 7.7   The equivalent circuit model of a generator injecting into a local utility.

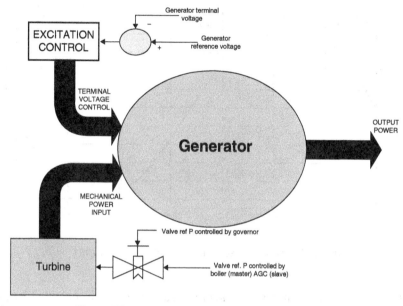

Figure 7.8 A generator as a three-terminal device.

To operate the synchronized generator at the unity power factor, the field current can be reduced such that the terminal current and terminal voltage is in phase (i.e., with zero angle). For unity power factor, the generator only produces active power and the reactive power production is zero. Therefore, by reducing field current I_f, the generator can be made to operate at unity power factor. If the field current, I_f, is reduced again, the terminal current leads the terminal voltage and the generator operates with the leading power factor while generating the same active power.[13,14]

Figure 7.8 depicts the operation of a generator as a three-terminal device. The mechanical power is supplied to the generator; the excitation voltage is controlled by the voltage regulator and the field current is set for the desired power factor of the generator. These conditions determine the active and reactive power supplied by the generator. The power supplied by the generators during the operation of power grid is controlled so that, the balance between total generation and total load is maintained as given by Equations (7.12) and (7.13).

AC grid generators are designed to produce a three-phase alternating current. The three sinusoidal distributed windings or coils are designed to carry the same current. Figure 7.9 depicts the voltage of each phase of the three-phase generator with respect to generator neutral.

Figure 7.9 represents the sinusoidal voltage (or current) as a function of time. In Figure 7.9, 0–360° (2π radians) is shown along the time axis. Power grids around the world each operate at a fixed frequency of 50 or 60 cycles per second. Based on a universal color-code convention,[16] black is used for one phase of

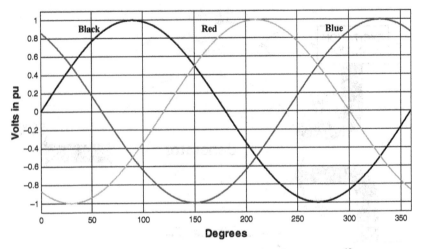

Figure 7.9 Three-phase generator voltage waveforms.[15]

the three-phase system; it denotes the ground as the reference phase with zero-degree angle. Red is used for the second phase, which is 120° out of phase with respect to the black phase. Blue is used for the third phase, which is also 120° out of phase with the black phase.

The three-phase AC system can be considered as three single-phase circuits. The first AC generators were single phase. However, it was recognized that the three-phase generators can produce three times as much power. However, the higher phase generators will not produce proportionally more power.[17] Based on Figure 7.9, the three phase voltages of three-phase generators can be expressed as

$$V_{ac} = \frac{460\sqrt{2}}{\sqrt{3}} \cdot \sin\left(2\pi 60 \cdot t\right)$$

$$V_{ab} = \frac{460\sqrt{2}}{\sqrt{3}} \cdot \sin\left(2\pi 60 \cdot t + 90\right)$$

$$V_{bc} = \frac{460\sqrt{2}}{\sqrt{3}} \cdot \sin\left(2\pi 60 \cdot t - 90\right)$$

In the above voltage expressions, the line-to-line voltage, V_{LL}, is 460 V and the line to neutral is $V_{L\text{-}N}$ is $\frac{460}{\sqrt{3}}$ and pick value of each phase voltage is $\frac{460\sqrt{2}}{\sqrt{3}}$.

7.6 THREE-PHASE POWER GRIDS

Consider the three-phase four-wire system given in Figure 7.10.

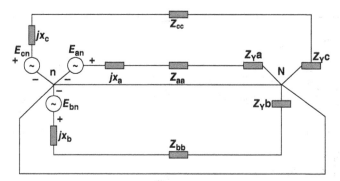

Figure 7.10 The three-phase four-wire distribution system.

In general, three-phase power grids[18,19] are designed as balanced grids. Therefore, the same structure for generators, distribution lines, and loads are used. Therefore, the balanced grid grids have the following specification.

$$X_a = X_b = X_c = X_s$$
$$Z_{aa} = Z_{bb} = Z_{cc} = Z_{line}$$
$$Z_{Ya} = Z_{Yb} = Z_{Yc} = Z_{load}$$

The power consumption in complex domain by Equation (7.14).

$$S_{aa} = S_{bb} = S_{cc} = P_L + jQ_L \tag{7.14}$$

The power consumption in complex domain.

The three phase voltages are 120 degrees out of phase as shown in Figure 7.11.

For balanced generators, we will usually have phase "a" selected as a reference and the other two phases are 120 degrees out of phase with respect to

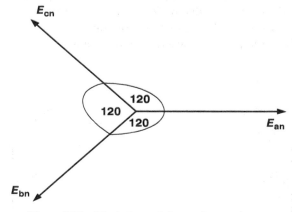

Figure 7.11 The balanced three-phase voltages.

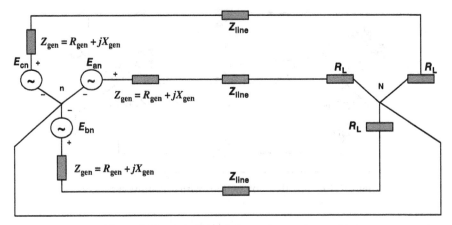

Figure 7.12   A balanced three-phase power grid.

phase "a." Figure 7.12 depicts a balanced three-phase grid which consists of balanced three-phase generators, transmission lines, and loads.

For a balanced three-phase grid, the currents are also balanced. The three-phase voltages are in a plane and they are out phase by 120° from each other.

$$E_{ab} + E_{bc} + E_{ca} = E_n \tag{7.15}$$

E_n is the voltage between neutral and ground.

Similarly, the three-phase currents are in a plane and they are out phase by 120° from each other. Therefore, we have:

$$I_a + I_b + I_c = I_n \tag{7.16}$$

If loads were unbalanced, the current I_n would flow. However, for a balanced three-phase load. The sum of $I_a + I_b + I_c = 0$, therefore $I_n = 0$.

In this case, the neutral conductor does not carry any current so it is omitted. Figure 7.13 shows a three-phase three-wire distribution system.

Consider the three-phase balanced system of Figure 7.14.

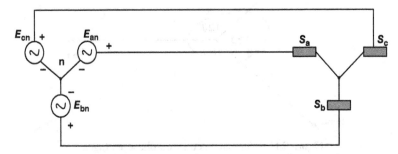

Figure 7.13   A balanced three-phase three-wire distribution system of a power grid.

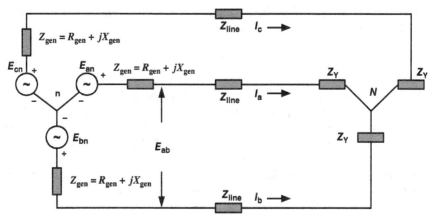

Figure 7.14 A balanced three-phase power grid.

The phase voltages and phase currents in Figure 7.15 are designated as

$$E_{an}, E_{bn}, E_{cn} = \text{Line-to-neutral voltages} = \text{Phase voltages}$$
$$I_a, I_b, I_c = \text{Line currents or phase currents}$$

Therefore, for line–line voltages, E_{ab} is 30° out of phase with the magnitude of $\sqrt{3}\, E$, E_{bc} is 90° out of phase with the magnitude of $\sqrt{3}\, E$. Because the source voltages are balanced and the loads are balanced, the resulting currents are also balanced and we will have:

$$I_a + I_b + I_c = 0 \tag{7.17}$$

7.7 CALCULATING POWER CONSUMPTION

The power consumption in a DC circuit is computed as:

$$P = V \cdot I = \frac{V^2}{R} = I^2 R \tag{7.18}$$

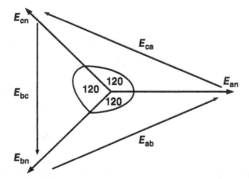

Figure 7.15 Phase voltages and line–line voltages.

The power consumption in a single-phase AC circuit is computed as described by Equation (7.18) or (7.19):

$$P = V \cdot I \cdot pf \tag{7.19}$$

In Equation (7.19), V is in volts and I is in ampere. The power factor (pf) is computed based on the phase angle between the voltage and current where the voltage is the reference phase. For purely resistive pf is equal to one. For most AC motors, the power factor is around 0.8 to 0.95. When the power factor of a motor is low, the motor consumes more reactive power. The low power factor causes the low voltage in power grids. Furthermore, the low power factor requires that grid generators produce reactive power. The reactive power can also be produced by placing capacitors in power grids.

$$Q_c = -.I^2 X_C \tag{7.20}$$
$$Q_L = +.I^2 X_L \tag{7.21}$$

The unit of reactive power is volt amp reactive, Var. In Equation (7.21), the negative sign indicates reactive power injected into power grid bus; and in Equation (7.21), the positive sign indicates reactive power is consumed by the load.

Figures 7.16a and 7.16b present the depiction of reactive power consumption (flow of reactive power to the load) by an inductive load and reactive power

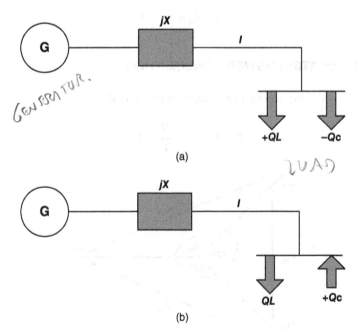

Figure 7.16 (a)Reactive power sign convention. (b) Reactive power flow direction

generation by a capacitive load negative convention, the depiction of reactive power consumption by an inductive load and reactive power generation by a capacitive load positive convention (flow of reactive power into the load bus).

A purely inductive grid element is presented by X_L.

$$XL = \omega L \tag{7.22}$$

And a purely capacitive element is given by

$$Xc = \frac{1}{\omega C} \tag{7.23}$$

In Equations (7.22) and (7.23) the units of XL and XC are in ohms. Therefore, pure capacitive loads are sources of reactive power generation.

An inductive load power consumption is given by

$$S = P + jQ \tag{7.24}$$

Equation (7.24) specifies that the load consumes active, P, and reactive power, Q.

Capacitive load power consumption is given by

$$S = P - jQ \tag{7.25}$$

Equation (7.25) specifies that the load consumes active, P, and produces the reactive power, Q.

7.8 ONE-LINE DIAGRAM REPRESENTATION OF THREE-PHASE POWER GRIDS

The three-phase power grids with three-phase loads can be represented by a one-line diagram. For example, Figure 7.17 depicts a balanced three-phase power grid with Y- and Δ-connected loads.

The power grid system depicted in Figure 7.17 is represented with a one-line diagram of Figure 7.18.

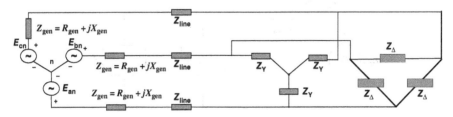

Figure 7.17 A three-phase power grid with two loads.

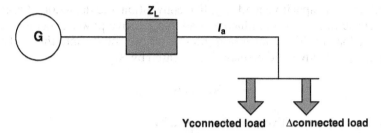

Ycconnected load   Δconnected load

Figure 7.18 A one-line diagram of Figure 7.17.

In a single-line diagram, the voltages are given as line-to-line voltage and power consumption is specified for all three phases. We can represent the one-phase equivalent circuit with line to neutral and power consumption per phase. Figures 7.17 and 7.18 are depicted in Figure 7.19 with Y equivalent of delta connected load as shown in Figure 7.19.

$$S_{3\phi} = 3S_{\phi} \tag{7.26}$$

However, the phase voltage is equal to the line-to-neutral voltage and line-to-line voltage is computed as

$$V_{\text{L-L}} = \sqrt{3}V_{\text{L-N}} = \sqrt{3}V_{\varphi} \tag{7.27}$$

Therefore, the three-phase power can be expressed as

$$S_{3\phi} = P_{3\phi} + jQ_{3\phi} \tag{7.28}$$

For three-phase Y-connected systems, we will have

$$P_{3\phi} = S_{3\phi}pf$$
$$S_{3\varphi} = \sqrt{3}V_{\text{L-L}}I_{\text{L}} \tag{7.29}$$

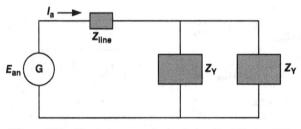

Figure 7.19 One-phase equivalent circuit of Figure 7.18.

And

$$|S_{3\phi}|^2 = |P_{3\phi}|^2 + |Q_{3\phi}|^2$$
$$Q_{3\phi} = \sqrt{|S_{3\phi}|^2 - |P_{3\phi}|^2} \tag{7.30}$$

Also, we can write complex power as

$$S_{3\phi} = \sqrt{\left(P_{3\phi}^2 + Q_{3\phi}^2\right)} \tag{7.31}$$

The power factor, pf is expressed as lagging or leading to indicate consumption or generation of reactive power. Also, pf can be expressed as

$$pf = \cos\theta = \frac{P}{|S|} \text{ leading or lagging} \tag{7.32}$$

To summarize for a lagging power factor, pf, the reactive power, Q, is positive. Therefore, the load consumes reactive power. Similarly, for a leading power factor, pf, the reactive power, Q, is negative. Therefore, the load generates reactive power and the load is a capacitive load.

Example 7.1   Consider a three-phase 480 V, 300 kVA load with pf of 0.9 lagging, what is the active, reactive power consumption of the load?

Solution
The following data is given about the load:

$$|S| = 300 \text{ kVA}$$
$$pf = 0.9 \text{ lagging}$$

In this example, we like to compute P, Q from S.

$$P_{3\phi} = |S_{3\phi}|.pf = 300 \times 0.9 = 270 \text{ kW}$$
$$\theta = \arccos(0.9) = 25.84 \text{ degrees}$$
$$\text{Sin}\theta = 0.43589$$
$$Q_{3\phi} = |S_{3\phi}| \cdot \text{Sin}\theta$$
$$Q_{3\phi} = 300*0.43589 = 130.77 \text{ kVAr}$$
$$Q_{3\phi} = \sqrt{|S_{3\phi}|^2 - |P_{3\phi}|^2}$$
$$= 130.77 \text{ kVAr}$$

$Q > 0$ because pf is lagging.

The load consumes 270 kW of active power and 130.77 kVAr of reactive power.

Example 7.2   Consider a three-phase 480 V, 240 kW load with $pf = 0.8$ lagging, what is the active, reactive, and complex power of the load?

Solution
We have the following known data:

$$P = 240 \text{ kW}$$
$$pf = \cos \theta = 0.8 \text{ lagging}$$

We can compute Q and S from P.

$$|S| = P / \cos \theta = 240/0.8 = 300 \text{ kVA}$$
$$Q_{3\phi} = |S_{3\phi}| \sin \theta = 300 \times 0.6 = 180 \text{ kVAr}$$

$Q > 0$ because pf is lagging.

$$S = 270 + j180 = 300\angle 36.8 \text{ kVA}$$

Example 7.3   Consider a three-phase 480 V, 180 kVA load with pf = 0.0 leading, what is the active, reactive, and complex power of the load?

Solution
We have the following data:

$$|S| = 180 \text{ kVA}$$
$$pf = \cos \theta = 0.0 \text{ leading}$$

To compute: P, Q, from S, we can compute $P_{3\phi}$ as

$$P_{3\phi} = |S_{3\phi}| \cos \theta = 180 \times 0.0 = 0$$
$$Q_{3\phi} = |S_{3\phi}| \sin \theta = 180 \times (-1) = -180 \text{ kVAr}$$

$Q < 0$ because pf is leading.

$$S = 0 - j\,180 = 180\angle -90° \text{ kVAr}$$

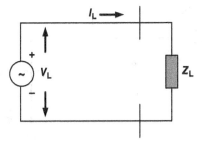

Figure 7.20   The inductive impedance load model.

7.9 LOAD MODELS

We can represent an inductive load by its impedance as shown in Figure 7.20.

The load impedance, Z_L is an inductive load. Most power system loads are inductive. The majority of industrial, commercial, and residential motors are of the induction type. In Figure 7.20, the load voltage, V_L is a line-to-neutral voltage and I_L is the phase current supplying the load.

$$Z_L = R_L + j\,\omega L = R + j\,X_L = |Z_L|\angle\theta \quad |Z_L| = \sqrt{R^2 + X_L^2}, \; \theta = \tan^{-1}\left(\frac{X_L}{R}\right)$$

The inductive load power representation is expressed by active and reactive power consumption of the load.

$$I_L = \frac{V_L\angle 0}{Z_L\angle\theta} = I_L\angle - \theta, \; I_L = \frac{|V_L|}{|Z_L|}\angle - \theta$$

With the load voltage as the reference (i.e., $V_L = |V_L|\angle 0$), the load current lags the voltage as shown in Figure 7.21.

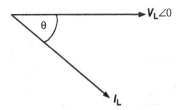

Figure 7.21   The load voltage and its lagging load current for the inductive load depicted in Figure 7.21.

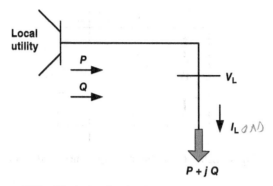

Figure 7.22   The inductive load–power representation.

The complex power absorbed by a load can be expressed as

$$S_L = V_L I_L^* = V_L(I_L\angle - \theta)^* = |V_L|\,|I_L|\cos\theta + j|V_L|\,|I_L|\sin\theta$$
$$S_L = |V_L||I_L|\text{VA}$$
$$P = |V_L||I_L|\cos\theta\text{W}$$
$$Q = |V_L||I_L|\sin\theta\text{ Vars}$$

And the complex power is expressed as

$$S = P + jQ, \text{ where } \theta = \tan^{-1}\frac{Q}{P} \text{ and } pf = \cos\theta, \text{ lagging}$$

An inductive load model power representation is shown in Figure 7.22.

Figure 7.22 depicts inductive load consumption of both active and reactive power.

Figure 7.23 depicts the capacitive impedance load model. Here, again, the load voltage is line-to-neutral and the reference and load current is the phase

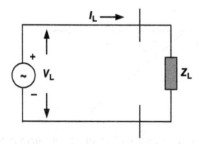

Figure 7.23   The capacitive impedance load model.

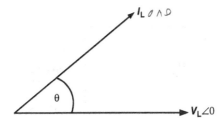

Figure 7.24   The load voltage and current of a capacitive load.

current supplied to the load.

$$Z_L = R - j X_c = |Z_L| \angle - \theta$$

$$|Z_L| = \sqrt{R^2 + X_c^2}, \theta = \tan^{-1} \left(\frac{X_c}{R} \right)$$

$$I_L = \frac{V_L \angle 0}{Z_L \angle - \theta} = I_L \angle \theta, |I_L| = \frac{|V_L|}{|Z_L|} \angle \theta$$

With the load voltage as the reference (i.e., $V_L = |V_L| \angle 0$), the load current leads the voltage as shown in Figure 7.24.

The complex power absorbed by the load is

$$S_L = V_L I_L^* = V_L (I_L \angle \theta)^* = V_L I_L \angle - \theta = |V_L| \, |I_L| \cos \theta - j \, |V_L| \, |I_L| \sin \theta$$
(7.33)
$$S = P - jQ, \theta = \tan^{-1} \left(\frac{Q}{P} \right) \text{ and power factor, } pf = \cos \theta, \text{ leading}$$

Therefore, the load model can be represented as shown in Figure 7.25.

Therefore, the active power is consumed by the load and reactive power is supplied by the capacitive load to the local power network as shown in Figure 7.25. Recently, more variable-speed drive systems are controlled by power converters, which are controlling many types of motors. In addition, more power electronic loads have penetrated the power systems. These types of loads act as nonlinear loads and can act both as inductive and capacitive loads

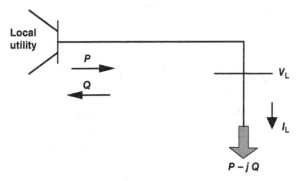

Figure 7.25   The power model for a capacitive load.

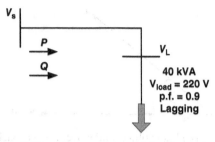

Figure 7.26 The power model for Example 7.4.

during their transient and steady-state operations. The *pf* corrections and voltage control and stability are active areas of research.

Example 7.4 For a single-phase inductive load, given below in Figure 7.26, compute the line current.

Solution

$$\text{kVA} = |V|\,|I_L| \times 10^3$$
$$|I_L| = 40 \times 10^3/220 = 181.8$$
$$I_L = 181.8\angle -25.8°\,\text{A}$$

Example 7.5 For a three-phase inductive load given by Figure 7.27, compute the line current.

Solution

$$\text{kVA}_{3\varphi} = 2000$$
$$V_{\text{L-L}} = 20\,\text{kV}$$
$$\text{kVA} = \sqrt{3}V_{\text{L-L}}I_L$$
$$|I_L| = 2000/\sqrt{3} \times 20 = 57.8\,\text{A}$$
$$I_L = 57.8\angle -25.8°$$
$$P_{3\phi} = \text{kW} = \text{kVA}\cos\theta = 2000 \times 0.9 = 1800\,\text{kW}$$
$$Q_{3\phi} = \text{kVar} = \text{kVA}\sin\theta = 2000 \times \sin(25.8°) = 870.46\,\text{kVAr}$$

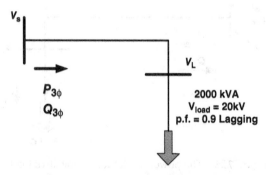

Figure 7.27 The power model for Example 7.5.

Figure 7.28   A generator operating with a lagging power factor.

Example 7.6   Consider the generator in Figure 7.28 that is operating with a lagging power factor. Compute the active and reactive power supplied to the system.

Figure 7.29 depicts the equivalent circuit model of Figure 7.28. Figure 7.30 presents the generator voltage and the generator current.

$$S_{3\varphi} = 3V_G I_G^* = P_{G3\varphi} + j\,Q_{G3\varphi}$$

7.10   TRANSFORMERS IN ELECTRIC POWER GRIDS

Two facts are clear by now. The transmitted power is the product of voltage times the current. Losses in a transmission line are the square of current through the lines times the line resistance. Therefore, the transmission of a large amount of power at low voltages would have a very large power loss. For high power transmission, we need to raise the voltage and lower the current. This problem was solved with the invention of transformers.

7.10.1   A Short History of Transformers

In the "War of Currents"[20] in the late 1880s, George Westinghouse[21] (1846–1914) and Thomas Edison (1847–1931) were at odds over Edison's promotion of direct current (DC) for electric power distribution and Westinghouse's advocacy of alternating current (AC), which was also Tesla's choice.[20] Because Edison's design was based on low-voltage DC, the power losses in distribution network were too high. Lucien Gaulard of France and John Gibbs of England[21] demonstrated the first AC power transformer in 1881 in London.[21]

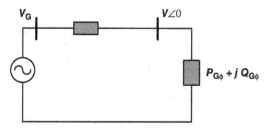

Figure 7.29   The equivalent circuit for Example 7.6.

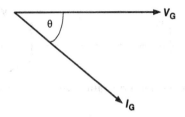

Figure 7.30   The phasor relationship of V_G and I_G.

This invention attracted the interest of Westinghouse, who then used Gaulard–Gibbs transformers and a Siemens AC generator in his design of an AC network in Pittsburgh. Westinghouse understood that transformers are essential for electric power transmission because they ensure acceptable power losses. By stepping up the voltage and reducing the current, power transmission losses are lessened. Ultimately, Westinghouse established that AC power was more economical for bulk power transmission and started the Westinghouse Company to manufacture AC power equipment.

7.10.2 Transmission Voltage

In early 1890, the power transmission voltage was at 3.3 kV. By 1970, power transmission voltage had reached a level of 765 kV. The standard operating voltages in the secondary distribution system are a low voltage range of 120–240 V for single phase and 208–600 V for three phase. The primary power distribution voltage has a range of 2.4–20 kV. The subtransmission voltage has a range of 23–69 kV. The high-voltage transmission has a range of 115–765 kV. The generating voltages are in the range of 3.2–22 kV. The generating voltage is stepped-up to the high voltage for bulk power transmission, then stepped-down to subtransmission voltage as the power system approaches the load centers of major metropolitan areas. The distribution system is used to distribute the power within the cities. The residential, commercial, and industrial loads are served at 120–600 V voltage levels.[18,19]

Figure 7.31 depicts a typical PV power source feeding a local distributed generation system. The PV panels are like batteries and are the source of DC power. They are connected in series and parallel. The maximum operating DC

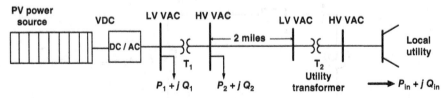

Figure 7.31   A photovoltaic (PV) power source feeding a radial distribution system.

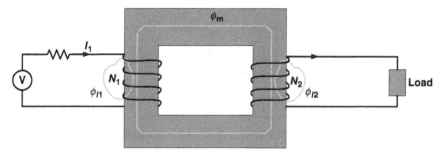

Figure 7.32   An ideal single-phase transformer.

voltage is 600 V. With the current DC technology of distribution, megawatts of power distribution are very expensive. If we distribute high DC power of a renewable source such as PV systems at low voltages, power losses through the distribution lines are quite high. We can step-up by using DC/DC converters (i.e., boost converters). However, the costs of associated DC/DC converters and the protection of a DC system are quite high. However, in the design of the system shown in Figure 7.31, the DC power is converted to AC using a DC/AC inverter. In Figure 7.32, the PV source is in a range of 3–kVA; it is connected to a 120 V residential system feeding the house loads. We have stepped-up the voltage before transferring the extra power to the local power grid company.

7.10.3 Transformers

A transformer is an element of a power grid that transfers electric power from one voltage to another voltage level through inductively coupled windings. For a single-phase transformer, each winding is wound around a single core. The two windings are magnetically coupled using the same core structure. As in an ideal transformer, the device input power is the same as its output power. This means that the input current times input voltage is equal to the product of the output voltage and output current. One of the windings is excited by an AC source. The time-varying magnetic field induces a voltage in the second winding by Faraday's fundamental law of induction.[9,17–19,22,23]

In a transformer, the volt per turn on the secondary winding is the same as in the primary winding. We can select the ratio of turns, to step up the voltage or to step it down. For high-voltage power transmission of power, transformers are needed to step up the voltage and therefore to step down the current and reduce the transmission line power losses. Figure 7.32 depicts an ideal transformer. An ideal transformer is assumed to have zero power losses in its core and winding.

$$N_1 I_1 = N_2 I_2 \text{ Ampere-Turn} \tag{7.34}$$

$$\frac{V_1}{N_1} = \frac{V_2}{N_2} \text{ Volt/Turn} \tag{7.35}$$

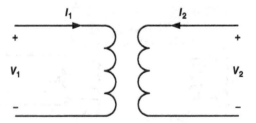

Figure 7.33   The schematic of an ideal transformer.

Figure 7.32, ϕ_m depicts the mutual flux linkage and ϕ_{l1} and ϕ_{l2} the leakage flux on either side of the transformer. The mutual flux linkage will result in the mutual inductance and the leakage flux will result in leakage inductance.

The schematic of an ideal transformer is depicted in Figure 7.33. Because the power losses in an ideal transformer are assumed to be zero, the input power and output power are the same.

$$S_1 = V_1 I_1^* = S_2 = V_2 I_2^* \tag{7.36}$$

A real transformer has both core and winding losses as shown in Figure 7.34. These losses are represented by R_1 and R_2. Here, in this representation, R_1 represents the primary-side ohmic loss and R_2 denotes the secondary-side ohmic loss.

It is customary to denote the side that is connected to the source as primary and the side connected to the load as secondary. For simplicity, we will number each side or call each side by their voltage levels, that is, one side of the transformer is the high-voltage side and the other side is the low-voltage side.

$$I_E = I_m + I_C$$

$$X_m = \omega L_m \tag{7.37}$$
$$N_1 I_2' = N_2 I_2 \tag{7.38}$$

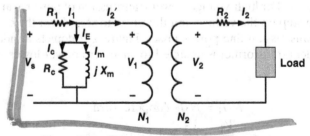

Figure 7.34   The schematic of a real transformer.

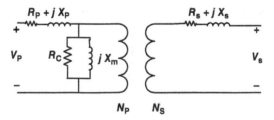

Figure 7.35 The complete schematic of a real transformer.

The current I_E is referred to as the excitation current. This current has two components, I_m and I_C. The current I_m is referred to as the magnetizing current and I_c, as the core current. In Equation (7.37), L_m denotes the magnetizing inductance. This inductance is computed from the mutual inductance and the inductance of each coil.

$$V_s = V_m \cos \omega t \qquad (7.39)$$

$$\overline{V}_s = \frac{V_m}{\sqrt{2}} \angle 0$$

$$I_m = \frac{V_s \angle 0°}{jX_m} = \frac{V_s}{X_m} \angle -90° \qquad (7.40)$$

In AC power distribution, the source voltage can be presented by Equation (7.39). The magnetizing current is computed as presented by Equation (7.40). Figure 7.35 depicts the complete transformer model. In this model, we have shown both sides of the transformer primary resistance and primary leakage reactance and on the secondary side, we have shown the secondary-side leakage reactance and winding resistance. The core losses and magnetizing reactance are represented by the resistance R_C and the reactance X_m and they are shown on the primary side.

$$X_p = \omega_e L_{lp}, \ X_s = \omega_e L_{ls} \text{ and } X_m = \omega_e L_m$$

where L_{lp}, L_{ls}, and L_m are referred to as leakage inductance of primary winding, leakage inductance of secondary winding, and the mutual inductance of the transformer, respectively. Because the transformers are designed to have very small exciting current—in the range of 2–5% of the load current, the magnetizing shunt elements of transformers are assumed to have very high impedance; therefore, the shunt elements are eliminated in the voltage calculation of transformers. Figure 7.36 depicts the transformer model used in voltage analysis.

In Figure 7.36, the resistance, R_1, denotes the winding resistance of winding number 1 and X_{l1} denotes the leakage reactance of the same winding. Similarly,

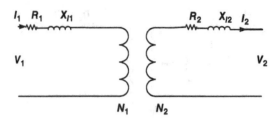

Figure 7.36 The equivalent model of a transformer for voltage analysis.

R_2 and X_{l2} denote the resistance and leakage reactance of the winding number 2. In this representation, designation of primary winding and secondary winding are omitted because, in general, either side can be connected to the load or source.

Figure 7.37 depicts the equivalent transformer model of Figure 7.36, where the impedance of both windings is referred to side 1. Let us assume side 1 of a transformer connected to the high-voltage side and side 2 to the low-voltage side. Figure 7.36 can be relabeled as Figure 7.37.

The above equivalent circuit model can be described by the following equations

$$\frac{V_{HV}}{N_{HV}} = \frac{V_{LV}}{N_{LV}} \text{ Volt per turn} \tag{7.41}$$

$$N_{HV}I_{HV} = N_{LV}I_{LV} \text{ Ampere-Turn} \tag{7.42}$$

The terms R_{sc} and X_{sc} are referred to as the short circuit resistance and the short circuit reactance of the transformer. These two terms are computed from the short circuit tests on the transformer. The short circuit test is performed by shorting one side of the transformer and applying a voltage in the range of 5–10% of the rated voltage and measuring the short circuit current. The applied voltage is adjusted such that the measured short circuit current is equal to the rated load current. We can think of the short circuit test as a load test because

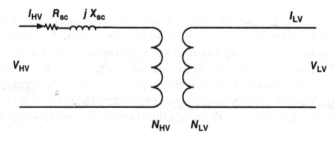

Figure 7.37 The equivalent model of a transformer.

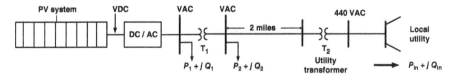

Figure 7.38   One-line diagram of a microgrid radial feeder.

the test reflects the rated load condition.

$$Z_{sc} = R_{sc} + jX_{sc} \qquad (7.43)$$

$$R_{sc} = R_1 + a^2R_2 \qquad (7.44)$$

$$X_{sc} = X_{l1} + a^2X_{l2} \qquad (7.45)$$

where $a = \dfrac{V_{HV}}{V_{LV}} = \dfrac{N_{HV}}{N_{LV}}$

7.11   MODELING A MICROGRID SYSTEM

As part of our objective in this chapter, we will now develop a model representing a microgrid system as depicted in Figure 7.38.

In Figure 7.39, we have presented the PV microgrid system and DC/AC invert by PV power source.

To calculate the voltage at the load using the model shown in Figure 7.39 requires a number of calculations because of the many transformers involved and the need to be aware of which sides of the transformers we are analyzing. First, we need to eliminate the transformers from the above model. To do this, we will normalize the system based on a common base voltage and current. To understand this normalizing method, we present the per unit system.

7.11.1   The Per Unit System

Before we present the concept of the per unit system (p.u.), we introduce the concept of the rated values or nominal values.[18,19] To understand the rated values, let us think of a light bulb. For example, consider a 50 W light bulb. The power and voltage ratings are stamped on the light bulb as shown in Figure 7.40.

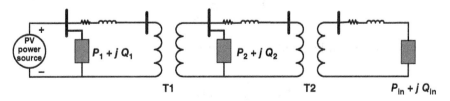

Figure 7.39   The impedance model diagram for a microgrid feeder.

Figure 7.40   The rated values of a light bulb.

The manufacturer of the light bulb is telling us that if we apply 120 V across the light bulb, we will consume 50 W. From the above values, we can calculate the impedance of the light bulb.

$$R_{\text{Light\_bulb}} = \frac{V^2}{P} = \frac{(120)^2}{50} = \frac{14400}{50} = 288.\Omega$$

We also know that if we apply 480 V to the light bulb, which is four times the rated value, then we will have a bright glow and possibly the light bulb will explode. Therefore, the rated values tell us the safe operating condition of an electrical device.

Let us set the following values:

$$P_{\text{base}} = 50 \text{ W} = P_{\text{rated}}$$
$$V_{\text{base}} = 120 \text{ V} = V_{\text{rated}}$$

Now let us use these values and normalize the operating condition of the light bulb:

$$P_{\text{p.u.}} = \frac{P_{\text{actual}}}{P_{\text{base}}} \tag{7.46}$$

$$V_{\text{p.u.}} = \frac{V_{\text{actual}}}{V_{\text{base}}} \tag{7.47}$$

Therefore, for our light bulb we will have

$$P_{\text{p.u.}} = \frac{50}{50} = 1.0 \text{ p.u. W}$$

$$V_{\text{p.u.}} = \frac{120}{120} = 1.0 \text{ p.u. V}$$

Therefore, one per unit represents the full load or rated load. Let us assume that we apply 480 V across the light bulb.

$$V_{\text{p.u.}} = \frac{480}{120} = 4.0 \text{ p.u. V}$$

The per unit (p.u.) voltage applied to the load is 4 p.u. or 4 times the rated voltage.

When we say a device is loaded at half load, the per unit load is 0.5 p.u. If we say the applied voltage is 10% above the rated voltage, we mean 1.10 p.u. volts.

In general the per unit system, the voltages, currents, powers, impedances, and other electrical quantities are expressed on a per unit basis.

$$\text{Quantity per unit} = \frac{\text{Actual value}}{\text{Base value of quantity}} \qquad (7.48)$$

It is customary to select two base quantities to define a given per unit system. We normally select voltage and power as base quantities.

Let us assume

$$V_{\text{b}} = V_{\text{rated}} \qquad (7.49)$$
$$S_{\text{b}} = S_{\text{rated}} \qquad (7.50)$$

Then, base values are computed for currents and impedances:

$$I_{\text{b}} = \frac{S_{\text{b}}}{V_{\text{b}}} \qquad (7.51)$$

$$Z_{\text{b}} = \frac{V_{\text{b}}}{I_{\text{b}}} = \frac{V_{\text{b}}^2}{S_{\text{b}}} \qquad (7.52)$$

And the per unit system values are

$$V_{\text{p.u.}} = \frac{V_{\text{actual}}}{V_{\text{b}}} \qquad (7.53)$$

$$I_{\text{p.u.}} = \frac{I_{\text{actual}}}{I_{\text{b}}} \qquad (7.54)$$

$$S_{\text{p.u.}} = \frac{S_{\text{actual}}}{S_{\text{b}}} \qquad (7.55)$$

$$Z_{\text{p.u.}} = \frac{Z_{\text{actual}}}{Z_{\text{b}}} \qquad (7.56)$$

$$Z\% = Z_{\text{p.u.}} \times 100\% \quad \text{Percent of base } Z \qquad (7.57)$$

Example 7.7   An electrical lamp is rated 120 V, 500 W. Compute the per unit and percent impedance of the lamp. Give the p.u. equivalent circuit.

Solution

We compute lamp resistance from the rate power consumption and rated voltage as

$$P = \frac{V^2}{R} \Rightarrow R = \frac{V^2}{P} = \frac{(120)^2}{500} = 28.8\Omega$$
$$pf = 1.0 \ Z = 28.8\angle 0\Omega$$

Select base quantities as:

$$S_b = 500 \text{ VA}$$
$$V_b = 120 \text{ V}$$

Compute base impedance

$$Z_b = \frac{V_b^2}{S_b} = \frac{(120)^2}{500} = 28.8 \ \Omega$$

The per unit impedance is

$$Z_{p.u.} = \frac{Z}{Z_b} = \frac{28.8\angle 0}{28.8} = 1\angle 0 \text{ p.u.}$$

Percent impedance:

$$Z\% = 100\%$$

Per unit equivalent circuit is given in Figure 7.41.

Example 7.8   An electrical lamp is rated 120 V, 500 W. If the voltage applied across the lamp is twice the rated value, compute the current that flows through the lamp. Use the per unit method.

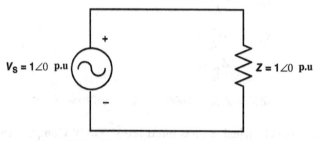

Figure 7.41   The equivalent circuit of Example 7.7.

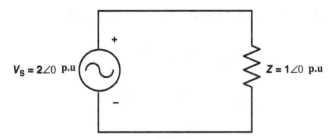

Figure 7.42 The per unit equivalent circuit of Example 7.8.

Solution

$$V_b = 120 \text{ V}$$

$$V_{p.u.} = \frac{V}{V_b} = \frac{240\angle0}{120} = 2\angle0 \text{ p.u.}$$

$$Z_{p.u.} = 1\angle0 \text{ p.u.}$$

The per unit equivalent circuit (see Figure 7.42) is as follows:

$$I_{p.u.} = \frac{V_{p.u.}}{Z_{p.u.}} = \frac{2\angle0}{1\angle0} = 2\angle0 \text{ p.u. A}$$

$$I_b = \frac{S_b}{V_b} = \frac{500}{120} = 4.167 \text{ A}$$

$$I_{actual} = I_{p.u.}I_b = 2\angle0 \times 4.167 = 8.334\angle0 \text{ A.}$$

The per unit system for a one-phase circuit is as follows:

$$S_b = S_{1-\phi} = V\phi I\phi \tag{7.58}$$

$$\searrow \text{ ONE PHASE}$$

where

$$V\phi = V_{\text{line-to-neutral}} \tag{7.59}$$
$$I\phi = I_{\text{line-current}} \tag{7.60}$$

For transformers, we select the bases as

MYSTERY SOLVED BAD EDITING?

$$V_{bLV} = V_{\phi LV} | V_{bHV} = V_{\phi HV} \tag{7.61}$$

$$I_{bLV} = \frac{S_b}{V_{bLV}} | I_{bHV} = \frac{S_b}{V_{bHV}} \tag{7.62}$$

The base impedances on the two sides of transformer are

$$Z_{bLV} = \frac{V_{bLV}}{I_{bLV}} = \frac{(V_{bLV})^2}{S_b} \tag{7.63}$$

$$Z_{bHV} = \frac{V_{bHV}}{I_{bHV}} = \frac{(V_{bHV})^2}{S_b} \tag{7.64}$$

$$S_{p.u.} = \frac{S}{S_b} = V_{p.u.} I_{p.u.}{}^* \tag{7.65}$$

$$P_{p.u.} = \frac{P}{S_b} = V_{p.u.} I_{p.u.} \cos\theta \tag{7.66}$$

$$Q_{p.u.} = \frac{Q}{S_b} = V_{p.u.} I_{p.u.} \sin\theta \tag{7.67}$$

When we have two or more transformers, we need to normalize the impedance model to a common base. We will make the selections as follows:
Selection 1:

$$S_{b1} = S_A \quad V_{b1} = V_A$$

then,

$$Z_{b1} = \frac{V_{b1}^2}{S_{b1}} \quad Z_{p.u.1} = \frac{Z_L}{Z_{b1}}$$

Selection 2:

$$S_{b2} = S_B \quad V_{b2} = V_B$$

then,

$$Z_{b2} = \frac{V_{b2}^2}{S_{b2}} \quad \text{and} \quad Z_{p.u.2} = \frac{Z_L}{Z_{b2}}$$

$$\frac{Z_{p.u.2}}{Z_{p.u.1}} = \frac{Z_L}{Z_{b2}} \times \frac{Z_{b1}}{Z_L} = \frac{Z_{b1}}{Z_{b2}} = \frac{V_{b1}^2}{S_{b1}} \times \frac{S_{b2}}{V_{b2}^2}$$

$$Z_{p.u.2} = Z_{p.u.1} \left(\frac{V_{b1}}{V_{b2}}\right)^2 \times \left(\frac{S_{b2}}{S_{b1}}\right) \tag{7.68}$$

In general, the values that are given as nominal values (rated values) are referred to as "old" values and new selection as common base as "new."

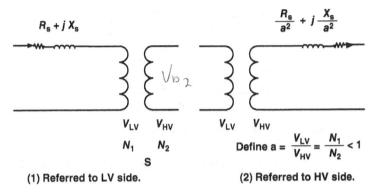

Figure 7.43   The equivalent circuit of a one-phase transformer.

Using this notation, the transformation between old and new bases can be written as:

$$Z_{\text{p.u.,new}} = Z_{\text{p.u.,old}} \left(\frac{V_{\text{b,old}}}{V_{\text{b,new}}} \right)^2 \times \left(\frac{S_{\text{b,new}}}{S_{\text{b,old}}} \right) \tag{7.69}$$

We can use the p.u. concept and define the p.u. model of a transformer. Consider the equivalent circuit of transformer model referred to LV side and HV side as shown by Figure 7.43.

We make the following selection for the transformer bases.

$$V_{\text{b1}} = V_{\text{LV,rated}} \tag{7.70}$$
$$S_{\text{b}} = S_{\text{rated}} \tag{7.71}$$

Then, we can compute the base voltage for the new common base.

$$V_{\text{b2}} = \frac{V_{\text{HV}}}{V_{\text{LV}}} V_{\text{b1}} = \frac{1}{a} V_{\text{b1}}$$

$$Z_{\text{b1}} = \frac{V_{\text{b1}}^2}{S_{\text{b}}} \quad Z_{\text{b2}} = \frac{V_{\text{b2}}^2}{S_{\text{b}}}$$

$$\frac{Z_{\text{b1}}}{Z_{\text{b2}}} = \frac{V_{\text{b1}}^2}{V_{\text{b2}}^2} = \frac{V_{\text{b1}}^2}{\left(\frac{1}{a}V_{\text{b1}}\right)^2} = a^2$$

Per unit impedances are

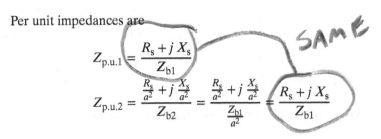

$$Z_{\text{p.u.1}} = \frac{R_{\text{s}} + j X_{\text{s}}}{Z_{\text{b1}}}$$

$$Z_{\text{p.u.2}} = \frac{\frac{R_{\text{s}}}{a^2} + j \frac{X_{\text{s}}}{a^2}}{Z_{\text{b2}}} = \frac{\frac{R_{\text{s}}}{a^2} + j \frac{X_{\text{s}}}{a^2}}{\frac{Z_{\text{b1}}}{a^2}} = \frac{R_{\text{s}} + j X_{\text{s}}}{Z_{\text{b1}}}$$

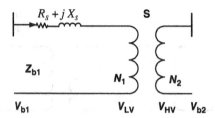

Figure 7.44 The equivalent circuit of a transformer.

From the above, we can see that the p.u. impedance of the transformer is identical.

$$Z_{\text{p.u.1}} = Z_{s2}$$

The per phase equivalent of a transformer is given by Figure 7.44, where R_s and X_s represents the parameters of a transformer referred to on the low-voltage side. $a = \dfrac{V_{\text{LV}}}{V_{\text{HV}}} = \dfrac{N_1}{N_2} < 1$

The per unit equivalent transformer is depicted in Figure 7.45.

Example 7.9 A single-phase transformer rated 200 kVA, 200/400 V, and 10% short-circuit reactance. Compute the voltage regulation when the transformer is fully loaded at unity pf and rated voltage 400 V.

Solution
Let

$$V_{b2} = 400 \text{ V}$$
$$S_b = 200 \text{ kVA}$$

Then, we will have

$$S_{\text{load,p.u.}} = 1\angle 0 \text{p.u. k VA}$$
$$X_{\text{s,p.u.}} = j0.1 \text{p.u.}$$

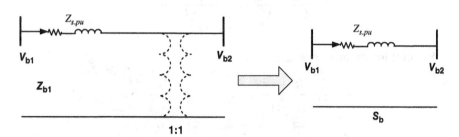

Figure 7.45 The per unit equivalent circuit of Figure 7.44.

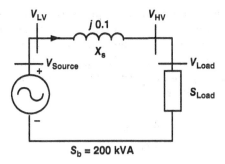

Figure 7.46   The per unit equivalent circuit of Example 7.9.

The per-unit equivalent model is given in Figure 7.46.
Rated voltage:

$$V_{\text{load,p.u.}} = 1\angle 0 \text{p.u. V}$$

$$I_{\text{load,p.u.}} = \left(\frac{S_{\text{load,p.u.}}}{V_{\text{load,p.u.}}}\right)^* = \left(\frac{1\angle 0}{1\angle 0}\right)^* = 1\angle 0 \text{ p.u. A}$$

where ∗ means conjugate.

$$\begin{aligned} V_{\text{source,p.u.}} &= V_{\text{load,p.u.}} + I_{\text{p.u.}} X_{\text{s,p.u.}} \\ &= 1.0\angle 0 + 1.0\angle 0 \times j0.1 = 1 + j0.1 = 1.001\angle 5.7 \text{ p.u. V} \end{aligned}$$

The p.u. load voltage under load is given as

$$V_{\text{p.u.,full-load}} = V_{\text{load,p.u.}} = 1\angle 0 \text{ p.u. V}$$

When we remove the load (no load) and set the source voltage at $1.001\angle 5.7$ p.u., the no-load voltage is given as:

$$V_{\text{p.u.,no-load}} = V_{P,\text{p.u.}} = 1.001\angle 5.7 \text{p.u. V}$$

Therefore, the voltage regulation (VR) by definition is given as

$$VR = \frac{|V_{\text{p.u.,no-load}}| - |V_{\text{p.u.,full-load}}|}{|V_{\text{p.u.,full-load}}|} \times 100\%$$

And we calculate:

$$VR = \frac{1.001 - 1.0}{1.0} \times 100\% = 0.1\%$$

In the following, we extend the concepts of per unit system to the three-phase systems. The extension is very simple. We will look at the three-phase system as three single-phase circuits that are connected as a three-phase system.

For three-phase systems, we select the S_b and V_b the same way as before, except the base values represent the three-phase power and the voltage base is equal to the line-to-line voltage.

$$S_b = S_{\text{three-phase}} = 3S_{1\text{phase}} \qquad (7.72)$$
$$S_b = 3V_\varphi I_\varphi \qquad (7.73)$$

where the phase voltage is given by:

$$V_\varphi = V_{\text{line-to-neutral}} = \frac{V_{L(\text{line})}}{\sqrt{3}} \qquad (7.74)$$

Therefore, we will have

$$V_\phi = \frac{V_L}{\sqrt{3}} \text{ and } I_\phi = I_{\text{line.current}} = I_L \qquad (7.75)$$

From Equations (7.73) and (7.75), we will have

$$S_b = 3\frac{V_L}{\sqrt{3}}I_L = \sqrt{3}V_L I_L \qquad (7.76)$$

For three-phase transformers, we select the voltage bases for high voltage (HV) and low voltage (LV) as:

$$V_{bLV} = V_{L(LV)} \quad V_{bHV} = V_{L(HV)} \qquad (7.77)$$

Therefore, the base power for the three-phase system can be expressed as:

$$S_b = \sqrt{3}V_{b(LV)}I_{b(LV)} = \sqrt{3}V_{b(HV)}I_{b(HV)} \qquad (7.78)$$

The current base for the low voltage (LV) and the current base for the high volt (HV) are given as:

$$I_{bLV} = \frac{S_b}{\sqrt{3}V_{b(LV)}} \quad I_{bHV} = \frac{S_b}{\sqrt{3}V_{b(HV)}} \qquad (7.79)$$

As before, we define the Z_b per phase for the three-phase system. Therefore, we will have

$$Z_{bLV} = \frac{V_{\varphi LV}}{I_{\varphi LV}} \qquad (7.80)$$

and

$$V_{\phi LV} = \frac{V_{bLV}}{\sqrt{3}} \quad I_{bLV} = \frac{S_b}{\sqrt{3}V_{b(LV)}}$$

From the above, we will have

$$Z_{bLV} = \frac{V_{bLV}}{\sqrt{3}} \times \frac{\sqrt{3}V_{bLV}}{S_b}$$

and finally, impedance bases as given by Equations (7.81) and (7.82)

$$Z_{bLV} = \frac{(V_{bLV})^2}{S_b} \tag{7.81}$$

$$Z_{bHV} = \frac{(V_{bHV})^2}{S_b} \tag{7.82}$$

The per unit power is defined as the ~~power of the three-phase system~~ divided by S_b as given by Equation (7.83)

$$S_{p.u.} = \frac{S_{3\varphi}}{S_b} = \frac{\sqrt{3}V_L I_L^*}{\sqrt{3}V_b I_b} = V_{p.u.} I_{p.u.}^* \tag{7.83}$$

Again, we restate the base impedance as

$$Z_b = \frac{V_{b\varphi}}{I_{b\varphi}} = \frac{V_{b\varphi}^2}{S_{b\varphi}} = \frac{3V_{b\varphi}^2}{3S_{b\varphi}} = \frac{3V_{b\varphi}^2}{S_{b3\varphi}}$$

By substituting for the phase voltage in terms of line-to-line voltage, we will have impedance base in terms of line-to-line voltages as given by Equation (7.84).

$$Z_b = \frac{\left(\frac{V_{bL-L}}{\sqrt{3}}\right)^2}{S_{b\varphi}} = \frac{V_{bL-L}^2}{3S_{b\varphi}} = \frac{V_{bL-L}^2}{S_{b3\varphi}} \tag{7.84}$$

7.12 MODELING THREE-PHASE TRANSFORMERS

The impedance diagram of a three-phase transformer is the same as three single-phase transformers. Let us first discuss how the three-phase transformers are constructed.[9,17–19,22,23]

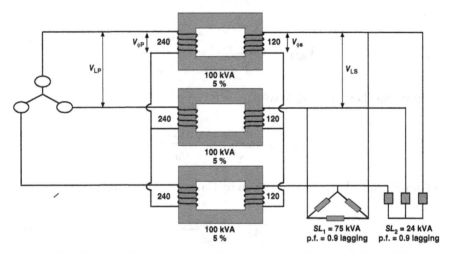

Figure 7.47 The three single-phase transformers connected as three-phase Y–Y transformer bank.

The three-phase transformers can be constructed from the three single-phase transformers.

Figure 7.47 depicts the description of a three-phase transformer that consists of three single-phase transformers.

In Figure 7.47, each single-phase transformer is rated at 240 V/120 V, 100 kVA with short-circuit reactance of 5% based on the rated transformer voltages and power rating. The three single-phase transformers are connected as Y–Y and makes a bank of three-phase transformers with rated line-to-line voltages of $\sqrt{3} \times 240$ V and $\sqrt{3} \times 120$ V, 300 kVA and the short-circuit reactance of 5%. The one-line diagram of Figure 7.47 is depicted in Figure 7.48 where the line-to-line voltages are calculated as

$$V_{LP} = 240 \times \sqrt{3} = 415.2V \quad \text{PRIMARY}$$
$$V_{LS} = 120 \times \sqrt{3} = 207.6V \quad \text{SECONDARY.}$$

Figure 7.49 depicts the one-phase equivalent circuit for Figure 7.48. In this figure, the short-circuit reactance is given in actual value in ohms (Ω) on the high-voltage side.

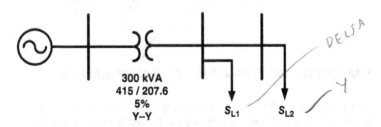

Figure 7.48 One-line diagram of Figure 7.47.

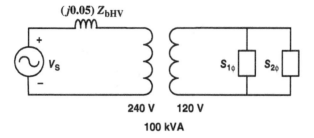

Figure 7.49   One-phase equivalent circuit model of Figure 7.48.

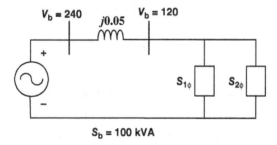

Figure 7.50   The per unit equivalent of Figure 7.49.

Figure 7.50 depicts the equivalent p.u. model of the Figure 7.49. In this figure, the power and voltage bases are given as shown.

Another important type of three-phase transformers is a Y–Δ connection with grounded neutral. Figure 7.51 depicts a Y–Δ connection. The figure shows the coupling windings. For example, the winding depicted with N_{p1} is coupled with the winding N_{s1} and similarly N_{p2} with N_{s2} and N_{p3} with N_{s3}.

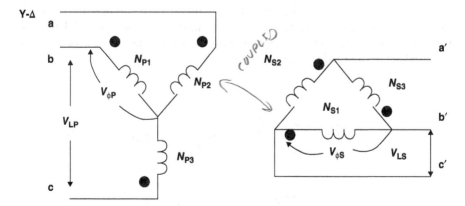

Figure 7.51   A three-phase Y–Δ connection.

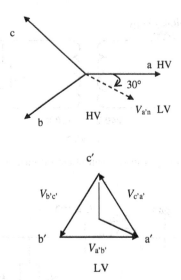

Figure 7.52 A 30° phase shift in Y–Δ transformers.

Figure 7.52 depicts the phase shift in Y–Δ transformers. Because of the phase shift in Y–Δ transformers, they cannot be connected in parallel unless they have proper phase sequence. Therefore, before connecting two Y–Δ and Y–Δ transformers, students are urged to study the IEEE standard on transformer paralleling or follow industry practice.[22] The dot dark circle identifies the relative positive terminal of each winding. Students are urged to review the dot notation from their circuit course.

7.13 TAP CHANGING TRANSFORMERS

Figure 7.53 presents a three-phase tap changing transformer. Load tap changing (LTC) transformers are of two types. If the taps are changed under loads, the LTC transformer is referred to as tap changing under load (TCUL) or on-load tap changing (OLTC). In off-load tap changing transformers the taps are changed when the loads are not connected.

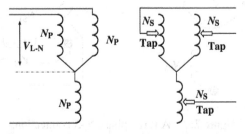

Figure 7.53 A schematic of a three-phase Y–Y connected tap changing transformer.

A tap changing transformer is constructed such that after a few turns a connection point is brought out. The taps can be on one of the windings or both. ~~These types of transformer facilitate voltage regulation.~~

Voltage stability of a power system is a function of load types. ~~An inductive load draws a lagging current that causes the load voltage to drop. A capacitive load draws a leading current and causes the load voltage to increase.~~ Voltage control elements such as capacitors, reactors, and under load tap changing (ULTC) transformers are used to regulate the power grid bus voltages. If a tap changer is used, tap points are usually made on the high voltage where the current is low because it minimizes the current handling requirements of the contacts. However, a transformer may include a tap changer on both windings.

In distribution networks, step-down transformers have an off-load tap changer on the primary winding and an on-load tap changer on the secondary winding. Normally, the high-voltage tap is controlled and adjusted to match the subtransmission voltage.

We need to define a few terms for LTC transformers. A nominal turns ratio is defined as the condition when all turns of both the primary and secondary windings are carrying currents.   *LOAD TAP CHANGING.*

$$\frac{V_{HV}}{N_{HV}} = \frac{V_{LV}}{N_{LV}} \text{ or } \frac{V_{HV}}{V_{LV}} = \frac{N_{HV}}{N_{LV}} = a$$

The off-nominal turns ratio refers to when parts of the turns of one or both windings are not carrying currents.

When nominal turns ratio is used, the base voltages are selected as:

$$V_{bHV} = \frac{V_{HV}}{V_{LV}} \times V_{bLV}$$

$$V_{bLV} = \frac{V_{LV}}{V_{HV}} \times V_{bHV}$$

Example 7.10   A single-phase tap changing transformer has 2000 turns on the primary side, and a variable number of turns on the secondary side ($N_{sec\ max} = 7300$, $N_{sec\ min} = 5300$) as shown in Figure 7.54. Compute the minimum and the maximum voltage that can be maintained on the secondary voltage. Assume the maximum primary voltage is equal to 36.4 kV.

Solution

$$\frac{V_p}{2000} = \frac{36.4 \times 10^3}{2000} = 18.2\ V/T$$

$$\frac{V_p}{2000} = \frac{V_{sec}}{N_{sec\ max}} = 18.2 = \frac{V_{sec}}{7300}, \ V_{sec} = 132.86\ \text{kV}$$

$$\frac{V_p}{2000} = \frac{V_{sec}}{N_{sec\ min}} = 18.2 = \frac{V_{sec}}{5300}, \ V_{sec} = 96.46\ \text{kV}$$

Thus, secondary voltage can be changed between 96.46 and 132.86 kV.

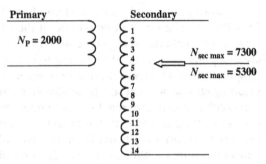

Figure 7.54 A schematic of a one winding of a three-phase Y–Y connected transformer.

7.14 MODELING TRANSMISSION LINES

As we know, electric power is mostly transmitted through three-phase systems.[9,17–19,22,23] The conductor used in each phase has resistance, inductance, and capacitance. In the modeling of a transmission system, we need to develop a model for the distributed resistance, inductance, and capacitance expressed in miles or kilometers (km) as a lumped model that can be used in the analysis of power grids. A transmission line is designed to carry current and a given designed voltage. The voltage of the transmission line is based on the amount of power, normally in MVA, that lines must carry. The current rating is based on the size of the conductor and its thermal rating. Normally, a conductor is sized based on its circular mil area (CM), which is a unit denoting the cross-sectional size of a conductor. The circular mil, CM is defined as

$$CM = d^2 \tag{7.85}$$

where d is the diameter of the conductor. Therefore, 1 mil is equal 0.001 inch and reversely, 1 inch = 1000 mil and the area A of the conductor is given as

$$A = d^2 \text{cmil}$$

The CM is also used as a designation for different classes of wires and cables by manufacturers.

The inductance and capacitance of transmission lines are the results of magnetic and electric fields generated from the current and voltage of the conductors. The shunt conductance ($G = 1/R$) is generally very small because the leakage current is very small due to leakage flux that may couple with the ground conductor or another metallic element around conductors. This conductance is normally ignored. The parameters R, L, and C are needed to construct the transmission line models. The inductance of the line is computed from the flux lines due the current flowing through the conductor. The capacitance of the line is computed from the voltage distribution along conductors.

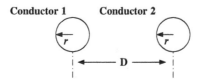

Figure 7.55 Two-wire transmission line.

The inductance of a solid cylindrical conduct has both internal and external inductance.[18,19]

Figure 7.55 depicts a two-wire transmission line. The total inductance is given as

$$L = 410^{-7} \ln \frac{D}{r'} \ H/m \qquad (7.86)$$

where $r' = re^{-1/4} = 0.7788r$

The term r' is defined as the geometric mean radius (GMR) of a conductor.

As we know, when two charged conductors are at different voltages they act as a charged capacitor. The capacitance can be calculated from the charge on conductors per unit potential difference between two conductors and earth. The neutral conductor is grounded to the earth. However, normally, by grounding the neutral conductor to ground, the earth conductor does not carry current unless a fault occurs, such as when a high-voltage conductor is accidently grounded. From a basic electromagnetic calculation,[17] we can compute the capacitance between two conductors as shown in Figure 7.56 below.

$$C_{12} = \frac{0.0388}{2.\ln \frac{D}{r}} \mu F/mile \qquad (7.87)$$

where D is the distance between the two conductors, $\mu F/mile$ is F times 10^{-6}, and r is the radius of the conductor.

$$C_{1n} = \frac{0.0388}{\ln \frac{D}{r}} \mu F/mile \qquad (7.88)$$

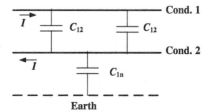

Figure 7.56 Capacitance between two conductors and ground.

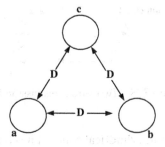

Figure 7.57 The inductance of a three-phase transmission line with equilateral spacing.

where C_{12} is the capacitance between the two conductors and C_{1n} is the capacitance of one conductor to neutral. In transmission line modeling, we are interested in equivalent capacitive reactance, X_c. The capacitive reactance in ohms per mile (Ω/mi) to neutral conductor is given as

$$X_C = \frac{1}{2\pi.f.C} = \frac{4.10}{f} \times 10^6 . \log\left(\frac{D}{r}\right) \ \Omega/\text{mi to neutral} \qquad (7.89)$$

where f is the frequency and C is capacitance in F.

Figure 7.57 depicts the inductance of a three-phase line with equilateral spacing.

Figure 7.57 presents a three-phase transmission line. The inductance of the three-phase transmission line is given as

$$L_a = 2 \times 10^{-7} . \ln\left(\frac{D}{r'}\right) \ \text{H/m} \qquad (7.90)$$

$$L_a = 0.7411 . \log\left(\frac{D}{r'}\right) \ m\text{H/mile} \qquad (7.91)$$

due to the symmetry of spacing between the conductors, $L_a = L_b = L_c$.

A three-phase transmission line with bundled conductor lines is used to carry high MVA power using high-voltage lines as depicted in Figure 7.58.

Inductances:

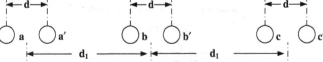

Figure 7.58 Three-phase transmission line with bundled conductor lines.

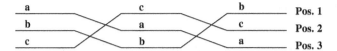

Figure 7.59 The transposed three-phase transmission line.

For bundled conductors, we use equivalent spacing, D_{eq}. Therefore, the inductance L, is given as

$$L = 2 \times 10^{-7} \times \ln \frac{D_{eq}}{D_s^6} \text{H/m} \qquad (7.92)$$

$$\text{where } D_{eq} = \sqrt[3]{d_1 \times d_1 \times 2d_1} = d_1\sqrt[3]{2} \qquad (7.93)$$

$$\text{and } D_s^6 = \sqrt[4]{D_{aa}.D_{aa'}.D_{a'a}.D_{a'a'}} \qquad (7.94)$$

$$D_s = D_{aa} = 0.7788r \qquad (7.95)$$

The term D_s defines the GMR of a bundle of conductors for a phase of a three-phase line.

Normally, in the bundled conductors, the spacing between the conductor of a phase is designated as $D_{a'a'} = D_s$. The value of D_s is given in a tabular form by many manufacturers.

$$D_s^6 = \sqrt[4]{D_s.D_s.d.d} = \sqrt{D_s.d} \qquad (7.96)$$

In the above, D_s defines the GMR of a bundle of conductors in a group of bundles in the three-phase line.

The capacitance of the three-phase transmission lines can be calculated in the same way:

$$C \text{ positive sequence} = C_{12} = \frac{2\pi\varepsilon}{\ln \dfrac{D_{eq}}{D_{sc}^6}} \qquad (7.97)$$

$$D_{eq} = \sqrt[3]{d_1 \times d_1 \times 2d_1} = d_1\sqrt[3]{2} \qquad (7.98)$$

$$D_{sc}^6 = \sqrt{rd} \qquad (7.99)$$

Due to the cost of having equal spacing of the conductors between transmission towers, the transmission lines are constructed with unsymmetrical spacing between the conductors. However, to create approximate spacing between phases, the phases are transposed. This means the position of phases are changed between towers or every few miles in a few towers. Figure 7.59 depicts a transposed three-phase transmission line.

Because the spacing between the phases of the transposed lines is approximately the same, we consider such a three-phase line to have equal resistance, inductance, and capacitance.

$$L_a = L_b = L_c = 2 \times 10^{-7} \times \log \frac{D_{eq}}{r'} \text{mH/mile} \qquad (7.100)$$

$$L = 2 \times 10^{-7} \times \ln \frac{D_{eq}}{D_s^6} \text{ H/m} \qquad (7.101)$$

$$D_{eq} = \sqrt[3]{D_{12} \times D_{23} \times D_{31}} \qquad (7.102)$$

$$X_L = 2\pi.f.L \qquad (7.103)$$

The capacitance to neutral for a three-phase line is

$$C_n = \frac{2\pi\varepsilon.\varepsilon_r}{\ln\left(\frac{D_{eq}}{D_s}\right)} \text{ F/m to neutral} \qquad (7.104)$$

where ε is the permittivity of free space, $\varepsilon = 8.85 \times 10^{-12}$, and ε_r is the relative permittivity of air which is equal to one.

The X_c at 60 hertz (Hz) is given as

$$X_c = 29.7 \times \ln\left(\frac{D_{eq}}{D_s}\right) \text{ k}\Omega\text{/mi to neutral} \qquad (7.105)$$

In the calculation of capacitance, D_s is the outside radius of the conductor.

Historically, the transmission line parameters are computed by conductor manufacturers and listed in tabular form for designers to use. However, now software packages are also available to calculate the transmission line parameters for different spacing and design of the lines. Here, we present how we can use a table to select a line parameter reactance and line-charging capacitive reactance for design problems in the book.

Generally, most manufacturers use values of GMR for calculating the line parameters in feet and miles or meters and kilometers. The inductive reactance can be calculated as

$$X_L = 2.02f \times 10^{-3} \ln\left(\frac{D_{eq}}{D_s}\right) \Omega\text{/mi} \qquad (7.106)$$

where D_{eq} is the distance between conductors. The GMR of tables are the same as D_s.

Some conductor manufacturers list the line data in terms of standard spacing of one and spacing of D as given below.

$$X_L = 2.02f \times 10^{-3} \ln\left(\frac{1}{GMR}\right) + 2.02f \times 10^{-3} \ln D \,\Omega\text{/mi} \qquad (7.107)$$

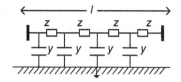

Figure 7.60   A distributed model of transmission lines.

Naturally, both GMR and D must be in feet and f in cycles per second in this formulation. However, because 1 mile is equal to 1.6 km, we can compute the inductive reactance; it can easily be converted to metric system. In the above equation, the first term is the inductive reactance for 1 foot and the second term is the inductive reactance for the spacing factor.

As can be expected, the inductance and capacitance is distributed along the line. Therefore, we can assume a distributed model as shown in Figure 7.60.

In a distributed model of transmission lines, we can define the following:

$z = R + j\omega L$ Ω/m seriesimpedanceunitlength

$y = G + j\omega C$ shunt admittance per unit length

$Z = zl$ Ω l = line length in meters

$$Y = yl\Omega^{-1}$$

Figure 7.61 depicts the lumped model of a transmission line. The parameters of this model can be computed from transmission line data provided by conductor manufacturers.

It should be noted that our discussion here is limited to overhead lines. We have presented very basic concepts related to the selection of overhead conductors and transmission line data for power grid analysis and design. Although this presentation is very brief, students can learn to design a microgrid by recognizing that the design of a microgrid is based on the selection of grid elements, and then analyse to obtain the desired performance. We will visit this concept in design throughout the book.

Finally, we must note that if we select insulated cables, the selection is more complex. In insulated cables, the heat generated within insulation is a major issue: the cable must be selected based not only on the load current it must carry, it also must be able to operate safely when there is a high transient or

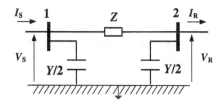

Figure 7.61   A general lumped model of a transmission line.

fault current due to accidental shorts. This topic will not be addressed further, but students may want to consult other sources.[19]

It is important to recognize that the current-carrying capacity is a thermal rating for the conductors. The power that can be transferred is based on the voltage rating of the lines. The voltage rating of the line is determined based on the insulation's requirement and how the spacing between the conductors is designed. Therefore, we need to select the voltage and then compute the power-carrying capacity of the line.

The current flow in the line is given by

$$I_L = \frac{S}{\sqrt{3} \times V_{L\text{-}L} \times pf}$$

where S is the power flow, $V_{L\text{-}L}$ is the line-to-line voltage and pf is the power factor.

The conductor selection is first based on the carrying capacity of the selected line; then we need to address the voltage drop and losses of the line. Therefore, depending on the design requirement, we may select a conductor that may have higher current-carrying capacity.

Example 7.11 A three-phase generator is supplying 1 MW load operating at 0.9 lagging power factor to a power grid. The generating station is 15 miles from the local power grid. Use the overhead line data for an aluminum steel conductor reinforced (ASCR) and perform the following:

(i) If the power is transferred at 460 V, select an overhead ASCR conductor to transfer the power. What is the generator voltage if the load voltage is specified at 460 V AC?

(ii) If the power is transferred at 3.3 kV, select an overhead ASCR conductor to transfer the power. What is the generator voltage if the load voltage is specified at 3.3 kV AC?

(iii) If the power is transferred at 11.3 kV, select an overhead ASCR conductor to transfer the power. What is the generator voltage if the load voltage is specified at 11.3 kV AC?

(iv) Compute active and reactive power losses for each design, put your results in a table, and discuss the results.

Solution
The single-phase equivalent diagram of the system is shown in Figure 7.62.
The load current is given by

$$I_{\text{Load}} = \frac{S}{\sqrt{3} \times V_{L\text{-}L} \times pf}$$

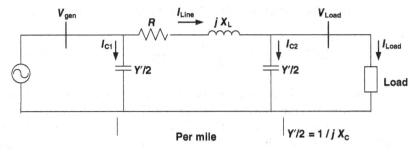

Figure 7.62 Single-phase equivalent diagram of the system in Example 7.11.

where S is the power flow, $V_{\text{L-L}}$ is the line-to-line voltage of the load and pf is the power factor of the load

(i) The load current at a load voltage of 460 is

$$I_{\text{Load}} = \frac{1 \times 10^6}{\sqrt{3} \times 460 \times 0.9} = 1394 \text{ A}$$

Because the conductor is carrying 1.39 kA we select the conductor with CM of 1590 kCmil. This conductor has a resistance of is 0.0591 Ω/mi at 25°C.

(ii) The reactance is 0.359 Ω/mi. The capacitive reactance is 0.0814 MΩ/mi. The half of the above value is used in the equivalent pie model of the transmission lines and is represented with their admittance as $Y'/2 = 1/jX_C$ in Figure 7.62. Therefore,

$$X_C = \frac{0.0814}{2} = 0.0407 \text{ MΩ/mi}$$

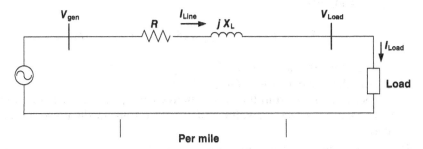

Figure 7.63 Simplified diagram of Figure 7.62 by ignoring the line-charging capacitances.

The line-charging current as shown in Figure 7.62 is given by

$$I_{C1} = \frac{V_{Load}}{jX_C} \times distance = \frac{460/\sqrt{3}}{j0.0407 \times 10^6} \times 15 = 97.81.82\angle 90° \, mA$$

It is seen that the line-charging current is in the milli-amp range and is negligible compared to the load current. Therefore, the single-phase equivalent circuit of Figure 7.62 reduces to a circuit as shown in Figure 7.63. We neglect the line-charging reactance as they are very large compared to the load impedance. It can be seen from Figure 7.63 that the line current I_{line} is approximately equal to the load current I_{Load}:

$$I_{Line} = I_{Load}$$

In the single-phase equivalent circuit, all the quantities are represented per phase. Wye-connected generator and load is assumed and their phase voltage is found from:

$$V_{ph} = V_{L-L}/\sqrt{3}$$

where V_{ph} is the phase voltage (line-to-neutral) and V_{L-L} is the line-to-line voltage.

From Kirchhoff's voltage law and Figure 7.63, the generator phase voltage is given as

$$
\begin{aligned}
V_{gen} &= V_{Load} + I_{Line}(R + jX_L) \\
&= \frac{460}{\sqrt{3}} + 1394\angle - \cos^{-1} 0.9 \times 15 \times (.0591 + j0.359) \\
&= 7767\angle 53.21° \, V
\end{aligned}
$$

The generator line voltage is $\sqrt{3}V_{gen} = \sqrt{3} \times 7763 = 13.4 \, kV$

The power loss is given by $S_{loss} = 3I_L^2.Z = 3 \times 1394^2 \times 15 \times (.0591 + j0.359)$

$$S_{loss} = P_{loss} + j \, Q_{loss} = (5.16 + j31.41) \times 10^6$$

Therefore, the active power loss, $P_{loss} = 5.16 \, MW$
And the reactive power, $Q_{loss} = 31.41 \, MVAr$.

As can be seen from the above analysis, the power losses are quite high. Next, we select line voltage of 3.3 kV to transfer the power to the load.

(iii) The load current at a load voltage of 3.3 kV $I_{Load} = \frac{1 \times 10^6}{\sqrt{3} \times 3.3 \times 10^3 \times 0.9} = 194 \, A$

Because the conductor is carrying 194 A, we select the conductor with CM of 266.8 kCmil. This conductor has a resistance of is 0.350 Ω/mi at 25°C. The reactance is 0.465 Ω/mi. The capacitive reactance is 0.1074 MΩ/mi.

Half of the above value is used in the equivalent pie model of the transmission lines and is represented with their admittance as $Y'/2 = 1/jX_C$ in Figure 7.62. Therefore,

$$X_C = \frac{0.1074}{2} = 0.0537 \text{ M}\Omega/\text{mi}$$

The line-charging current as shown in Figure 7.62 is given by

$$I_{C1} = \frac{V_{\text{Load}}}{jX_C} \times \text{distance}$$

$$= \frac{3.3 \times 10^3 \times /\sqrt{3}}{j0.0537 \times 10^6} \times 15 = 0.531.82\angle 90° \text{ A}$$

We can safely neglect the line-charging current because it is very small as compared to the load. Applying Kirchhoff's voltage law to the circuit in Figure 7.62 the generator phase voltage is given as:

$$V_{\text{gen}} = V_{\text{Load}} + I_{\text{Line}}(R + j X_L)$$

$$= \frac{3300}{\sqrt{3}} + 194\angle - \cos^{-1} 0.9 \times 15 \times (0.350 + j0.465)$$

$$= 3.5\angle 53.21° \text{kV}$$

The generator line voltage is

$$V_{\text{L-L}} = \sqrt{3}V_{\text{gen}} = \sqrt{3} \times 3.5 = 6.06 \text{ kV}$$

The power loss is given as

$$S_{\text{loss}} = 3I_L^2.Z = 3 \times 194^2 \times 15 \times (0.350 + j\,0.465)$$

$$= (0.57 + j0.78) \times 10^6$$

$$S_{\text{loss}} = P_{\text{loss}} + j\,Q_{\text{loss}}$$

Therefore, the active power loss $P_{\text{loss}} = 0.57$ MW and the reactive power loss $Q_{\text{loss}} = 0.78$ MVar.

As we can see by increasing the transmission line, we have reduced the power losses. Next, we evaluate the power losses if the line voltage is selected at 11.3 kV.

(iv) The load current at a load voltage of 11.3 kV,

$$I_{Load} = \frac{1 \times 10^6}{\sqrt{3} \times 11.3 \times 10^3 \times 0.9} = 57 A$$

The same conductor is selected as in part (ii). The resistance of the line for 57 A is 0.350 Ω/mi. The series reactance of the line is 0.465 Ω/mi. The capacitive reactance is 0.1074 MΩ/mi.

Therefore,

$$X_C = \frac{0.1074}{2} = 0.0537 \ M\Omega/mi$$

The line-charging current as shown in Figure 7.62 is given by $I_{C1} = \frac{V_{Load}}{jX_C} \times$ distance $= \frac{11.3 \times 10^3/\sqrt{3} \times 15}{0.0537 \times 10^6} = 1.82\angle 90° A$

In this case, the line-charging current is of comparable magnitude to that of the load current. Hence, we do not neglect the line-charging reactance. Applying Kirchhoff's current law in Figure 7.62, the line current is given by:

$$
\begin{aligned}
I_{Line} &= I_{Load} + I_{C1} \\
&= 57\angle - \cos^{-1} 0.9 + 1.82\angle 90° = 56.2\angle - 24.2° \ A
\end{aligned}
$$

The generator phase voltage is given as

$$
\begin{aligned}
V_{gen} &= V_{Load} + I_{Line} (R + j X_L) \\
&= \frac{11300}{\sqrt{3}} + 56.2\angle - 24.2° \times 15 \times (0.350 + j0.465) \\
&= 6.96\angle 1.95° kV
\end{aligned}
$$

The generator line voltage is given as

$$V_{L-L} = \sqrt{3}V_{gen} = \sqrt{3} \times 6.96 = 12.1 \ kV$$

The power loss is given by

$$
\begin{aligned}
S_{loss} &= 3I_L^2.Z = 3 \times 56.2^2 \times 15 \times (0.350 + j\,0.465) \\
S_{loss} &= P_{loss} + jQ_{loss} = (49.8 + j66) \times 10^3
\end{aligned}
$$

TABLE 7.1   Active and Reactive Power Losses at Different Line Voltage Levels

Voltage (kV)	Active Power Loss (MW)	Reactive Power Loss (MVA$_r$)
0.460	5.16	31.41
3.3	0.57	0.78
11.3	0.050	0.066

Therefore, the active power loss P_{loss}= 49.8 kW and the reactive power loss Q_{loss}= 66.0 kVAr.

With line voltage of 11.3 kV, the power losses are quite low.

(v) The active and reactive power loss for each voltage level is shown in Table 7.1.

From Table 7.1 it can be seen that the transmission line losses can be reduced substantially by increasing the transmission voltage.

Example 7.12   If the generator of Example 7.11 is a PV farm and we want to operate the PV bus at unity power factor, how much reactive power will we need to place on the PV bus or load bus if the power is transferred at 3.3 kV?

Solution
In Example 7.11, the load is 1 MW at 0.9 power factor lagging.

Therefore, the power factor angle is $\cos^{-1} 0.9 = 25.84°$

The MVA rating of the load is given as

$$S_{load} = \frac{P}{pf} = \frac{1}{0.9} = 1.11 \text{ MVA}$$

Therefore, the reactive power consumption of the load is given as

$$Q = 1.11 \sin 0.9 = 0.48 \text{ MVAr}$$

From Table 7.1, the reactive power loss of transmission line is

$$Q = 0.78 \text{ MVAr}$$

Therefore, the total reactive power consumed

$$Q_{Consumed} = 0.78 + 0.48 = 1.26 \text{ MVAr}$$

The line-charging capacitances will supply some reactive power to the system. The reactive power supplied by the line-charging reactance is given by (see Figure 7.65):

$$Q_{produced} = \left(\frac{V_{gen}^2}{X_C} + \frac{V_{Load}^2}{X_C} \right) \times distance = \frac{\left(6.06 \times 10^3\right)^2 + \left(3.3 \times 10^3\right)^2}{0.0537 \times 10^6} \times 15$$

$$= 13.3 \text{ kVAr}$$

The net reactive power supplied by the generator is $Q_{gen} = Q_{consumed} - Q_{produced}$

$$= 1.26 - 0.013 = 1.25 \text{ MVAr}$$

Therefore, a reactive power of 1.25 MVAr must be supplied by the PV bus. In other words, a capacitive load of 1.25 MVAr must be connected to the PV bus.

7.15 THE CONSTRUCTION OF A POWER GRID SYSTEM

A power system grid is a network of transmission and distribution systems for delivering electric power from suppliers to consumers. The power grids use many methods of energy generation, transmission, and distribution. Following the energy crisis of the 1970s, the federal Public Utility Regulatory Policies Act (PURPA)[24,25] of 1978 aimed at improving energy efficiency and increasing the reliability of electric power supplies. PURPA required open access to the power grid network for small, IPPs. After the deregulation of the power industry, the power generation units of many power grid companies began operating as a separate business. New power generation companies entered the power market as IPPs: IPPs generate power that is purchased by electric utilities at wholesale prices. Today, power grid generating stations are owned by IPPs, power companies, and municipalities. The end-use customers are connected to the distribution systems of power grid companies who can buy power at a retail price.

Power companies are tied together by transmission lines referred to as *interconnections*. An *interconnected network* is used for power transfer between power companies. Interconnected networks are also used by power companies to support and increase the reliability of the power grid for stable operation and to reduce costs. If one company is short of power due to unforeseen events, it can buy power from its neighbors through the interconnected transmission systems.

The construction of a power-generating station with a high capacity, say in the range of 500 MW may take 5–10 years. Before constructing such a power-generating station, a permit must be obtained from the government. Stakeholders, the local power company, and IPPs will have to undertake an economic evaluation to determine the cost of electric energy over the life of the plan

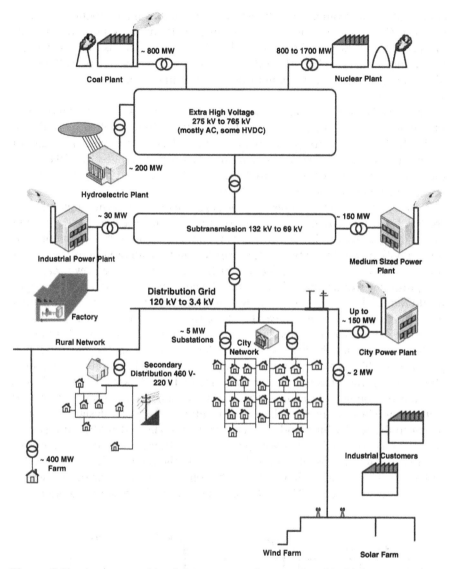

Figure 7.64 A power grid of interconnected network with high green energy penetration.[16]

as compared with the price of power from other producers before deciding to build the plant.

Figure 7.64 presents a power grid of interconnected network with high green energy penetration. Under a deregulated power industry, power grid generation and the cost of electric power is determined by supply and demand. In the United States and most countries around the world, the interconnected network power grid is deregulated and is open for all power producers. The control of an interconnected network is maintained by an independent system operator (ISO). The ISO is mainly concerned with maintaining instantaneous balance of the power grid system load and generation to ensure that the system would remain stable. The ISO performs its function by controlling and dispatching the least costly generating units to match power generated with system loads.

Historically, power plants are located away from heavily populated areas. The plants are constructed where water and fuel (often supplied by coal) are available. Large-capacity power plants are constructed to take advantage of economies of scale. The power is generated in a voltage range of 11–20 kV and then the voltage is stepped up to a higher voltage before connection to the interconnected bulk transmission network.

High-voltage (HV) transmission lines are constructed in the range of 138–765 kV. These lines are mostly overhead. However, in large cities, underground cables are also used. The lines consist of copper or aluminum. A major concern in bulk power transmission is power loss in transmission lines that is dissipated as heat due to the resistance of the conductors. The power capacity is expressed as voltage magnitude times the current magnitude. High voltages would require less current for the same amount of power and less surface area of conductor, resulting in reduced line loss. The distribution lines are normally considered lines that are rated less than 69 kV. Bulk power transmission lines are like the interstate highway systems of the energy industry, transferring bulk power along high-voltage lines that are interconnected at strategic locations. High-voltage transmission lines in the range of 110–132 kV are referred to as subtransmission lines. In Figure 7.65, the subtransmission lines supply power to factories and large industrial plants. Distribution systems are designed to carry power to the feeder lines and end-use customers. The distribution transformers are connected to the high-voltage side of the transmission or the subtransmission system. The distribution voltages are in the range of 120, 208, 240, 277, and 480 V. The service voltage of distribution systems depends on the size of service in terms of loads. The higher commercial loads are served at 480 V. In

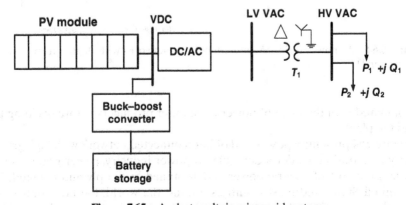

Figure 7.65   A photovoltaic microgrid system.

TABLE 7.2   Variation of AC Bus Voltage with Modulation Index

$V_{dc} = 120$ V		
Modulation Index, M_a	AC Bus Voltage (Low-Voltage Side), V	AC Bus Voltage (High-Voltage Side), V
0.70	51.4	110
0.75	55.1	110
0.80	58.8	110
0.85	62.5	110
0.90	66.1	110

smart grids, power generation sources are supplied to the grids from all voltage levels by independent power producers, commercial, industrial, and residential energy users.

Example 7.13   Consider the three-phase radial PV distribution system given below.

Assume load number 1 is a three-phase load of 5 kW at a power factor of 0.85 leading and load number 2 is a three-phase load of 10 kW at a power factor of 0.9 lagging at rated voltage of 110 V with a 10% voltage variation. Assume a PV source voltage is rated 120 V DC. Compute the transformer ratings. Assume an ideal transformer.

If the amplitude modulation index, M_a, can be controlled from 0.7 to 0.9 in steps of 0.05, compute the following:

  (i) The transformer low-voltage-side rating (LV)
 (ii) The transformer high-voltage-side rating (HV)
(iii) The PV system per unit model if the transformer ratings are selected as the base values
(iv) The per-unit model of this PV system

Solution
For the DC/AC inverter:

$$V_{\text{L-L,rms}} = \frac{\sqrt{3}}{\sqrt{2}} . M_a \frac{V_{dc}}{2}$$

Let us vary the modulation index from 0.7 to 0.9 to find the suitable voltage. From Table 7.2, we see that the modulation index of 0.9 is most suitable. Therefore,

$$V_{\text{L-l,rms}} = \frac{\sqrt{3}}{\sqrt{2}} . M_a \frac{V_{dc}}{2} = \frac{\sqrt{3}}{\sqrt{2}} \times 0.9 \times \frac{120}{2} = 66.13 \text{ V}$$

$$\text{kVA for load 1} = \frac{\text{kW}_1}{pf\_1} = \frac{5}{0.85} = 5.88 \text{ kVA}$$

$$pf \text{ angle} = \cos^{-1} 0.85 = 31.78°, \text{leading}$$

$$\text{kVA for load 1} = \frac{\text{kW}_1}{pf\_1} = \frac{10}{0.9} = 11.11 \text{ kVA}$$

$$pf \text{ angle} = \cos^{-1} 0.9 = 25.84°, \text{lagging}$$

Let an ideal transformer be selected. It is rated at 70/110. The kVA of the transformer should be greater than or equal to the total load it supplies. A 20 kVA has adequate capacity.

$$\text{Therefore, the p.u. value of load 1} = S_{\text{p.u.,1}} = \frac{\text{kVA}_1}{\text{kVA}_{\text{base}}} = \frac{5.88}{20} = 0.29\angle 31.78°$$

$$\text{Therefore, the p.u. value of load 2} = S_{\text{p.u.,2}} = \frac{\text{kVA}_2}{\text{kVA}_{\text{base}}} = \frac{11.11}{20} = 0.556\angle -25.84°$$

Now, for the DC–AC inverter:

$$\text{The base value of the DC side is} = \frac{\text{DC side voltage}}{\text{AC side voltage}} \times \text{Base voltage of AC side}$$

$$= \frac{120}{66.13} \times 70 = 127.02 \text{ V}$$

$$\text{Therefore, the p.u. value of the DC side of inverter} = \frac{120}{127.02} = 0.94 \text{ p.u.}$$

Therefore, the p.u. value of the AC side of inverter $= \frac{66.13}{70} = 0.94$ p.u.

The microgrid of Figure 7.65 can operate as standalone grid since storage system is connected for storing PV power. Figure 7.66 presents the p.u. model of Figure 7.65. Also, if the microgrid is connected to local power provider, it can operate in parallel with power grid of local power provider.

Example 7.14 Consider the three-phase radial microgrid PV depicted in Figure 7.67.

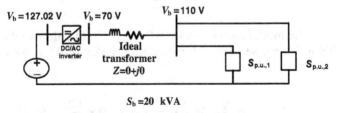

$S_b = 20$ kVA

Figure 7.66 Per unit model of Figure 7.65.

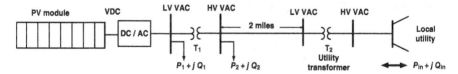

Figure 7.67 A radial microgrid photovoltaic system operating in parallel with the local utility.

Assume a PV is providing power to two submersible pumps on a farm. The load number 1 is rated at 5 kVA at 120 V AC and load number 2 is rated at 7 kVA at 240 V AC. Assume a load voltage variation of 10%, and an amplitude modulation index of 0.9. The distribution line has impedance of 0.04 Ω and reactance of 0.8 Ω per mile. Assume a utility voltage of 460 V.

Compute the following:

- (i) The transformer T_1 rating at the (LV) voltage side
- (ii) The transformer T_2 rating at the high (HV) voltage side
- (iii) The PV system per unit model with the transformer T_2 ratings selected as the base values and assuming the transformers have an impedance of 10% based on selected ratings

Solution

The transformer T_2 is selected as base with voltage of 240/460 V. It must supply the loads on the low-voltage side and must have a kVA rating greater than the loads connected to it. We choose a transformer of rating 15 kVA ($> 5 + 7 = 12$) and select this rating as the base.

The base on the transmission line side is 240V and 15 kVA.

The impedance of the transmission line is equal to its length times the impedance per unit length.

$$Z_{\text{line}} = 2 \times (0.04 + j0.8) = 0.08 + j1.6\Omega$$

The base impedance for the transmission line is

$$Z_{\text{b,trans}} = \frac{V^2_{\text{base}}}{VA_{\text{base}}} = \frac{240^2}{15 \times 10^3} = 3.84$$

The p.u. value of the transmission line is equal to

$$Z_{\text{p.u.,trans}} = \frac{Z_{\text{trans}}}{Z_{\text{b,trans}}} = \frac{0.08 + j1.6}{3.84} = 0.02 + j0.42\text{p.u.}\Omega$$

The p.u. value of load 2 is given as

$$S_2 = \frac{kVA_2}{kVA_{base}} = \frac{7}{15} = 0.467 \text{ p.u. kVA}$$

Let us select a transformer T_1 of 120/240 V, 15 kVA. The base on the low-voltage side of T_1 is 120 V.

Therefore, the p.u. value of the AC side of inverter $= \frac{120}{120} = 1$ p.u. V and the p.u. value of load 1 is given as

$$S_1 = \frac{kVA_1}{kVA_{base}} = \frac{5}{15} = 0.33 \text{p.u. kVA}$$

The transformer impedances are as given: $j0.10$ p.u. Ω
For the DC–AC inverter:

$$V_{L,rms} = \frac{\sqrt{3}}{\sqrt{2}}.M_a\frac{V_{dc}}{2}$$

Let us select a modulation index of 0.9.

$$V_{dc} = \frac{\sqrt{2}}{\sqrt{3}}.\frac{2.V_{L,rms}}{M_a} = \frac{\sqrt{2}}{\sqrt{3}}.\frac{2 \times 120}{0.9} = 217.7 \text{ V}$$

The base value of the DC side is $= \dfrac{\text{DC side voltage}}{\text{AC side voltage}} \times$ Base voltage of AC side

V base DC side$= \dfrac{217.7}{120} \times 120 = 217.7$ V

Therefore the p.u. value of the DC side of inverter $= \dfrac{217.7}{217.7} = 1$ p.u. V
The p.u. model of Figure 7.67 is presented in Figure 7.68.

Example 7.15   Compute the minimum DC input voltage if the switching frequency is set at 5 kHz. Assume the inverter is rated 207.6 V AC, 60 Hz, 100 kVA.

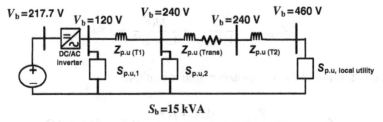

S_b =15 kVA

Figure 7.68   Per unit model of Figure 7.67.

Solution

$$V_{\text{rated}} = 207.6 \text{ V, kVAr} = 100 \text{ kVA}, f = 60 \text{ Hz}$$

The AC-side voltage of the inverter is given as

$$V_{\text{peak}} \sin(2.\pi.f.t) = 207.6\sqrt{2}. \sin(2.\pi.60.t) = 293.59 \sin(2.\pi.60.t)$$

Here, $V_{\text{peak}} = 293.59$ V

For the three-phase inverter using sine PWM, the line–line peak voltage is given as $V_{\text{L-L,peak}} = M_a.\sqrt{3}.\frac{V_{dc}}{2}$

Therefore,

$$V_{dc} = \frac{2}{\sqrt{3}} . \frac{V_{\text{L-L,peak}}}{M_a}$$

For the inverter to operate at maximum value of M_a is 1, the minimum DC voltage is given as

$$V_{dc,min} = \frac{2}{\sqrt{3}} . \frac{V_{\text{L-L,peak}}}{M_{a,max}} = \frac{2}{\sqrt{3}} . \frac{293.59}{1} = 339.01 \text{ V}$$

Therefore, the minimum $V_{dc} = 339.01$ V

7.16 MICROGRID OF RENEWABLE ENERGY SYSTEMS

Figure 7.69 depicts a community microgrid system operating in parallel with the local utility system. The transformer T_1 voltage is in the range of 240 VAC/600 VAC and the transformer T_2 that connects the PV system to the utility bus system is rated to step up the voltage to the utility local AC bus substation.

Example 7.16 Consider a microgrid that is fed from a PV generating station with AC bus voltage of 120 V as shown in Figure 7.70.

The modulation index for the inverter is 0.9.

Compute the following:

  (i) The per unit value of line impedance.
 (ii) The per unit impedance of transformers T_1 and T_2.
(iii) The per unit model of the load.
(iv) The per unit equivalent impedance and give the impedance diagram. Assume the base voltage, V_b of 240 V in the load circuit and the S_b of 500 kVA.

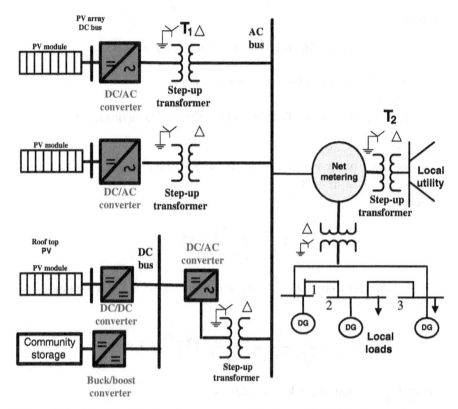

Figure 7.69   One-line diagram for a community microgrid distribution system connected to a local utility power grid.

Solution
We select the voltage base and power base as specified. For the utility side, we will have

$$V_b = 240 \text{ V } S_b = 500 \text{ kVA}$$

The p.u. load, $S_{\text{p.u.(Load)}} = \frac{400}{500} \angle -\cos^{-1} 0.8 = 0.8 \angle -36.87° \text{ p.u. kVA}$

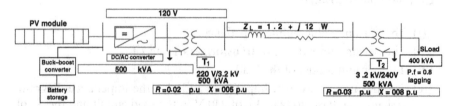

Figure 7.70   The microgrid of Example 7.15.

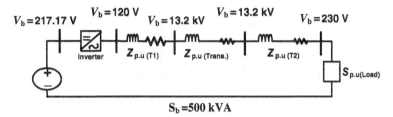

Figure 7.71   The per unit model of Figure 7.70.

The p.u. impedance of T_2 will remain the same as its rated value of $0.03 + j0.08$.

The base impedance of the transmission line can be computed as

$$Z_{b\_3200} = \frac{V_b^2}{S_b} = \frac{(3200)^2}{500 \times 10^3} = 20.48 \ \Omega$$

The per unit transmission line impedance can be computed as

$$Z_{trans.line,p.u.} = \frac{Z_{tran.actual}}{Z_{b\_3200V}} = \frac{1.2 + j12}{20.48} = 0.058 + j0.586 \ p.u.\Omega$$

The p.u. impedance of T_1 at its rated value also will remain the same as $0.02 + j0.05$.

For the inverter:

$$V_{L,rms} = \frac{\sqrt{3}}{\sqrt{2}}.M_a \frac{V_{dc}}{2}$$

$$V_{dc} = \frac{\sqrt{2}}{\sqrt{3}}.\frac{2.V_{L,rms}}{M_a} = \frac{\sqrt{2}}{\sqrt{3}}.\frac{2 \times 120}{0.9} = 217.7 \ V$$

The base value of the DC value is $= \dfrac{\text{DC side voltage}}{\text{AC side voltage}} \times$ Base voltage of AC side

$$= \frac{217.17}{120} \times 120 = 217.17 \ V$$

Therefore, the p.u. value of the DC side of inverter $= \dfrac{217.17}{217.17} = 1 \ p.u.$
The per unit impedance diagram is given in Figure 7.71.

Example 7.17   Consider the microgrid of Figure 7.72. Assume the PV generating station can produce 175 kW of power, transformer T_1 is rated at 5% impedance, 240 V/120 V and 75 kVA, and transformer T_2 is rated at 10%

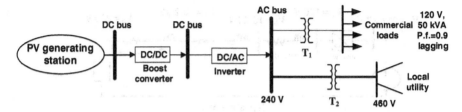

Figure 7.72 One-line diagram of Example 7.16.

impedance, 240 V/460 V, and 150 kVA. Assume the power base of 1000 kVA and the voltage base of 460 V in transformer T_2. Assume the three-phase inverter has a modulation index of 0.8 and the boost converter has a duty ratio of 0.6.

Compute the following:

(i) The per unit value of transformers T_1 and T_2, and per unit model of the load

(ii) The per unit equivalent impedance and give the impedance diagram

Solution

The power base is selected as 1000 kVA and voltage base on the utility side is selected as 460 V.

We will have

$$V_{bnew} = 460 \text{ V} \quad S_{bnew} = 1000 \text{ kVA}$$

Therefore, the new impedance base on the 460 V is given as

$$Z_{b460V} = \frac{V_b^2}{S_b} = \frac{(460)^2}{1000 \times 10^3} = 0.2116\Omega$$

For new base, the transformer T_2, we will have

$$Z_{p.u.\_new} = Z_{p.u.\_old} \times \left(\frac{V_{b\_old}}{V_{b\_new}}\right)^2 \frac{S_{b\_new}}{S_{b\_old}}$$

$$Z_{p.u.\_new-T2} = 0.1 \times \left(\frac{460}{460}\right)^2 \times \frac{1000}{150} = 0.67 \text{p.u.}\Omega$$

The old values of the transformer T_1 are given on its nameplate rating. We recalculate the new per unit impedance of T_1 based on the new selected bases.

For the transformer T_1, we will have

$$Z_{\text{p.u.\_new-T1}} = 0.05 \times \left(\frac{240}{240}\right)^2 \times \frac{1000}{75} = 0.67 \text{ p.u.}\Omega$$

The base voltage on the low side of the T_1 transformer is given by the ratio of the old base voltage to the new base voltage (i.e., both on the LV side or HV side).

$$V_{\text{b\_new(LV)}} = \frac{V_{\text{b\_new(HV)}}}{V_{\text{b\_old(HV)}}} V_{\text{b\_old(LV)}}$$

Using the transformer, T_1, nameplate ratings and the new base voltage on the transmission can be computed as

$$V_{\text{b\_new(LV)}} = \left(\frac{240}{240}\right)(120) = 120 \text{ V}$$

The per unit load can be calculated as

$$S_{\text{p.u.(Load)}} = \frac{50}{1000} \angle - \cos^{-1} 0.9 = 0.05\angle - 25.84°\text{p.u. kVA}$$

For the inverter:

$$V_{\text{L,rms}} = \frac{\sqrt{3}}{\sqrt{2}}.M_{\text{a}}\frac{V_{\text{dc}}}{2}$$

$$V_{\text{dc}} = \frac{\sqrt{2}}{\sqrt{3}}.\frac{2.V_{\text{L,rms}}}{M_{\text{a}}} = \frac{\sqrt{2}}{\sqrt{3}}.\frac{2 \times 240}{0.8} = 489.9 \text{ V}$$

The base value of the DC side is $= \dfrac{\text{DC side voltage}}{\text{AC side voltage}} \times$ Base voltage of AC side

$$= \frac{489.9}{240} \times 240 = 489.9 \text{ V}$$

Therefore, the p.u. value of the DC side of the inverter $= \dfrac{489.9}{489.9} = 1.\text{p.u. V}$

Now, for the DC/DC converter:

$$V_{\text{o}} = \frac{V_{\text{in}}}{1 - \text{Duty ratio}}$$

$$V_{\text{in}} = (1 - \text{Duty ratio}) \times V_{\text{o}} = (1 - 0.6) \times 489.9 = 195.96 \text{ V}$$

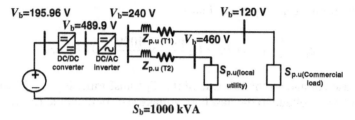

V_b=195.96 V V_b=240 V V_b=120 V
   V_b=489.9 V

DC/DC DC/AC $Z_{p.u \, (T1)}$ V_b=460 V
converter inverter $Z_{p.u \, (T2)}$

$S_{p.u(local \, utility)}$ $S_{p.u(Commercial \, load)}$

S_b=1000 kVA

Figure 7.73   The per unit model of Figure 7.72.

The base value of the low-voltage side is given as

$$\text{V base on DC side} = \frac{\text{Low voltage side}}{\text{High voltage side}} \times \text{Base voltage of high voltage side}$$

$$\text{V base on DC side} = \frac{195.96}{489.9} \times 489.9 = 195.96 \text{ V}$$

Therefore, the p.u. value of the DC side of the inverter $= \dfrac{195.96}{195.96} = 1$ p.u. V
The p.u. model of power grid of Figure 7.72 is presented in Figure 7.73.

This chapter presents the basic knowledge of circuit theory and power electronics to modeling and design power grids. We have studied the meaning of impedance loads, a power consumption model for loads, the rated values of electrical devices, and the modeling of single- and three-phase transformers. We also explored the basic concept of per unit system and how to develop a per unit model—a three-phase systems distribution feeder that can be fed from a renewable energy source.

In the following chapters, we will study the concept of operation of power grids and explore the operation of smart grid system, the design of smart microgrid rentable energy systems, and energy monitoring and energy efficiency.

PROBLEMS

7.1 Assume $a = 120 + j\,100$, $b = 100 + j\,150$ and $c = 50 + j\,80$. Compute the following complex numbers:
   (i) $(a \times b)/c$
   (ii) $(a/b) \times c$
   (iii) $(a - b)/(c - a)$

7.2 $V = 120\cos(377t + 5°)$. If a voltmeter is used to measure this voltage, what value should the voltmeter read? Express this voltage in polar form.

7.3 The operation of AC machines (in particular, transformers and induction machines) can be studied with the aid of the T-circuit shown in Figure 7.74 below.

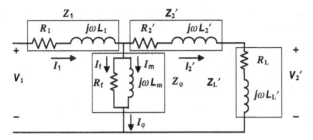

Figure 7.74   Equivalent circuit for Problem 7.3.

Assume the frequency is 60 Hz. The circuit elements are given in the table below.

	R_1	L_1	R_f	L_m	R_2'	L_2'	R_L	L_L	V_1	V_2	I_1	I_2'	I_f
1	1	0.01	1000	8	1	0.01		Open	480	?	?	?	
2	1	0.01	1000	8	1	0.01	200	0	480	?	?	?	?
3	0.02	0.00265	Open		0	0		Open	1	?	?		
4	0.02	0.00265	Open		0	0	1	0	1	?	?	?	
5	0.02	0.00265	Open		0	0	0.707	1.875×10⁻³	1	?	?	?	
6	0	0	100	0.1	0.01	106×10⁻⁶	1	0	1	?	?	?	?
7	0	0	100	0.01	0.01	106×10⁻⁶	1.414	3.75×10⁻³	1	?	?	?	?
8	0.3	1.33×10⁻³	Open	3.45×10⁻²	0.15	0.56×10⁻³	7.35	0	127	?	?	?	?
9	10	5.2×10⁻²	Open	0	0	200	0.4	0	500	?	?	?	?
10	0.15	2.54×10⁻³	Open	1.57	6.24×10⁻³	98.5	0.178	0	240	?	?	?	?
11	0.3	0.003	1	4.25×10⁻²	0.2	0.003	10	0	440	?	?	?	?
12	0.3	0.003	0	4.25×10⁻²	0.2	0.003	1	0	380	?	?	?	?

Use polar form for all complex numbers and solve set 1 through set 12. Show your calculations separately.

7.4 Assume a balanced three-phase load. Each phase load is rated 120 Ω with a phase angle of 10° and the load is connected as a Y bank. A three-phase balance source rated 240 V is applied to this three-phase load. Perform the following:

(i) Give the three-phase circuit of this load.

(ii) Compute the active and reactive power consumed by this load.

7.5 Assume a single-phase load has inductance of 10 millihenry (mH) and resistance 3.77 Ω and it is connected to a 60 Hz source at 120 V. Compute the active and reactive power consumed or generated by the load.

7.6 Assume a single-phase load has a capacitance of 10 millifarad (mF) and resistance of 2.0 Ω and it is connected to a 60 Hz source at 120 V. Compute active and reactive power consumed or generated by the loads.

7.7 A single-load consumes 10 kW at 0.9 power factor, *pf*, lagging at 220 V. What is the magnitude and phase angle of current drawn from the source? If another single-phase load rated at 10 kW and 0.8 leading power factor is connected in parallel to the same sources, what is the magnitude of current drawn? What is the current drawn from the source? Give your answer in polar form.

7.8 A single-phase transformer is rated 200 kVA, 120/220 V, 5% reactance. Compute the reactance of the transformer from the high-voltage side. Give the single-phase equivalent circuit.

7.9 Compute the per unit model of Problem 7.7 based on 400 kVA and a base voltage of 440 V.

7.10 Three single-phase transformers are each rated 460 V/13.2 kV 400 kVA, 5% short-circuit reactance. The three single-phase transformers are connected as a three-phase Y–Y transformer. Perform the following:

(i) Compute the three-phase Y–Y transformer high-voltage-side and low-voltage-side voltage and kVA ratings.

(ii) Compute the per unit model of the three-phase transformer.

7.11 A bank of three-phase Y–Y resistive load of 30 Ω is connected to the low-voltage side of the transformer of Problem 7.9 using a feeder with resistance of $1 + j10$ Ω. Perform the following:

(i) Give the three-phase equivalent model.

(ii) Give a one-line diagram.

(iii) Compute the per unit model based on the transformer rating.

7.12 Three single transformers are each rated 460 V/13.2 kV 400 kVA, 5% short circuit reactance. The three single-phase transformers are connected as a three-phase Y–Δ transformer. Perform the following:

(i) Compute the three-phase Y–Δ transformer high-voltage-side and low-voltage-side voltage and kVA ratings.

(ii) Compute the per unit model of the three-phase transformer.

7.13 A bank of three-phase Y–Y resistive load of 30 Ω is connected to the low-voltage side of the transformer of Problem 7.11 using a feeder with resistance of $1 + j10$ Ω. Perform the following:

(i) Give the three-phase equivalent model.

(ii) Give the one phase of a Y-connected equivalent circuit.

(iii) Give a one-line diagram.

(iv) Compute the per unit model based on the transformer rating.

7.14 Three single-phase transformers are each rated 460 V/13.2 kV 400 kVA, 5% short-circuit reactance. The three single-phase transformers are connected as a three-phase Δ–Δ transformer. Perform the following:

(i) Compute the three-phase Δ–Δ transformer high-voltage side and low-voltage side voltage and kVA ratings.

(ii) Give the three-phase equivalent model.

(iii) Give the one phase of a Y-connected equivalent circuit.

(iv) Compute the per unit model based on the transformer rating.

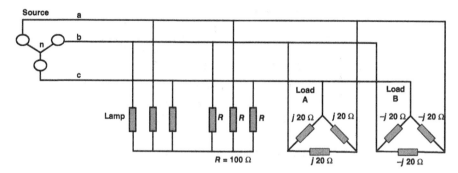

Figure 7.75   The three-phase diagram of Problem 7.16.

7.15   A bank of three-phase Δ–Δ resistive load of 30 Ω is connected to the low-voltage side of the transformer of Problem 7.13 using a feeder with resistance of $1 + j10$ Ω. Perform the following:

  (i) Give the three-phase equivalent model.

  (ii) Give the one phase of a Y-connected equivalent circuit.

  (iii) Give a one-line diagram.

  (iv) Compute the per unit model based on the transformer rating.

7.16   A balanced three-phase, three-wire feeder has three balanced loads as shown in Figure 7.75.

  Each lamp is rated at 100 W and 120 V. The line-to-line voltage on the feeder is 240 V and remains constant under the loads. Find the source current in the feeder lines and the power delivered by the source.

  (i) Compute the line current in phase a

  (ii) Compute active and reactive power supplied by the sources.

7.17   Consider a three-phase distribution feeder as shown by Figure 7.76.

  For the three-phase system, all voltages are given line-to-line and the complex power is given as three-phase power.

  Compute the following:

  (i) The source voltage V_s, if V_R is to be maintained at 4.4 kV ($V_R = $ 4.4 kV V line value)

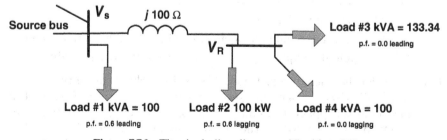

Figure 7.76   The single-line diagram of Problem 7.17.

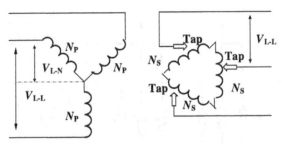

Figure 7.77   Figure for Problem 7.18.

(ii) The source current and the power factor at the source

(iii) The total complex power supplied by the source

(iv) How much reactive power should be connected to the source bus to obtain a unity power factor at the source bus?

7.18   Assume a three-phase transformer is Y-grounded on the low-voltage side and Δ-connected winding on the high-voltage side as shown in Figure 7.77 Each winding of tap changing winding has 2000 turns on the low-voltage side, and a variable number of turns on the high-voltage winding side. Assume the high-voltage winding side has a maximum number of turns equal to 7300 turns and a minimum number of turns equal to 5300. Compute the minimum and maximum line-to-line voltage that can be maintained on the high-voltage side. Assume the winding voltage across the low-voltage side is set equal to 36.4 kV.

7.19   Write a term paper on the capital cost of a one megawatt solar system and its payback over 25 years. Assume the cost of a coal-fired power plant is $10 per kWh and PV, to generate electricity the cost is about $0.25–0.40 per kWh.

7.20   A three-phase generator rated 480 V and 400 kVA connected to a three-phase transformer rated 3.2 kV/480 V, 200 kVA, with reactance of 10% is used to serve a load over a line of $1 + j20$ Ω. At the load site, the voltage is stepped down to 220 V from 3.2 kV using a three-phase transformer rated 150 kVA with reactance of 7%. The voltage at the load is to be maintained at 220 V. Perform the following:

(i) Give a one-line diagram.

(ii) Compute a per unit equivalent circuit based at 200 kVA and 480 V.

(iii) Compute the required generator voltage. Is this design viable?

7.21   Consider the microgrid of Figure 7.78. A three-phase transformer, T1, is rated 500 kVA, 220/440, transformer with the reactance of 3.5%. The microgrid is supplied from an AC bus of a PV generating station with its DC bus rated at 540 V. The distribution line is 10 miles long and has a series impedance of $0.1 + j\,1.0$ Ω per mile and local load of 100 kVA

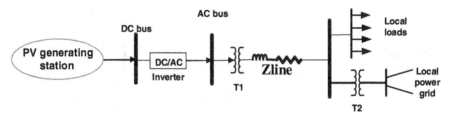

Figure 7.78   A single-line diagram for Problem 7.21.

at 440 V. The microgrid is connected to the local power grid using a three-phase transformer T2, rated at 440 V/13.2 kV with the reactance of 8%. Compute the per unit impedance diagram of the microgrid system. Assume the voltage base of 13.2 kV on the local power grid side and kVA base of 500.

7.22   Consider the microgrid of Figure 7.79. Assume the inverter AC bus voltage of 240 V and transformer T1 is rated 5% impedance, 240 V/120 V and 150 kVA. The transformer T2 is rated at 10% impedance, 240 V/3.2 kV and 500 kVA. The local loads are rated at 100 kVA and a power factor of 0.9 lagging. The inverter modulation index is 0.9.

Compute the following:

(i) The DC bus voltage and inverter rating.

(ii) The boost converter PV bus input voltage and input current ratings for the required DC bus voltage of the inverter.

(iii) The size of the microgrid PV station.

(iv) The per unit model of the microgrid.

7.23   For Problem 7.22 depicted in Figure 7.79, assume the DC bus voltage of inverter is equal to 800 V DC. The transformer T1 is rated at 9% impedance, 400 V/220 V and 250 kVA. The transformer T2 is rated at 10% impedance, 400 V/13.2 kV and 500 kVA. The load number 1 is rated at 150 kVA with a power factor of 0.85 lagging. The load number 2 is rated at 270 kVA with a power factor of .95 leading.

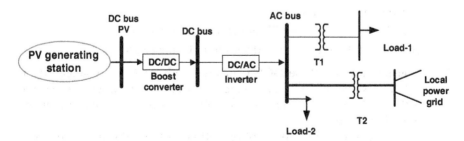

Figure 7.79   A one-line diagram of Problem 7.22.

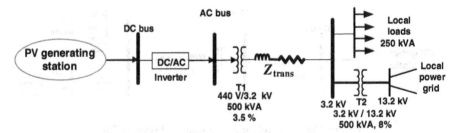

Figure 7.80 A one-line diagram of Problem 7.25.

Compute the following:

(i) The modulation index and inverter rating.

(ii) The boost converter PV bus voltage and input current for the required DC bus voltage of the inverter.

(iii) The minimum size of the microgrid PV station.

7.24 For Problem 7.22 depicted in Figure 7.79, assume transformer T1 is rated at 5% impedance, 440 V/120 V and 150 kVA. The transformer T2 is rated at 10% impedance, 240 V/3.2 kV and 500 kVA. The inverter modulation index is 0.9. Select the power base of 500 kVA and the voltage base of 600 V. The local power grid internal reactance is 0.2 per unit base on 10 MVA and 3.2 kV. Assume the microgrid is not loaded.

Compute the following.

(i) The per unit equivalent impedance model.

(ii) The fault current if a balance three-phase fault occurs on the inverter bus.

7.25 Consider the microgrid of Figure 7.80. A three-phase 500 kVA, 440/3.2 kV transformer with the per unit reactance of 3.5% feeds from an AC source of a PV generating station. The distribution line is 10 miles long and has a series impedance of $0.01 + j\,0.09\ \Omega$ per mile. The local load is 250 kVA. Balance of power can be injected into the local utility using a 3.2/13.2 kV transformer and with the per unit reactance of 8%. Assume the voltage base of 13.8 kV on the local power grid side, kVA base of 500, and the DC bus voltage of 800 V.

Compute the following:

(i) The inverter and the PV generating station ratings.

(ii) The per unit impedance diagram of the microgrid.

7.26 Consider a three-phase DC/AC inverter with a DC bus voltage rate at 600 V. Assume the DC bus voltage can drop during a discharge cycle to 350 V DC. Determine the AC bus voltage range and its corresponding modulation index.

Consider a rectifier mode of operation for the following system data of a three-phase rectifier. Assume the data given in Table 7.3.

TABLE 7.3   Data for Problem 7.26

V_{AN}	f_1	P	X	pf	M_a
120 V RMS	60 Hz	1500 W	15 Ω	1	0.85

Perform the following:

(i) Find the angle δ and the DC link voltage. Ignore the resulting harmonics.

(ii) Write a MATLAB program to simulate the rectifier operation to plot the current I_a and the voltage V_{AN} considering V_{PWMa} and δ are known and have values same as the result of part (i). Assume the PWM voltage is calculated from the incoming AC power. Ignore the resulting harmonics.

7.27 For a three-phase rectifier with the supply-phase voltage of 120 V, the reactance between the AC source and the rectifier is 11 Ω. Write a MATLAB program to plot the modulation index necessary to keep the V_{DC} constant at 1200 V when the input power factor angle is varied from − 60° to 60°. Assume that the active power supplied by the system remains fixed at 5.5 kW.

7.28 Develop MATLAB testbed for the DC/AC inverter given in Figure 7.81. Assume V_{idc} is equal to 560 V (DC), $V_{C\,(max)}$ is equal to 220 V, f_e is equal to 60 Hz and modulation index of $5 \le M_a \le 1.0$, and change M_f from 2 to 20. Perform the following:

(i) The ratio of the fundamental frequency of the line-line (RMS) to the input DC voltage, V_{idc}, $(V_{(lin\text{-}line\,(RMS))}/V_{idc}.)$

(ii) Plot the $V_{(lin\text{-}line\,(RMS))}/V_{idc}$ as a function of M_a for $M_f = 2$ and $M_f = 20$.

7.29 Consider a PV source of 60 V. A single-phase inverter with four switches is used to convert DC to 50 Hz AC using a unipolar scheme. Select the following modulation indices:

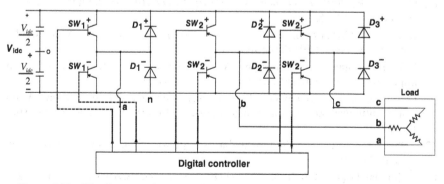

Figure 7.81   The linear and overmodulated operation of a three-phase converter.

(i) $M_a = 0.5$ and $M_f = 7$.

(ii) $M_a = 0.5$ and $M_f = 10$.

(iii) $M_a = 0.9$ and $M_f = 4$.

Write a MATLAB program to generate the waveforms of the inverter showing V_{an}, V_{bn}, V_{cn} voltages. Make tables and discuss your results.

REFERENCES

1. http://edisontechcenter.org/generators.html#history. Accessed October 18, 2013.

2. Enjeti, P. (2010) An advanced PWM strategy to improve efficiency and voltage transfer ratio of three-phase isolated boost dc/dc converter. Paper presented at: 2008 Twenty-Third Annual IEEE Applied Power Electronics; February 24–28, 2010; Austin, TX.

3. Dowell, L.J., Drozda, M., Henderson, D.B., Loose, V.W., Marathe, M.V., Roberts, D.J. ELISIMS: comprehensive detailed simulation of the electric power industry, Technical Rep. No. LA-UR-98-1739. Los Alamos National Laboratory, Los Alamos, NM.

4. Chapman, S. (2003) *Electric Machinery and Power System Fundamentals,* McGraw Hill, New York.

5. Complex numbers. Available at http://en.wikipedia.org/wiki/Complex_number#History_in_brief. Accessed December 20, 2003.

6. Keyhani, A., Marwali, M.N., and Dai, M. (2010) *Integration Of Green And Renewable Energy in Electric Power Systems,* John Wiley & Sons, Hoboken, NJ.

7. Mohan, N., Undeland, T., and Robbins, W. (1995) *Power Electronics*, John Wiley & Sons, New York.

8. Rashid, M.H. (2003) *Power Electronics, Circuits, Devices and Applications*, 3rd edn, Pearson Prentice Hall, Englewood Cliffs, NJ.

9. Grainger, J. and Stevenson, W.D. (2008) *Power Systems Analysis,* McGraw Hill, New York .

10. http://www.greatachievements.org. Accessed September 18, 2009.

11. http://edisontechcenter.org/HistElectPowTrans.html. Accessed October 25, 2013.

12. http://en.wikipedia.org/. Accessed September 28, 2009.

13. Clayton, P.R. (2001) *Fundamentals of Electric Circuit Analysis*, John Wiley & Sons. ISBN 0-471-37195-5.

14. Keyhani, A. (2011) *Design of Smart Power Grid Renewable Energy Systems*, John Wiley & Sons and IEEE Publication.

15. U.S. Environmental Protection Agency. Clean Air Act. Available at http://www.epa.gov/air/caa/. Accessed January 25, 2009.

16. California Energy Commission. Energy Quest. Glossary of Energy Terms. Available at http://www.energyquest.ca.gov/glossary/glossary-i.html#i. Accessed April 18, 2009.

17. 3-Phase Power Resource Site. Available at http://www.3phasepower.org. Accessed January 28, 2009.

18. Majmudar, H. (1965) *Electromechanical Energy Converters*, Allyn & Bacon, Boston, MA.

19. El-Hawary, M.E. (1983) *Electric Power Systems: Design and Analysis*, Reston Publishing, Reston, VA.

20. Nikola Tesla. Available at http://en.wikipedia.org/. Accessed September 28, 2009.

21. A Century of Innovation-Electrification National Academy of Engineering. Available at http://www.greatachievements.org. Accessed September 18, 2009.

22. Elgerd, O.I. (1982) *Electric Energy System Theory: An Introduction,* 2nd edn, McGraw-Hill, New York.

23. IEEE Brown Book. (1980) *IEEE Recommended Practice For Power System Analysis*, Wiley-Interscience, New York.

24. War of Currents. Available at wikipedia.org/wiki/. Accessed January 28, 2009.

25. U.S. Energy Information Administration. Independent Statistics and Analysis. Available at www.eia.doe.gov. Accessed September 18, 2009.

CHAPTER 8

MICROGRID WIND ENERGY SYSTEMS

8.1 INTRODUCTION

Historians estimate that wind energy has been utilized to sail ships since about 3200 B.C.[1] The first windmills were developed in Iran (Persia) for pumping water and grinding grain.[1] Denmark developed the first wind turbine for electricity generation in 1891.[2] Today over 20% of Denmark's electricity comes from wind energy and Denmark's wind energy industry has a 27% share of the global market.

Wind energy is captured from the mechanical power of wind and converted to electric power using the classical process of Faraday's law of induction.[3] Figure 8.1 depicts the distribution of energy consumption by energy source from 2004 to 2008 as determined by the U.S. Department of Energy (DOE).[4,5]

Although renewable energy production is small, the renewable electricity production in the United States (excluding hydropower) has nearly tripled since 2000. In 2008, it represented 41.9 GW of installed capacity. Wind energy installed capacity in 2008 (26 GW) is almost 10 times that of 2000 (2.6 GW). Wind energy and solar PV cumulative capacity grew 61% and 44%, respectively, from 2007 to 2008.

The National Renewable Energy Laboratory (NREL) of the Department of Energy (DOE)[6] of the United States is dedicated to the research, development, commercialization, and deployment of renewable energy sources. The NREL provides wind data by US location; students are encouraged to visit the NREL website to obtain data for wind power in their geographic area.

Design of Smart Power Grid Renewable Energy Systems, Second Edition. Ali Keyhani.
© 2017 John Wiley & Sons, Inc. Published 2017 by John Wiley & Sons, Inc.
Companion website: www.wiley.com/go/smartpowergrid2e

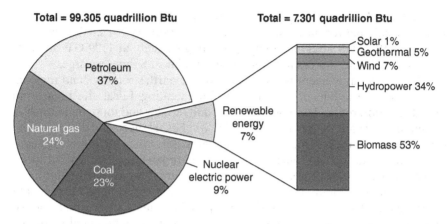

Total = 99.305 quadrillion Btu

Total = 7.301 quadrillion Btu

Petroleum 37%

Natural gas 24%

Coal 23%

Renewable energy 7%

Nuclear electric power 9%

Solar 1%
Geothermal 5%
Wind 7%
Hydropower 34%
Biomass 53%

Figure 8.1 The distribution of energy consumption by energy source from 2004 to 2008 as determined by the U.S. Department of Energy (DOE).

8.2 WIND POWER

Wind energy,[6–14] as one of our most abundant resources, is the fastest growing renewable energy technology worldwide as shown in Figure 8.2. Improved turbine and power converter designs have promoted a significant drop in wind energy generation cost making it the least-expensive source of electricity— from 37 cents/kWh in 1980 down to 4 cents/kWh in 2008. In 2008, wind energy systems worldwide generated 331,600 million kWh, which is 1.6% of total electricity generation—making wind the second highest resource after hydroelectric power (16.6%), while the photovoltaic (PV) technology contribution was only 0.1%.

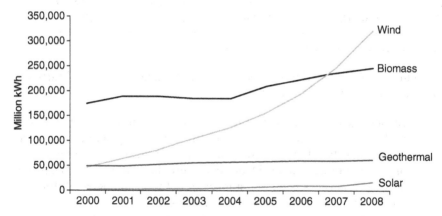

Figure 8.2 Worldwide electricity generation for wind, biomass, geothermal, and solar resources, years 2000–2008.[4]

The United States alone possesses more than 8000 GW of land-based wind resources suitable for harnessing, and an extra 2000 GW of shallow offshore resources. With US total electricity generating capacity at 1109 GW in 2008, the untapped wind sources in United States is almost 8 times as large.

Global wind movement is predicated on the earth's rotation, and regional and seasonal variations of sun irradiance and heating. Local effects on wind include the differential heating of the land and the sea, and topography such as mountains and valleys. We always describe wind by its speed and direction. The speed of the wind is determined by an anemometer, which measures the angular speed of rotation and translates it into its corresponding linear wind speed in meters per second or miles per hour. The average wind speed determines the wind energy potential at a particular site. Wind speed measurements are recorded for a 1-year period and then compared to a nearby site with available long-term data to forecast wind speed and the location's potential to supply wind energy.

A wind energy resource map for the contiguous United States from the NREL is shown in Figure 8.3.[6] This map is based on a location's annual average wind speed and wind power density (in W/m^2) at 60-m (164 ft) tower height; it can be used for initial site assessment. The coastal areas have high wind energy potential; nevertheless, 90% of the US usable wind resources lie in the wind belt spanning the eight Great Plains states.[6]

Wind Power Classification

Wind Power Class	Resource Potential	Wind Power Density at 50 m (W/m^2)	Wind Speed[a] at 50 m (m/s)	Wind Speed[a] at 50 m (mph)
3	Fair	300–400	6.4–7.0	14.3–15.7
4	Good	400–500	7.0–7.5	15.7–16.8
5	Excellent	500–600	7.5–8.0	16.8–17.9
6	Outstanding	600–800	8.0–8.8	17.9–19.7
7	Superb	800–1800	8.8–11.1	19.7–24.8

[a]Wind speeds are based on a Weibull's[15] value of 2.0.

Figure 8.4 depicts a wind turbine system. A horizontal-axis wind turbine is made of the following subsystems.

1. Blades: Two or three blades are attached to the hub of the shaft (rotor) of the wind generator; they are made of high-density epoxy or fiberglass composites. Wind exerts a drag force perpendicular to the blades and produces lift forces on the blades that cause the rotor to turn. The blades' cross section is designed to minimize drag forces and boost lift forces to increase turbine output power at various speeds.

2. Rotor: The rotor transfers the mechanical power of wind and acts to power the generator.

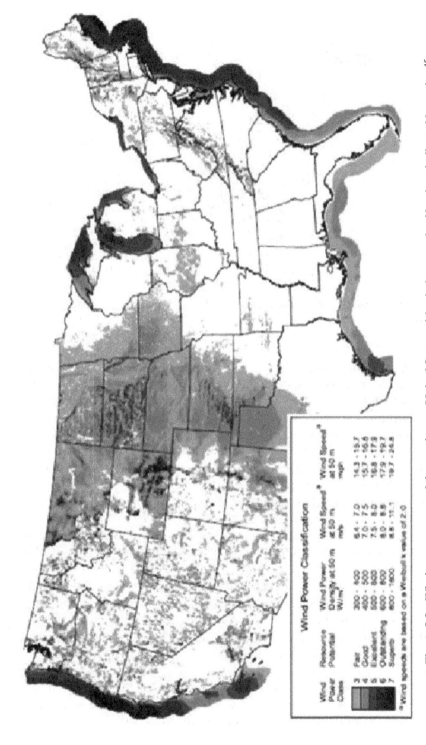

Figure 8.3 Wind resource map of the contiguous United States with wind power classifications indicated by region.[15]

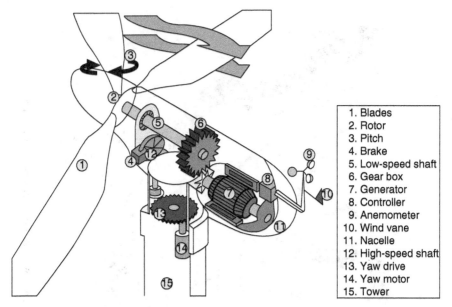

1.	Blades
2.	Rotor
3.	Pitch
4.	Brake
5.	Low-speed shaft
6.	Gear box
7.	Generator
8.	Controller
9.	Anemometer
10.	Wind vane
11.	Nacelle
12.	High-speed shaft
13.	Yaw drive
14.	Yaw motor
15.	Tower

Figure 8.4   A wind turbine system (Adapted from the National Renewable Energy Laboratory).[6–8]

3. Pitch control uses an electric motor or hydraulic mechanism. It is used to turn (or pitch) the blades to maximize the power capture of the turbine or to reduce the rotor's rotational speed in high winds.

4. A rotor brake system is used to stop the rotor for maintenance. Some advanced turbines use hydraulic brakes for the cut-in and the cut-out wind speeds when turbine power output is either too low or too high.

5. Low-speed shafts are designed to transfer mechanical power of the rotor at a speed of 30–60 rpm to the gearbox.

6. A gearbox is used to couple low-speed and high-speed shafts and steps-up the rotational speed to 1200–1600 rpm suitable for electric generators. Gearboxes have disadvantages such as noise, high cost, frictional losses, and maintenance requirements, which preclude their use in some turbine designs.

7. Induction and permanent magnet generators are used for wind turbines.

8. Wind controllers are used to regulate and control the turbine's electrical and mechanical operations.

9. An anemometer is used to measure the wind speed and sends the measurements to the controller.

10. A weather vane is an instrument for showing the direction of the wind. The vane is used to measure wind direction and sends it to the controller, which in turn commands the yaw drive to aim the turbine nose cone in the proper orientation as the wind direction changes.

11. A nacelle is used as a weatherproof streamlined enclosure for housing shafts, a generator, a controller, and rotor brakes.

12. A high-speed shaft is used that mechanically couples the gearbox and rotor of the electric generator.

13. A yaw drive is used to orient the nacelle and the rotor using the yaw motor or hydraulic mechanism.

14. A yaw motor is used to move the nacelle with its components.

15. A tower is used to raise the turbine and hold the rotor blades and the nacelle. Turbine towers are tubular with heights approximately equal to the rotor diameter; however, the minimum height is 26 meters to avoid turbulence.

Appendix D gives a summary of estimating the mechanical power of wind.

8.3 WIND TURBINE GENERATORS

Wind turbine generators (WTGs) are rapidly advancing in both technology and installed capacity.[15] Conventional geared wind generator systems have dominated the wind market for many years. Wind turbine technologies are classified based on their speed characteristics. They are either at fixed or variable speed. The speed of a wind turbine is usually low. The classical WTGs are of two types: (1) wound rotor winding, and (2) squirrel-cage induction. These systems use multistage gear systems coupled to a fixed-speed squirrel-cage induction generator (SCIG), which are directly connected to the power grid.

Figure 8.5a depicts the distribution of stator and rotor windings. The concentrated representations of stator and rotor windings are represented in Figure 8.5a. However, in practice, the windings of a stator and rotor are approximately distributed sinusoidal windings. The axes of these windings are displaced by 120°. The sinusoidal distribution of stator windings are measured with angle φ_s and sinusoidal distribution of rotor windings are measured with angle φ_r. The angle θ_r represents the rotor angle as it rotates around the air gap.

When the stator windings are excited with balanced three-phase sinusoidal currents, each phase winding produces a pulsating sinusoidal vector field along the winding axis and pointing to where the field is maximum positive as shown in Figure 8.5a. The effect of the three-phase winding vector field distribution is equivalent to having a single sinusoidal distributed vector field. For a two-pole machine, if the stator windings are excited by a 60-Hz source, the synchronous speed of the two-pole vector field is 3600 rpm. Figure 8.6 depicts the schematic of a squirrel cage induction generator (SCIG) system microgrid.

An electronic switch is used to provide a soft start to smooth the connection and disconnection of the generator to the grid by limiting the unwanted inrush current to about 1.6 times the nominal current. The circuit breaker can be automatically controlled by a microcontroller or manually controlled. A zero crossing is the instantaneous point at which there is no voltage (current) present. In

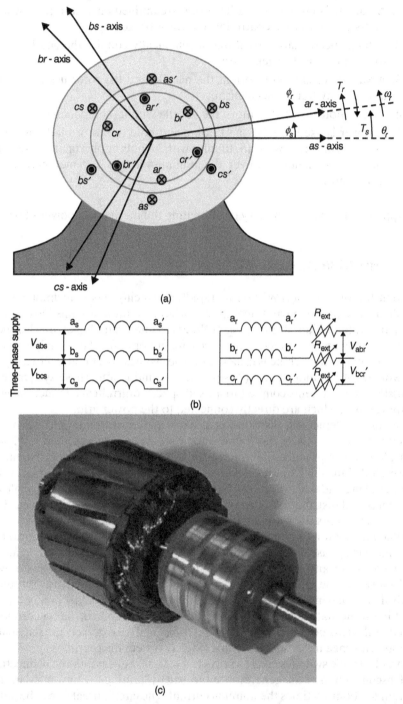

Figure 8.5 Induction machine types. (a) Wound machine stator and rotor windings. (b) Equivalent circuit.[14,16] (c) Wound field rotor.[17]

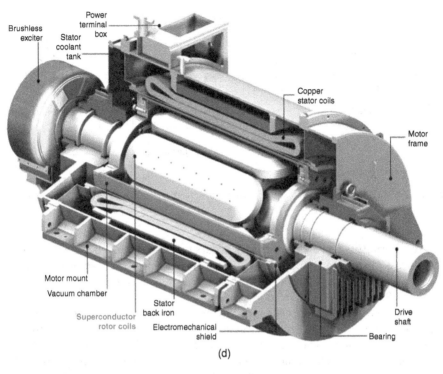

(d)

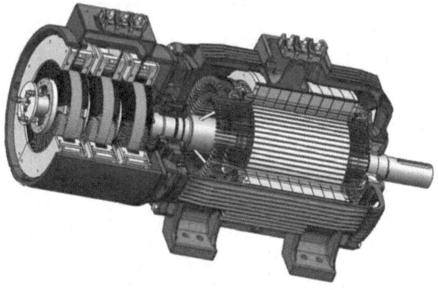

(e)

Figure 8.5 (*Coninued*)   (d) Superconducting ship propulsion motor.[18] (e) Cutaway of a doubly fed induction generator with a rotary transformer.[19]

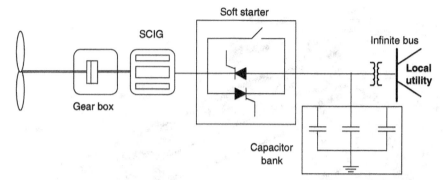

Figure 8.6   The schematic of a squirrel cage induction generator (SCIG) system microgrid.

an AC alternating wave, this normally occurs twice during each cycle as shown in Figure 8.7. Assuming the system operates at a lagging power factor, for example, at a lagging power factor of 0.86, the current then lags the voltage as shown in Figure 8.7.

The soft-switch circuit breaker is controlled by a microcontroller or a digital signal processor. The soft-switch circuit breaker is closed at a zero crossing of voltage and it is open at zero crossing of current. The soft switch is also protecting the mechanical parts of the turbines such as a gearbox and shaft against high forces. The soft switch uses controllable thyristors to connect the generator to

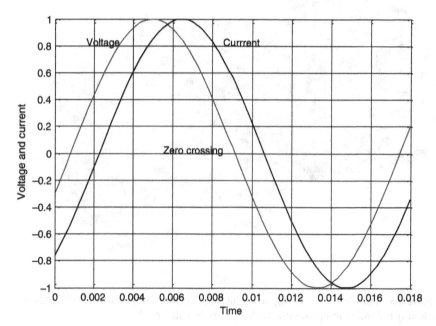

Figure 8.7   The zero crossing of an AC source supplying power at 0.86 power factor lagging.

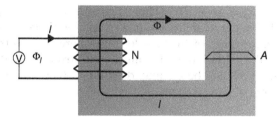

Figure 8.8   An inductor winding.

the induction generator at zero crossing of the sinusoidal voltage of the generator 2 seconds after the induction machine operates above the synchronous speed.

8.4 THE MODELING OF INDUCTION MACHINES

In the three-phase round induction machines with wound windings on the rotor, the stator windings consist of three-phase windings. The rotor winding also consists of three-phase wound windings. For a squirrel cage induction machine, the rotor of the machine does not have winding; instead it has a cage that is normally constructed from aluminum.

For mathematical modeling of the machine's steady-state model,[20,21] it is assumed that the stator and rotor windings are balanced. In the case of a squirrel cage induction machine, it is assumed the rotor is represented by an equivalent winding. Each phase of the stator winding is designed using the same wire size and occupies the same space on the stator of the machine. The same is true for rotor winding. To understand the basic relationship between the current flowing in the windings and the resulting field distribution, we need to recall the relationship of field distribution in the machine stator and rotor and machine air gap. Let us first review the fundamentals of field distribution in a rectangular structure of an inductor depicted in Figure 8.8. These discussions facilitate our understanding of the basic concept of inductance and induction, which is used later in this chapter.

According to the fundamentals of electromagnetics, the field intensity can be expressed as the product of H in A/m, times the mean length of the magnetic core is equal to number of turns, N times the current, I flowing in the winding as given below.

$$Hl = NI \tag{8.1}$$

From the fundamental relationship the field intensity, H and flux density, B in weber per square meter (Wb/m²), and the permeability, μ of the core material are related as:

$$H = \frac{B}{\mu} \tag{8.2}$$

where $\mu = \mu_r\,\mu_o$; $\mu_o = 4\pi \times 10^{-7}$ (H/m), and μ_r is relative permeability. For example, the relative permeability is equal to one for air. For iron, the relative permeability is in the range of 26,000–360,000. Therefore, μ is the slope of the $B = \mu H$ and H of the field winding. The flux lines distribution in the rectangular structure of Figure 8.8 can be computed by multiplying the field density by cross-sectional area of core, A expressed as:

$$\Phi = BA \tag{8.3}$$

The flux linkage Φ represents the flux lines that link the winding; they are expressed in weber. The flux leakage, Φl represents the flux lines that are leaked in the surrounding medium and do not remain in the core structure.

If we assume the leakage flux lines are zero, then we can express the product of field intensity, H, and the mean length of the core to be equal to magneto-motive force, abbreviated as mmf, and shown by Equation (8.4).

$$Hl = \text{mmf} \tag{8.4}$$

Therefore, the mmf is the force of the number of turn times of the current flowing through the winding, NI, which creates the force that produces the magnetic flux that is the result of the field intensity H.

By substituting for H in Equation (8.4), we obtain the flux flow in the stator structure, rotor structure, and the air gap as:

$$\Phi \cdot \frac{l}{\mu \cdot A} = \text{mmf} \tag{8.5}$$

$$\mathfrak{R} = \frac{l}{\mu \cdot A} \tag{8.6}$$

Equation (8.6) is defined as reluctance. The reluctance is analogous to resistance. It is a function of the dimension of the medium through which the flux flows.

$$\Phi \cdot \mathfrak{R} = \text{mmf} \tag{8.7}$$

The reluctance is inversely proportional to inductance, which is directly proportional to the square of the number of turns. The inductance, L, represents the inductance of the inductor of Figure 8.8

$$L = \frac{N^2}{\mathfrak{R}} \text{ or } L = \mu\frac{N^2 A}{l} \tag{8.8}$$

To understand the permeability, μ, we need to understand the hysteresis phenomena. A hysteresis loop shows the magnetic characteristics of a ferromagnetic material under test. The relationship shows the relationship between the voltage applied to a winding with a core structure made from a ferromagnetic

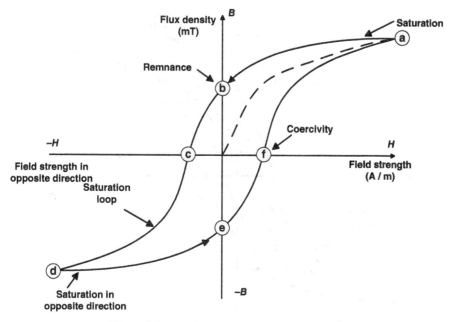

Figure 8.9 A schematic of a hysteresis loop.[22]

material and resulting current flow in the winding. The induced magnetic flux density (B) is calculated from the applied voltage and the magnetizing field intensity (H) from the current flow through the winding. It is often referred to as the B–H loop. An example of a hysteresis loop is shown in Figure 8.9. The loop is produced by measuring the magnetic flux of a ferromagnetic material while the applied voltage is changed.

A ferromagnetic core that has not been previously magnetized will follow the dashed line as the applied voltage is increased, resulting in higher current and higher magnetic intensity, H is increased. The greater the applied excitation current to produce, $H+$, the stronger is the magnetic field density, $B+$. For a Point a, depending on the type of magnetic material, however, a higher applied voltage will not produce an increased value in magnetic flux. At the point of magnetic saturation the process is changed in the opposite direction as shown in Figure 8.9 and recorded voltage and current representing flux density and field intensity moves from Point a to Point b. At this point, some residual flux remains in the material even after the applied voltage is reduced to zero. This point is called retentivity on the hysteresis loop and shows the residual magnetism (remanence) in the material. As the applied voltage direction is changed, the recorded flux density and field intensity moves to Point c. At this point the flux is reduced to zero. This is called the point of coercivity on the hysteresis loop. When the voltage is applied in a negative direction, the material is saturated magnetically in the opposite direction and reaches Point d. As the applied voltage is increased in positive direction, again H crosses zero at

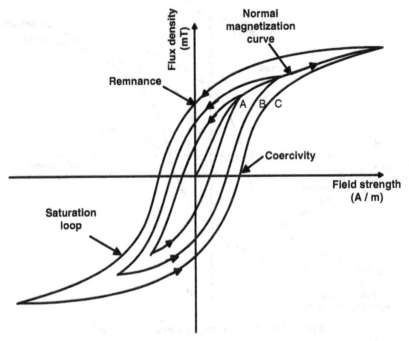

Figure 8.10   A family of hysteresis loops as applied voltage to the winding is varied.[22]

some value of the applied voltage and the hysteresis loop reaches the Point e. As the applied voltage is increased further, H is increased back in the positive direction, the flux density, B returns to zero. If the applied voltage repeats the same cycle, the process is repeated. If the process is reversed, the loop does not return to original position. Some force is required to remove the residual magnetism and to demagnetize the material.[22]

Figure 8.10 shows the traces of a family of hysteresis loops as applied voltage to the winding is varied. When the tips of the hysteresis loops, A, B, C, and D are connected, the result is a normalized magnetizing curve. Figure 8.11 depicts the normalized magnetization curve.

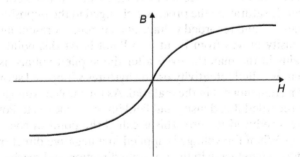

Figure 8.11   The normalized magnetization curve.

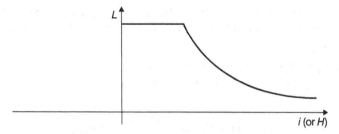

Figure 8.12   Schematic of inductance as a core structure saturated.

Because $B = \mu H$, the permeability, μ is the slope of hysteresis as expressed by Equation (8.9).

$$\mu = \frac{B}{H} \tag{8.9}$$

In the linear region of magnetization curve, the inductance is linear as given by Equation (8.8). Figure 8.12 shows the change in inductance of the rectangular core, as the core is saturated.

Figure 8.8 represents a simple rectangular structure. However, a three-phase induction machine has three windings on the stator and three windings on the rotor as shown in Figure 8.5b. The three windings on the stator are placed on the circumference of the stators at 120° apart. The three-phase rotor windings are also located on the circumference of the rotor at 120° apart. As a motor operation, the stator is supplied from the three-phase AC voltage source. The time-varying AC source would magnetize the stator structure and give rise to time-varying flux density and time-varying inductance for each phase of the stator. Although the cylindrical structure is more complex to calculate these inductances, we can use advanced finite element calculation and make discrete elements of the stator structure to calculate the flux in these elements. However, finite element calculation is an advanced topic that is not the subject of this book. From our study of the inductor, we can conclude that the inductances of the stator are a function of the diameter, number of turns, permeability of the stator material, and cross-sectional area and excitation current. Therefore, we recognize that the flux density variation in winding a-a' when it is energized from the AC alternating source can be expressed as:

$$B_a(\theta_r, t) = B(t) \cos\left(\frac{P}{D}\theta_r\right) \tag{8.10}$$

where $B(t)$ is time-varying flux density produced by the phase a applied voltage, $v_a(t)$.

$$B(t) = B_{max} \cos \omega_s t \tag{8.11}$$

In Equation (8.11), the $\omega_s = 2\pi f$ is the source frequency and θ_r is the angle around the circumference of the stator; P is number of poles, D is diameter and $\frac{P}{D}\theta_r$ angle in radians.

For example, $B_a(x, t) = B_{max}$ when $\theta_r = 0$.

And $\theta_r = \frac{2\pi D}{P}$, Equation (8.10) can be expressed as

$$B_a(x, t) = B(t) \cos\left(\frac{P}{D}\frac{2\pi D}{P}\right) = B_{max} \tag{8.12}$$

Substituting Equation (8.12) in Equation (8.10), we can obtain Equation (8.13).

$$B_a(x, t) = \hat{B} \cos \omega_s t \cos\left(\frac{P}{D}\theta_r\right) \tag{8.13}$$

Using the cosine identity in Equation (8.13), we can obtain Equation (8.14).

$$\cos \alpha \cos \beta = \frac{1}{2}[\cos(\alpha - \beta) + \cos(\alpha + \beta)]$$

$$B_a(x, t) = \frac{\hat{B}}{2} \cos\left(\frac{P\theta_r}{D} - \omega_s t\right) + \frac{\hat{B}}{2} \cos\left(\frac{P\theta_r}{D} + \omega_s t\right) \tag{8.14}$$

We can obtain the same expression for windings b-b' and c-c' as expressed by Equations (8.15) and (8.16).

$$B_b(x, t) = \hat{B} \cos\left(\omega_s t - \frac{2\pi}{3}\right) \cos\left(\frac{P\theta_r}{D} - \frac{2\pi}{3}\right) \tag{8.15}$$

$$B_c(x, t) = \hat{B} \cos\left(\omega_s t - \frac{4\pi}{3}\right) \cos\left(\frac{P\theta_r}{D} - \frac{4\pi}{3}\right) \tag{8.16}$$

Recall that the flux density is a vector quantity, we can add flux density generated by phase a, phase b, and phase c, and we can obtain the total flux generated as expressed by Equation (8.17).

$$B_{tot}(x, t) = B_a(\theta_r, t) + B_b(\theta_r, t) + B_c(\theta_r, t) \tag{8.17}$$

$$B_{tot}(x, t) = \frac{3}{2}\hat{B} \cos\left(\frac{P\theta_r}{D} - \omega_s t\right) \tag{8.18}$$

The peak value of flux density crossing the air gap of the machine is given by Equation (8.19).

$$B_{max} = \frac{3}{2}\hat{B} \tag{8.19}$$

The rotor windings are distributed on the rotor structure. The total time-varying flux density of stator windings crossing the air gap links the rotor windings and induces voltage in the rotor windings. Let us analyze the rotor induced voltage assuming two conditions: (1) the rotor is at standstill, and (2) the rotor is free to rotate.

1. The rotor is at standstill.
   $B_{tot}(\theta_r, t)$ is total flux distribution in the machine and frequency of flux wave is expressed as

$$\omega_s = 2\pi \cdot f_s \tag{8.20}$$

   f_s = stator frequency.
   This flux distribution can be regarded as a flux wave that is distributed in the air gap machine with mechanical equivalent speed as

$$\omega_{mech} \text{ (equivalent)} = \frac{2}{P} \cdot \omega_s \tag{8.21}$$

   The mechanical speed expressed by Equation (8.21) is also called the synchronous speed.

$$\omega_{syn} = \frac{2}{P} \cdot \omega_s \tag{8.22}$$

   Let us assume that we short the rotor windings and also assume that the rotor is restrained and remains at standstill. The stator flux wave distribution of stator crosses the machine air gap and links the rotor windings and induces the voltage in phases a, b, and c of the rotor windings. The resulting current flow in rotor windings gives rise to the rotor flux of rotor windings. Adding flux waves produced by phases a, b, and c of rotor windings, we can obtain,

$$B_R(x,t) = B_{aR}(x,t) + B_{bR}(x,t) + B_{cR}(x,t)$$
$$= \frac{3}{2}\hat{B}\cos\left(\frac{P\theta_r}{D} - \omega_{stator}t\right) \tag{8.23}$$

   Because the rotor is at standstill, that shaft speed is zero, the rotor frequency is the same as stator frequency. In fact, the induction machine is acting like a three-phase transformer; however, in a cylindrical structure with the rotor supported on a bearing system that separates the stator from the rotor with a very small air gap, the stator and rotor windings are coupled.

2. The rotor is free to rotate.

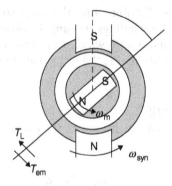

Figure 8.13   The schematic flux distributions for two-pole machines.

Suppose the rotor is free to rotate. Let us assume rotor is rotating at a speed of ω_m, then the rotor total flux wave distribution, $B_R(x,t)$ has the mechanical speed as expressed by Equation (8.24).

$$\omega_r \text{ (mech)} = \omega_{syn} - \omega_m \tag{8.24}$$

ω_r (mech) is the rotor mechanical speed that is the difference between the synchronous speed and the shaft speed because the shaft is free to rotate. The electrical speed of the rotor flux wave is expressed as given by Equation (8.25).

$$\omega_r = \frac{P}{2}\omega_r(\text{mech}) = \frac{P}{2}(\omega_{syn} - \omega_m) \tag{8.25}$$
$$\omega_r = 2\pi f_r$$

f_r = frequency of induced voltage (current) in the rotor.

The total flux distribution of stator flux, $B_s(\theta_r, t)$ has the equivalent mechanical speed at ω_{syn}. This flux distribution for a two-pole machine has two poles as shown in Figure 8.13. The total flux distribution of the rotor, $B_R(\theta_r, t)$ of the rotor rotates at $\omega_{syn} - \omega_m$ for a two-pole machine with respect to the rotor structure and $B_R(\theta_r, t)$ of the rotor structure rotates at ω_m (mechanical speed) with respect to the stator.

The two flux waves are distributed in the machine at an angle with respect to each other as shown in Figure 8.13. Motor action ω_{syn} and ω_m are rotating in the same direction and $\omega_{syn} > \omega_m$.

8.4.1 Calculation of Slip

$$\omega_r = \frac{P}{2}(\omega_{syn} - \omega_m) \tag{8.26}$$

Let us multiply and divide by ω_{syn} (stator electrical speed):

$$\omega_r = \frac{\omega_{syn} - \omega_m}{\frac{2}{P}\omega_{syn}}\omega_{syn} \qquad (8.27)$$

Let us define

$$s = \frac{\omega_{syn} - \omega_m}{\omega_{syn}} = \text{slip} \qquad (8.28)$$

$$\omega_r = s\omega_{syn} \text{ or } f_r = sf \qquad (8.29)$$

where ω_{syn} rad/sec, ω_m rad/sec. The slip can be calculated as presented in Equation (8.30).

$$s = \frac{n_{syn} - n_m}{n_{syn}} \qquad (8.30)$$

$$n_{syn} = \frac{120 f_s}{P} \qquad (8.31)$$

where n_{syn} is the synchronous speed in rpm and n_m is the rotor speed in rpm.

8.4.2 The Equivalent Circuit of an Induction Machine

Figure 8.14 depicts a three-phase Y-connected stator and rotor windings. As a motor operation, the stator is fed from a three-phase AC source or through a PWM inverter with a variable frequency. The rotor is short circuited or connected to an external resistance. As a motor, the machine operates at a shaft speed below the synchronous frequency and supports a mechanical load. As a wind generator, the mechanical wind power is supplied and the generator operates above the synchronous speed. To describe how the machine operates, we need to develop an equivalent circuit model for the machine. In Figures 8.15 and 8.16, we refer to stator winding with subscript "1" and rotor as subscript

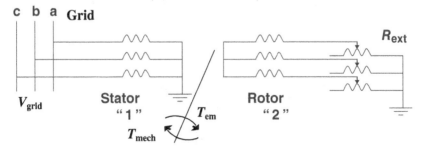

Figure 8.14 The equivalent circuit of a wound rotor induction machine with external resistance, $R_{ext,}$ inserted in the rotor circuit.

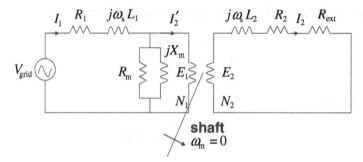

Figure 8.15   The equivalent circuit model of an induction machine at standstill.

"2." We look at two operating conditions: (1) the rotor is at standstill, and (2) the rotor is free to rotate.

1. The rotor is at standstill.

   Suppose the rotor is at standstill and the shaft speed is zero ($\omega_m = 0$). Using the approach used in modeling the coupled winding, the equivalent circuit for one phase to ground can be depicted by Figure 8.15.

   The machine is designed to use a small amount of current to magnetize it. Normally, the magnetizing current is less than 5% of the rated load current. This requires that magnetizing reactance, X_m, and resistance, R_m, which is net magnetizing impedance, acts as high impedance. Therefore, we can ignore the shunt elements. With this assumption, we can obtain the following:

$$I_1 = I_2' \tag{8.32}$$
$$\tilde{V}_{grid} = \tilde{I}_1(R_1 + j\omega_s L_1) + \tilde{E}_1 \tag{8.33}$$
$$\tilde{E}_2 = \tilde{I}_2(R_2 + R_{ext} + j\omega_s L_2) \tag{8.34}$$

2. The rotor is free to rotate.

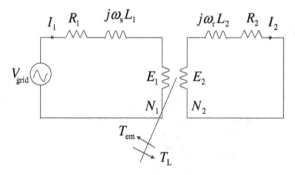

Figure 8.16   The equivalent circuit model of an induction machine when the rotor is free to rotate.

Suppose rotor speed is running at ω_m. Let us express $R_2 = R_{2,\text{rotor}} + R_{\text{ext}}$. Figure 8.16 depicts the equivalent circuit model of an induction machine when the rotor is free to rotate.

$$\tilde{V}_{\text{grid}} = \tilde{I}_1(R_1 + jX_1) + \tilde{E}_1 \tag{8.35}$$
$$\tilde{E}_2 = \tilde{I}_2(R_2 + j\omega_r L_2) \tag{8.36}$$

where R_1 is stator resistance/phase and $X_1 = \omega_s L_1$ is stator reactance/phase, I_1 is stator current (line) and E_1 is induced emf/phase.

Let us define

$$a = \frac{N_1}{N_2} \tag{8.37}$$

$$I_1 = \frac{I_2}{a} \tag{8.38}$$

Based on Faraday's law of induction, the following expression defines the induced voltage, e_1 and e_2.

$$e_1 = N_1 \frac{d\Phi}{dt} = N_1 \Phi_{\text{max}} \cos(\omega t + \theta_1) \tag{8.39}$$

$$e_2 = N_2 \frac{d\Phi}{dt} = sN_2 \Phi_{\text{max}} \cos(\omega t + \theta_2) \tag{8.40}$$

Then, the RMS values E_1 and E_2 can be expressed as

$$E_1 = \frac{N_1 \Phi_{\text{max}}}{\sqrt{2}} \text{ and } E_2 = \frac{sN_2 \Phi_{\text{max}}}{\sqrt{2}} \tag{8.41}$$

The ratio of E_1 and E_2 can be computed as

$$E_2 = sN_2 \frac{E_1}{N_1} = s\frac{E_1}{a} \tag{8.42}$$

And the I_2 and I_1 can be expressed as

$$\tilde{I}_2 = a\tilde{I}_1 \tag{8.43}$$

Using the above relationship, we can compute the equivalent model of induction machine from the stator side by referring the rotor variables to the stator sides.

$$\tilde{E}_2 = \tilde{I}_2(R_2 + j\omega_r L_2) \tag{8.44}$$

$$s\frac{\tilde{E}_1}{a} = a\tilde{I}_1(R_2 + j\omega_r L_2) \tag{8.45}$$

Equation (8.50) can by recomputed by multiplying by a and dividing by s.

$$s\tilde{E}_1 = \tilde{I}_1(a^2 R_2 + jsa^2 X_2) \tag{8.46}$$

$$\tilde{E}_1 = \tilde{I}_1\left(a^2\frac{R_2}{s} + ja^2 X_2\right) \tag{8.47}$$

Equation (8.47) represents the rotor variables from the stator side. The one-phase stator variables are given by Equation (8.48).

$$\tilde{V}_{\text{grid}} = \tilde{I}_1(R_1 + jX_1) + \tilde{E}_1 \tag{8.48}$$

Combining Equation (8.47) with Equation (8.48), we obtain the following:

$$\tilde{V}_{\text{grid}} = \tilde{I}_1(R_1 + jX_1) + \tilde{I}_1\left(a^2\frac{R_2}{s} + ja^2 X_2\right) \tag{8.49}$$

We can rewrite Equation (8.49) as defined by Equation (8.50):

$$R'_2 = a^2 R_2 X'_2 = a^2 X_2 \tag{8.50}$$

$$\tilde{V}_{\text{grid}} = \tilde{I}_1\left[\left(R_1 + \frac{R'_2}{s}\right) + j(X_1 + X'_2)\right] \tag{8.51}$$

The induction motor equivalent circuit from the stator side is given in Figures 8.17 and 8.18.

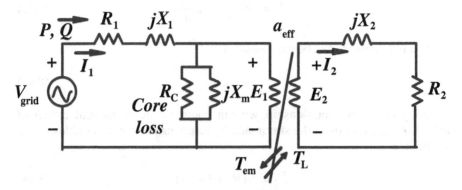

Figure 8.17 The equivalent circuit model of an induction machine with magnetizing inductance represented on the stator side.

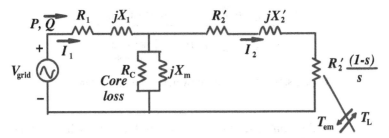

Figure 8.18   The equivalent circuit model of an induction machine with rotor variables referred to the stator side.

8.5 POWER FLOW ANALYSIS OF AN INDUCTION MACHINE

The input power can be calculated from the input voltage and current drawn by the machine as given by Equation (8.52).

$$P_i = 3\text{Re}\{V_1 I_1^*\} \tag{8.52}$$

The power crossing the air gap of the machine (see Figure 8.17 can be calculated by subtracting stator wire resistance losses and core losses as given by Equation (8.53).

$$P_{AG} = 3P_i - 3\left(I_1^2 R_1 + P_c\right) = 3\,|I_2|^2 \frac{R_2'}{s} \tag{8.53}$$

where P_c is the core loss shown by equivalent resistance R_c in Figure 8.18.

As can be observed from Equation (8.53), the air gap power can also be computed by the square of rotor current, I_2 and $\frac{R_2'}{s}$.

When the machine is operating as a motor, the power delivered to the shaft can be calculated by accounting for rotor losses.

$$P_{em} = P_{AG} - 3I_2^2 R_2' \tag{8.54}$$

By substituting for the air gap power in Equation (8.54), we can obtain Equation (8.55).

$$P_{em} = 3I_2'^2\frac{R_2'}{s} - 3I_2'^2 R_2' = 3I_2'^2\left(\frac{R_2'}{s} - R_2'\right) \tag{8.55}$$

$$P_{conv} = P_{em} = 3I_2'^2 R_2'\left(\frac{1-s}{s}\right) \tag{8.56}$$

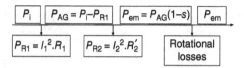

Figure 8.19   The power flow in induction machines.

Therefore, the electromagnetic power delivered to the machine shaft can be expressed as a function of air gap power by Equation (8.59).

$$P_{em} = 3I_2'^2 \frac{R_2'}{s} (1-s) = P_{AG} (1-s) \tag{8.57}$$

The flow of power from the stator to the machine shaft is depicted by power flow in Figure 8.19. The power flow is given from input power P_i, the air gap power, P_{AG}, electromagnetic power, P_{em}, and output power, P_o to the shaft of the machine.

An induction machine has three regions of operations. It can operate as a motor, as a generator, or as a brake. To describe these regions of operation, we need to study the machine torque as a function of speed.

Using Equation (8.58) and the machine model of Figure 8.15, we first need to calculate the machine torque as a function of input voltage. Figure 8.20 depicts the equivalent circuit of the induction machines neglecting the magnetizing element.

$$P_{em} = T_{em}\omega_m \tag{8.58}$$

In Equation (8.58), ω_m is the shaft speed and T_{em} is torque driving the shaft and P_{em} is the power supplied to the machine. We can calculate P_{em} by calculating the current supplied to the machine from the grid. Because the magnetizing element is ignored, the current, I_1 is equal to I_2'.

$$V_{grid} = I_1\{(R_1 + R_2) + j(X_1 + X_2')\} \tag{8.59}$$

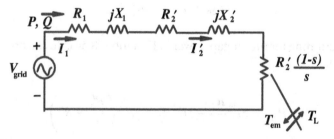

Figure 8.20   The equivalent circuit model of an induction machine, operating as a motor, from stator terminals omitting the magnetizing elements.

From the above equation, we can calculate the current I_1 as expressed by Equation (8.60).

$$|I_1|^2 = \frac{|V_{grid}|^2}{\left(R_1 + \frac{R_2'}{s}\right)^2 + (X_1 + X_2')^2} \tag{8.60}$$

The torque supplied to the rotor is expressed by Equation (8.61).

$$T_{em} = \frac{3}{\omega_m}\left(\frac{1-s}{s}\right)R_2'|I_1|^2 \tag{8.61}$$

The shaft speed can also be expressed as given by Equation (8.62).

$$\omega_m = \omega_{sync}(1-s) \tag{8.62}$$

By substituting Equations (8.61) and (8.62), we can express the shaft torque as a function of input voltage as given by Equation (8.63).

$$T_{em} = \frac{3}{\omega_{sync}}\frac{R_2'}{s}\frac{|V_{grid}|^2}{\left(R_1 + \frac{R_2'}{s}\right)^2 + (X_1 + X_2')^2} \tag{8.63}$$

It is instructive to study the machine performance for various values of external resistance for a constant input voltage. From careful examination of Equation (8.83), we can conclude the following:

a. When the slip is zero, the shaft speed, ω_m is equal the synchronous speed, ω_{sync} and the produced electromagnetic torque is zero.
b. When the slip, s is equal to one, that is, the shaft speed is zero (starting), the standstill torque (starting torque) can be obtained.
c. As the external resistance is changed using a controller, the value of maximum torque occurs at different values of the shaft speed.

The control of shaft speed over a wide range would provide the capability to operate the machine at various speeds. When the machine is used as an induction generator, we can capture the wind power and inject the generated power into the local grid.

There are three distinct points in an induction machine torque versus speed curve: (1) the starting torque when the shaft speed is zero; (2) when the shaft

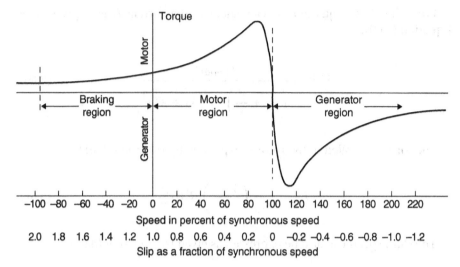

Figure 8.21   An induction machine's various regions of operation.

speed is equal the synchronous speed of the flux waves, the generated torque is zero (see Figure 8.21), and (3) the point of maximum torque production. To calculate the maximum torque, we compute the derivative of torque expression as given by Equation (8.63), with respect to slip and set it to zero.

$$\frac{dT_{\mathrm{m}}}{ds} = 0 \qquad\qquad (8.64)$$

The maximum slip point is given by Equation (8.65).

$$s_{\max} = \pm\frac{R_2'}{\sqrt{R_1^2 + \left(X_1 + X_2'\right)^2}} \qquad\qquad (8.65)$$

The resulting maximum torque is given by Equation (8.66).

$$T_{\max} = \pm\frac{3}{2\omega_{\mathrm{sync}}}\frac{|V_{\mathrm{grid}}|^2}{\left[R_1 \pm \sqrt{R_1^2 + \left(X_1 + X_2'\right)^2}\right]} \qquad\qquad (8.66)$$

Equation (8.65) shows that the slip at which the maximum torque occurs is directly proportional to the rotor resistance. Equation (8.66) shows that the maximum torque is independent of the rotor resistance. However, it is directly proportional with the square of input voltage.

8.6 THE OPERATION OF AN INDUCTION GENERATOR

Wound rotor induction machines have stators like the squirrel cage machine. However, their rotors' winding terminals are brought out via slip rings and brushes for torque and speed control. For torque control no power is applied to the slip rings. External resistances are placed in series with the rotor windings during starting to limit the starting current. Without the external resistances, the starting currents are many times the rated currents. Depending on the size of the machine, it can draw 300% to over 900% of full-load current. The resistances are shorted out once a motor is started. The external resistances are also used to control the machine speed and its starting torque. The negative slip operation of an induction machine indicates that the machine is operating as a generator and power is injected into the local power grid.[23] Induction generators required reactive power for magnetizing the rotor, and this power is supplied by the following different methods.

1. The machine is magnetized and started as an induction. Then, the wind turbine is engaged to supply mechanical power and shaft speed is increased above synchronous speed. As a standalone, we can use any induction machine and by adding capacitors in parallel with the machine terminals, then driving it above synchronous speed, the machine would operate as an induction generator. The capacitance helps to induce currents into the rotor conductors and AC current is produced. The loads can be connected to the capacitor leads because the capacitors are in parallel. This method of starting a standalone machine is assured, if the residual magnetism in the rotor exits. However, the machine can be magnetized by using a direct current source such as a 12 V battery for a very short time.

2. Rotor resistance of the induction generator is varied instantly using a fast power electronics controller. Variable rotor speed (consequently variable slip) provides the capability to increase the power captured from the wind at different wind speeds. This can be achieved if rotor winding terminals can be accessed by changing slip via external resistors connected to the generator rotor winding. The external resistors will only be connected to produce the desired slip when the load on the wind turbine becomes high.

Figure 8.22 presents equivalent circuits of induction machines. The relationships between the variation of external resistance and the location of maximum and standstill torques can be studied by varying the external resistance in the rotor circuit. We can study this relationship by writing a MATLAB M-file as described in Example 8.1.

The power from the rotation of the wind turbine rotor is transferred to the induction generator through a transmission train consisting of the main turbine shaft, the gearbox, and the high-speed generator shaft. The normal range of wind speed is not high and changes during the day and with the seasons. For

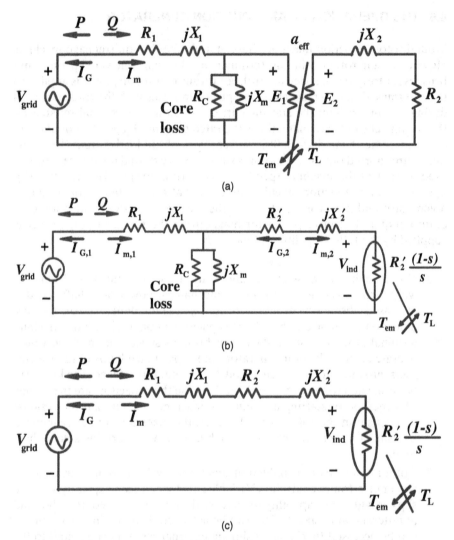

Figure 8.22 (a) Equivalent circuit of an induction generator. (b) Equivalent circuit of an induction generator with all quantities referred to the stator side. (c) Equivalent circuit of an induction generator referred to the stator neglecting magnetizing and loss components.

example, a three-phase generator with two poles directly connected to a local power grid operating at 60 Hz is running at 3600 rpm. We can reduce the speed of the generator rotor, if the generator has a higher number of poles. For example, if the number of poles is four, six, or eight, the shaft speed reduces to 1800, 1200, and 900 rpm, respectively. A power grid supplied by a wind generator requires much higher wind speed in the range of 900–3600 rpm. To reduce the speed of the generator rotor, we can increase the number of poles. However, designing a generator with a very high number of poles requires a large diameter and will result in high volume and weight. We can increase the low speed and high torque of the wind turbine to low torque and high speed using gear systems. Before discussing the gear concepts, we need to understand how to convert linear velocity to rotational velocity.

From fundamental physics, we know the linear velocity and rotational velocity can be expressed as

$$V = r \cdot \omega \tag{8.67}$$

where V is the linear velocity in meters per second, r is the radius in meter, and rotational velocity, ω is in radians per second.

We can convert the rotational velocity, ω to revolutions per minute as

$$\omega = N\frac{2\pi}{60}$$

In the above, N is in revolutions per minute. Equation (8.67) can be restated as

$$V = N\frac{2\pi}{60}r \tag{8.68}$$

where V is in meters per second, N is in rotations per minute, and r is in meters. Conversely, we can rewrite the above as

$$N = V \cdot \frac{60}{2\pi \cdot r} \tag{8.69}$$

In Equation (8.69), V is in meters per second, r is in meters, and N is in rotations per minute. Conversely, the unit in Equation (8.69) can be in feet per second for V, r in feet, and N is in rotations per minute.

If we want to express the velocity in miles per hour (mph), we rewrite Equation (8.69) as

$$\frac{5280}{3600}V = \frac{\pi r \cdot N}{30} \tag{8.70}$$

where V is in miles per hour, r is in feet, and N is in rotations per minute. In Equation (8.70), we have substituted the value of 5280 feet for one mile and 3600 seconds for an hour. Equation (8.70) can be restated as

$$N = \frac{14.01 \cdot V}{r} \tag{8.71}$$

The wind speed in the range of 10 miles per hour (14.67 ft/sec) will result in 124.2 rotations per minute.

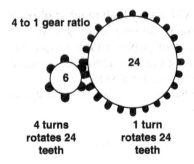

4 to 1 gear ratio

24

6

4 turns	1 turn
rotates 24	rotates 24
teeth	teeth

Figure 8.23 A schematic of a gear mechanism.

The wind generators are designed to capture the wind power in the range less than 120 rpm using a gear box transmission system. The gear box operates like a transformer. The gearbox provides speed and torque conversions.

$$T_{input}\omega_{input} = T_{output}\omega_{output} \tag{8.72}$$

where T_{input} and T_{output} are torque in newton-meter and ω_{input} and ω_{output} in radian per second. To provide torque speed conversion, a gearbox transmission is used. A gearbox is designed with a number of teeth.

$$\text{Gear ratio} = \frac{\text{Input number of teeth}}{\text{Output number of teeth}} = \frac{T_{input}}{T_{output}} \tag{8.73}$$

The gearbox converts low speed and high torque powers of a wind turbine rotor to low torque power and high speed of the generator rotor. The gear ratio is in the range of 1–100 for a wind generator in the range of 600 kW to 1.5 MW. Figure 8.23 depicts a schematic of a gear mechanism.

Example 8.1   Consider a three-phase Y-wound rotor-connected induction machine operating as a generator in parallel with a local power grid. The machine is rated at 220 V, 60 Hz, and 14 kW, with eight poles and the following parameters:

Stator resistance (R_1) of 0.2 ohms per phase (Ω/phase) and reactance of (X_1) of 0.8 Ω/phase.
Rotor resistance (R_2') of 0.13 Ω/phase and reactance (X_2) of 0.8 Ω/phase.

Ignore magnetizing reactance and core losses.
Perform the following:

(i) Develop a one-line diagram and one-phase equivalent model.
(ii) If the prime mover speed is 1000 rpm, determine the active and reactive power between the local grid and wind generator. How much capacitor

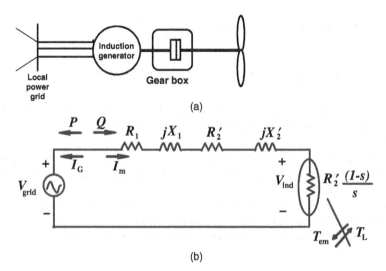

Figure 8.24   (a) A one-line diagram. (b) A one-phase equivalent circuit of an induction generator.

must be placed at the machine stator terminal for unity power factor operation?

(iii) Plot the torque speed characteristics of the machine at different values of external resistances.

Solution

The synchronous speed, $N_s = \dfrac{120f}{P} = \dfrac{120 \times 60}{8} = 900$ rpm

The rotor speed, $N_r = 1000$ rpm

The slip, $s = \dfrac{N_s - N_r}{N_s} = \dfrac{900 - 1000}{900} = -0.111$

Writing Kirchhoff's voltage law (KVL) for the circuit of Figure 8.24b, the current in motor convention,

$$I_m = \frac{V_{grid}}{R_1 + R_2'/s + j\left(X_1 + X_2'\right)}$$

$$= \frac{220/\sqrt{3}}{0.2 - 0.13/0.111 + j(0.8 + 0.8)} = 67.88\angle - 121.22°$$

For motor operation, the power flows from the grid to the induction machine. For generator operation, the direction of active power is reversed. Using motor convention and calculating the current, the angle of the current is greater than 90°. This means that the direction of current is opposite to the direction as shown by generator convention in Figure 8.24b.

With current following the generator convention and flowing from induction generator to the grid, the current $I_G = 67.88\angle 180 - 121.22° = 67.88\angle 58.77°$.

The induced voltage is given as $V_{\text{ind}} = \dfrac{1-s}{s} R_2' \cdot I_G$

$$= \frac{1-(-0.111)}{-0.111} \times 0.13 \times 67.88\angle 58.77° = 88.25\angle 58.77°$$

Neglecting the mechanical losses, the electrical power generated from the supplied wind mechanical power is given as $S_{\text{ind}} = 3V_{\text{ind}} \cdot I_G^* = 3 \times 88.25\angle 58.77° \times 67.88\angle -58.77° = 17.93 + j0$ kVA.

From S_{ind}, it is seen that the induction generator produces active power. However, it does not produce any reactive power. The reactive power is supplied by the grid.

The power produced by the induction generator is injected to the local grid and some power is lost in stator and rotor resistance.

The complex power injected to the grid is given by $S_{\text{grid}} = 3V_{\text{grid}} \cdot I_G^*$

$$= 3 \times \frac{220}{\sqrt{3}}\angle 0° \times 67.88\angle -58.77° = 13.41 - j22.12 \text{ kVA}$$

The induction generator feeds 13.41 kW of active power to the grid, but consumes 22.12 kVAr of reactive power from the grid.

Let a three-phase Y-connected capacitor bank be connected at the terminals of the induction generator. For unity power factor operation, the capacitor bank must supply the reactive power demand of the induction generator.

$$C_P = \frac{3V_{\text{grid}}^2}{2\pi f \cdot Q} = \frac{3 \times \left(220/\sqrt{3}\right)^2}{2\pi \times 60 \times 22.12 \times 10^3} = 5.8 \text{ mF}$$

Figure 8.25 depicts an induction generator with its reactive power supplied locally by a capacitor. Let us assume that the value of external resistance in varied from 0 to 0.75 in steps of 0.25 and the speed of the machine is varied from zero to the synchronous speed. In the following MATLAB testbed, a plot

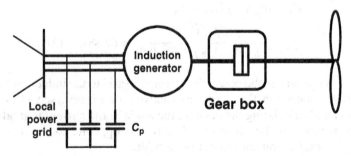

Figure 8.25 A squirrel-cage induction generator with reactive power supplied locally by a capacitor bank.

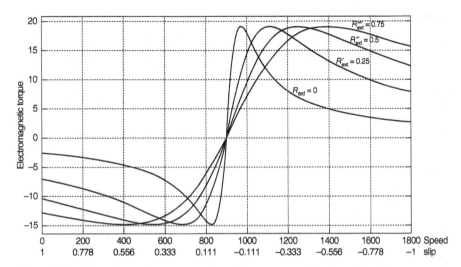

Figure 8.26 The schematic presentation of torque as a function of various external resistances with generator convention.

is made for the different values of R_{ext}. Figure 8.26 depicts the schematic presentation of torque as a function of various external resistances with generator convention.

```
%TORQUE vs SPEED
clc; clear all;
v1=220/sqrt(3);
f=60;
P=8;
r1=0.2;
x1=0.8;
r2d=0.13;
x2d=0.8;                  %The electrical quantities are defined
ws=120*f/P;
Tmax=-(3/2/ws)*v1^2/(r1+sqrt(r1^2+(x1+x2d)^2))
Tmax_gen=-(3/2/ws)*v1^2/(r1-sqrt(r1^2+(x1+x2d)^2))
w=0:1:2*ws;
for r_ext=0:0.25:0.75    %the value of external resistance is varied
    Tstart=-(3/ws)*((r2d+r_ext)/1)*v1^2/((r1+(r2d+r_ext)/1)^2
            +(x1+x2d)^2)
    smax=(r2d+r_ext)/sqrt(r1^2+(x1+x2d)^2)
    for j = 1:length(w)
        s(j)=(ws-w(j))/ws;
        Tem(j)=-(3/ws)*((r2d+r_ext)/s(j))*v1^2/((r1+(r2d+r_ext)/
            s(j))^2+(x1+x2d)^2);
    end
    plot(w,Tem,'k','linewidth',2)
    hold on;
end
```

TABLE 8.1   The Results of Example 8.1

External Resistance (Ω)	Starting Torque (N-m)	Slip at Maximum Torque (N-m)	Maximum Torque (N-m)
0.00	−2.62	0.08	
0.25	−7.06	0.24	−14.84 (Motor)
0.50	−10.43	0.39	19.04 (Generator)
0.75	−12.70	0.55	

```
grid on;
xlabel('Speed')
ylabel('Electromagnetic Torque')
axis([0 2*ws 1.1*Tmax_gen 1.1*Tmax])
gtext('R_e_x_t=0')
gtext('R_e_x_t^,=0.25')
gtext('R_e_x_t^,^,=0.5')
gtext('R_e_x_t^,^,^,=0.75')
```

The results are tabulated in Table 8.1.

The operation of an induction generator is the same as an induction motor except the direction of power flow is from wind power driving the shaft of the machine. Therefore, an induction generator injects or supplies power to the local power grid. For motor convention, the positive current flows from the source to the motor. For generator convention, the positive current flows from the motor terminal voltage to the local power grid. This means the sign of the current with motor convention will be negative for generator operation.

Example 8.2   For the machine of Example 8.1 with the same supply voltage connected to the local utility, plot the torque-slip characteristics in the speed range of 1000–2000 rpm with motor convention. Write a MATLAB M-file testbed and plot. Figure 8.27 depicts the plot of torque versus speed of the induction machine of Example 8.2.

Solution
The MATLAB M-file testbed for Example 8.2 is given below.

```
%TORQUE vs SPEED
clc; clear all;
v1=220/sqrt(3);
f=60;
P=8;
r1=0.2;
x1=0.8;
r2d=0.13;
x2d=0.8;              %The electrical quantities are defined
ws=120*f/P;
```

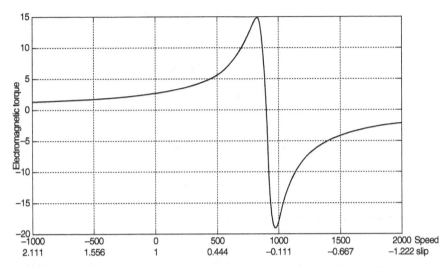

Figure 8.27   The plot of torque versus speed of the induction machine of Example 8.2.

```
w=-1000:0.2:2000;
for j = 1:length(w)
   s(j)=(ws-w(j))/ws;
   Tem(j)=(3/ws)*(r2d/s(j))*v1^2/((r1+r2d/s(j))^2+(x1+x2d)^2);
end
plot(w,Tem,'k','linewidth',2)
hold on;
grid on;
xlabel("Speed")
ylabel("Electromagnetic Torque")
```

The following observations can be made from the above examples:

When ω_{syn} and ω_m are rotating in the same direction and ω_{syn} is rotating faster than ω_m: this condition describes the normal operation of the induction machine as a motor.

In this region, the slip as given by Equation (8.70) is positive because both are rotating in the same direction.

$$s = \frac{\omega_{syn} - \omega_m}{\omega_{syn}} \tag{8.74}$$

$$R_{eff} = \frac{1-s}{s}R_2' \tag{8.75}$$

When ω_{syn} and ω_m are rotating in the same direction and ω_{syn} is rotating slower than ω_m, this operating condition describes the operation of a machine in generator mode. In this region, ω_{syn} and ω_m are rotating in the same direction. However ω_{syn} is rotating slower than ω_m. Therefore, in this region, mechanical power is supplied to the shaft by an external source and ω_m is greater than

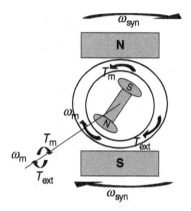

Figure 8.28   The operation of an induction machine as a generator.

ω_{syn} and slip is negative ($s < 0$). This region ($s < 0$) corresponds to the generator operation. For this region, the equivalent effective resistance as given by Equation (8.71) of the rotor is negative (see Figure 8.22) and the corresponding power (torque) is also negative. This means that the mechanical power is driving the machine and the machine in turn delivers electric power at its stator terminals to the source. Figure 8.28 depicts the operation of an induction machine as a generator.

The generator operation can be summarized as follows:

(i) $\omega_m > \omega_{syn}$.
(ii) ω_m and ω_{syn} are rotating in the same direction.
(iii) Generator action: Electrical power is supplied to the network via stator terminals.
(iv) $s = \dfrac{\omega_{syn} - \omega_m}{\omega_{syn}} < 0$

The negative slip indicates generator action.

When ω_{syn} and ω_m are rotating in opposite directions, the machine is operating in the brake region. The brake mode of operation can be implemented when an induction motor is operating under normal conditions at some value of positive slip in a stable region ($0 < s < s_{max}$); then we interchange any two terminals of the stator. This operation reverses the direction of the stator rotating field. The rotor speed ω_m may now be considered as negative with respect to that of the stator field. For this case, $s > 1$ and power loss is negative, indicating that mechanical energy is being converted to electric energy. The power fed from the stator and the power fed from the rotor are both lost as heat in the rotor resistance. This region is called the braking region.

Example 8.3   The air-gap power of an eight-pole 60-Hz induction machine, running at 1000 rpm is 3 kW. What are the rotor copper losses?

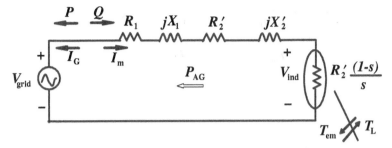

Figure 8.29   A single-phase equivalent circuit of an induction generator.

Solution
Figure 8.29 depicts the flow of currents in single-phase equivalent circuit of an induction generator.
   The air-gap power is

$P(\text{Input power to rotor}) = P_{AG_\varphi} = 3|I_2|^2 \frac{R_2'}{s}$
The rotor copper loss is given by

$$P_{\text{rotor loss}} = 3|I_2|^2 R_2'$$

Therefore,

$$\frac{P_{\text{rotor loss}}}{P_{\text{rotor in}} = P_{AG}} = \frac{3|I_2|^2 R_2'}{3|I_2|^2 \frac{R_2'}{s}} = s$$

The rotor power loss, $P_{\text{rotorloss}}$, is slip times the air-gap power:

$$P_{\text{rotorloss}} = sP_{AG}$$

For motor convention, $P > 0$ indicates power is being consumed by the machine.
   For generator convention, $P > 0$ indicates power is being generated by the machine.
   Synchronous speed is given by $N_{\text{syn}} = 120\frac{f}{P}$

$$= 120\frac{60}{8} = 900 \text{ rpm}$$

Slip is given by $s = \frac{N_{\text{syn}} - N_m}{N_{\text{syn}}}$

$$= \frac{900 - 1000}{900} = -0.11$$

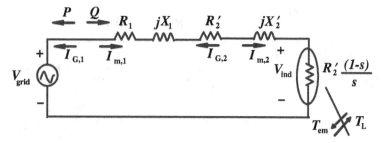

Figure 8.30 The single-phase equivalent circuit of induction generator.

The machine is operating as an induction generator.

Therefore, following generator convention, reversing the direction of air-gap power, the rotor power loss is given by

$$P_{rotorloss} = -sP_{AG} = 0.111 \times 3000 = 333 \text{ kW}$$

Example 8.4 A three-phase, six-pole, Y-connected induction generator rated at 400 V, 60 Hz is running at 1500 rpm supplying a current of 60 A at a power factor of 0.866 leading. It is operating in parallel with a local power grid. The stator copper losses are 2700 W, rotational losses are 3600 W.

Perform the following:

   (i) Determine the active and reactive power flow between induction generator and the power grid.

  (ii) Determine how much reactive power must be supplied at the induction generator to operate the induction generator at unity power factor.

 (iii) Calculate the efficiency of the generator.

 (iv) Find the value of stator resistance, rotor resistance, and the sum of the stator and rotor reactances.

  (v) Compute the electromechanical power developed by the rotor.

Solution

Figure 8.30 depicts active and reactive power flows in the single-phase equivalent circuit of induction generator.

Figure 8.31 depicts the power flow diagram for the generator mode of the operation of an induction machine.

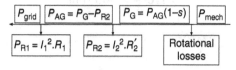

Figure 8.31 A power flow diagram for the generator mode of the operation of an induction machine.

Synchronous speed is given by $N_{syn} = 120\frac{f}{P}$

$$= 120\frac{60}{6} = 1200 \text{ rpm}$$

Slip is given by

$$s = \frac{N_{syn} - N_m}{N_{syn}} = \frac{1200 - 1500}{1500} = -0.25$$

The active power at the stator terminals is given by $P_{grid} = 3V_{L-N} \cdot I_s \cdot \cos\theta$

$$= 3 \times \frac{400}{\sqrt{3}} \times 60 \times 0.866 = 36,000 \text{ W}$$

The reactive power at the stator terminals is given by $Q_{grid} = 3V_{L-N} \cdot I_s \cdot \sin(\cos^{-1}\theta)$

$$= 3 \times \frac{400}{\sqrt{3}} \times 60 \times \sin(\cos^{-1} 0.866) = 20,786 \text{ VAr}$$

For unity power factor operation, the reactive power that needs to be supplied locally is the same as

$$Q_{grid} = 20,786 \text{ VAr}$$

Stator copper loss, $P_{R1} = 2700 \text{ W}$
The air-gap power is $P_{AG} = P_{grid} + P_{R1}$

$$= 36000 + 2700 = 38,700 \text{ W } (3\varphi)$$

The fixed loss, $P_{rotational\ loss} = 3600 \text{ W}$
Let P_G is the electromechanical power developed in the rotor

$$P_{mech} = P_G + P_{rotational} = (1-s)P_{AG} + P_{rotational}$$
$$= (1-(-0.25))38700 + 3600 = 51,975 \text{ W}$$

The efficiency, $\eta = \frac{P_{elec}}{P_{mech}} = \frac{36,000}{51,975} = 0.6276 = 69.26 \%$

The stator resistance is given by $R_1 = \frac{P_{R1}}{3I^2}$

$$= \frac{2700}{3 \times 60^2} = 0.25 \text{ }\Omega$$

Following generator convention,

Rotor copper loss is given by $P_{R2} = -sP_{AG}$

$$= -(-0.25) \times 38700 = 9675 \text{ W}$$

The rotor resistance is given by $R_2' = \dfrac{P_{R2}}{3I^2}$

$$= \frac{9675}{3 \times 60^2} = 0.90 \ \Omega$$

The sum of reactances of the rotor and the stator is given by $X = X_1 + X'_2 = \dfrac{Q_{\text{grid}}}{3I^2}$

$$= \frac{20,786}{3 \times 60^2} = 1.92 \ \Omega$$

The electromechanical power developed by the rotor $= -3I^2 \frac{1-s}{s} R_2'$

$$= -3 \times 60^2 \times \frac{1 - (-0.25)}{-0.25} \times 0.9 = 48,600 \text{ W}$$

8.7 DYNAMIC PERFORMANCE

In the previous sections, we analyzed the steady state of induction machines. For dynamic analysis, we must model the machines by a set of differential equations. For stator windings, we have three coupled windings that are sinusoidally distributed around the stator. When these coupled windings are represented in terms of self-inductances and mutual inductances, they give rise to a set of three time-varying differential equations. Similarly, we can obtain three time-varying differential equations for the rotor windings. The electromagnetic torque can be expressed by a nonlinear algebraic equation and the rotor s dynamic can be represented by a differential equation expensing the rotational speed of the motor. Therefore, the induction machine dynamic performance can be expressed by seven differential equations and one algebraic equation. The dynamic modeling of an induction machine is an advanced concept that students need to study with additional coursework.[20,21] Here it is instructive to study the results of a dynamic analysis as depicted by Figures 8.32 through 8.34.

Figure 8.32 depicts the start-up condition of an induction machine. As expected, the machine stator current has many cycles of transient oscillation before it reaches the steady-state current. The steady-state current supplied by the source magnetizes the machine and is lost as heat because the machine is operating at no load.

Figure 8.33 depicts the machine shaft speed from rest (start-up) to no load speed that is just below the synchronous speed.

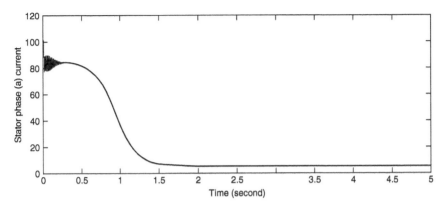

Figure 8.32   The induction machine stator current for a no load start-up.

Figure 8.34 depicts the transient oscillations of a machine. As can be seen from Figure 8.34, the machine goes through 0.4 seconds of oscillations and reaches it maximum torque. The normal region of the induction machine operation is below maximum speed. In Figure 8.34, the motor operation of the machine is presented. Because the machine is simulated at no load from rest, the machine develops enough torque to support the machine resistive and rotational losses.

The schematic of a microgrid of an induction machine controlled as a generator is presented in Figure 8.35.

Example 8.5   Consider the microgrid given by Figure 8.36 with the following data:

(i) Transformer rated at 440 V/11 kV, with reactance of 0.16 Ω and resistance of 0.02 Ω, and rated 60 kVA.

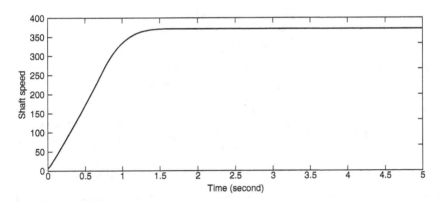

Figure 8.33   Induction machine shaft speed for no load start-up.

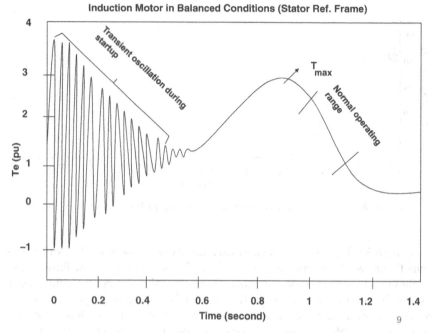

Figure 8.34 The dynamic performance of an induction machine.

(ii) The induction machine rated at 440 V, 60 Hz, three-phase, eight-pole; 50 kVA, 440 V, 60 Hz; stator resistance of 0.2 Ω/phase; rotor referred resistance in the stator side of 0.2 Ω/phase, stator reactance of 1.6 Ω/phase; rotor referred reactance of 0.8 Ω/phase. The generator speed is 1200 rpm. Perform the following:

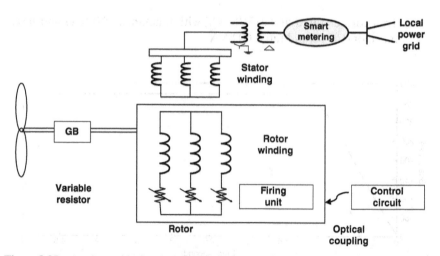

Figure 8.35 A microgrid of an induction machine controlled as an induction generator supplying power to the local power grid.

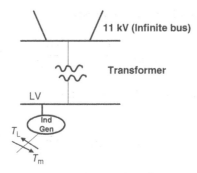

Figure 8.36   The microgrid connected to a local power grid for Example 8.5.

(iii) Give the per unit (p.u) model of the system.
(iv) Compute the power delivered to local power grid.
 (v) Compute the reactive power flow between grid and induction genera-
     tor. Assume base values equal to the rating of the induction machine.

Solution
Figure 8.37 depicts one-phase per unit equivalent circuit for example 8.5

 (i) Selecting the induction machine's rating as base,

$$S_b = 50 \text{ KVA}$$
$$V_b = 440 \text{ V}$$

The base impedance of the system

$$Z_b = \frac{V_b^2}{S_b} = \frac{440^2}{50 \times 10^3} = 3.872$$

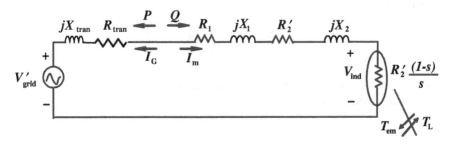

Figure 8.37   One-phase per unit equivalent circuit for Example 8.5.

The p.u value of transformer resistance

$$R_{tran,p.u} = \frac{R_{tran}}{Z_b} = \frac{0.02}{3.872} = 0.005$$

The p.u value of transformer reactance

$$X_{tran,p.u} = \frac{X_{tran}}{Z_b} = \frac{0.15}{3.872} = 0.04$$

The p.u value of stator resistance

$$R_{s,p.u} = \frac{R_s}{Z_b} = \frac{0.2}{3.872} = 0.052$$

The p.u value of rotor resistance referred to primary

$$R'_{r,p.u} = \frac{R'_r}{Z_b} = \frac{0.2}{3.872} = 0.052$$

The p.u value of stator reactance

$$X_{s,p.u} = \frac{X_s}{Z_b} = \frac{1.6}{3.572} = 0.413$$

The p.u value of rotor resistance referred to primary

$$X'_{r,p.u} = \frac{X'_r}{Z_b} = \frac{0.8}{3.572} = 0.207$$

(ii) Synchronous speed, N_s,

$$N_s = \frac{120f}{P} = \frac{120 \times 60}{8} = 900 \text{ rpm}$$

Slip, s is

$$s = \frac{N_s - N}{N_s} = \frac{900 - 1200}{900} = -0.333$$

where N is the motor shaft (rotor) speed in rpm.
The rotor voltage frequency

$$f_r = s.f_s$$
$$f_r = 0.333 \times 60 = 20 \text{ Hz}$$

The supply voltage is 440 V = 1 p.u

The base current

$$I_b = \frac{VA_b}{\sqrt{3}V_b} = \frac{50 \times 10^3}{\sqrt{3} \times 440} = 65.61 \text{ A}$$

p.u impedance

$$Z_{p.u} = \sqrt{\left(R_{s,p.u} + R_{tran,p.u} + R'_{r,p.u}/s\right)^2 + \left(X_{s,p.u} + X_{tran,p.u} + X'_{r,p.u}\right)^2}$$

$$= \sqrt{(0.052 + 0.005 - 0.052/0.333)^2 + (0.413 + 0.04 + 0.207)^2} = 0.667$$

The power factor angle

$$\tan^{-1}\left(\frac{X_{s,p.u} + X_{tran,p.u} + X_{r,p.u}}{R_{s,p.u} + R_{tran,p.u} + R_{r,p.u}/s}\right) = \tan^{-1}\left(\frac{0.413 + 0.04 + 0.207}{0.052 + 0.005 - 0.052/0.333}\right)$$

$$= 98.54°$$

$$Z_{p.u} = |Z_{p.u}|\angle\theta = 0.667\angle98.54°$$

The p.u stator current in motor convention is

$$I_{p.u} = \frac{V_b}{Z_{p.u}} = \frac{1}{0.667\angle98.54°} = 1.499\angle - 98.54°$$

The actual value of current in motor convention, $I_m = I_b \times I_{p.u}$

$$= 65.61 \times 1.499 = 98.35\angle - 98.54° \text{ A}$$

Therefore, $I_m = 98.35\angle - 98.54°$ A.

Because this angle is more than 90°, the power flow is from the induction generator to the local power grid.

Using generator convention, the direction of current is reversed to represent I_G as in Figure 8.38.

Therefore, the current in generator convention, $I_G = 98.35\angle180 - 98.54° = 98.35\angle81.46°$ A

The active power input to the grid,

$$P_{grid} = \sqrt{3} \cdot V_{grid} \cdot I_G \cdot \cos\theta = \sqrt{3} \times 440 \times 98.35 \times \cos 81.46 = 11,130 \text{ W}$$

Active power lost in the transformer $P_{loss} = 3I_G^2 R_{trans} = 3 \times 98.35^2 \times 0.02 = 580$ W

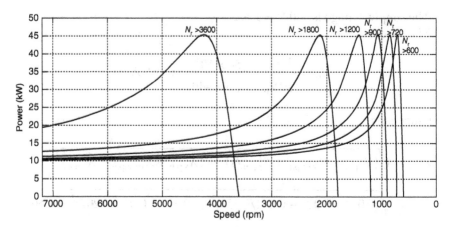

Figure 8.38 Power versus speed of a variable pole induction generator for various rotor speeds.

Therefore, the active power injected by the induction generator to the transformer, $P = P_{grid} + P_{loss}$

$$= 11,130 + 580 = 11,710 \text{ W}$$

The reactive power input to the grid,

$$Q_{grid} = \sqrt{3} \cdot V_{grid} \cdot I_G \cdot \sin\theta = \sqrt{3} \times 440 \times 98.35 \times \sin 81.46 = 74,121 \text{W}$$

Reactive power lost in the transformer $Q_{loss} = 3I_G^2 X_{trans} = 3 \times 98.35^2 \times 0.15 = 4355 \text{ W}$.

Therefore, the reactive power consumed by the induction generator is given by $Q = Q_{grid} + Q_{loss}$

$$= 74121 - 4355 = 69,766 \text{ W}$$

Example 8.6 Consider a three-phase Y-wound rotor connected induction generator rated 220 V, 60 Hz, and 16 hp; the poles of the machine can be changed from 2 to 12 to control the wind speed. The machine has the following parameters:

$R_1 = 0.2 \, \Omega$/phase and $X_1 = 0.4 \, \Omega$/phase
$R_2' = 0.13 \, \Omega$/phase and $X_2 = 0.4 \, \Omega$/phase

Plot the power versus speed curve of the machine controlling the wind speed by changing the number of poles.

The following MATLAB M-file testbed depicts the operation of the machine.

```
%POWER vs SPEED
clc;
v1=220/sqrt(3);
f=60;
P=2;
r1=0.2;
x1=0.4;
r2d=0.13;
x2d=0.4; % The electrical quantities are defined
for P=2:2:12 % The no. of poles is varied from 2 to 12
ws=120*f/P;
w=0:.2:7200; % The value of speed is varied till synchronous speed
for i=1:length(w)
s(i)=(ws-w(i))/ws;
Tem(i)=(3/ws)*(r2d/s(i))*v1^2/((r1+r2d/s(i))^2
        +(x1+x2d)^2);
Po(i)=-Tem(i)*w(i)/1000;   % Power in kW
end
plot(w,Po)
hold on;
end
axis([0 7200 0 50])
grid on;
set(gca,'XDir','reverse')
xlabel('Speed (rpm)')
ylabel('Power (kW)')
```

Figure 8.38 depicts the power versus speed of a variable pole induction generator for various rotor speeds.

8.8 THE DOUBLY FED INDUCTION GENERATOR

Electric machines are classified based on the number of windings in the conversion of mechanical power to electric power. A singly fed machine has one winding. The SCIG-type machines have one winding, which contributes to the energy conversion process. Doubly fed machines have two windings that likewise are instrumental in energy conversion. The wound rotor doubly fed induction generator (DFIG)[23] is the only electric machine that can operate with rated torque to twice the synchronous speed for a given frequency of operation. In a DFIG, the current flowing in the magnetizing branch and torque current are orthogonal vectors.[23] It is not desirable to design machines that are magnetized from the rotor because commutation systems, supporting slip rings, and brushes are needed to inject current into the rotor winding. These types of machines have high maintenance costs. However, in these machines, the stators can have a unity power factor. The frequency and the magnitude of the rotor voltage are proportional to the slip as expressed by Equation (8.30). In principle, the DFIG is a transformer at standstill.

If the DFIG is producing torque and operating as a motor, the rotor is consuming power. At standstill all power fed into the stator is consumed as heat in the stator and the rotor. Therefore, at low speeds, the efficiency of the DFIG is very low because the supplied current is mainly used to produce magnetizing current and the power conversion processes as a function of a motor or a generator do not take place. If the DFIG is operating at above the synchronous speed, the mechanical power is fed in both through the stator and rotor. Therefore, the machine has higher efficiency and the machine can produce twice the power as a singly fed electric machine.

With the DFIG at below synchronous speeds, the stator winding is producing electric power and part of its power is fed back to the rotor. At speeds above synchronous speeds, the rotor winding and stator winding are supplying electric power to the grid. However, DFIGs do not produce higher torque per volume than singly fed machines. The higher power rating can be obtained because of the higher speed and without weakening the magnetic flux.

A DFIG configuration system is depicted in Figure 8.38. This DFIG is a wound rotor induction generator (WRIG) with the stator windings directly connected to the power grid. The DFIG has two parallel AC–DC converter units. Although these converters act together, they are not necessarily fully identical with regard to their power rating.

The rotor windings are connected to an AC–DC power converter on the grid side and a DC–AC power converter on the rotor side. The back-to-back converter operates as a bidirectional power converter with a common DC bus. The transformer in Figure 8.39 has two secondary windings; one winding connecting the stator and the other connecting the rotor. The converter on the rotor side makes it possible to operate the rotor excitation at a lower DC bus

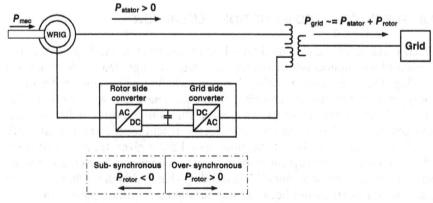

Figure 8.39 A microgrid of a doubly fed wound induction generator.

voltage. This DFIG provides reactive power control through its power converter because it decouples active and reactive power control by independently controlling the rotor excitation current. A DFIG can supply (absorb) reactive power to and from the power grid. The control method is based on the variable speed/variable pitch wind.[24] Two hierarchical control levels are used. These controllers are designed to track the wind turbine operation point, to limit the turbine operation in the case of high wind speeds. In addition, the controller controls the reactive power injected into the power grid and the reactive power consumed by the WTG. The power supervisory controller controls the pitch angle to keep the wind turbine operating at the rated power. At the same time, the speed controller controls the shaft speed of the generator to ensure that it remains within safe range. However, at low wind speeds, the speed controller attempts to maximize the generated power and generator efficiency. Therefore, the change in generator speed follows the slow change in wind speed.

As an inherent part of DFIG systems, the generator stator feeds up to 70% of the generated power directly into the grid, usually at low voltage, using a step-up transformer. A well-established disadvantage of DFIG systems is the occurrence of internal stray currents in the generator. These currents accelerate generator bearing failure. Protective counter-methods include the design of special generator bearings and/or seals that shield the bearings against the negative impact of stray currents.

A wound rotor DFIG has several advantages over a conventional induction generator. Because the rotor winding is actively controlled by a power converter, the induction generator can generate and consume reactive power. Therefore, the DFIG can support the power system stability during severe voltage disturbances by providing reactive power support. As an inherent part of DFIG systems, the generator stator feeds the remaining 70–76% of total power directly into the grid, usually for up to 690 V. The wind microgrid is connected to the local grid using a step-up transformer. The DFIG systems are designed for both a 60-Hz and a 50-Hz system, but each grid situation requires a generator operation adapted to the specific operational circumstances. However, the control of DC–AC converters is an advanced topic that is not the subject of this book. Students who are interested in the control of converters in green energy systems should refer to reference 24. SCIGs have a lower cost of manufacturing than WRIGs and are relatively simple to design; they are robust, cost effective, and widely used in wind microgrids.

8.9 BRUSHLESS DOUBLY FED INDUCTION GENERATOR SYSTEMS

Brushless DFIG systems[25] are designed by placing two multiphase winding sets with a different number of pole pairs on the stator structure. One of the stator winding sets is designated as power winding and is connected to the power grid. The second winding control is supplied from a converter and controls the energy conversion process. The generator is controlled by varying the frequency of the winding connected to the power converter. Because the pole

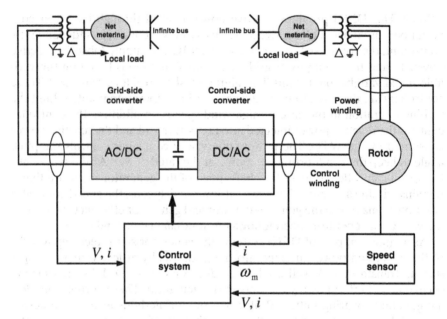

Figure 8.40   A microgrid of brushless doubly fed induction generator.

pairs of two windings are not identical, the low-frequency magnetic induction is created in the winding that connects to the power grid over a speed range that the rotor supplies by wind power. Brushless DFIGs do not utilize the magnetic core efficiently. The dual winding set stator area is physically larger than that of other electric machines of comparable power rating. Figure 8.40 presents a microgrid of brushless doubly fed induction generator.

8.10 VARIABLE-SPEED PERMANENT MAGNET GENERATORS

These types of wind generators operate with variable wind speed.[26] They use a "full" AC–DC and DC–AC power converter.[8] The DC power is inverted using a DC–AC inverter and is connected to a step-up transformer, and then connected to the local power grid as shown in Figure 8.41.

The variable-speed permanent magnet generator of Figure 8.41 produces a variable AC voltage with a variable frequency. Because the power generated is not at the frequency of the local power grid, the output power of the variable frequency generator cannot be injected into local power grid. Therefore, the variable AC frequency voltage is rectified by the AC–DC rectifier (see Figure 8.41). The DC bus of Figure 8.40 can be used to charge a storage system using a boost–buck converter. Figure 8.42 depicts a variable-speed generator. This type of generator is studied in Chapter 4. The field winding of the generator of Figure 8.42 is supplied from DC power using an AC–DC rectifier from an AC bus of the microgrid. Because the supplied wind mechanical power has a

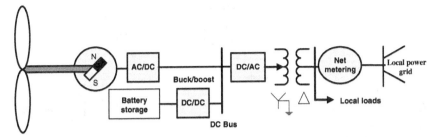

Figure 8.41 A microgrid of a variable-speed permanent magnet for a wind generator.

variable speed, the generator output power would also contain variable frequency.[9] The DC–AC inverter is used to convert the DC power to AC power at the local power grid frequency and voltage as shown in Figures 8.41 and 8.42. The microgrid of wind power can be connected to the local power grid of an AC bus because it is operating at the frequency of the local power grid. However, the coordinated control of converters of Figures 8.41 and 8.42 is an advanced topic that is not the subject of this book. Students who are interested in the control of converters in green energy systems should refer to Reference 24.

8.11 A VARIABLE-SPEED SYNCHRONOUS GENERATOR

The rotor of a synchronous generator rotates at synchronous speed.[22,25] For a synchronous generator, the frequency of the voltage induced in the stator windings is given by the expression below.

$$\omega_{syn} = \frac{2}{P}\omega_s$$

If $P = 2$, then $\omega_{syn} = \omega_s$

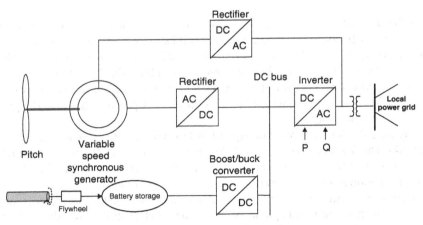

Figure 8.42 A microgrid of a multipole synchronous generator.

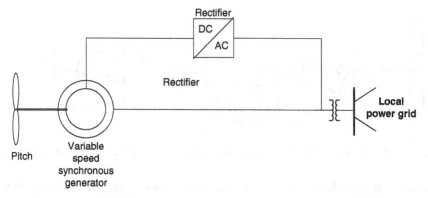

Figure 8.43 A variable-speed synchronous generator.

And the frequency of induced voltage is given by

$$\omega_s = 2 \cdot \pi \cdot f_e$$

If the wind speed is variable, the induced voltage will also be time varying, and it will be of multiple frequencies. Figure 8.43 depicts such a wind-based microgrid. However, before connecting the generator to the local power grid, we must operate the generator at the synchronous speed. Depending on the expected wind speed for a given location, the generator can be designed with a gear system. The gear ratio is adjusted such that the speed of the rotor of the generator is at synchronous speed.[10] Therefore, the induced voltage of the stator of the generator is at the same frequency as the local power grid.

8.12 A VARIABLE-SPEED GENERATOR WITH A CONVERTER ISOLATED FROM THE GRID

Another type of WTG system consists of an electrical exciter machine together with a DFIG. In comparison to a normal DFIG system, this wind generator has one converter.[11] By including the excitation machine, it is possible to isolate the power converter so it is not directly connected to the grid. That is, the stator is the only grid-connected output. This is a solution that is different from a normal DFIG grid connection, in which the generator rotor power is fed into the grid via a power converter.

Figure 8.44 depicts a variable-speed WTG with a converter isolated from the grid. The first converter is a DC–AC inverter that feeds the rotor of DGIF. However, in this topology, the second converter unit is an AC–DC rectifier, which is supplied from the exciter machine (see Figure 8.44).

Furthermore, in contrast to both synchronous and asynchronous generators, DFIGs are inherently incapable of acting as electric brakes for the sudden separation of a wind turbine from the grid or a sudden high wind speed. However, in the above topology, the exciter machine power can be used to

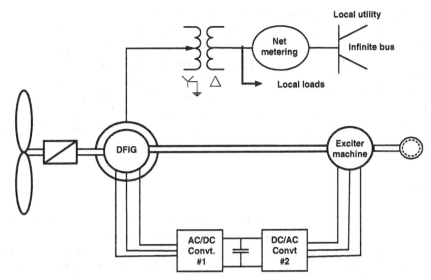

Figure 8.44 A microgrid of variable-speed wind turbine generator with the converter isolated from the grid.

drive an electric brake.[12] The electric brake may also be used together with aerodynamic braking, minimizing peak torque loads.

These wind turbines are characterized by lower inertia than classical power plants; therefore, they cannot participate in power system load–frequency regulation. When the wind generators are equipped with a storage system, they can participate in load–frequency control. The variable-speed turbines are designed based on the use of back-to-back power electronic converters. The intermediate DC voltage bus creates an electrical decoupling between the machine and the grid. Such decoupling creates a new opportunity to use these types of wind generating systems for load–frequency control.[13]

Example 8.7   Select AC–DC rectifiers and DC–AC inverters for a 600 kW variable-speed wind generator operating at 690 V AC. The utility-side voltage is 1000 V.

Solution
The peak value of the instantaneous line-to-line as voltage: $V_{L\text{-}L,\text{peak}} = \sqrt{2} \cdot V_{L\text{-}L,\text{rms}} = \sqrt{2} \times 690 = 975.8$ V.

Therefore, the DC-side voltage rating of the rectifier is $\geq$976.8 V.

Let the voltage rating of the rectifier be 1000 V on the DC side and 690 V on the AC side. The inverter also can be selected with a rating of 1000 V on the AC side. Both the rectifier and the inverter should be rated 600 kW.

In this chapter, we have studied the modeling of induction machines and their operation as motors and generators. The use of induction generators as a source of power in microgrids requires an excitation current for generator operation to be provided from local microgrids. If the wind-based microgrids are connected to the local power grids, the power grids will provide the excitation currents (VAr). Therefore, the local power grids must be planned to provide the reactive power (VAr) requirements of the wind microgrids.

We have also reviewed DFIGs, variable-speed induction generators, and variable-speed permanent magnet generators.[23,24] The coordinated control of converter is an advanced topic that is not the subject of this book. Students who are interested in control of converters in green energy systems should refer to Reference 24.

PROBLEMS

8.1   Consider a wind microgrid given in Figure 8.45. The system has a local load rated 100 kVA at a power factor rated 0.8 lagging.

The three-phase transformer is rated 11 kV/0.44 kV; 300 kVA; $X = 0.06$ p.u. The induction generator is rated as 440 V, 60 Hz, three-phase, eight-pole; stator resistance of 0.08 Ω/phase; rotor referred resistance in the stator side of 0.07 Ω/phase, stator reactance of 0.2 Ω/phase; rotor referred reactance of X_2 0.1 Ω/phase. Compute the following:

(i) The p.u equivalent model based on a kVA base of 300 kVA and 440 V.

(ii) The shaft mechanical power if the shaft speed is at 1200 rpm.

(iii) The amount of power injected into the local utility.

(iv) The flow of reactive power between grid and local microgrid.

(v) How much reactive power must be placed at the local grid to have unity power factor at the local power grid.

8.2   The microgrid of Figure 8.46 is supplied by an induction generator. The system has a local load rated 100 kVA at power factor rated 0.8 lagging. The three-phase transformer is rated 11 kV/0.44 kV; 300 kVA, and reactance of 6%. The induction machine rated at 440 V, 60 Hz, three-phase,

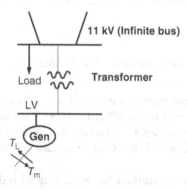

Figure 8.45   The schematic of Problem 8.1.

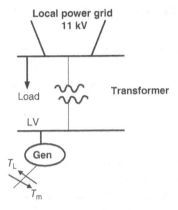

Local power grid
11 kV

Load   Transformer

LV

Gen

T_L

T_m

Figure 8.46   System of Problem 8.4.

eight-pole; 500 kVA, 440 V, 60 Hz; stator resistance of 0.1 Ω/phase; rotor referred resistance in the stator side of 0.1 Ω/phase, stator reactance of 0.8 Ω/phase; rotor referred reactance of 0.4 Ω/phase. Compute the following:

  (i) Compute the per unit power flow model and short circuit model based on a base of 500 kVA and 440 V.

 (ii) If the speed of the induction generator is 1000 rpm, what is the rotor frequency?

(iii) The flow of active and reactive power between microgrid and the local power grid.

8.3 A six-pole wound rotor induction machine rated at 60 Hz, 380 V, 160 kVA. The induction machine has a stator and referred rotor resistance of 0.8 Ω/phase and stator and rotor reactance of 0.6 Ω/phase. The generator shaft speed is at 1500 rpm. Determine how much resistance must be inserted in the rotor circuit to operate the generator at 1800 rpm.

8.4 A 400 V, three-phase Y-connected induction generator has the following data.

    The generator is connected to a local power grid. Perform the following:

  (i) The maximum active power that generator can supply.

 (ii) The reactive power flow between the induction generator and local power grid.

8.5 Design a 15 kW wind power generator that is supplied from a variable wind speed. The designed system must provide 220 V_{AC}, single-phase AC power. Compute the DC bus voltage.

8.6 The same as Problem 8.7, except the wind generating system must provide three-phase AC nominal voltage of 210 V_{AC}. Compute the DC bus voltage.

TABLE 8.2   13.2–132 kV Class One Phase–Neutral Return Line Model

Conductor	DC Resistance (Ω/km)	Inductance (Ω/km) L	Susceptance (S/km) C	Current Ratings (Amp)
Magpie	1.646	$j\,0.755$	$j\,1.45\text{e-}7$	100
Squirrel	1.3677	$j\,0.78$	$j\,6.9\text{e-}7$	130
Gopher	1.0933	$j\,0.711$	$j\,7.7\text{e-}7$	150

8.7   Design a 2 MW wind systems using a variable-speed system. The DC bus voltage is to be at nominal value of 600 V_{DC}. The panels are located at 5 miles from the local utility. The utility voltage is three-phase AC rated at 34.5 kV. Assume the rated power factor is 0.8 leading.

The data for transmission line is given by Table 8.2. The data for transformers are given as:

460 V/13.2 kV 250 kVA 10% impedance and 13.2–34.5 kV, 1 MVA 8.5% impedance.

Perform the following:

(i) Give the one-line diagram.

(ii) Per unit model based on rated wing generator.

8.8   A wound rotor six-pole 60 Hz induction generator has stator resistance of 1.1 Ω/phase and rotor resistance of 0.8 Ω and runs at 1350 rpm. The prime mover torque remains constant at all speeds. How much resistance must be inserted in the rotor circuit to change the speed to 1800 rpm. Neglect the motor leakage reactance, X_1 and X_2.

8.9   Consider a three-phase Y-wound rotor connected induction generator rated 220 V, 60 Hz, 16 hp, eight pole with the following parameters:

$R_1 = 1$ Ω/phase and $X_1 = 1.6$ Ω/phase

$R_2{}' = 0.36$ Ω/phase and $X_2 = 1.8$ Ω/phase

The generator is connected to the local power grid.

Write a MATLAB simulation testbed to plot slip and speed as a function of machine torque and various external inserted resistance in rotor circuit. Make the plot for a value of external resistance of 0.0, 0.4, 0.8, and 1.2 Ω.

REFERENCES

1. Wikipedia. History of Wind Power. Available at http://en.wikipedia.org/wiki/Wind_Energy}History. Accessed August 10, 2009.

2. Technical University of Denmark. National Laboratory for Sustainable Energy. Wind Energy. Available at http://www.risoe.dk/Research/sustainable_energy/wind_energy.aspx. Accessed July 11, 2009.

3. California Energy Commission. Energy Quest. Chapter 6: Wind energy. Available at http://www.energyquest.ca.gov/story/chapter16.html. Accessed June 10, 2009.

4. U.S. Department of Energy, Energy Information Administration. Official Energy Statistics from the US Government. Available at http://www.eia.doe.gov/energyexplained/index.cfm?page=us_energy. Accessed September 10, 2009.

5. U.S. Department of Energy. U.S. installed wind capacity and wind project locations. Available at http://www.windpoweringamerica.gov/wind_installed_capacity.asp. Accessed December 5, 2010.

6. Justus, C.G., Hargraves, W.R., Mikhail, A., and Graber, D. (1978) Methods of estimating wind speed frequency distribution. *Journal of Applied Meteorology*, **17**(3), 350–353.

7. Jangamshetti, S.H. and Guruprasada Rau, V. (1999) Site matching of wind turbine generators: a case study. *IEEE Transactions in Energy Conversion*, **14**(4), 1537–1543.

8. Jangamshetti, S.H. and Guruprasada Rau, V. (2001) Optimum siting of wind turbine generators. *IEEE Transactions of Energy Conversion*, **16**(1), 8–13.

9. Quaschning, V. (2006) *Understanding Renewable Energy Systems*, Earthscan, London.

10. Freris, L. and Infield, D. (2008) *Renewable Energy in Power Systems*, Wiley, Hoboken, NJ.

11. Patel, M.K. (2006) *Wind and Solar Power Systems: Design, Analysis, and Operation*, CRC Press, Boca Raton, FL.

12. Hau, E. (2006) *Wind Turbines: Fundamentals, Technologies, Application and Economics*, Springer, Heidelberg.

13. Simoes, M.G., and Farrat, F.A. (2008) *Alternative Energy Systems: Design and Analysis with Induction Generators*, CRC Press, Boca Raton, FL.

14. AC Motor Theory. Available at http://www.pdftop.com/ebook/ac+motor+theory/. Accessed December 5, 2010.

15. U.S. Department of Energy, National Renewable Energy Laboratory. Available at http://www.nrel.gov/. Accessed October 10, 2010.

16. HSL Automation Ltd. Basic motor theory: squirrel cage induction motor. Available at http://www.hslautomation.com/downloads/tech_notes/HSL_Basic_Motor_Theory.pdf. Accessed October 10, 2010.

17. Tatsuya, K. and Takashi, K. (1997) IEEE: A Unique Desk-top Electrical Machinery Laboratory for the Mechatronics Age. Available at http://www.ewh.ieee.org/soc/es/Nov1997/09/INDEX.HTM. Accessed December 10, 2010.

18. Bretz, E. (2004) IEEE Spectrum: Superconductors on the High Seas. Available at http://spectrum.ieee.org/energy/renewables/winner-superconductors-on-the-high-seas. Accessed December 19, 2010

19. Wenping, C., Ying, X., and Zheng, T. (2012) Wind Turbine Generator Technologies, Advances in Wind Power, Dr. Rupp Carriveau (Ed.), ISBN: 978-953-51-0863-4, InTech, DOI: 10.5772/51780. Available at http://www.intechopen.com/books/advances-in-wind-power/wind-turbine-generator-technologies. Accessed December 15, 2010.

20. Krause, P., and Wasynczuk, O. (1989) *Electromechanical Motion Devices*, McGraw-Hill, New York.

21. Majmudar, H. (1966) *Electromechanical Energy Converters*, Allyn & Bacon, Reading, MA.

22. Sung, S.W.F., and Rudowicz, C. A closer look at the hysteresis loop for ferromagnets Available at http://arxiv.org/ftp/cond-mat/papers/0210/0210657.pdf. Accessed December 5, 2010.

23. Pena, R., Clare, J.C., and Asher, G.M. (1996) Doubly fed induction generator using back-to-back PWM converters and its application to variable-speed wind-energy generation. *IEE Proceedings of Electric Power Applications*, **143**(3), 231–241.

24. Keyhani, A., Marwali, M., and Dai, M. (2010) *Integration of Green and Renewable Energy in Electric Power Systems*, Wiley, Hoboken, NJ.

25. 1999 European Wind Energy Conference: Wind energy for the next millenium. Proceedings of the European Wind Energy Conference, Nice, France, 1–5 March 1999. London: Earthscan; 1999.

26. Boldea, I. (2005) The electric generators handbook. *Variable Speed Generators*, CRC Press, Boca Raton, FL.

CHAPTER 9

MARKET OPERATION OF SMART POWER GRIDS

This chapter introduces the smart grid elements and their functions from a systems approach, and provides an overview of the complexity of smart power grid operations. Topics covered in the chapter include the basic system concepts of sensing, measurement, integrated communications, and smart meters; real-time pricing; cyber control of smart grids; high green energy penetration into the bulk interconnected power grids; intermittent generation sources; and the electricity market.

9.1 INTRODUCTION—CLASSICAL POWER GRIDS

Initially designed in the early 1900s, today's power grid has evolved to become a large interconnected network that connects thousands of generating stations and load centers through a system of power transmission lines.[1-3] A power grid system is designed based on the long-term load forecast of the power grid load centers, which is developed according to the anticipated needs of the community it serves. Then, an analytical model of the system is developed to project the grid's real-time operation. In a smart power grid system, a large number of microgrids operate as part of an interconnected power grid. For example, a photovoltaic (PV)-based residential system with its local storage system and load would be one of the smallest microgrids in the smart power grid system.[4] To understand the new paradigm of tomorrow's smart power grid design and operation, we need to understand today's electric power grid operation and costs of design.[4,5]

Design of Smart Power Grid Renewable Energy Systems, Second Edition. Ali Keyhani.
© 2017 John Wiley & Sons, Inc. Published 2017 by John Wiley & Sons, Inc.
Companion website: www.wiley.com/go/smartpowergrid2e

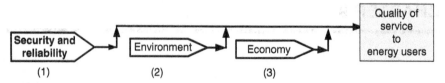

Figure 9.1 The interrelated operation objectives of power grids.

This chapter introduces the basic concepts of generator operation, load-frequency control, power flow, the limit of power flow on transmission lines, and load factor calculation and its impact on the operation of a smart grid and microgrids. A basic understanding of a power grid's operation will facilitate how to design a microgrid to operate as a standalone system when it is separated from its local power grid.

9.2 POWER GRID OPERATION

The operational objectives of a power grid are to provide continuous quality service at an acceptable voltage and frequency with adequate security, reliability, and an acceptable impact upon the environment—without damage to the power grid equipment—all at a minimum cost.[5,6]

In Figure 9.1, the direction of the arrows indicates the priority in which the objectives are implemented. The quality service that is environmentally acceptable, secure, and reliable, and entails minimum cost is the main objective of power grid operation. However, during emergency conditions, the grid may be operated without regard for economy and, instead concentrating on the security and reliability of the service for energy users, while maintaining power grid stability.

The term *continuous service* means "secure and reliable service." The term *security*, as it is used here, means that upon occurrence of a contingency, the power grid could recover to its original state and supply the same quality electric energy as before.

For example, if in the power grid of Figure 9.2, the line connecting bus 2 and bus 4 is out of service, the power grid is *secure* if it still can serve all loads. However, the power grid is *reliable* if it has adequate reserves to face increased load demands. In addition, the power grid of Figure 9.2 is reliable if it is subjected to scheduled or unscheduled energy source outages; it is still able to supply quality electric energy to the users.

To ensure security and reliability, power plant facilities and resources must first be planned then managed effectively. A large power grid comprises many elements including generating units, transmission lines, transformers, and circuit breakers. As new green energy sources are adopted into the power grid and a smart power grid is put in place, additional equipment such as DC/DC converters and DC/AC converters must be integrated and scheduled for power

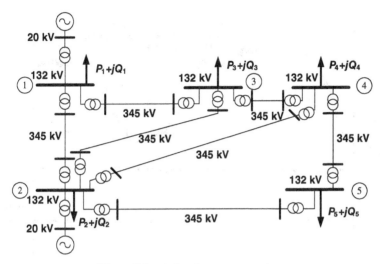

Figure 9.2   A five-bus power system.

grid operation. In addition, market structure and real-time pricing[5,6] of energy need to be evaluated.

At the outset, the main objective is to schedule power generation to supply the system loads for every second of the grid's operation.[6–8] The energy resources of a large power system consist of hydro and nuclear energy, fossil fuel, renewable energy sources such as wind and solar energy, as well as green energy sources such as fuel cells, combined heat and power (CHP; also known as cogeneration), and microturbines. These resources must be managed and synchronized to satisfy the load demand of the power grid. The load demand of a power grid is cyclic in nature and has a daily peak demand over a week, a weekly peak demand over a month, and a monthly peak demand over a year. Energy resources must be optimized to satisfy the peak demand of each load cycle, such that the total cost of production and distribution of electric energy is minimized.

Figure 9.3 depicts a 24-hour load variation sampled every 5 minutes. From Figure 9.3, it can be seen that peak demand is twice the minimum power demand. Figure 9.4 shows that the peak grid demand occurs on Monday and the minimum grid demand occurs on Sunday. The grid operator must plan the grid energy resources and facilities to satisfy the varying load conditions.

Operations planning is divided into three tasks—long-, medium-, and short-term operations—as shown in Figure 9.5. Operations control deals with controlling the system as it is operated minute by minute. Operations accounting system records the events occurring on a grid and by analyzing recorded data attempts to account for various events that affected the grid. Operations accounting data is also used in the planning of future power grids.

The decision time involved in operations planning and control of the power grid is depicted in Figure 9.6. The vertical axis of Figure 9.6 shows the

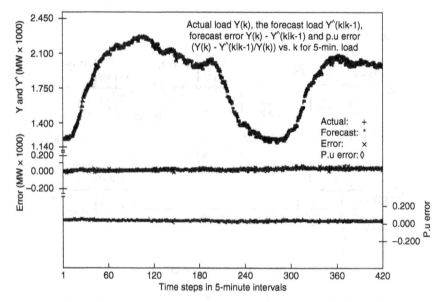

Figure 9.3 A 24-hour load variation sampled every 5 minutes.[7]

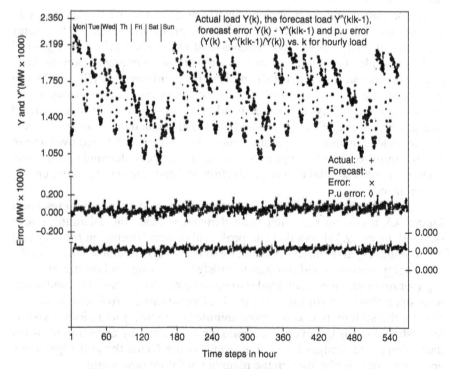

Figure 9.4 A weekly load variation sampled hourly.

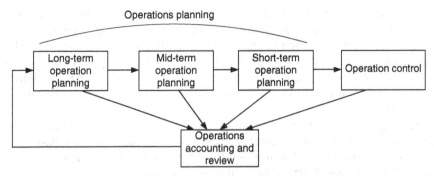

Figure 9.5 The interrelated tasks of planned scheduling operation.[7]

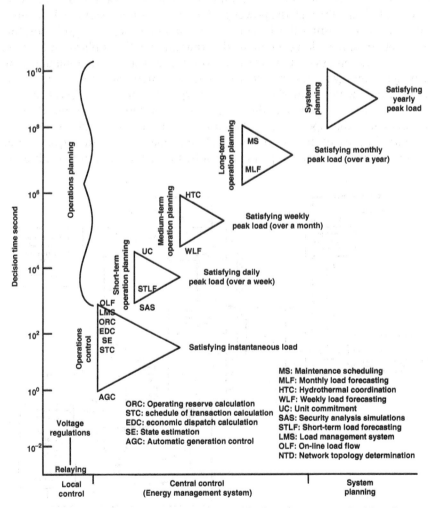

Figure 9.6 Energy management system and its functions versus decision time.

decision time for implementing a function. The horizontal axis indicates where the control of a function takes place. These functions are programmed into the computer software. The hardware for implementing these functions includes two computers, one operating in real time in charge of the grid; the second is on-line as a backup if the first computer encounters problems. The computer system is referred to as an energy management system (EMS) or the energy control center.

The Supervisory Control and Data Acquisition (SCADA) system consists of data acquisition and control hardware and software, man–machine and interface software systems, and dual computer systems with real-time operating systems. Therefore, the EMS consists of a SCADA system plus the application functions used to operate and control the power grid. The primary functions of SCADA are (1) to collect information throughout the power grid, (2) to send the collected data through the power grid communication system to the control center, and (3) to display the data in the control center for power grid operators to use for decision making and in the determination of the application function for grid operation. As part of the smart power grid design, additional data concerning the energy resources such wind, solar, PV energy sources, and real-time pricing from the power market must be incorporated into the SCADA system. Distributed over a wide area, the smart power grid must be optimized for its efficient and stable operation—yet another task for the SCADA system.

Figure 9.7 presents the interrelated tasks for a planned scheduling operation. The problems of operations planning can be broken down into four different

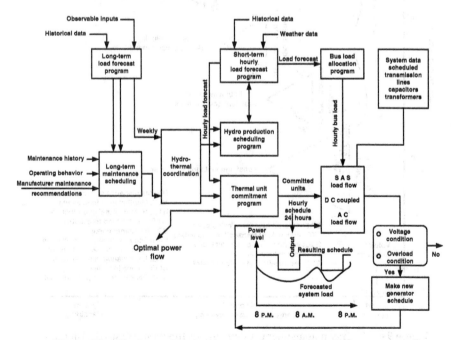

Figure 9.7 The interrelated tasks for a planned scheduling operation.

tasks: (1) to schedule all resources and facilities yearly, (2) on a monthly basis for satisfying the forecasted monthly peak load, (3) then utilizing the weekly results to produce a daily schedule, and (4) finally using the daily schedule to prepare a feasible and secure hourly schedule.

The long-term operation planning consists of two functions: a monthly load-forecasting program and a maintenance scheduling program based on estimates of the peak load demand of every month. The maintenance scheduling program schedules units for maintenance according to many criteria, for example, the manufacturer's maintenance recommendations, experience with particular equipment such as generators, transformers, transmission lines, such that the peak load demand of every month is satisfied within a reasonable risk.

The scheduling of resources and facilities monthly on a weekly basis is accomplished through medium-term operation planning. This task consists of two functions: a weekly load-forecasting program and a hydrothermal coordination program. The weekly load-forecasting program estimates the peak demand of every week over a period of 1 month. The hydrothermal coordination program determines the best schedule of operation of the hydro and thermal units such that the amount of fuel consumed in the thermal units is minimized, and the weekly load demand of the system is satisfied. As renewable energy sources are increasingly used in the power grid, operation planning will become highly complex due to the intermittent nature of renewable energy sources. Furthermore, as real-time pricing is put into effect,[8] the task of operation planning will become more complex as energy users react to daily hourly pricing. These problems are being addressed in the planned operation of a smart power grid.

The weekly scheduling of facilities and resources on a daily basis is performed by short-term operation planning, which consists of a short-term load-forecasting program, a security analysis simulations program, and a unit commitment program. The short-term load-forecasting estimates the hourly load demands of the next 168 hours. The unit commitment or economic generation scheduling, which is also referred to as the ordering method or pre-dispatching, determines the generating units to be committed to operation based on the availability of different units, as specified by long- and medium-term operation planning so that the hourly forecasted load for the next 24–168 hours is satisfied. In addition, the unit commitment function makes a sequence of decisions leading to the starting and stopping of thermal units to assure an adequate but not excessive amount of generating capacity is available to meet the hourly forecasted grid load. It should be mentioned that due to uncertainty in the load forecast, it is customary to plan for additional capacity either synchronized or ready to be synchronized within a short period. This excess capacity is called the operating reserve. The security analysis simulation function is a set of dispatcher-oriented, interactive programs that are used to compute the system bus-load voltages. The last activity of short-term operation planning is a security analysis simulation function. The security analysis simulation function computes the system bus voltages based on the hourly generation schedule from unit commitment program and hourly bus-load forecast. If for the expected hourly loading, generation scheduling, and the expected or scheduled system configuration, the resulting operation condition is not acceptable, then changes in the system configuration (e.g., changing the tap setting of the transformers or the scheduled generation) will be made and the computation will be repeated. Once a feasible economical hourly schedule of generation is determined, the planned operation of the system will be passed to operations control, which will attempt to satisfy the minute-to-minute demands of the system. Bus-load voltage calculation is important in power grid planning and operation; it is called power flow or load flow. We will discuss this function in Chapter 10.

In the deregulated power market, many contracts are executed between buyers and sellers in advance. For example, the day-ahead market is developed based on the forecasted load demands of the system to supply electrical power at least 24 hours before delivery to buyers and end users.[9,10]

9.3 VERTICALLY AND MARKET-STRUCTURED POWER GRID

Figure 9.8 depicts a vertically integrated power grid, which dates back to Thomas Edison.[1,2] It is essentially a network structure, with a large power station located where coal or hydropower were available. The electric energy users were first served at large cities and then through radial distribution

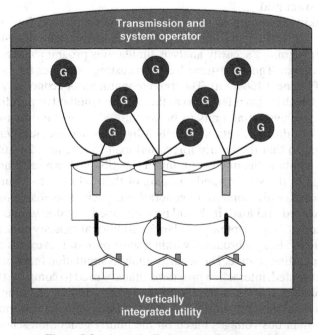

Figure 9.8   A vertically integrated power grid.

system to the rural areas. In today's power grid, energy flows one way—essentially, from power stations to the residential, commercial, and industrial users. Because most power stations are remotely located where coal resources and hydropower sources are available, the power must be transmitted through high voltage transmission systems and subtransmission systems to end users. The power losses to the load during transmission can be estimated by studying the power flow through the system. For the United States, in 2004, the losses were estimated to be approximately over 265,180,000,000 million kWh.[3]

If a power grid is subject to a sudden outage of a power generating station or the loss of a critical line carrying large megawatts of power from generating stations to the load centers, such a loss creates a sudden deficiency of generating power. With the result that system operators cannot match energy production to energy demand; hence, the power grid is subject to power oscillation. If the grid operators are not able to match energy production to energy consumption, the grid becomes unstable, and protective relays remove the equipment from service prompting a blackout.

Figure 9.9 depicts a market-operated power grid. In this structure, the independent system operator (ISO) is in charge of power grid operation. ISO energy management computer systems compute the operating reserve that is necessary to maintain reliable interconnected power grid operation.[7–9] The ISO operators operate the power grid based on the North American Electric Reliability Council (NERC)[9,10] Policy 1 generation control and performance standard. This document specifies the operating reserve requirements.[11–19] According to NERC, "Each control area shall operate its power resources to provide for a level of operating reserve sufficient to account for such factors as errors in forecasting, generation and transmission equipment unavailability, number and size of generating units, system equipment forced outage rates, maintenance schedules, regulating requirements, and regional and system load diversity."[2] Following loss of resources or loads, the ISO operator takes appropriate steps to stabilize the grid. The additional generation is in the form of a spinning reserve.[7–9] The spinning reserve is defined as synchronized power that is ready to be dispatched by the system operator. The spinning reserve is normally from a 5- to 10-minute reserve. The reserve power can also be in the form of standby off-line reserves, and contracted customer-interrupted loads.

9.3.1 Who Controls the Power Grids?

The first question that must come to mind is who controls the power grids? What is the meaning of control of grid? Control means to initiate an action that will lead to the desired outcome. The power grids have three stakeholders: (1)One energy user. (2) Energy producers. (3) Facilities (grid) that deliver the energy to energy users.

When energy users turn lights off at nights, this control action acts to control the energy consumption and reduce energy cost. This small change in use of energy is taken care of by kinetic energy stored in high mass of rotating generators. However, the result is reduction of speed of generators that is

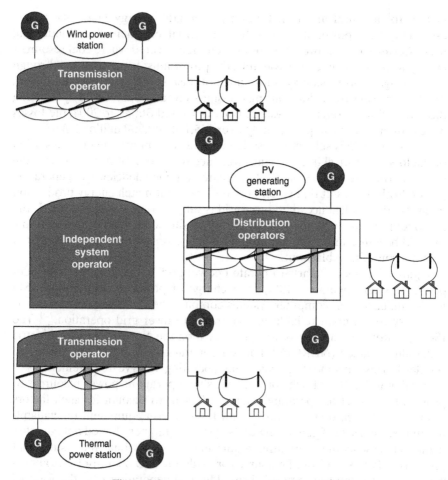

Figure 9.9 A market-structured power grid.

reflected in reduction in frequency. In the United States, the ISO through its computer control systems sets the frequency control set point at 60 cps. This action is accomplished by control action to increase the generation and restore the speed of generators to synchronous speed and grid frequency to its set point. When frequency returns to its set point and grid the loads plus losses in lines become equal to grid generation. This is a fundamental law of power grid stable operation:

Power produced (injected into grids) = Power used by energy users + power loss in lines.

When the power balance (the above law) is not obeyed, the power grid's synchronized operation is no longer possible and some generators may slowdown and other generators may speed up and protection of grid facilities take the affected equipment that may be subjected to damage out of service by opening the circuit breakers. Results in black out of power grids.

These concepts can be summarized in the following grid operation. As load on the grid increases, the system frequency drops. The drop in frequency is reflected in drop of generators' synchronized speed. So as more loads are put on grids' generators, the synchronized speed and frequency drops. Just like a car going up a hill. Going up a hill puts more load on the car.

Therefore, in order to maintain the speed of the car, more fuel must be used by pushing the gas paddle. As frequency drops, ISO operator increases power inputs of the cheapest generator and inject more power into grid. The grid frequency is constantly changing. Energy user controls the grid frequency by varying power demands. For example, when power demand is reduced at night, the input power to power grid must be reduced to keep the system frequency at its set point. ISO operators' through computer control follow the load changes by adjusting input power into the grid. To summarize:

- Load goes up, then the grid frequency goes down
- Load goes down, then the grid frequency goes up

Therefore, it is apparent that the end-energy users are collectively controlling grid loads. When energy users turn on or turn off power devices, these actions impact the grid instantaneously. The grid control system must respond to the load changes and produce or reduce generation to balance total generated power with total loads. This action is referred to as load following and it takes place under load-frequency control of computer control of ISO.[11]

In a market-structured power grid system, the ISO has the responsibility of controlling the power grid. All stakeholders, power companies, independent power producers (IPP), and municipal producers are operating under power market rules to inject the generated power into grid for highest payoff economically. To obtain the best price all stakeholders must study the system load profile and expected demand to position themselves for maximum profit. However, ISO still operates the power grid such that its stability is assured, while it provides efficient utilization of power in the power market.

In this book, the functions that are important for the understanding of a power grid consisting of many microgrids where every bus (node) of the grid has both load and generation are presented. To understand these concepts, understanding of the operation control of a power grid in more detail is essential.

9.4 THE OPERATION CONTROL OF A POWER GRID

The primary functions of operations control are satisfying the instantaneous load on a second-to-second and minute-to-minute basis.[6-8] Some of these functions are

1. Load-frequency control (LFC)
2. Automatic-generation control (AGC)

3. Network topology determination (NTD)
4. State estimation (SE)
5. On-line load flow and contingency studies
6. Schedule of transactions (ST)
7. Economic dispatch calculation (EDC)
8. Operating reserve calculation (ORC)
9. Load management system (LMS)

The decision time of operations control is from dynamic response in a fraction of a cycle in LFC, to 1–10 seconds for automatic-generation control, to 5–10 minutes for economic-dispatch calculations, and from a second up to 30 minutes for a load management system.[8–10] However, with the implementation of a smart grid system with a high penetration of renewable and green energy sources and a smart metering system, we will have a more-complex power grid. In the following sections, we will study the function of LFC and automatic-generation control. Other functions are left for students to study on their own using the reference section at the end of this chapter and the rich sources on the Internet.[1–4]

9.5 LOAD-FREQUENCY CONTROL

LFC is also referred to as the governor response control loop as shown in Figure 9.10. As the load demand of the power system increases, the speed of the generators decreases and this reduces the system frequency. Similarly, as the system load-demand decreases, the speed of the system generators increases

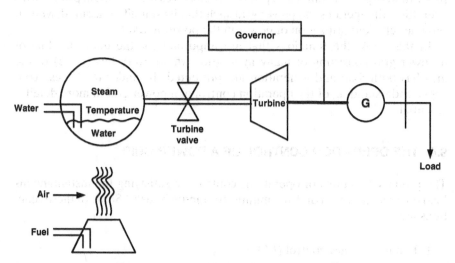

Figure 9.10 The governor control systems.[11]

and this increases the system frequency. The power system-frequency control must be maintained for the power grid to remain stable.

In the AC power grids, all generating sources are operating in parallel and all (inject) supply power to the power grid. This means that all power sources are operating at the same system frequency. The system operating frequency in the United States is 60 Hz and 50 Hz in the rest of the world. The generators are operating at the system frequency; they are all synchronized and operating at the same synchronized speed: all are (injecting) supplying power to the power grid. The synchronized speed can be computed as

$$\omega_{syn} = \frac{2}{P}\omega_s \qquad (9.1)$$

where $\omega_s = 2\pi f_s$ and f_s is the system frequency. In revolutions per minute (rpm), we have

$$n_{syn} = \frac{120 f_s}{P} \text{ rpm} \qquad (9.2)$$

In the above equation, P is the number of poles and f_s is the generator frequency. Therefore, for a two-pole machine, operating at 60 Hz ($f = 60$ Hz), the shaft of the machine is rotating at 3600 rpm. If the prime mover power has a slower speed, such as the hydropower unit, the generator has more poles. For example, if $P = 12$, the prime mover speed is 600 rpm and still the unit operates at 60 Hz.

Synchronized operation means that all generators of the power grid are operating at the same frequency and all generating sources are operating in parallel. This also means that all generating units are operating at the system frequency regardless of the speed of each prime mover. In AC grids, the energy cannot be stored; it can only be exchanged between inductors and capacitors of the system and is consumed by loads. Therefore, for an AC grid to operate at a stable frequency, the power generated by AC sources must be equal to the grid loads. However, the loads on the grid are controlled by the energy users. As discussed before when the lights turned off, the grid loads are reduced; when the lights are turned on, the grid loads are increased. In response to load changes, the energy is supplied from the inertial energy stored in the massive mass of a rotor. However, at every instant, the balance between energy supplied to the grid and the energy consumed by loads plus losses must be maintained for stable operation. This concept can be expressed as

$$\sum_{i=1}^{n_1} P_{G_i} = \sum_{i=1}^{n_2} P_{L_i} + P_{losses} \qquad (9.3)$$

where P_{G_i} is the power generated by generator I, P_{L_i} is the power consumed by load i and the n_1 is the number of the system generators and n_2 is the number of the system loads. The transmission line losses are designated by P_{losses}.

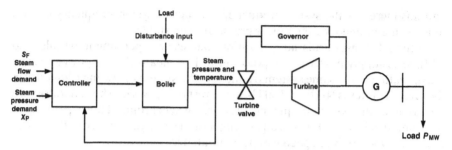

Figure 9.11 A boiler control system.

As can be expected, as the grid load demand increases at time t, the grid frequency decreases because the power grid at that instant has more loads than at the instant $t - k$ where k is the time step. In fact, this is precisely what happens at first. However, the grid has a feedback loop that is called the load-speed control and as the system frequency drops, that is, the prime mover shaft speed decreases, the feedback loop increases the input power to match the grid total generation with the grid total load. This is called governing system control: the governor opens the turbine valves to increase the input power that in turn speeds up the shaft of the generators under speed control. Therefore, with an increase in the grid loads, the additional power generated matches the grid generated power to the grid total load and the grid operates at the system synchronized speed.

The governor control keeps the turbine shaft speed constant at the desired synchronized speed to generate power at a synchronized grid frequency. To ensure the safety of the boiler and turbine, the boiler control system controls the condition of steam that is expressed by steam pressure and steam temperature. The boiler control system controls the turbine valve in the desired position such that the steam pressure and temperature are within their specified range. Figure 9.11 depicts the boiler control system. The governor control feedback controls the turbine shaft speed—as the system load changes; the governor feedback opens or closes the turbine valves. However, the opening and closing turbine valves are dependent on steam conditions. The turbine valves can be opened or closed as long as the boiler steam conditions are within the desired range.

To match the grid generated power to the grid total load, two control methods are implemented. These methods are turbine-following control and boiler follow-up control. In the turbine-following control, the turbine generator is assigned the responsibility of throttle pressure. The turbine valves are controlled within specified range that ensures that steam conditions, steam pressure, and temperature are within the safe range. The MW load demand corresponds to steam flow demand and it is the responsibility of the boiler. When the step increase in load control command is issued, the control command is sent to the boiler. The boiler control system then increases the fuel rate, feed water, and airflow, which increases the throttle pressure. The change in the throttle pressure is measured by the turbine control system. The turbine valves are controlled by the turbine control system. The turbine valves are opened to increase the steam flow and MW output of the generator. Note that when the steam flow increases due to the opening of the turbine valves, the turbine shaft accelerates. However, because the generator is synchronized to the power grid and the grid load has increased, the MW power generated is injected into the power grid and a new balance between the grid total load and total generated power is established. Therefore, the system frequency is maintained and all connected generators operate at a synchronized speed.

In boiler follow-up control,[11] the boiler is assigned the responsibility of throttle pressure. The MW load demand is the responsibility of the turbine generator. In this mode of operation, a step increase in generation due to a step change in load demand goes directly to the turbine valves. The load demand increases, the turbine valves open, and hence the steam flow and MW output of the generator increase. However, the boiler is controlling the throttle pressure, and if the pressure drops out of the range assigned by the boiler, the boiler control system overrules the turbine control action to maintain the pressure. Both the proposed control systems can provide satisfactory control performance. The boiler follow-up control has a faster response and is widely used. The turbine-following control system has slower response; however, it protects the boiler and assures that steam is conditioned before energy is extracted from the boiler.

For the stable operation of a power grid, in addition to frequency control, the terminal voltages of generators and power factors must be controlled. Figure 9.2 depicts the voltage regulator for a steam turbine generator. As was indicated, the governor controls the main steam valves of the turbine and controls the steam flow to the turbine. The steam flow to the turbine is the primary mechanical power on the shaft of the generator. In Figure 9.12, the generator excitation system that is located on the rotor of the machine is also depicted. The generator's terminal voltage is controlled by the voltage regulator. The field voltage is applied to the generator excitation winding based on the regulator set point (V_{ref}).

By applying the mechanical power to the rotor winding that is supplied with DC current, a time-varying field is established in the air gap of the machine. Based on Faraday's law of induction, the voltage is included on the stator windings. Again, because the generator is synchronized to the power grid, the power is injected into the system. Figure 9.13 depicts the main concepts that we must understand from the operation of a steam power plant. A power generator is a three-terminal device. The field current of the generator can be set to control the generator's terminal voltage.

$$E = K \cdot I_{\text{f}} \cdot \omega \qquad (9.4)$$

The open-circuit-induced voltage, E is a function of machine dimensions that is depicted by constant K and field current, I_{f} and shaft speed, ω.

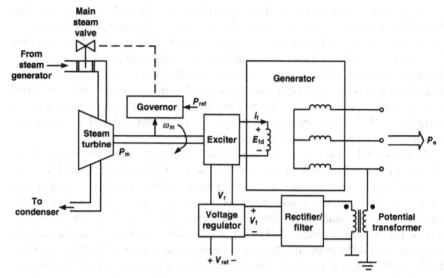

Figure 9.12 Voltage regulator and turbine-governor controls for a steam turbine generator.

By adjusting the field current, a generator can operate at leading or lagging power factor. The reactive power, Q_G generated by generators must be equal to the total reactive loads and transmission lines' reactive losses.

$$\sum_{i=1}^{n_1} Q_{Gi} = \sum_{i=1}^{n_2} Q_{Li} + Q_{\text{losses}} \tag{9.5}$$

where Q_G is the reactive power generated, Q_L is the reactive power of load, Q_{losses} is the reactive power loss.

Let us formally introduce two important analysis studies in power grids, planning, design, and operation.

1. Power Flow Studies. Given the schedule system generation, system load, and schedule system elements such as transmission lines and transformers, we compute the system bus voltages and power flow on transmission lines. These conditions are expressed by Equations (9.3) and (9.5). We often refer to bus voltages as system states that represent the voltage magnitude and phase angle at each bus. For power flow studies, we are interested in the system injection model: we do not include the generator impedance in the power grid injection model that describes the injected power at the terminal of the generator into the network model of the transmission system.

2. Short-Circuit Studies. Given the system model, the bus voltages, and load, we compute balanced and unbalanced fault currents that can flow on the

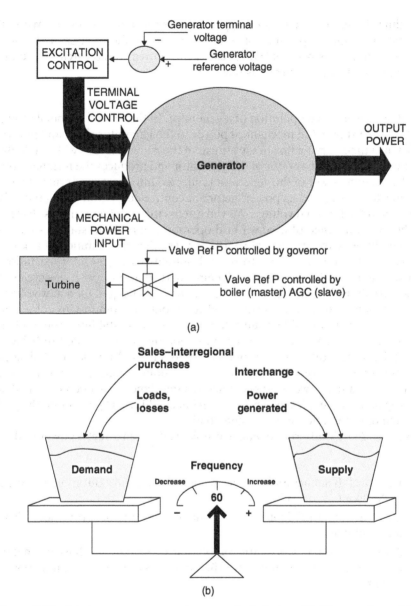

Figure 9.13 (a) A generator as a three-terminal device. (b) The supply and load demand balance for synchronized operation.[24]

system if a fault happens. Based on this study, we calculate the short-circuit currents that the breakers may experience upon occurrence of a fault. This study also provides the level of fault current throughout the system for setting relays of the protection system. In the short-circuit studies, the internal input impedance of generating sources must be included because they limit the fault current as it happens upon occurrence of a fault. Without internal input impedance of generating sources, the fault current would be infinite; this is unrealistic because the source would catch fire before an extremely high current is reached.

Let us return to the operation of a generator. In the second terminal of Figure 9.13a, we supply the mechanical power to the generator shaft, and in turn, we set up a time-varying flux in the air gap of the generator that will couple the windings located on the stator of the generator and produces the terminal voltage. The output power of the generator is injected into the power grid network. The injected power and its power factor are controlled by controlling the field current and the terminal voltage. We will discuss this point later in this chapter.

The dynamic range of a power grid operation starts from startup—a transient condition to the steady-state operation. The dynamic duration of a power grid can be from a few cycles to several minutes. The generator excitation-control system can be subjected to dynamic perturbation from a few cycles to a few seconds as the field current of the generator is changed for a new voltage setting. When a power grid is subjected to an outage from the loss of a generator, the power grid will be subjected to the dynamic stability problem that can be stabilized if the power grid can provide the power needed to balance the total grid generation to the total grid load. Figure 9.13b depicts balanced operation of power grids. For example, for a generator outage, the governors of all units within the power grid will react to a deficiency in needed power (that is a drop in the system frequency) and will inject additional power into the grid to match the generation to the system load.

We can identify different dynamic problems that can affect a power grid:

1. Electrical dynamics and excitation controls may have duration of several cycles to a few seconds.
2. Governing and LFC may have a dynamic duration of several seconds to a few minutes.
3. A prime mover and an energy supply control system may have a dynamic duration of several minutes. A prime mover is a steam-generating power system.

9.6 AUTOMATIC GENERATION CONTROL

In the preceding sections, we defined the function of a boiler and a governor in the LFC. As we pointed out, as the load changes, the supply of power to the grid must be adjusted such that at all times, the system-generated power balances the system loads and losses to keep the grid operating at rated voltages and rated frequency.

The grid load has a general pattern of increasing slowly during the day and then decreasing at night. The cost of generated power is not the same for all generating units. Therefore, more power generation is assigned to the least costly units. In addition, a few lines connect one power grid to another neighboring power grid. These lines are referred to as tie lines. Tie lines are controlled to import or export power according to set agreed contracts. When power is exported from a power system to a neighboring power system through the tie lines, the exported power is considered as load; conversely when imported, it is considered as power generation. The control of the power flow through these lines is prespecified on agreed schedules and they are based on secure operation and economic transactions. To control both the power flow through transmission tie lines and the grid frequency, the concept of area control error (ACE) is defined as

$$ACE = \Delta P_{TL} - \beta \Delta f \tag{9.6}$$

where

$$\Delta P_{TL} = P_{Sch} - P_{Actual}$$

$$\Delta f = f_S - f_{Actual} \tag{9.7}$$

P_{Sch}: The scheduled power flow between two power networks
P_{Actual}: The actual power flow between two power networks
f_s: The reference frequency, that is, the rated frequency
f_{Actual}: The actual measured system frequency
β: The frequency bias

The AGC software control is designed to accomplish the following objectives:

1. Match area generation to area load, that is, match the tie-line interchanges with the schedules and control the system frequency.
2. Distribute the changing loads among generators to minimize the operating costs.

The above condition is also subject to additional constraints that might be introduced by power grid security considerations such as loss of a line or a generating station.

Figure 9.14 presents the automatic generation control (AGC) block diagram. The first objective involves the supplementary controller and the concept of tie-line bias. The term β is defined as bias and it is a tuning factor that is set when AGC is implemented. A small change in the system load produces proportional changes in the system frequency. Hence, the area control error

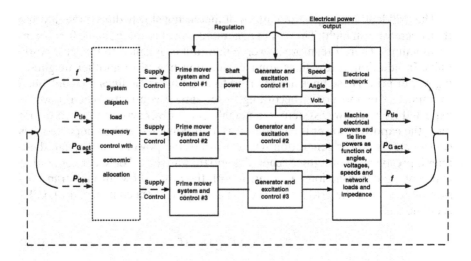

Figure 9.14   The automatic generation control (AGC).

$(ACE = \Delta P_{TL} - \beta \Delta f)$ provides each area with approximate knowledge of the load change and directs the supplementary controller for the area to manipulate the turbine valves of the regulating units. To obtain a meaningful regulation (i.e., reducing the ACE to zero), the load demands of the grid are sampled every few seconds. The second objective is met by sampling the load every few minutes (1–5 minutes) and allocating the changing load among different units to minimize the operating costs. This assumes that the load demand remains constant during each period of economic dispatch. To implement the above objectives, nearly all AGC software is based on unit control. For unit i, the desired generation at time instant K is normally sampled every 2 or 4 seconds and is given by Equation (9.8).

$$P_D^i(K) = P_E^i(K) + P_R^i(K) + P_{EA}^i(K) \tag{9.8}$$

where $P_E^i(K)$, $P_R^i(K)$, and $P_{EA}^i(K)$ are the economic, regulating, and emergency assist components of desired generation for unit i at time instant K, respectively.

Figure 9.15b depicts the concept of stored inertial energy and the energy management time scale of power grid control. The stored inertial energy in the rotor of generating units provides energy to the high-frequency load changes. The high-frequency loads are shown in Figure 9.3. Simply, when an energy user turns off a light, the load drop creates a high-frequency load fluctuation. Of course, when a large number of energy users turn their lights off, they create high- and low-frequency load fluctuations. The low-frequency load fluctuation has a clear load rise or a load drop trend. This change of load is controlled by AGC as shown in Figure 9.15c. The AGC and the system operator follow the

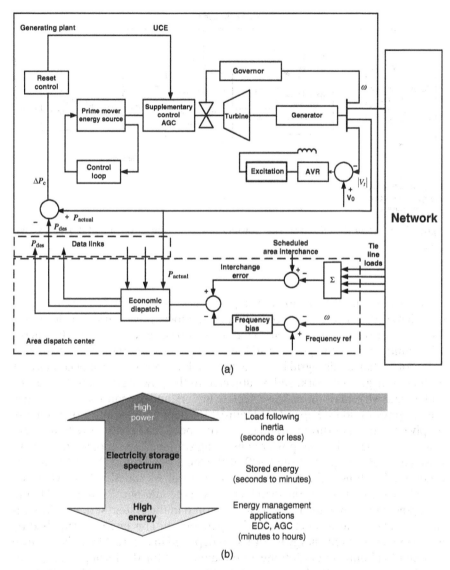

Figure 9.15 (a) Schematic diagram of load-frequency control system with economic dispatch. (b) The energy management time scale of power grid control.[12] (Adapted from IEEE Power and Energy Magazine, IEEE 2009.)

grid load. However, the AGC controls the input energy into the power grid in response to load changes.

The AGC also controls the connected microgrids in a large intercon-nected power grid. The microgrid concept assumes a cluster of loads and its microsources, such as photovoltaic, wind, and CHP are operating as a single controllable power grid. To the local power grid, this cluster becomes a single

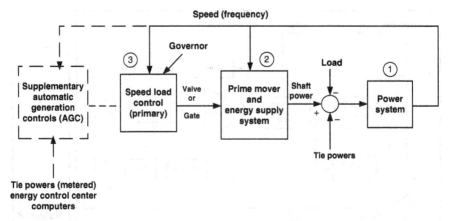

Figure 9.15 (*Continued*) (c) The automatic generation control (AGC) block diagram.

dispatchable load. When a microgrid power grid is connected to a power grid, the microgrid bus voltage is controlled by the local power grid. Furthermore, the power grid frequency is controlled by the power grid operator. The microgrid cannot change the power grid bus voltage and the power grid frequency. Therefore, when a microgrid is connected to a local power grid, it becomes part of the power grid network, and is subjected to the power grid disturbances. To understand why this is the case, we need to understand the control system that is used by power grid ISO operators. Figure 9.15a depicts the prime mover, energy supply (steam or gas turbine), and governor (speed load control) system. These systems are located in the power-generating station. The supplementary controls and AGC are part of the EMS of the local power grid. The LFC system is designed to follow the system load fluctuation. As stated before, when the load changes, let us say as the load increases in the microgrids connected to the local power grid, then the inertial energy stored in the system supplies the deficiency in energy, to balance the load to generation. This energy is supplied by prime movers (stored energy in rotors) as depicted in Figure 9.15b. The balance between load and generation must be maintained for the local power grid to remain stable. When the balance between generation and load is disturbed, the dynamics of the generators and loads can cause the system frequency and/or voltages to vary, and if this oscillation persists, it will lead to system collapse of local power grid and connected microgrids. If the load increases rapidly and the power grid frequency drops, then steam units open the steam valves and hydro unit control loops will open the hydro gates, to supply energy to stabilize the grid frequency. This action takes place regardless of the cost of energy from generating units. All units that are under LFC participate in the regulation of the power grid frequency. This is called the governor speed control, as shown in Figure 9.15c. Every 1–2 minutes, the supplementary control loop, under AGC, will economically dispatch all units to match load to generation, and at the same time, minimize the total operating cost. Therefore, the AGC will change the set points of the generators under its control. This timing of the cycle can fall within one to several minutes. In Figure 9.15c, the dotted line section encompasses the AGC, which is located at the local power grid energy control center under control of ISO. For LFC control, ISO receive the following information for all units under AGC control from power plant operators: (1) generator upper and lower limits; (2) rate limit; (3) economic participation factor.[23] However, when a microgrid is disconnected from the local power grid, the microgrid must be designed to control its voltage and frequency. A smart grid system and smart metering will facilitate load control.

9.7 OPERATING RESERVE CALCULATION

As discussed, the power grid operation remains stable as long as a balance exists between the system loads and system generation. The operating reserve decision is made based on the security and the necessary reliability. A stable frequency response is essential to stabilize the operation of an interconnected system upon the loss of load or generation outage.[9,10]

The spinning reserve is the amount of additional power that is distributed in the form of a few megawatts among many generators operating in the power grid. These units are under AGC control and can dispatch power to ensure the balance of grid loads and grid generation. The cost of additional power will add to the cost of proving electric energy services. The real-time pricing and smart meters will empower many energy end users to participate in improving the spinning reserves in the future operation of power systems, increasing overall efficiency, and reducing the cost of operation of power grids.

9.8 BASIC CONCEPTS OF A SMART POWER GRID

In a classical power grid, a fixed price is charged to energy users. However, the cost of energy is highest during the daily peak load operation.[15,16] The classical power system operation has no control over the loads except in an emergency situation when a portion of the loads can be dropped as needed to balance the power grid generation with its loads. Therefore, many grid facilities are used for a short time during the peak power demand and they remain idle during the daily operation.

For an efficient smart power grid system design and operation, substantial infrastructure investment in the form of a communication system, cyber network, sensors, and smart meters must be installed to curtail the system peak loads when the cost of electric energy is highest. The smart power grid introduces a sensing, monitoring, and control system that provides end users with the cost of energy at any moment through real-time pricing. It also provides monitoring capabilities and incipient failure tracking, advanced protection for management of grid facilities. The advanced control systems of smart metering provide the energy users with the ability to respond to real-time pricing. Furthermore, the smart power grid supplies the platform for the use of renewable and green energy sources and adequate emergency power for major metropolitan load centers. It safeguards against a complete blackout of the interconnected power grids due to man-made events or environmental calamity. It also allows for the break-up of the interconnected power grid into smaller, regional clusters. In addition, the smart power grid enables every energy user to become an energy producer by giving the user the choice of PV or wind energy, fuel cells, and CHP energy sources and to participate in the energy market by buying or selling energy through the smart meter connection.

The bulk power grid of the United States and many other countries are already operating as a large interconnected network. The mission of the North American Electric Reliability Corporation (NERC)[9] is to ensure the reliability and security of America's bulk power grid. Figure 9.16 depicts the North American electric reliability centers.

To increase the reliability of North American power grid, the power grid networks of North America is interconnected as show in Figure 9.16b.[24] Each grid network of an area owned by a utility company, from technical operation view point, acts as an equivalent large generator. All equivalent generators operate in nearly steady-state synchronism. As discussed, if the total interconnection loads exceed the supplied generation, the frequency drops from the set frequency of 60 Hz. The balanced operation of power grid is accomplished by load frequency control and with frequency variable frequency operation of induction motors and stored inertial energy in rotors of all generators. Those areas that are deficient in generation in supplying their loads are assisted by other areas through high voltage transmissions (called tie lines) that interconnect North America grids. For accounting purposes there are a number of balancing authorizers in North America overseeing the balancing for schedule buying and selling of power. These authorizers are called Reliability Coordinators.[24] The inadvertent exchange of power during emergency is accounted for by Reliability Coordinators. The frequency error is typically expressed in megawatts per 0.1 Hz (MW/0.1 Hz).[24]

Similarly, it is natural to expect that future cyber-controlled smart grid systems will be developed for the NERC-mandated reliability centers of the US grid. A future cyber-controlled system is depicted in Figure 9.17.

The cyber-fusion point (CFP) represents a node of the smart grid system where the renewable and green energy system is connected to large-scale interconnected systems. The US interconnected system has eight regional reliability centers as shown in Figure 9.16. It is expected that renewable microgrids will be connections of regional reliability centers such as Reliability *First* Corporation (RFC) transmission systems. The CFP is the node in the system that receives data from upstream, that is, from the interconnected network, and downstream, that is, from the microgrid renewable and green energy (MRG) system and its associated smart metering systems. The CFP node is the smart node of the system where the status of the network is evaluated and controlled, and where economic decisions are made as to how to operate the local MRG system. A CFP

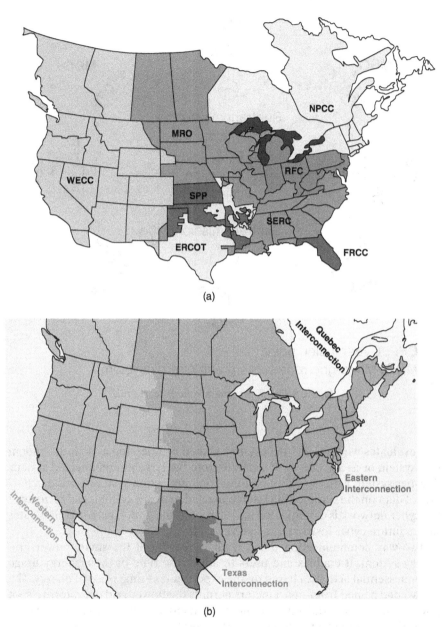

(a)

(b)

Figure 9.16   (a) Adapted from North American Electric Reliability Centers (NERC). ERCOT, Electric Reliability Council of Texas; FRCC, Florida Reliability Coordinating Council; MRO, Midwest Reliability Organization; NPCC, Northeast Power Coordinating Council, Inc.; RFC, Reliability First Corporation; SERC, SERC Reliability Corporation; SPP, Southwest Power Pool, Inc.; WECC, Western Electricity Coordinating Council.[1–3] (b) Adapted from North American Electric Reliability Corporation (NRC).[24]

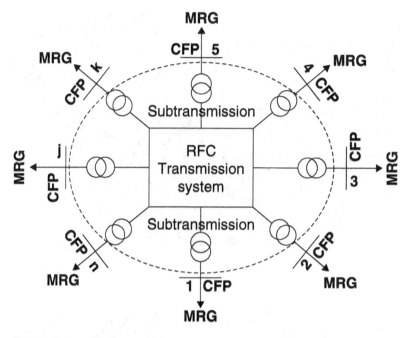

CFP: Cyber fusion point
MRG: Micro-grid renewable green energy system

Figure 9.17   A cyber-controlled smart grid.

also evaluates whether its MRG system should be operated as an independent grid system or as a grid system separate from the large interconnected system. A cyber system is the backbone of the communication system for the collection of data on the status of the interconnected network system. The security of the cyber network is essential for the security of the grid. Figure 9.17 illustrates such a future cyber-instrumented power grid.

Two-way communication is a key characteristic of the smart power grid energy system. It enables end users to adjust the time of their energy usage for nonessential activities based on the expected real-time price of energy. The knowledge gained from smart meters permits the power grid operators to spot power outages more quickly and smooth demand in response to real-time pricing as the cost of power varies during the day.[10–15]

The cyber control of a smart grid is the subject of research by many disciplines in electrical and computer engineering. It requires a control system that analyzes the performance of the power grid using distributed, autonomous, and intelligent controllers. The cyber system will learn on-line from the sensors, the smart grid, and microgrid states. The control system analyzes the system for possible impeding failure. By sensor measurements and monitoring, the cyber

TABLE 9.1   The Cost of Electric Energy in 2009[3]

Energy Source	Cost per Kilowatt Hour (Cents)	Typical Uses	Typical Installation Size
Solar energy (photovoltaics)	20–40	Base load power source	1–10,000 kW
Microturbines	10–15	Can be used in base load, peaking, or cogeneration applications	30–300 kW
Fuel cells	10–15	Rural (off-grid) power transportation appropriate for base load applications	1–200 kW
Wind turbines	5–10		5–10 MW
Internal combustion engines	1.5–3.5	Well-established, long history as back up or in peaking applications	50 kW–5 MW
Central power generation	1.7–3.7	Base load/peaking electricity generation	500–3000 MW

control system governs grid behavior based on real-time data in the face of ever-changing operating conditions and new equipment. The system uses electronic switches that control multiple MRG systems with varying costs of generation and reliability.

As a result, a cyber-controlled smart grid requires consumers to pay the real-time price of produced electric power. Table 9.1 presents the cost of electric energy as of 2009 from different sources.[3]

Figure 9.18 shows that the feeder maximum load and minimum load is changing by a factor of 2 over 24 hours. As we have learned in the previous chapter, the local power company must use many types of electric power sources to match the system generation to the system load.

The power flow into this load center is supplied by 345 kV and 138 kV transmission systems. The area load demands are satisfied by the secondary and tertiary windings of transformers rated at 138/69/12. Industrial loads are served at 138/69/23 kV. The bus load is the power flowing into the primary windings of the transformers connected to 23 kV. The power flows from higher voltage systems to lower voltage systems. Therefore, the bus load can be defined as 138 kV and/or 345 kV transformer loads.

An important factor affecting the load demands and real-time pricing is the effect of weather and its ensuing rapid increase in load demands. To separate weather-induced bus load, we use the average bus load when weather conditions are normal and subtract when weather conditions are above normal. In Figure 9.19, a composite hourly load sequence was generated by a

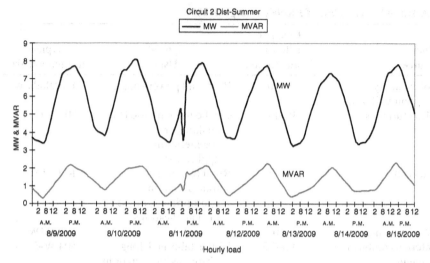

Figure 9.18 The hourly loads of a distribution feeder of a midwestern power company.[7]

midwestern power grid company.[16] The recursive mean and variance of $\{Y(.)\}$ can be computed based on Equations (9.9) and (9.10).

$$\bar{Y}(K+1) = \bar{Y}(K) + \frac{Y(K+1) - \bar{Y}(K)}{K+1} \tag{9.9}$$

$$\sigma y^2(K+1) = \frac{K}{K+1}\sigma y^2(K) + \frac{[Y(K+1) - \bar{Y}(K+1)]^2}{K} \tag{9.10}$$

The weather effect on the load sequence, holidays, and unforeseen conditions, which cause the load demand sequence to be higher or lower than a given nominal mean load profile, is subtracted from the recorded data. By removing the component of $\{Y_R(.)\}$, weather-induced load demand due to the weather effect, we generate a new sequence designated as the nominal load sequence $\{Y_N(.)\}$. The effect of weather conditions on the load depends on temperature, humidity, wind speed, and illumination. However, to demonstrate the basic concept, only the weighted average, maximum and minimum temperature that were recorded, were used. Therefore, the effect of weather condition on the load is expressed in terms of temperature. A weather-sensitive load is present when the daily temperature, T, is outside of the comfort range of $T_{min} < T < T_{max}$ where T_{min} and T_{max} are the lower and upper limits of the comfort range. This suggests that the nominal nonweather-sensitive load sequence $\{Y_N(.)\}$ is assumed to be equal to the sequence $\{Y_R(.)\}$ when the temperature is within the comfort range, and sudden load changes due to outages or special events have not occurred.

Figure 9.19 shows the plots of $\{Y_R(.)\}$ and $\{Y_N(.)\}$ load sequences for week 10, 11, 12, and 13 bus loads. It can be seen that the general weekly

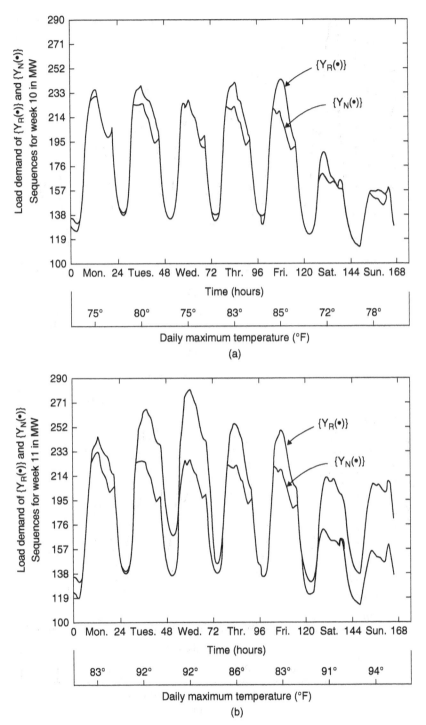

Figure 9.19 (a–d) Plot of $\{Y_R(.)\}$ and $\{Y_N(.)\}$ for weeks 10, 11, 12, and 13.[16]

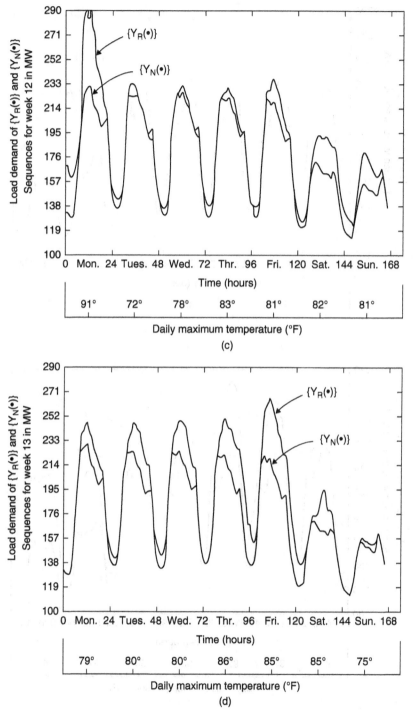

Figure 9.19 (*Continued*)

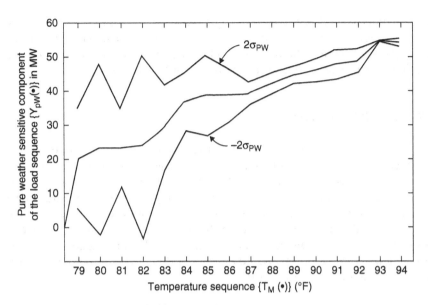

Figure 9.20   The mean and standard deviation of a pure weather-sensitive load versus temperature.[7]

profile of weather-sensitive load sequence $\{Y_R(.)\}$ and nominal load (nonweather-sensitive) sequence $\{Y_N(.)\}$ are essentially the same when the daily temperature is normal. However, when the daily temperature is high, a weather-induced load is superimposed on the $\{Y_N(.)\}$ sequence.

The procedure is based on computing an average relationship between the temperature and the pure weather-induced component of the load, which is designated as Y_{PW}. The sequence $\{Y_{PW}(.)\}$ is generated where each member of $\{Y_{PW}(.)\}$ is a mean value of the pure weather-sensitive component of the load at a given temperature.

Figure 9.20 shows the mean and standard deviation ($\pm 2\sigma_{PW}$) of the pure weather-induced component of the load at different temperatures.

Figure 9.20 depicts the pure weather-sensitive component of the load. The weather-sensitive load sequence has saturated at 80°–82°, 84°–87°, and above 93°F.

Historically, power grid companies have operated the power system as a public service. They have provided reliable electric power at a constant price regardless of changing conditions. Their systems used additional spinning reserve units to serve the unexpected loading and outages due to the loss of equipment. However, in an age of global climate change, this kind of service cannot be provided without severe environmental degradation.[5]

A power grid operator has to schedule its generation sources based on the cost of energy. However, the weather-sensitive load component adds substantial uncertainties in planning load–generation balance. As can be expected, the least costly units are scheduled to satisfy the base loads. The more costly units are scheduled to satisfy the time-changing loads. Therefore, the price of electric energy is continuously changing as load demands are changing. If real-time pricing is implemented, the variable electric rates must be used for the privilege of reliable electrical service during high-demand conditions.

9.9 THE LOAD FACTOR

The load factor is one of the key factors that determine the price of electricity. The load factor is the ratio of a customer's average power demand to its peak demand. As has been observed, the daily load demands have a daily variation. The cost of peak power demand is substantially higher than the average power demand. Therefore, the cost of power demand changes with the time of day. The term "real-time pricing" refers to the minute-by-minute price of electric power as the energy control center commits the scheduled generators to the production of electric power. The load factor as defined below defines the price of power in an electric power grid.

$$\text{Load factor} (\%) = \frac{\text{Average power}}{\text{Peak power}} \times 100 \qquad (9.11)$$

The average power is defined as the amount of power consumption during a period. The peak power is defined as the amount of maximum power consumption during the same period. The load factor can be calculated based on daily, monthly, or yearly cycle. For system planning, the load factor is calculated on a monthly or annual basis. The facility investments must be made so that the system can handle the maximum demand. Therefore, it is desirable to have a low maximum demand. On the other hand, because revenues are generated in proportion to the average demand, it is desirable to have a high average demand. Therefore, a desirable load factor is close to one, that is, when the peak demand and average demand are close to each other.

We should understand the difference between "power factor" and "load factor." As we know, the load power factor determines the active and reactive power consumption of the load. However, induction motors that are widely used in a power grid have lagging power factors. The load factor is power consumption in kW divided by the worst-case consumption over a period. Therefore, the load factor defines the cost to the supplier per unit of energy delivered in that period. In other words, the load factor indicates how efficiently the power grid is operating.

Example 9.1 An industrial site has a constant power demand of 100 kW over a year of energy consumption. Compute the customer load factor over 1 year of providing energy to this site.

Solution:
Total energy = 8760 h/yr × 100 kW = 876,000 kWh/yr

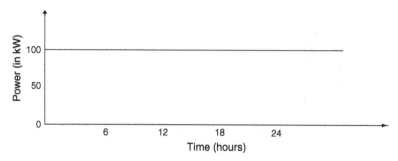

Figure 9.21   The load for 24 hours operation for Example 9.1.

Because the power demand is constant the average and peak is the same. Therefore, the load factor of this customer is 100%.

Figure 9.21 presents the load for 24 hours operation for Example 9.1.

Example 9.2   A commercial site has peak demand of 200 kW during 12 hours a day and an average demand of 50 kW demand the rest of a day. Compute the customer load factor over 1 year of providing energy to this site. Explain the associated cost of providing energy to the industrial site (Example 9.1) and the commercial site.

Solution

$$\text{The average power} = \frac{\sum \text{Power}_i \times \text{Time}_i}{\sum \text{Time}_i} = \frac{200 \times 12 + 50 \times 12}{12 + 12} = 125\,\text{kW}$$

$$\text{Load factor} = \frac{\text{Average power}}{\text{Peak power}} = \frac{125}{200} \times 100 = 62.5\%$$

When the load factor is close to unity (100%), the generating plant is efficiently used. The cost of supplying power to the load is more when the load factor is low.

At a commercial site with a low load factor, say in the range of 50%, the power grid would need twice as much installed equipment and resources to serve the site. The lower load factor means that the price must be adjusted to recuperate the extra costs. Because the industrial site and commercial site use the same amount of kW, the same price is charged for the two sites. However, the smart meter, in conjunction with real-time pricing can provide an incentive for efficiency and load demand control. The users as stakeholders would be encouraged to control the loads during peak power demands by shifting the usage at times when prices are favorable. Furthermore, the end users have a

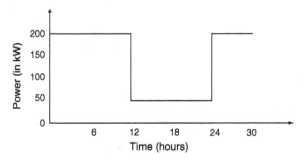

Figure 9.22 The load for 24 hours operation for Example 9.2.

high incentive to participate by installing local green energy sources such as wind and PV.

Figure 9.22 presents the load for 24 hours operation for Example 9.2.

9.10 THE LOAD FACTOR AND REAL-TIME PRICING

Real-time pricing was introduced by F. Schweppes[8] in July 1978 during an energy crisis. A simple analysis of the cost to the supplier per unit of energy delivered explains the relationship between cost and plant utilization. The real-time price of electricity is a function of load factor, load demand, and unexpected events. The first cost is the utilization of a power plant and its operating costs. To build a large plant, many issues must be addressed. In a regulated market, large plants take years to build and are located far away from load centers; electric power is transferred by long transmission lines. Normally, coal-fired power plants are built close to a coal mine. From an operational viewpoint, the sudden loss of a large plant creates instant real-time price change in the power market because the allocated real-time reserve is limited due to cost. Small power plants are normally gas fired; they are built over a short time and their construction costs can be accurately estimated. Gas-fired plants can be placed close to load centers because these plants need limited space. Furthermore, when plants are close to load centers and their power does not need to be shipped over long-distance transmission lines, these plants have lower system losses. These plants also tend to have good system security and are generally more reliable and less consequential when they are subjected to sudden outage. Combined cycle units are highly desirable because of their high efficiency. Cogeneration facilities are also attractive because they typically have lower ratings. Plants fueled by renewable energy sources are attractive as well because of their low operating cost. Due to many sources of power and their associated cost, the cost of real-time power is a variable and needs to be determined as sources of electric energy to supply the system load changes.

Example 9.3   Suppose a PV plant of 1000 kW capacity is constructed for $500 per kW. Compute the cost of energy per kWh to the end users for 1 year of operation at full capacity if the total cost on investment is to be recovered in 2 years when PV plant operates 6 hours a day on an average for 2 years and the cost of production is negligible.

Solution
The energy consumed in 1 year = Power capacity × Time in hours

$$= 1000 \times 365 \times 6 = 2190 \text{ MWh}$$

Let the price of 1 kWh of energy be = $x per kWh
The investment cost = Capacity × Cost per unit capacity = 1000 × 500 = 500,000
Therefore, the energy consumed for 2 years 2190 × 2 = 4380 MWh

$$4380 \times 10^3 \times x = 500,000$$

$$x = \frac{500,000}{4380 \times 10^3} = \$0.1142 \text{ kWh}$$

Let us introduce the cost of fuel, labor, and maintenance in load factor calculation.

$$\text{EUC} = \text{VC} + \frac{\text{Amortized fixed cost}}{\text{LF}} \tag{9.12}$$

In Equation (9.12), the term VC defines variable cost associated with fuel and other cost of plant operation and EUC represents the energy unit cost in cents per kWh.

Example 9.4   Suppose a natural gas plant of 1000 kW capacity is constructed for $300 per kW. Assume the variable cost, VC, is 2 cents per kWh.
Perform the following:

(i) Compute the cost of electric energy to end users if 100% of installed capacity is used 24 hours a day over 5 years.
(ii) Compute the cost of electric energy to end users if 100% of installed capacity is used 12 hours and 50% of installed capacity is used for the rest of a day over 5 years.
(iii) Plot the energy unit cost versus the load factor, LF from 0 to 1.

Solution
The investment cost = Capacity × Cost per unit capacity = 1000 × 300 = $300,000
Energy consumed over a period of 5 years at full capacity = 1000 × 24 × 365 × 5 = 43,800 MWh

If this cost is distributed over 5 years, the amortized fixed cost $= \frac{300,000}{43,800\times10^3} =$ 0.007

(i) Load factor $= 1$

$$EUC = VC + \frac{\text{Amortized fixed cost}}{LF}$$

$$EUC = 0.02 + \frac{0.007}{1} = \$0.027/\text{kWh}$$

(ii) The average power $= \dfrac{\sum \text{Power}_i \times \text{Time}_i}{\sum \text{Time}_i} =$

$$\frac{1000 \times 1 \times 12 + 1000 \times 0.5 \times 12}{12 + 12} = 750 \text{ kW}$$

$$\text{Load factor} = \frac{\text{Average power}}{\text{Peak power}} = \frac{750}{1000} \times 100 = 75\%$$

$$EUC = 0.02 + \frac{0.007}{0.75} = \$0.029 \text{ kWh}$$

(iii) To compute the energy unit cost versus the load factor, LF from 0 to 1, a MATLAB m-file is developed as presented below:

```
% PLOT OF EUC
clc; clear all;
LF=0.01:.01:1;      % defining the range of load factor
VC=0.02;            % variable cost
A_FC=0.007;         % amortized fixed cost
EUC=VC+A_FC./LF;    % defining EUC in $/kWh
EUC=EUC*100;        % converting into cents/kWh
plot(LF,EUC)
grid on;            % labeling the axes
xlabel('Load Factor');
ylabel('EUC (in cent/kWh)');
title('EUC vs Load Factor');
Plot:
```

From the solution and plot, it can be seen that the EUC increases rapidly as the load factor is decreased. This is because the capacity of the plant is under-utilized when the load factor is small. The real-time pricing takes into account the power grid load factors and the cost of spinning reserve that are sitting on-line to keep the system operating at stable operating conditions if the system is subjected to contingencies. Figure 9.23 presents the cost in cents per kwh as a function of load factor for Example 9.4.

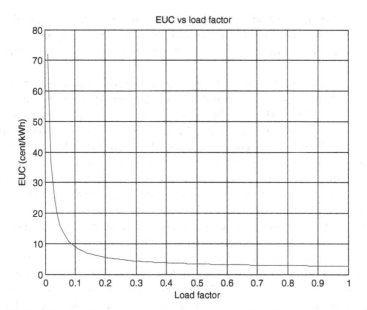

Figure 9.23   The cost in cents per kwh as a function of load factor for Example 9.4.

9.11   A CYBER-CONTROLLED SMART GRID

A cyber-controlled smart grid consists of many distributed generation stations in the form of microgrids. The microgrids incorporate intelligent load control equipment in its design, operation, and communication.[13,14] This enables the energy end users and the microgrids serving them to better control energy usage. Smart appliances such as refrigerators, washing machines, dishwashers, and microwaves can be turned off if the energy end user elects to reduce energy use. This is done by connecting the smart appliances to the EMSs in smart buildings. This technology will enable the energy end users to control their energy costs. Advanced communication capabilities in conjunction with smart meters and smart appliances enable the energy end users with the tools to take advantage of real-time electricity pricing and incentive-based load control. Furthermore, the emergency load reduction can be achieved by turning off millions of air conditioners on a rotation basis for a few minutes. With real-time pricing, the energy end users would have a very high incentive to become energy producers and install green energy sources. As real-time prices take hold, commercial and industry units are expected to generate their own energy and sell their extra power back to the power grid.

Cyber-controlled smart grid technology has three important elements: sensing and measurement tools, a smart transducer, and an integrated communication system.[13,14] These elements monitor the state of the power system by measuring line flows, bus voltages, magnitude, and phase angle using phasor measurement technology[15,16] and state estimation.[17] The technology is based on advanced digital technology such as microcontrollers/digital signal processors. The digital technology facilitates wide-area monitoring systems, real-time line rating, and temperature monitoring combined with real-time thermal rating systems.

Transducers are sensors and actuators play a central role in automatic computerized data acquisition and monitoring of smart grid power systems. A smart transducer is a device that combines a digital sensor, a processing unit, and a communication interface. The smart/controller transducer accepts standardized commands and issues control signals. The smart transducer/controller is also able to locally implement the control action based on feedback at the transducer interface. The utilization of low-cost smart transducers is rapidly increasing their penetration for building embedded control systems in smart grid monitoring and control.

Real-time, two-way communication is enabling a new paradigm in the smart grid system. It enables the end users to install green energy sources and to sell energy back to the grid through net metering. The customers can sign up for different classes of service. The smart meters will facilitate the communication between the customers by providing the real price by the supplier. The customers can track energy use via Internet accounts, where the expected price of energy can be announced a day ahead for planning purposes and the real-time price of energy can be provided to end users so they may be aware of the savings that can be realized by curtailing their energy use when the energy system is under stress.[16,17]

A smart meter allows the system operator to control the system loads. Load control ultimately provides new markets for local generation in the form of renewable and green energy sources. With the installation of smart meters (i.e., a net metering system), end users can produce their own electric power from renewable sources and sell their extra power to their local power grid.

As more customers use a net metering system, a substantial change in energy demand will result. Residential, commercial, and industrial concerns will install PV systems, use wind farm and micro generation technologies and store energy as IPP.[18] The EMSs of smart buildings with their own renewable power sources and CHP is likely the trend of the future. With the installation of an advanced net metering system, every node of the system will be able to buy and sell electric power. The use of real-time prices will facilitate the control of frequency and tie-line deviations in a smart grid electric power system. Under the grid emergency operation, the real-time pricing will provide a feedback signal as the basis of an economic load/shedding policy to assist the direct stabilizing control for a smart grid. Real-time pricing can be integrated with demand response to match the system load demand and generation in real time. This will facilitate coordinating demand to flatten a sudden change in energy use. If the sudden surge in demand is not satisfied, it will result in the cascading collapse of the power grid. In demand response control, these spikes can be eliminated without the cost of adding spinning reserve generators. It will also reduce maintenance and extend the life of equipment. The energy users can reduce their energy bills

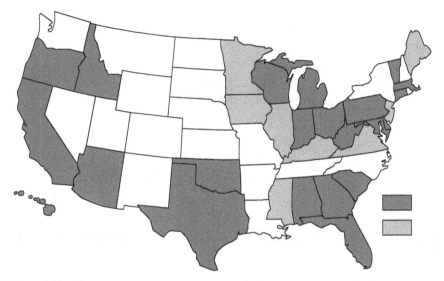

Figure 9.24   The map represents smart meter deployments, planned deployments, and proposed deployments by investor-owned utilities and some public power utilities. (Adapted from The Edison Foundation Institute for Electric Innovation; as of this writing, approximately 60 million customers have been equipped with a smart meter.[1]) The shaded area indicates that deployment is for more than 50% of end users. The dotted area indicates that deployment is for less than 40% of end users.

by using their smart meters to program and operate their low-priority household appliances only when energy is at its cheapest.

Figure 9.24 shows the state of smart metering installation in the United States as of April 2009.

Figure 9.17 depicts the MRG system. The MRG system's EMS through ISO communicates with individual smart meters located at residential, commercial, and industrial customer sites. The smart meters can control loads, such as air-conditioning systems, electric ranges, electric water heaters, electric space heaters, refrigerators, washers, and dryers using Ethernet TCP/IP sensors, transducers, and communication protocol, as shown in Figure 9.25.

The intelligence nodule of the EMS of the MRG system will receive information on the status of connected loads from local smart meters. The EMS of the MRG system will control various customer loads, based on real-time pricing signals and normal, alert, or emergency grid signals. In general, the EMS takes information from the power grid and the open access same-time information system (OASIS).[1–6] Based on real-time pricing, smart meters are programmed to control loads on the customer's sites. The EMS's control of loads will depend on input signals from its EMS and the customer's pre-established contract criteria. The EMS of the MRG microgrids would have the capability to shed customer load and respond to local power grid operating conditions.

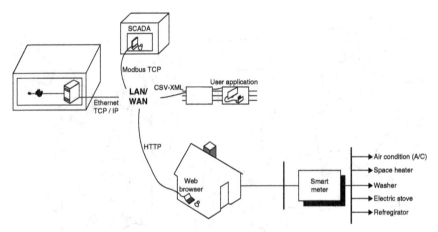

Figure 9.25 Ethernet TCP/IP sensors, transducers, and communication protocol for load control.

Smart microgrid systems comprises green and renewable energy sources with their associated power converters, efficient transformers, and storage systems.[18]

9.12 SMART GRID DEVELOPMENT

Global warming and the environmental impact of coal-based power generation are changing the design and operation of the power grid. The industry is experiencing a gradual transformation that will have a long-term effect on the development of the infrastructure for generating, transmitting, and distributing power. This fundamental change will incorporate renewable generation and green energy sources in a new distributed generation program based on increased levels of distributed monitoring, automation, and control as well as new sensors. The power grid control will rely on data and information collected on each microgrid for decentralized control. In return, the microgrids and interconnected power grid will be able to operate as a more reliable, efficient, and secure energy supplier.

The technology of the power grid and microgrids has a number of key elements. Adaptive and autonomous decentralized controls respond to changing conditions. Predictive algorithms capture the power grid state (phasor measurements)[16] for a wide area and are able to identify potential outages. The system also provides market structure for real-time pricing and interaction between customers, grid networks, and power markets. Furthermore, the smart grid provides a platform to maximize reliability, availability, efficiency, economic performance, and higher security from attack and naturally occurring power disruptions.

TABLE 9.2   A Comparison of the Current Grid and the Smart Grid

	Current Grid	Smart Grid
System communications	Limited to power companies	Expanded, real time
Interaction with energy users	Limited to large energy users	Extensive two-way communications
Operation and maintenance	Manual and dispatching	Distributed monitoring and diagnostics, predictive
Generation	Centralized	Centralized and distributed, substantial renewable resources, energy storage
Power flow control	Limited	More extensive
Reliability	Based on static, off-line models and simulations	Proactive, real-time predictions, more actual system data
Restoration	Manual	Decentralized control
Topology	Mainly radial	Network

The implementation of an advanced metering infrastructure provides real-time pricing to the energy end users. In parallel, the penetration of renewable energy sources is providing a platform for autonomous control or local control of connected microgrids to the local power grid. A distributed autonomous control will provide reliability through fault detection, isolation, and restoration. The autonomous control and real-time pricing also delivers efficiency in feeder voltage to minimize feeder losses and to reduce feeder peak demand of plug-in electric vehicles. The maturing storage technology will provide community energy storage, which becomes yet another important element for microgrid control and allows the energy user to become an energy producer. These interrelated technologies require a coordinated modeling, simulation, and analysis system to achieve the benefits of a smart power grid. Table 9.2 presents a comparison of the current grid and the smart grid.

9.13 SMART MICROGRID RENEWABLE AND GREEN ENERGY SYSTEMS

Figures 9.26 and 9.27 present the DC and AC architectures of MRG systems. The MRG systems will also include cyber communication systems consisting of smart sensors for monitoring, controlling, and tracking the normal, alert, emergency, and restorative states of systems.

The smart meters of MRG system are connected to a large interconnected power grid (see Figure 9.17). The MRG system is also designed to provide an intelligent grid optimization manager that would allow control of various customer loads based on pricing signals and grid stress. The smart meters will

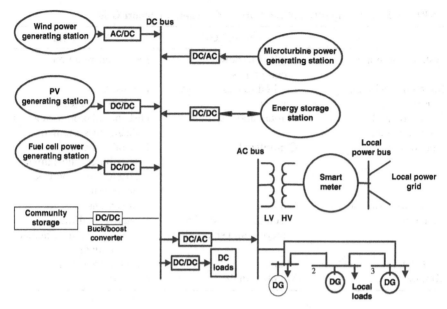

Figure 9.26 The DC architecture of a microgrid renewable and green energy (MRG) distributed generation (DG) system.

control devices at the customer location by changing their use of power. The smart meters would have the capability to shed customer load and allow distributed generation to come on-line, when the price of power is above a specified level. The EMS through ISO has two-way communication with the smart meters under its control. The EMS of microgrid receives status and power signals from all of the modules (loads and generating sources). The EMS is able to control power flow into and out of the microgrid system from its host local power grid based on variables such as weather forecasts, load forecasts, unit availability, and power sales transactions.

The MRG systems provide a new paradigm for defining the operation of distributed generation (DG). The MRG systems are designed as clusters of loads and microsources, operating as a single controllable system. To the local power grid, this cluster becomes a single dispatchable load, which can respond in seconds. The point of interconnection in the smart microgrid is represented by a node where the microgrid is connected to the local power grid, as shown in Figures 9.26 and 9.27. This node is referred to as the locational marginal pricing (LMP),[5,6] where the node price (cost) represents the locational value of energy. Power grids (energy serving entities) try to provide reliable supplies of electric energy to their energy users. Maximum benefits require low cost, sufficient supply, and stable operation. The architectures shown in Figures 9.26 and 9.27 are of interest to smart grid technology because they facilitate "plug and play" capabilities. In these architectures, green energy sources, such as fuel cells, microturbines, or renewable sources, such as PV-generating stations and wind

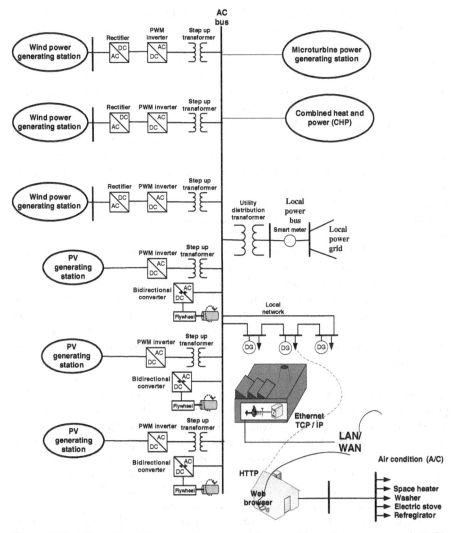

Figure 9.27   The AC architecture of a microgrid renewable and green energy (MRG) distributed generation (DG) system.

farms, can be connected to a DC bus or an AC bus, using uniform interchangeable converters. The MRG systems must be able to operate in two modes of operation: (1) in synchronized operation with the local power grid system, and (2) in the island mode of operation, upon the loss of the power grid system. In the island mode of operation, the MRG operates as an autonomous microgrid controlling its frequency and bus voltage. When an MRG system is connected to a power grid, the MRG system operates using a master and slave control technology. The master referred to is the EMS of the local power grid and slave is the MRG microgrid. If the MRG system is suddenly separated from its local power grid and MRG stability is maintained, then the slave controller takes over LFC and voltage control. For MRG systems with high power capacity, there is a purchasing agreement between the power grid and MRG systems, regarding active and reactive power transfer. The MRG system can control its loads and accept a "price signal" and/or an "emergency operation signal" from its local power grid to adjust its active and reactive power generation. The MRG system has hardware in place to shed loads in response to a price signal, and it can rotate nonessential loads to keep on critical loads. However, because disturbances in a local power grid cannot be predicted with current technology, it is quite possible upon the loss of the local power grid that an MRG system would not be rapidly disconnected from the local power grid; hence, the stability of the MRG systems would not be maintained. Future research in cyber-monitoring and control MRG systems is being conducted to provide predictive models to track states in the system and to provide distributed intelligence control technology.

In Figures 9.26 and 9.27, the EMS controls the infinite bus voltage and the system frequency. The slave controller controls the AC bus voltage of the inverter and the inverter current. Therefore, the slave controller of the MRG system inverter must be able to control active and reactive power, at leading or lagging power factor, or operate at the unity power factor. In small, renewable energy systems, the inverter is controlled at the unity power factor, and it leaves the voltage control, that is, the reactive power (Vars) control, to the EMS of the local power grid.

The MRG system of Figures 9.26 or 9.27 has a number of distribution generation sources such as wind-generating stations and PV-generating stations. The power capacity of such a system is in a range of 1–10 MW. We are using step-up transformers to step up the AC voltage to reduce the power losses. Depending on the size of distributed generators (DGs), we will step up the voltage to a range of 34.5–69 kV, before injecting the power into the local power grid network. As shown in Figure 9.28, the storage system facilitates storing the renewable energy. The stored energy can be used as load demand fluctuates during the daily load cycle of 24 hours. Because the investment in developing a power grid is quite high and its construction takes many years, the power system is planned years in advance. However, if renewable and green energy sources are installed in the distribution system, the lead time is from a few months to a few years depending on the size of facilities to be installed.

The residential MRG system depicted in Figure 9.28 consists of rooftop photovoltaics with a capacity in the range of 5–25 kVA depending on the available roof surface area. A DC/DC boost converter is used to boost the voltage of the DC bus. The maximum power point tracking (MPPT) system is designed to track and operate the PV power generator at the maximum power point. The DC/AC inverter converts the DC bus voltage to AC voltage at the operating frequency and rated residential voltage. The MRG system of Figure 9.28 is connected to the local power grid by stepping up the voltage to the local distribution voltage. The MRG system can also store DC power during the day for use during the night.

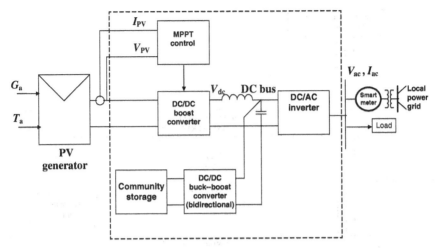

Figure 9.28 A residential microgrid renewable and green energy (MRG) system with a local storage system.

Figure 9.27 depicts the injection of electric energy into the network at all voltage levels in interconnected power grid systems. These injection nodes are part of an MRG system of a cyber-controlled smart grid as shown in Figure 9.17. The loads are under the control of smart meters. Therefore, every node in the cyber-controlled power grid is price sensitive.

The high penetration of renewable energy sources can be placed at all voltage levels as shown in Figure 9.29. Many states have mandated the use of renewable energy. It requires that investor-owned utilities obtain 25% of their electricity from renewable and advanced energy sources by 2025. In some states, a 0.5% set aside has been put in for solar PV energy sources. There are penalties of $450 per MWh for noncompliance with solar benchmarks and $45 per MWh for noncompliance with other benchmarks that began in 2009. These benchmarks can be met by purchasing renewable energy credits (REC) each equal to 1 MWh of renewable energy. Purchasers of the product could sell RECs into the REC market. As of this writing, their current value is around $37.50. It is expected that solar RECs will probably sell at much higher rates.[1-5]

There are a number of advantages to smart MRG systems. It empowers the individual to provide his or her own energy needs. This participation by energy end users has been called the "democratization of energy." This allows matching the characteristics' of loads to their generators. This makes these types of microgrids independent from the interconnected grid power failures. Smart MRG technologies have the ability to identify potential stability problem. Real-time information enables MRG systems to separate themselves from the interconnected grid and operate as an "island." During normal synchronized operation, the interconnected smart grid with its smart monitoring of power grids can control and manage the interconnected power grid to control the system loads and avoid system-wide black out.

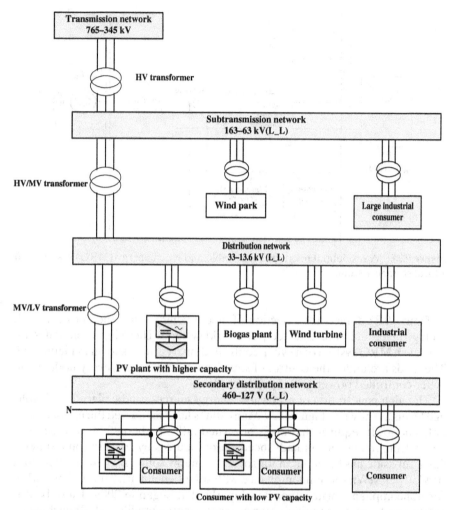

Figure 9.29 A smart grid with high penetration of renewable energy sources.

When a smart microgrid DG is connected to a power grid, then the smart microgrid DG generating station should operate, using a master and slave control technology. The master is the EMS of the power grid system. The EMS controls the infinite bus voltage located at the connection node of one microgrid to its local power grid. In Figure 9.27 the smart grid DG is connected to the power grid system by a transformer. The slave controller controls the AC bus voltage of the microgrid inverter (bus voltage magnitude and phase angle) and the inverter current. Therefore, the slave controller of the microgrid DG inverter also controls active and reactive power. The microgrid inverter is designed to operate as a unity power factor and to leave the voltage control, that is, the reactive power (Vars) control, to the EMS of the power grid system. Or, the inverter of DG operates with a leading power factor, a lagging power factor. If the smart grid is suddenly separated from its local power grid and system stability is maintained, then the slave controller takes over LFC and voltage control.

The word "smart" refers to the requirement that a microgrid controls its loads, and that it accepts the "price signal" and/or the "emergency operation signal" from its local power grid to adjust its active and reactive power generation. Other designs are also possible such as a net smart metering communication between the EMS of the local power grid and the EMS of the microgrid system. The smart microgrids should have hardware in place to shed loads, in response to the price signal, or they should have hardware to rotate nonessential loads and to supply power critical loads. However, because disturbances in a power grid system cannot be predicted, the DG microgrid system should be designed to rapidly disconnect from its power grid system to maintain the stability of the microgrid DG systems. To ensure stable operation, the storage system of the microgrid, with its inverter, has to be able to participate in the ancillary market.[5] The architecture of Figure 9.27 is of interest to distributed generation technology because it facilitates plug and play capabilities. In this architecture, green energy sources, such as fuel cells, microturbines, or renewable sources, such as PV-generating stations and wind farms, can be connected to a DC bus, using uniform interchangeable converters. The architecture of Figure 9.27 satisfies the "Renewable Portfolio" laws that have been mandated by many states. This architecture allows for the selling of and buying of energy to and from local utilities. Note that the microgrid can offer real-time pricing or a tariff-based contract by facilitating load control through a cyber-controlled smart meter or net metering, if a customer has local generation at the customer site to control loads. The smart meter and its cyber control are also shown in Figure 9.27. In addition, the microgrid can participate in system control at a bulk power level, through its own smart metering that is interfaced with its local power grid EMS.[20]

The microgrid DG system is designed such that upon separation from the local power grid, the microgrid DG system is restored by shedding noncritical loads if there has been a generation deficiency, so local microgrid is able to return to normal operation. This problem of stability is a function of how strongly a microgrid DG system is connected to a power grid system. This problem must be addressed as system parameters are defined, and as the level of connecting a microgrid DG system to a local power grid is defined. It also requires the microgrid DG system to have other green energy sources, such as CHP and microturbines, to ensure quality and stable service if some power sources have forced outage.[21]

The classical interconnected power network is designed for one-way flow of power. However, when the subtransmission and distribution and residential systems have the ability to generate electric power that can reverse the flow of power from low voltages to higher voltages and create reverse power flow, it can raise safety and reliability issues. These protection problems of a smart grid are another important topic of current research.[22]

The design of a power grid requires major investment. However, a smart grid can minimize costs of operations and maintenance by the rational pricing of electric power. The optimized generation and transmission lines planning facilitate efficient power flows that will reduce power losses and promote the efficient use of lowest-cost generators. Currently, the interconnected power networks have a number of communication systems within control systems for their generating plants, transmission lines, substations, and major load centers. The energy flow is directed from generators to the load centers. The power system operators, through metering of power flows and measurement of bus voltages and frequency, monitor the system. The system demands are met through load frequency and AGC. When the system demands are not met, the system frequency drops and a brownout, a rolling blackout, or an uncontrolled blackout may occur. The total amount of power demand by the users can vary widely as a function of environmental conditions. The spinning reserve at spare generating plants is kept in standby mode on-line to respond to the fluctuating power demands. The smart grid with its independent MRG system eliminates the cost of a spinning reserve because its operating system is based on real-time pricing.

In this chapter, we have learned about the fundamental operation of a power grid and how to model the power grid for the analysis and design of a smart grid. We have also presented the important elements of a smart grid and load dynamics including how load variation during daily operation affects the price of electric energy. The importance of generator and motor internal impedance for limiting fault current power grid network was emphasized. Finally, the problems of power flow studies and short-circuit studies for design of a power grid were presented. In the following chapter, we will study the design of smart microgrid renewable energy systems.

9.14   THE IMPACT OF RENEWABLE POWER ON VOLTAGE STABILITY AND REACTIVE POWER SUPPLY

As penetration of renewable power, wind, and PV power has increased, the voltage stability and reactive power supply has impacted the grid operation.[23] Inverters are employed to control and schedule the renewable distributed energy sources (DER) in smart grid. The intermittent nature of renewables and the large-scale utilization of inverter control bring forth numerous challenges to system operation and design. Inverters are controlled to schedule unity power factor in smart DER to grid leaving voltage control for local power producers. When the penetration of renewable was low, the reactive power requirement was supplied by the local power producers to keep the grid bus voltage within acceptable range by placing capacitors in distribution systems. For residential PV, in United States, the reactive power is supplied by local power producers. Essentially, local power providers are providing a service to end energy users without being compensated for the service. However, industrial and commercial energy users, the local power providers have charged the energy users for inadequate power factor resulting in low voltage due to excessive reactive power requirement. Most of the wind power is produced by IPP. The IPPs have been installing capacitors to compensate for power factor (PF) especially when the technology used wind induction generators without inverters.

In European Union (EU), "11,895 MW of wind power capacity (worth between €12.8bn and €17.2bn) was installed in the EU during 2012. The National Renewable Energy Action Plans forecast a net increase in 2012 of 11,360 MW, 328 MW less than the actual net annual increase of 11,688 MW." (http://www.ewea.org/fileadmin/files/library/publications/statistics/Wind_in_power_annual_statistics_2012.pdf. Accessed June, 2014.) In 2014, wind power accounted for 33% of all the electric energy generated in Portugal. When the gird networks provide reactive power in distribution system and subtransmission systems, it contributes to excessive losses, imbalances in voltage levels, flows upward from the distribution to the transmission systems. In EU, it is required; IPPs must provide the reactive power at the point of interconnection to the grid by placing capacitors at IPPs power generation station.

The design of DER must integrate the dispersed generation and participate in the voltage/reactive support of the operation of the distribution systems. One approach is to establish market mechanism to buy and sell reactive power. For smart PV residential microgrids, the trend is requiring a fee for connection to local power providers.

PROBLEMS

9.1 Write a 3000 word report on the Cost of Electric Energy for the last year similar to Table 9.1. Use the DOE sources and state the type power, PV, wind, peaking unit, and real-time price. Give your references using the citation of sources as used in the reference section of the book.

9.2 Write a 3000 word report on balancing and frequency control technical report prepared by NREC. Give your references using the citation of sources as used in the reference section of the book.

9.3 Assume ACE is equal to 100 MW between to two areas and frequency of grid has dropped to 59.8 Hz while the tie line is on schedule. Calculate the bias factor.

9.4 If in Problem 9.3, area 1 has emergency due to outage of a generator, determine the inadvertent power flow of each area.

9.5 Compute the load factor for a feeder assuming that the maximum load is 8 MW and the average power is 6 MW.

9.6 Compute the load factor for a feeder for daily operation for 1 month assuming the same daily profile. Assume the average power is 170 MW and the peak is 240 MW.

9.7   If the feeder of Example 9.13 is supplied from a wind source rated 80 MW and central power-generating station rated 500 MW, assume the capital cost of wind power is $500 per KW and the central station $100 per KW. Compute the EUC if the maintenance cost for the wind source is free, except maintenance 1 cent per kWh and central power-generating station fuel and maintenance cost is 3.2 cents per kWh. Give a figure for EUC from zero load factors to unity over 5 year's utilization.

9.8   If the feeder of Example 9.13 is supplied from 10 fuel cell sources rated for a total of 2 MW and 20 microturbines rated for total of 6 MW, assume the capital cost of fuel cell is $1000 per kW and microturbine is $200 per kW. Compute the fuel source EUC if the variable cost for the fuel cell is 15 cents per kWh and the microturbine is 2.2 cents per kWh, assume 5 years of operation. Show EUC as a function of time from zero load factors to unity in a figure.

REFERENCES

1. Institute for Electric Energy. Homepage. Available at http://www.edison-foundation.net/IEE. Accessed October 7, 2010.
2. MISO-MAPP Tariff Administration. Transition Guide.
3. Energy Information Administration. Official energy statistics from the US government. Available at http://www.eia.doe.gov. Accessed October 29, 2010.
4. Carbon Dioxide Information Analysis Center. Frequently asked global change questions. Available at http://www cdiac.ornl.gov/pns/faq.html. Accessed September 29, 2009.
5. Shahidehpour, M. and Yamin, H. (2002) *Market Operations in Electric Power Systems: Forecasting, Scheduling, and Risk Management,* Wiley/IEEE, New York/Piscataway, NJ.
6. Hirst, E. and Kirby, B. Technical and market issues for operating reserves. Available at http://www.ornl.gov/sci/btc/apps/Restructuring/Operating_Reserves.pdf. Accessed January 10, 2009.
7. Keyhani, A. and Miri, S.M. (1983) On-line weather-sensitive and industrial group bus load forecasting for microprocessor-based applications. *IEEE Transactions on Power Apparatus and Systems,* **102**(12), 3868–3876.
8. Schweppe, F.C. (1978) Power systems 2000: hierarchical control strategies. *IEEE Spectrum,* **14**(7), 42–47.
9. North American Electric Reliability Council. NERC 2008 long term reliability assessment 2008–2017.
10. Wood, A.J. and Wollenberg, B.F. (1996) *Power Generation, Operation, and Control,* Wiley, New York.
11. Babcock & Wilcox Company. (1975) *Steam its Generation & Use,* 38th edn, Babcock & Wilcox, Charlotte, NC.
12. Nourai, A. and Schafer, C. (2009) Changing the electricity game. *IEEE Power and Energy Magazine,* **7**(4), 42–47.

13. Ko, W.H. and Fung, C.D. (1982) VLSI and intelligent transducers. *Sensors and Actuators*, **2**, 239–250.

14. De Almeida, A.T. and Vine, E.L. (1994) Advanced monitoring technologies for the evaluation of demand-side management programs. *IEEE Transactions on Power Systems*, **9**(3), 1691–1697.

15. Adamiak, M. Phasor measurement overview.

16. Phadke, A.G. (1993) Computer applications in power. *IEEE Power and Energy Society*, **1**(2), 10–15.

17. Schweppe, F.C. and Wildes, J. (1970) Power system static-state estimation, part I: exact model power apparatus and systems. *IEEE Transactions on Volume*, **PAS-89**(1), 120–125.

18. Ducey, R., Chapman, R., and Edwards, S. The U.S. Army Yuma Proving Ground 900-kVA Photovoltaic Power Station. Available at http://photovoltaics.sandia.gov/docs/PDF/YUMADOC.PDF. Accessed October 10, 2010.

19. Grainger, J., and Stevenson, W.D. (2008) *Power Systems Analysis*, McGraw Hill, New York.

20. Gross, A.C. (1986) *Power System Analysis*, Wiley, New York.

21. Keyhani, A., Marwali, M., and Dai, M. (2010) *Integration of Green and Renewable Energy in Electric Power Systems*, Wiley, Hoboken, NJ.

22. Majmudar, H. (1965) *Electromechanical Energy Converters*, Allyn and Bacon, Boston, MA.

23. Sakis Meliopoulos, A.P., Cokkinides, G.J., and Bakirtzis, A.G. (1999) *Load-Frequency Control Service in a Deregulated Environment*, Elsevier Publication, Decision Support Systems 24 _1999, pp. 243–250.

24. North American Electric Reliability Corporation (NRC). http://www.nerc.com/docs/oc/rs/NERC%20Balancing%20and%20Frequency%20Control%200405201-11.pdf. Accessed January 7, 2014.

ADDITIONAL RESOURCES

Anderson, P.M. and Fouad, A.A. (1977) *Power System Control and Stability*, 1st edn, Iowa State University Press, Ames, IA.

Anderson, R., Boulanger, A., Johnson, J.A. and Kressner, A. (2008) *Computer-Aided Load Management for the Energy Industry*, Pennwell, Tulsa, OK, p. 333.

Berger, A.W. and Schweppe, F.C. (1989) Real time pricing to assist in load frequency control. *IEEE Transactions on Power Systems*, 4(3): 920–926.

Bohn, R., Caramanis, M. and Schweppe, F. (1984) Optimal pricing in electrical networks over space and time. *Rand Journal on Economics*, 18(3):360–376.

Caramanis, M., Bohn, R. and Schweppe, F. (1982) Optimal spot pricing: practice & theory. *IEEE Transactions on Power Apparatus and Systems*, PAS-101(9): 3234–3245.

Chapman, S. (2003) *Electric Machinery and Power System Fundamentals*, McGraw Hill, New York.

Dowell, L.J., Drozda, M., Henderson, D.B., Loose, V.W., Marathe, M.V. and Roberts, D.J. ELISIMS: comprehensive detailed simulation of the electric power industry, Technical Rep. No. LA-UR-98-1739, Los Alamos National Laboratory, Los Alamos, NM.

Electricity Storage Association. Technologies and applications: Flywheels. Available at http://electricitystorage.org/tech/technologies_technologies_flywheels.htm. Accessed October 10, 2010.

Elgerd, O.I. (1982) *Electric Energy System Theory: An Introduction*, 2nd edn, McGraw-Hill, New York.

El-Hawary, M.E. (1983) *Electric Power Systems: Design and Analysis*, Reston Publishing, Reston, VA.

IEEE Brown Book. (1980) *IEEE Recommended Practice for Power System Analysis*, Wiley-Interscience, New York.

Institute of Electrical and Electronics Engineers, Inc. (1999) *IEEE Std 1451.1-1999, Standard for a Smart Transducer Interface for Sensors and Actuators–Network Capable Application Processor (NCAP) Information Model*, IEEE, Piscataway, NJ.

Masters, G.M. (2004) *Renewable and Efficient Electric Power Systems*, Wiley, New York.

Nourai, A. (2002) Large-scale electricity storage technologies for energy management, in Proceedings of the Power Engineering Society Summer Meeting, Vol. 1; July 25, 2002; Chicago, IL. IEEE, Piscataway, NJ, pp. 310–315.

Schweppe, F., Caramanis, M., Tabors, R. and Bohn, R. (1988) *Spot Pricing of Electricity*, Kluwer, Alphen aan den Rijn, The Netherlands.

Schweppes, F., Tabors, R., Kirtley, J., Outhred, H., Pickel, F. and Cox, A. (1980) Homeostatic utility control. *IEEE Transactions on Power Applications and Systems*, **PAS-99**(3): 1151–1163.

CHAPTER 10

LOAD FLOW ANALYSIS OF POWER GRIDS AND MICROGRIDS

10.1 INTRODUCTION

We have established that the main objectives of electric energy distribution are to provide rated voltage and rated frequency at specified loads. In Chapter 4, we discussed load frequency control; here we will focus on voltage control. To ensure the rated voltage at each bus in a power grid, the system must be modeled in steady state. Hence, the integration of renewable energy sources as part of a microgrid that is connected to the local power grid as well as the asynchronous operation of a microgrid is also examined. The calculation of bus load voltage is formulated as a power flow problem. Once the load voltages are established—this includes the bus voltage magnitude and phase angle—the power flow on lines can be computed from line impedance and voltages on the two ends of a line. This takes us to the use of power flow analysis as an engineering tool. In the formulation of the power flow problem, we will study how interconnected transmission systems are modeled. Both the bus admittance matrix and bus impedance matrix models are presented in this chapter. At the end of the chapter, we will review three methods for solving power flow problems: the Gauss–Seidel, Newton–Raphson, and fast-decoupled load flow (FDLF) solutions. Several solved examples and homework problems are provided to further students' understanding of microgrid design.

Design of Smart Power Grid Renewable Energy Systems, Second Edition. Ali Keyhani.
© 2017 John Wiley & Sons, Inc. Published 2017 by John Wiley & Sons, Inc.
Companion website: www.wiley.com/go/smartpowergrid2e

10.2 VOLTAGE CALCULATION IN POWER GRID ANALYSIS

In a circuit problem, the impedance of loads and the source voltage are given, and the problem is to find the current flow in the circuit and to calculate the voltage across each load. In voltage calculation in a power problem, the loads are given in terms of active and reactive power consumption. We can study this problem via two methods: (1) assume the voltage across the load and calculate the source voltage and (2) assume the source voltage and compute the bus load voltage (this is known as a power flow or load flow problem). Example 10.1 illustrates the first method.

Example 10.1 A three-phase feeder is connected through two cables with equal impedance of $4 + j15$ Ω in series to 2 three-phase loads. The first load is a Y-connected load rated 440 V, 8 KVA, p.f. $= 0.9$ (lagging) and the second load is a Δ-connected motor load rated 440 V, 6 KVA, p.f. $= 0.85$ (lagging). The motor requires a load voltage of 440 V at the end of the line on the Δ-connected loads. Perform the following:

(i) Give the one-line diagram.
(ii) Find the required feeder voltage.

Solution
Figure 10.1 depicts the one line diagram of Example 10.1. The line voltage at bus 3 is equal to 440 V.

The rated current drawn by a motor on bus 3 is $I_3 = \dfrac{kVA_{r3}}{\sqrt{3}.V_3} = \dfrac{6000}{\sqrt{3}\times 440}\angle -$

$\cos^{-1} 0.85 = 7.87\angle - 31.77° \, A$

The voltage at bus 2 is given by $V_{2,ph} = V_{3,ph} + z_{2\text{-}3} \times I_3 = 440/\sqrt{3} +$
$(4 + j15) \times 7.87\angle - 31.77 = 353.04\angle 13.7° \, V$

The rated current drawn by a load on bus 2 is $I_2 = \dfrac{kVA_{r2}}{3.V_2} = \dfrac{8000}{3\times 353.04}\angle -$

$\cos^{-1} 0.9 = 7.55\angle - 25.84° \, A$

The supply current of the generator is given by $I_1 = I_2 + I_3 = 7.55\angle - 25.84 + 7.87\angle - 31.77 = 15.39\angle - 28.87° \, A$

The generator phase voltage is given by $V_1 = V_2 + z_{1\text{-}2} \times I_1 = 353.04\angle 13.7 + (4 + j15) \times 15.39\angle - 28.87$

$$= 569.21\angle 26.73° \, V$$

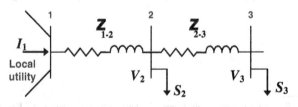

Figure 10.1 The one-line diagram of Example 10.1.

The line voltage of the generator is given by $V_1\sqrt{3} = 569.21 \times \sqrt{3} = 985.90$ V.

Example 10.1 is not a practical problem. The generator voltage is controlled by its excitation system. In practice, the field current is set to obtain the rated generator voltage. If the generator has two poles and the generator is operating at synchronous speed, that is, for a 60-Hz system, it operates at 3600 rpm.

Therefore, in a practical problem, we need to compute the voltage at load buses, given the generator voltage and load power consumption. The solution to this latter problem—known as the power flow problem—is more complex.

Example 10.2   Consider a distributed feeder presented in Figure 10.2. Assume the following:

a. Feeder line impedances, that is, $z_{1\text{-}2}$ and $z_{2\text{-}3}$, are known.
b. The active and reactive power consumed, that is, S_2 and S_3, by loads are known.
c. The local power grid bus voltage V_1 is known and all data are in per unit.

Solution
Let us write the Kirchhoff's current law for each node (bus) of Figure 10.2 and assume that the sum of the currents away from the bus is equal to zero. That is, for buses 1–3, we have

$$(v_1 - v_2)y_{12} - I_1 = 0(v_2 - v_1)y_{12} + (v_2 - v_3)y_{23} + I_2$$
$$= 0(v_3 - v_2)y_{23} + I_3 = 0 \qquad (10.1)$$

where $y_{12} = 1/z_{1\text{-}2}$ and $y_{23} = 1/z_{2\text{-}3}$.

$$I_1 = \left(\frac{S_1}{V_1}\right)^*, \ I_2 = \left(\frac{S_2}{V_2}\right)^*, \text{ and } I_3 = \left(\frac{S_3}{V_3}\right)^* \qquad (10.2)$$

We can rewrite Equation (10.1) as

$$y_{12}v_1 - y_{12}v_2 = I_1$$
$$-y_{12}v_1 + (y_{12} + y_{23})v_2 - y_{23}v_3 = -I_2$$
$$-y_{23}v_2 + y_{23}v_3 = -I_3$$

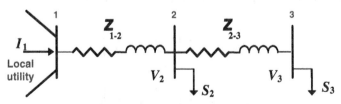

Figure 10.2   A distribution feeder.

The above can be written as:

$$\begin{bmatrix} Y_{11} & Y_{12} & 0 \\ Y_{21} & Y_{22} & Y_{23} \\ 0 & Y_{23} & Y_{33} \end{bmatrix} \cdot \begin{bmatrix} V_1 \\ V_2 \\ V_3 \end{bmatrix} = \begin{bmatrix} I_1 \\ -I_2 \\ -I_3 \end{bmatrix} \tag{10.3}$$

where $Y_{11} = y_{12}$, $Y_{12} = -y_{12}$, $Y_{21} = -y_{12}$, $Y_{22} = y_{12} + y_{23}$, $Y_{23} = -y_{23}$, $Y_{32} = -y_{23}$, and $Y_{33} = y_{23}$.

The matrix Equation (10.3) represents the bus admittance matrix; it is also the Y_{Bus} model for Example 10.2.

The Y_{Bus} matrix is described as

$$I_{\text{Bus}} = Y_{\text{Bus}} \cdot V_{\text{Bus}} \tag{10.4}$$

If the system has n buses, I_{Bus} is a vector of $n \times 1$ current injection, V_{Bus} is a voltage vector of $n \times 1$, and Y_{Bus} is a matrix of $n \times n$.

For Example 10.2, we have three buses. Therefore, the Y_{Bus} matrix is 3×3 as shown in Equation (10.3) or (10.4).

Let us continue our discussion for a general case of a power grid with n buses. For each bus, k, we have

$$S_k = V_k I_k^* \quad k = 1, 2, \dots, n \tag{10.5}$$

And I_k is the current injection into the power grid at bus k. Therefore, from the kth row of Y_{Bus} matrix, we have

$$I_k = \sum_{j=1}^{n} Y_{kj} V_j \tag{10.6}$$

Substituting Equation (10.6) in Equation (10.5), we have

$$S_k = V_k \left(\sum_{j=1}^{n} Y_{kj} V_j \right)^* \quad k = 1, 2, \dots, n \tag{10.7}$$

For each bus k, we have a complex equation of the form given by Equation (10.7). Therefore, we have n nonlinear complex equations.

$$P_k = \text{Re} \left\{ V_k \sum_{j=1}^{n} Y_{kj}^* V_j^* \right\}$$

$$Q_k = \text{Im} \left\{ V_k \sum_{j=1}^{n} Y_{kj}^* V_j^* \right\}$$

$$P_k = V_k \sum_{j=1}^{n} V_j \left(G_{kj} \cos \theta_{kj} + B_{kj} \sin \theta_{kj} \right) \qquad (10.8)$$

$$Q_k = V_k \sum_{j=1}^{n} V_j \left(G_{kj} \sin \theta_{kj} - B_{kj} \cos \theta_{kj} \right) \qquad (10.9)$$

where $Y_{kj} = G_{kj} + jB_{kj}$, $\theta_{kj} = \theta_k - \theta_j$, $V_j = V_j \left(\cos \theta_j + j \sin \theta_j \right)$, $V_k = V_k \left(\cos \theta_k + j \sin \theta_k \right)$.

For Example 10.2, $n = 3$, we have six nonlinear equations. However, because the power grid bus voltage magnitude is given and used as a reference with a phase angle of zero, we have four nonlinear equations.

In Example 10.2, we are given the feeder impedances ($z_{1\text{-}2}$ and $z_{2\text{-}3}$) and loads (S_2 and S_3). To find the bus load voltages, we need to solve the four nonlinear equations for $V_2, V_3, \theta_2,$ and θ_3. After calculating the bus voltages, we can calculate the complex power ($S_1 = P_1 + jQ_1$) injected by the local power feeder. The four nonlinear equations are

$$P_2 = V_2 \sum_{j=1}^{n} V_j \left(G_{2j} \cos \theta_{2j} + B_{2j} \sin \theta_{2j} \right)$$

$$P_3 = V_3 \sum_{j=1}^{n} V_j \left(G_{3j} \cos \theta_{3j} + B_{3j} \sin \theta_{3j} \right)$$

$$Q_2 = V_2 \sum_{j=1}^{n} V_j \left(G_{2j} \sin \theta_{2j} - B_{2j} \cos \theta_{2j} \right)$$

$$Q_3 = V_3 \sum_{j=1}^{n} V_j \left(G_{3j} \sin \theta_{3j} - B_{3j} \cos \theta_{3j} \right)$$

The above expressions present the basic concepts of bus active and reactive power injections of a power grid. If we know the bus-injected power, then we can solve for the load bus voltages. Voltage calculation is an important step in the design of a power grid network.

We should understand that Example 10.2 is not the same as Example 10.1. Example 10.1 is not a realistic problem because we cannot expect the local power grid to provide the voltage at the point of interconnection of a microgrid or a feeder. However, in Example 10.2 we know the local power grid bus voltage and our objective is the design of a feeder to provide the rated voltage to its loads.

10.3 THE POWER FLOW PROBLEM

In the design of a power grid, a fundamental problem is the power flow analysis.[1-9] The solution of the power flow ensures that the designed power grid can deliver adequate electric energy to the power grid loads at acceptable voltage and frequency (acceptable voltage is defined as the rated load voltage). For example, for a light bulb rated at 50 W and 120 V, the voltage provided across the load should be 120, with deviation of no more than 5% under normal operating conditions and 10% under emergency operating conditions. In per unit (p.u) value, we seek to provide 1 p.u voltage to the loads within the range 0.95–1.05 p.u. That is, once we have specified the schedule of generation to satisfy the system load demand, the solution for bus voltages must be 1 p.u $\pm$ 5%.

The acceptable frequency can be ensured if a balance between the system loads and generation is maintained on a second-by-second basis as controlled by load and frequency control and automatic generation control. In terms of loads, the frequency deviation from the rated frequency—60 Hz in the United States and 50 Hz in the rest of the world—affects the machinery loads such as that of the induction motors and pumps. If the frequency drops, the speed of the induction motors will drop and result in excessive heating and failure of the induction motors. For example, consider a power system supported by a few diesel generators. If the operating frequency drops due to heavy load demands on the system, the pumps of a diesel generating station slow down; as a result, the diesel generators are not cooled enough. The overheated diesel engines are then removed from service by the system's override control system, causing a cascade failure of that power system. Therefore, to maintain stable operation, system bus load voltages are maintained at 1 p.u with a deviation of no more than 10% in emergency operating conditions.

The terms *power flow studies* and *load flow studies* are used interchangeably to refer to the flow of power from the generating units to the loads. To solve a power flow problem, the question is "Given the system model and the schedule of generation and loads, find the voltage of the load buses."

10.4   LOAD FLOW STUDY AS A POWER SYSTEM ENGINEERING TOOL

As we discussed in the preceding section, the power grid must be designed to provide rated voltage to the grid's loads. We need to calculate bus voltages from the scheduled transmission system, scheduled generation system, and scheduled bus loads. In power grid system planning, the grid is planned based on projected future loads. The main objective of a power flow study is to determine whether a specific system design with a lower cost can produce bus voltages within an acceptable limit. In general, power grid planning entails many studies addressing (1) power generation planning, (2) transmission system planning, and (3) reactive power supply planning. The objective of planning studies is to ensure that all power grids are operating within their operating limits and bus load voltages are within acceptable limits.

In the operation of a power grid, the following questions are addressed. Over the next 24 hours, for all scheduled busloads, transmission systems,

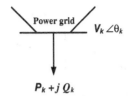

Figure 10.3 A load bus.

transformers, and generation will the bus voltages be within the limits of their rated values? If a transformer has oil leaks, can the transformer be taken out of service without affecting load voltages? In the sudden loss of a line, can the power grid load demand be satisfied without any lines being overloaded?

Load flow studies in power grid and operational planning, outage control, as well as power system optimization and stability studies are performed to provide the needed answers. In Section 10.5, we present the power flow problem formulation by first introducing bus types in a power grid.

10.5 BUS TYPES

In a power flow problem, several bus types are defined. The three most important types are a load bus, a generator bus, and a swing bus. Figure 10.3 depicts a load bus.

A power system bus has four variables. These variables are (1) the active power at the bus, (2) the reactive power at the bus, (3) the voltage magnitude, and (4) the phase angle. For a load bus, the active and reactive power consumptions are given as a scheduled load for a given time. The time can be specified as the day ahead forecasted peak load. If the system is being planned for 10 years ahead, then the forecasted peak load is used at the bus.

Figure 10.4 depicts a generator bus. This type of bus is referred to as a constant P-V bus.

A P-V (voltage-controlled) bus models a generator bus. For this bus type, the power injected into the bus by the connected generator is given in addition to the magnitude of bus voltage. The reactive power injected into the network and phase angle must be computed from the solution of the power flow problem. However, the reactive power must be within the limit (minimum and maximum) of what the P-V bus can provide.

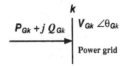

Figure 10.4 A constant voltage-controlled (P-V) bus.

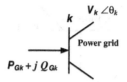

Figure 10.5   A constant $P_G - Q_G$ bus.

Figure 10.5 depicts a constant $P_G - Q_G$ bus. This bus type represents a generator with known active and reactive power injection into the power system. However, the generator magnitude of the voltage and the phase angle must be computed from the solution of power flow problems.

Figure 10.6 defines a swing bus or slack bus. The swing bus is identical to a P-V bus except the bus voltage is set to 1 p.u and its phase angle to zero. For a swing bus, the net-injected active power and reactive power into the network are not known. The generator connected to the swing bus is called a swing generator or slack generator. The function of a swing bus is to balance power consumption and power loss with net-injected generated power. The swing bus can also be considered as an infinite bus, that is, it can theoretically provide an infinite amount of power. Hence, the swing bus is considered an ideal voltage source: it can provide an infinite amount of power and its voltage remains constant. Note that all the above definitions are identical and are used interchangeably.

Let us consider the bus type for microgrids of photovoltaic (PV) or wind generating stations. Because the energy captured from the sun or wind source is free, these types of generating units are operated to produce active power. This means that a PV or wind bus is operating at unity power factor. Therefore, the PV or wind bus can be modeled as given in Figure 10.7.

Figure 10.7 depicts the modeling of a PV or wind generating station connected to a bus when a microgrid is connected to the local power grid. In this model, for the voltage analysis of the microgrid, the PV or wind bus active power generation is given and the reactive power generation is assumed to be zero. The bus voltage and phase angle are computed from the solution of the power flow problem subject to a minimum and maximum limitation as specified by the modulation index setting of a PV generating station. Therefore, we can summarize the P-V generating bus model for bus k as P_{GK} and $V_{min} < V_k < V_{max}$ as specified. At the same time, the reactive power to be provided by

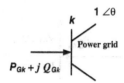

Figure 10.6   Swing bus or slack bus.

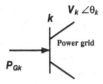

Figure 10.7   A photovoltaic or wind generating station bus model.

a PV or wind generator must be within the limits (minimum and maximum) of the generating station.

Let us now consider the case when a microgrid is separated from the local power grid. In this case, the local microgrid must control its own frequency and bus voltages. When the microgrid of a PV and wind generating system is separated from the local power grid, the PV or wind generating bus can be modeled as given in Figure 10.8.

In the above model, the magnitude of bus voltage is specified with a minimum and maximum as defined by the modulation index of the inverter; active power generation and reactive powers are also specified. The phase angle and voltage magnitude are to be computed from the power flow solution. However, a PV generating station without a storage system has very limited control over reactive power. To control an inverter power factor, a storage system is essential. To make an inverter with its supporting storage system operate like a steam unit and be able to provide active and reactive power is the subject of ongoing research on modeling and inverter control modeling. In case a wind generating station is connected to the microgrid directly, the reactive power injection control is limited within the acceptable voltage range of a connected wind bus.

For an isolated microgrid to operate at a stable frequency and voltage, it must be able to balance its loads and generation at all times. Because the load variation is continuous and renewable energy sources are intermittent, it is essential that a storage system and/or a fast-acting generating source such as high-speed microturbines, and/or a combined heat and power generating station be part of the generation mix of the microgrid.

Students are urged to read Chapter 4 again and study the factors that must be considered for the stable operation of a power grid.

In a load flow problem, all buses within the network have a designation. In general, the load buses are modeled as a constant P and Q model, where the active power, P, and reactive power, Q, are given and bus voltages are to be calculated. It is assumed that power flowing toward loads is represented as a

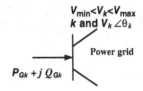

Figure 10.8   A photovoltaic or wind generating bus model.

negative injection into the power system network. The generator buses can be modeled as a constant P_G and Q_G or as PV bus type. The generators inject algebraically positive active and reactive power into the power system network. For formulation of the power flow problem, we are interested in injected power into the power system network; the internal impedance of generators is not included in the power system model. However, for short-circuit studies, the internal impedance of the generators is included in the system model. The internal impedance limits the fault current flow from the generators. For viable power flow, the balance of the system loads and generation must be maintained at all times. This balance can be expressed as

$$\sum_{k=1}^{n_1} P_{Gk} = \sum_{k=1}^{n_2} P_{Lk} + P_{\text{losses}} \tag{10.10}$$

where P_{Gk} is the active power generated by generator k, P_{Lk} is the active power consumed by the load k, n_1 is the number of the system generators, and n_2 is the number of the system loads. Similarly,

$$\sum_{k=1}^{n_1} Q_{Gk} = \sum_{k=1}^{n_2} Q_{Lk} + Q_{\text{losses}} \tag{10.11}$$

where Q_{Gk} is the reactive power generated by generator bus k, Q_{Lk} is the reactive power consumed by load k, n_1 is the number of system generators, and n_2 is the number of system loads.

Let us consider the system depicted in Figure 10.9.

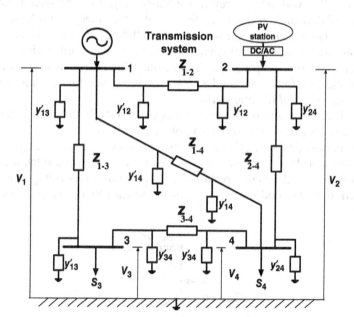

Figure 10.9 The schematic presentation of a three-bus microgrid system.

In the system given in Figure 10.9, we will need to balance the three-bus power system loads and generation.

$$P_{G1} + P_{G2} = P_{L4} + P_{L3} + P_{\text{losses}} \; P_{G1} + P_{G2}$$

$$- P_{L4} - P_{L3} - P_{\text{losses}} = 0 \tag{10.12}$$

$$Q_{G1} + Q_{G2} = Q_{L4} + Q_{L3} + Q_{\text{losses}} \; Q_{G1} + Q_{G2}$$

$$- Q_{L4} - Q_{L3} - Q_{\text{losses}} = 0 \tag{10.13}$$

In the above formulation, we assume inductive loads consume reactive power $Q_{\text{Ind}} > 0$ and capacitive loads supply reactive power $Q_{\text{Cap}} < 0$.

To ensure the balance between load and generation, we must calculate the active and reactive power losses. However, to calculate power losses, we need the bus voltages. The bus voltages are the unknown values to be calculated from the power flow formulation. This problem is resolved by defining a bus of the power grid — a swing bus and the generator behind it as a swing generator as defined earlier. By definition, the swing bus is an ideal voltage source. As an ideal voltage source, it provides both active and reactive power while the bus voltage remains constant. Therefore, a swing generator is a source of infinite active and reactive power in a power flow problem formulation The swing bus voltage is set to 1 p.u and its phase angle as the reference angle set to zero degree, $V_s = 1 \angle 0$. With this assignment, the generator behind the swing bus can provide the required power to the loads of the power grid and its voltage will not be subject to fluctuations. Of course, in practice, such a constant voltage source with an infinite power source does not exist. However, if the connected loads are much smaller than the power behind a bus then it can be approximated as an ideal voltage source. The swing bus allows balancing the system loads plus system losses to the system's supply generation; thus, the balance of energy in the network is ensured.

10.6 GENERAL FORMULATION OF THE POWER FLOW PROBLEM

Let us now formulate the same problem, for a network of a power grid. We should keep in mind the following assumptions.

a. The generators are supplying balanced three-phase voltages.
b. The transmission lines are balanced.
c. The loads are assumed to be balanced.
d. The PV or wind generating stations are presented by a *PV* bus with the bus voltage having a minimum and maximum limit.

Consider the injection model of a power grid given in Figure 10.10.

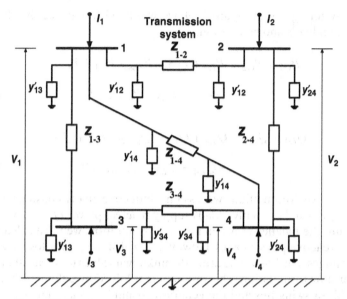

Figure 10.10   A current injection model for power flow studies.

In Figure 10.10, the current injections at each bus are presented based on the known power injection and the bus voltage that is calculated from the mathematical model of the system.

In Figure 10.10, the following definitions are implied:

- The bus voltages are actual bus-to-ground voltages in per unit.
- The bus currents are net-injected currents in per unit flowing into the transmission system from generators and loads.
- All currents are assigned a positive direction into their respective buses. This means that all generators inject positive currents and all loads inject negative currents.
- The $z_{i\text{-}j}$ is the one-phase primitive impedance, also called the positive sequence impedance between buses i and j. We will study sequence impedances in Chapter 11. However, the positive sequence impedance is the same balanced line impedance.
- The y'_{ij} is the half of the shunt admittance between buses i and j.

Representing the series primitive impedance at the lines by their corresponding primitive admittance form where

$$y_{12} = 1/Z_{1-2}; \ y_{13} = 1/Z_{1-3}; \ y_{14} = 1/Z_{1-4}; y_{24}$$
$$= 1/Z_{2-4}; \ y_{34} = 1/Z_{3-4} \tag{10.14}$$

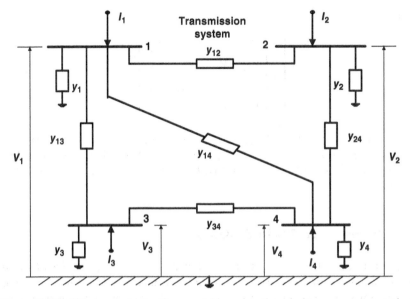

Figure 10.11 The current injection model by using the admittance representation.

The power system shown in Figure 10.10 can be redrawn as Figure 10.11.

where $y_1 = y'_{12} + y'_{13} + y'_{14}$ is the total shunt admittance connected to bus 1, $y_2 = y'_{12} + y'_{24}$ is the total shunt admittance connected to bus 2, $y_3 = y'_{13} + y'_{34}$ is the total shunt admittance connected to bus 3, and $y_4 = y'_{14} + y'_{24} + y'_{34}$ is the total shunt admittance connected to bus 4.

Assuming the ground bus as the reference bus, Kirchhoff's current law for each bus (node) gives

$$
\begin{aligned}
I_1 &= V_1 y_1 + (V_1 - V_2)\, y_{12} + (V_1 - V_3)\, y_{13} + (V_1 - V_4)\, y_{14} \\
I_2 &= V_2 y_2 + (V_2 - V_1)\, y_{12} + (V_2 - V_4)\, y_{24} \\
I_3 &= V_3 y_3 + (V_3 - V_1)\, y_{13} + (V_3 - V_4)\, y_{34} \\
I_4 &= V_4\, y_4 + (V_4 - V_1)\, y_{14} + (V_4 - V_2)\, y_{24} + (V_4 - V_3)\, y_{34}
\end{aligned}
\tag{10.15}
$$

The above equations can be written, respectively, as

$$
\begin{aligned}
I_1 &= V_1\,(y_1 + y_{12} + y_{13} + y_{14}) + V_2\,(-y_{12}) + V_3\,(-y_{13}) + V_4\,(-y_{14}) \\
I_2 &= V_1\,(-y_{12}) + V_2(y_2 + y_{12} + y_{24}) + V_3\,(0) + V_4\,(-y_{24}) \\
I_3 &= V_1\,(-y_{13}) + V_2\,(0) + V_3\,(y_3 + y_{13} + y_{34}) + V_4\,(-y_{34}) \\
I_4 &= V_1\,(-y_{14}) + V_2\,(-y_{24}) + V_3\,(-y_{34}) + V_4\,(y_4 + y_{14} + y_{24} + y_{34})
\end{aligned}
\tag{10.16}
$$

In matrix form

$$
\begin{bmatrix} I_1 \\ I_2 \\ I_3 \\ I_4 \end{bmatrix} = \begin{bmatrix} Y_{11} & Y_{12} & Y_{13} & Y_{14} \\ Y_{21} & Y_{22} & Y_{23} & Y_{24} \\ Y_{31} & Y_{32} & Y_{33} & Y_{34} \\ Y_{41} & Y_{42} & Y_{43} & Y_{44} \end{bmatrix} \begin{bmatrix} V_1 \\ V_2 \\ V_3 \\ V_4 \end{bmatrix}
\tag{10.17}
$$

where $Y_{11} = y_1 + y_{12} + y_{13} + y_{14}$; $Y_{12} = -y_{12}$; $Y_{13} = -y_{13}$; $Y_{14} = -y_{14}$; $Y_{21} = Y_{12}$; $Y_{22} = y_2 + y_{12} + y_{24}$; $Y_{23} = 0$; $Y_{24} = -y_{24}$; $Y_{31} = Y_{13}$; $Y_{32} = Y_{23}$; $Y_{33} = y_3 + y_{13} + y_{34}$; $Y_{34} = -y_{34}$; $Y_{41} = Y_{14}$; $Y_{42} = Y_{24}$; $Y_{43} = Y_{34}$; $Y_{44} = y_4 + y_{14} + y_{24} + y_{34}$.

10.7 THE BUS ADMITTANCE MODEL

We can formalize the formulation of a bus admittance matrix. This formulation is known as an "algorithm."[10] An algorithm is used to solve a problem using a finite sequence of steps. In 825 AD, Al-Khwārizmī, a Persian astronomer and mathematician wrote, *On Calculation with Hindu Numerals*. His work was translated into Latin as *Algoritmi de Numero Indorum* in the 12th century. The words *algebra* and *algorithm* are derived from Al-Khwārizmī's treatise.[10] Later, Omar Khayyam (1048–1122)[11] the renowned poet, mathematician, and astronomer wrote *Demonstrations of Problems of Algebra* (1070), which laid down the principles of algebra. He also developed algorithms for the root extraction of arbitrarily high-degree polynomials.[10] Since 825 AD, the word algorithm has been used by mathematicians to formulate and solve complex problems. Here, we will formulate an algorithm for the determination of the Y_{Bus} matrix and solve a power flow problem.

The elements of Y_{Bus} matrix can be calculated from the following algorithm.

Step 1. If $i = j$, $Y_{ii} = \sum y$, that is, the $\sum$ of admittances connected to bus i.
Step 2. If $i \neq j$, and bus i is not connected to bus j then the element $Y_{ij} = 0$.

$$\tag{10.18}$$

Step 3. If $i \neq j$, and bus i is connected to bus j through the admittance y_{ij}, then the element is $Y_{ij} = -y_{ij}$.

The above algorithm can be easily programmed for the solution of a power flow problem encompassing an eastern US power grid.

In a more compact form, we can express the bus current injection vector into a power grid in terms of a bus admittance matrix and a bus voltage vector:

$$[I_{Bus}] = [Y_{Bus}] [V_{Bus}]$$

where I_{Bus} is the bus-injected current vector; Y_{Bus} is the bus admittance matrix; V_{Bus} is the bus voltage profile vector.

The Y_{Bus} matrix model of the power grid is a symmetric, complex, and sparse matrix. The row sum (or column sum) corresponding to each bus is equal to the admittance to the reference bus. If there is no connection to a reference bus, every row sum is zero. For this case, the Y_{Bus} matrix is singular and det $[Y_{Bus}] = 0$, and such a Y_{Bus} matrix cannot be inverted.

At this time, we should recall that if we formulate the Y_{Bus} matrix model for short-circuit studies, we will include the internal impedance of generators and motors. However, for power flow studies, we represent the power grid with injection models. We should also note that in general, a power grid is normally grounded through the capacitance of transmission lines.

10.8   THE BUS IMPEDANCE MATRIX MODEL

From Figure 10.11, it can be seen that the bus current injections are related to bus voltages by the bus admittance matrix as given below:

$$[I_{Bus}] = [Y_{Bus}][V_{Bus}]$$

$$[V_{Bus}] = [Z_{Bus}][I_{Bus}] \tag{10.19}$$

$$[Z_{Bus}] = [Y_{Bus}]^{-1} \tag{10.20}$$

Therefore, the Z_{Bus} matrix is the inverse of the Y_{Bus} matrix. Now the bus voltage vector is expressed in terms of Z_{Bus}, which is the bus impedance matrix, and I_{Bus} is the bus-injected current vector.

For the system of Figure 10.11, the impedance matrix of Equation (10.20) can be expressed as

$$\begin{bmatrix} V_1 \\ V_2 \\ V_3 \\ V_4 \end{bmatrix} = \begin{bmatrix} Z_{11} & Z_{12} & Z_{13} & Z_{14} \\ Z_{21} & Z_{22} & Z_{23} & Z_{24} \\ Z_{31} & Z_{32} & Z_{33} & Z_{34} \\ Z_{41} & Z_{42} & Z_{43} & Z_{44} \end{bmatrix} \begin{bmatrix} I_1 \\ I_2 \\ I_3 \\ I_4 \end{bmatrix} \tag{10.21}$$

Example 10.3   For the power grid given in Figure 10.12, compute the bus admittance and bus impedance models.

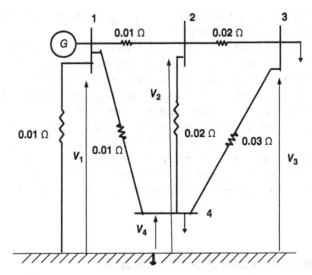

Figure 10.12   The power grid for Example 10.3.

Solution

The admittance matrix is calculated according to Equation (10.16).

$$Y_{11} = y_1 + y_{12} + y_{14} = \frac{1}{0.01} + \frac{1}{0.01} + \frac{1}{0.01} = 300,$$

$$Y_{12} = -y_{12} = -\frac{1}{0.01} = -100, \quad Y_{14} = -y_{14} = -\frac{1}{0.01} = -100$$

$$Y_{21} = -y_{21} = -\frac{1}{0.01} = -100,$$

$$Y_{22} = y_{12} + y_{23} + y_{24} = \frac{1}{0.01} + \frac{1}{0.02} + \frac{1}{0.02} = 200,$$

$$Y_{23} = -y_{23} = -\frac{1}{0.02} = -50, \quad Y_{24} = -y_{24} = -\frac{1}{0.02} = -50,$$

$$Y_{32} = -y_{23} = -\frac{1}{0.02} = -50,$$

$$Y_{33} = y_{32} + y_{34} = \frac{1}{0.02} + \frac{1}{0.03} = 83.33,$$

$$Y_{34} = -y_{34} = -\frac{1}{0.03} = -33.33, \quad Y_{41} = -y_{14} = -\frac{1}{0.01} = -100,$$

$$Y_{42} = -y_{24} = -\frac{1}{0.02} = -50, \quad Y_{43} = -y_{34} = -\frac{1}{0.03} = -33.33,$$

$$Y_{44} = y_{41} + y_{42} + y_{43} = \frac{1}{0.01} + \frac{1}{0.02} + \frac{1}{0.03} = 183.33.$$

The rest of the elements of the admittance matrix elements are zero if there are no direct connections between the buses.

$$Y_{\text{Bus}} = \begin{bmatrix} Y_{11} & Y_{12} & Y_{13} & Y_{14} \\ Y_{21} & Y_{22} & Y_{23} & Y_{24} \\ Y_{31} & Y_{32} & Y_{33} & Y_{34} \\ Y_{41} & Y_{42} & Y_{43} & Y_{44} \end{bmatrix} = \begin{bmatrix} 300 & -100 & 0 & -100 \\ -100 & 200 & -50 & -50 \\ 0 & -50 & 83.33 & -33.33 \\ -100 & -50 & -33.33 & 183.33 \end{bmatrix}$$

$$Z_{\text{Bus}} = Y_{\text{Bus}}^{-1} = \begin{bmatrix} 0.010 & 0.010 & 0.010 & 0.010 \\ 0.010 & 0.017 & 0.015 & 0.013 \\ 0.010 & 0.015 & 0.027 & 0.015 \\ 0.010 & 0.013 & 0.015 & 0.017 \end{bmatrix}$$

Z_{Bus} is the bus impedance model for the power grid in Example 10.3.

10.9 FORMULATION OF THE LOAD FLOW PROBLEM

Consider the power grid presented by Figure 10.11. The power flow problem can mathematically be stated as given by a bus admittance matrix.

$$[I_{\text{Bus}}] = [Y_{\text{Bus}}][V_{\text{Bus}}] \tag{10.22}$$

The vector of the current injection represents the net injection where the injected current is algebraically a positive injection for power generation and a negative injection for loads. Therefore, if the generation at a bus is larger than the load connected to the bus, then there is a positive net injection into the power grid. Otherwise, it will be negative if there are more loads connected to the bus than generating power. Therefore, for each bus k we have

$$S_k = V_k I_k^* \quad k = 1, 2, \ldots, n \tag{10.23}$$

and I_k is the current injection into the power grid at bus k. Therefore, from row k of Y_{Bus} matrix, we have

$$I_k = \sum_{j=1}^{n} Y_{kj} V_j \tag{10.24}$$

Substituting Equation (10.24) in Equation (10.23), we have

$$S_k = V_k \left(\sum_{j=1}^{n} Y_{kj} V_k \right)^* \quad k = 1, 2, \ldots, n \qquad (10.25)$$

For each bus k we have a complex equation of the form given by Equation (10.25). Therefore, we have n nonlinear complex equations.where $Y_{kj} = G_{kj} + jB_{kj}, \theta_{kj} = \theta_k - \theta_j \, V_j = V_j(\cos\theta_j + j\sin\theta_j), V_k = V_k(\cos\theta_k + j\sin\theta_k), I_k^* = \left(\frac{S_k}{V_k} \right)^* = \frac{(P_k + jQ_k)^*}{V_k^*} = \frac{P_k - jQ_k}{V_k . \angle - \theta_k}$, where n is the total number of buses in the power grid network.

From the Y_{Bus} model, we have the relationship of injected current into the power grid as it relates to the network admittance model, as well as how the power will flow in the transmission system based on the bus voltages. Therefore, for each bus k based on the bus admittance model, we have the following expressions:

$$\frac{P_1 - jQ_1}{V_1^*} = Y_{11} \, V_1 + Y_{12} \, V_2 + Y_{13} \, V_3 + Y_{14} \, V_4 \qquad (10.26)$$

$$\frac{P_2 - jQ_2}{V_2^*} = Y_{21} \, V_1 + Y_{22} \, V_2 + Y_{23} \, V_3 + Y_{24} \, V_4 \qquad (10.27)$$

$$\frac{P_3 - jQ_3}{V_3^*} = Y_{31} \, V_1 + Y_{32} \, V_2 + Y_{33} \, V_3 + Y_{34} \, V_4 \qquad (10.28)$$

$$\frac{P_4 - jQ_4}{V_4^*} = Y_{41} \, V_1 + Y_{42} \, V_2 + Y_{43} \, V_3 + Y_{44} \, V_4 \qquad (10.29)$$

We can rewrite Equations (10.26) through (10.29) and express them as

$$P_1 - jQ_1 = Y_{11} V_1^2 + Y_{12} V_1^* V_2 + Y_{13} V_1^* V_3 + Y_{14} V_1^* V_4$$

$$P_2 - jQ_2 = Y_{21} V_1 V_2^* + Y_{22} V_2^2 + Y_{23} V_2^* V_3 + Y_{24} V_2^* V_4$$

$$P_3 - jQ_3 = Y_{31} V_1 V_3^* + Y_{32} V_2 V_3^* + Y_{33} V_3^2 + Y_{34} V_3^* V_4 \qquad (10.30)$$

$$P_4 - jQ_4 = Y_{41} V_1 V_4^* + Y_{42} V_2 V_4^* + Y_{43} V_4^* + Y_{44} V_{44}^2$$

The above systems of equations are complex and nonlinear. As we stated, one bus of the system is selected as a swing bus and its voltage magnitude is set to 1 p.u; its phase angle is set to zero as the reference phasor. The swing bus will ensure the balance of power between the system loads and the system generations. In a power flow problem, the load bus voltages are the unknown variables and all the injected powers are known variables. Here, there are three nonlinear complex equations to be solved for bus load voltages.

In general, as we discussed earlier, for each bus k a complex equation can be written as two equations in terms of real numbers. Using the above expressions in general formulation, we have

$$P_k = V_k \sum_{j=1}^{n} V_j (G_{kj} \cos \theta_{kj} + B_{kj} \sin \theta_{kj})$$

$$Q_k = V_k \sum_{j=1}^{n} V_j (G_{kj} \sin \theta_{kj} - B_{kj} \cos \theta_{kj})$$

where $Y_{kj} = G_{kj} + jB_{kj}, \theta_{kj} = \theta_k - \theta_j, V_k = V_k(\cos \theta_k + j \sin \theta_k), V_j = V_j(\cos \theta_j + j \sin \theta_j), I_k = \left(\frac{S_k}{V_k} \right)^* = \frac{P_k - jQ_k}{V_k^*} = \frac{P_k - jQ_k}{V_k \angle -\theta_k}$.

If the system has n buses, the above equations can be expressed as $2n$ equations:

$$f_1(V_1 \ldots \ldots V_n, \theta_1 \ldots \ldots \ldots \ldots \ldots \theta_n) = 0$$

$$f_2(V_1 \ldots \ldots \ldots V_n, \theta_1 \ldots \ldots \ldots \ldots \theta_n) = 0$$

$$f_n(V_1 \ldots \ldots \ldots V_n, \theta_1 \ldots \ldots \ldots \ldots \theta_n) = 0$$

$$f_{2n}(V_1 \ldots \ldots \ldots V_n, \theta_1 \ldots \ldots \ldots \ldots \theta_n) = 0$$

$$(10.31)$$

The above $2n$ equations can be expressed as

$$F(x) = \begin{bmatrix} f_1(x) \\ f_2(x) \\ . \\ . \\ . \\ f_{2n}(x) \end{bmatrix} \qquad (10.32)$$

where the elements of vector X represent the magnitude of voltage and phase angle. In Equation (10.32), we have $2n$ nonlinear equations to be solved and vector X can be presented as

$$[X]^t = [V_1 \ldots \ldots \ldots V_n, \theta_1 \ldots \ldots \ldots \theta_n]$$

$$= [X_1 \ldots \ldots \ldots X_n, X_{n+1} \ldots \ldots \ldots X_{2n}]$$

10.10 THE GAUSS–SEIDEL Y_{BUS} ALGORITHM

The Gauss–Seidel algorithm is an iterative process. In this method, the objective is to satisfy the set of nonlinear equations by repeated approximation. The solution is reached when all nonlinear equations are satisfied at an acceptable accuracy level. In the case of power flow problems, the solution is reached when all bus voltages are converging to around 1 p.u within 5% of the rated voltage and all nonlinear equations are satisfied at an acceptable tolerance.

Let us now restate the main equations of power flow problems:

$$V_1 = 1 \angle 0 \tag{10.33}$$

$$I_{Bus} = Y_{Bus} \cdot V_{Bus} \tag{10.34}$$

$$S_k = V_k I_k^* \tag{10.35}$$

where $k = 1, 2, 3, \ldots, n$.

Equation (10.33) is the model for the swing bus (also called a *slack bus*) that creates balance between the system loads and system generation by providing the system transmission losses. When a microgrid is connected to a local power grid, the power grid bus is selected as the swing bus; this ensures that the balance between loads and local generation and losses are maintained. If the microgrid has deficiencies in generation, the balance is maintained by the power grid bus. If the microgrid has excess generation, the balance is injected into the local power grid. Equation (10.34) describes the current flow or power flow through transmission lines for a set of bus voltages. Equation (10.35) presents the net-injected power at each bus of the system.

The Gauss–Seidel Y_{Bus} algorithm can be stated as depicted in Figure 10.13.

In the Gauss–Seidel method, we repeatedly solve the fundamental load flow equation expressed by Equation (10.31).

$$V_k = \frac{\frac{P_k - jQ_k}{V_k^*} - \sum_{\substack{j=1 \\ j \neq k}}^{n} Y_{kj} V_j}{Y_{kk}} \quad k = 2, \ldots, n \tag{10.36}$$

Equation (10.36) represents the load bus voltages with the power grid depicted by the bus admittance matrix. The diagonal element for a bus for which a voltage approximation is being computed appears in the denominator. Because the diagonal elements of the bus admittance matrix are never zero, an approximation of bus voltages is assured. If the power grid is correctly designed, the convergence can be obtained.

In Equation (10.37), we check to see if the original nonlinear load flow equations are satisfied with the last approximated bus voltages.

$$\Delta P_k = P_{k(\text{Scheduled})} - P_{k(\text{Calculated})} \leq c_P$$

$$\Delta Q_k = Q_{k(\text{Scheduled})} - Q_{k(\text{Calculated})} \leq c_Q \tag{10.37}$$

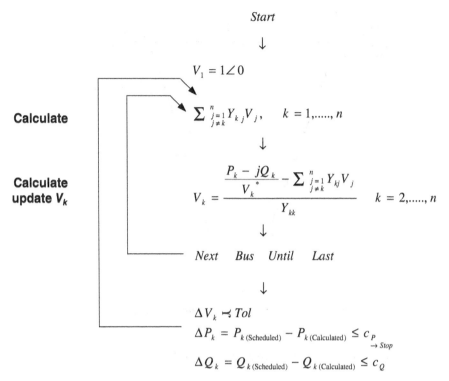

Figure 10.13   The Gauss–Seidel algorithm for iterative approximation.

Example 10.4   For the system below, use the Gauss–Seidel Y_{Bus} method and solve for the bus voltages.

For the example depicted in Figure 10.14, bus 1 is the swing bus and its voltage is $V_1 = 1 \angle 0$. The scheduled power at bus 2 is 1.2 p.u. The load at bus 3 is 1.5 p.u. Compute the bus 2 and bus 3 voltages.

Solution

To solve the above problem, we need to formulate the bus admittance matrix.

$$
Y_{Bus} = \begin{bmatrix} 14 & -4 & -10 \\ -4 & 9 & -5 \\ -10 & -5 & 15 \end{bmatrix}
$$

The power flow models are

$$
V_1 = 1 \angle 0
$$

$$
I_{Bus} = V_{Bus} \cdot Y_{Bus}
$$

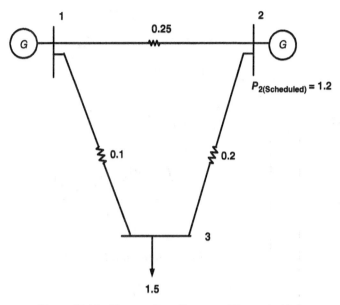

Figure 10.14   The one-line diagram of Example 10.4.

For a DC system, we only have active power flow.

$$I_k^* = I_k, \ V_k^* = V_k$$

$$S_k = V_k \cdot I_k^* = V_k I_k = P_k,$$

and

$$Q_k = 0$$

Therefore, for bus 2 we have

$$1.2 = V_2 \cdot I_2$$

For bus 3, we have

$$-1.5 = V_3 \cdot I_3$$

We can start the iterative approximation for $i = 0$ iteration by assuming bus 2 and bus 3 voltages are equal to 1 p.u.

$$V_{\text{Bus}}^{(0)} = \begin{bmatrix} 1 \\ 1 \end{bmatrix}$$

For bus 2, we have

$$\sum_{\substack{j=1 \\ j\neq 2}}^{3} Y_{2j}V_j = Y_{21}V_1 + Y_{23}V_3 = (-4)(1) + (-5)(1)$$

Next, we update V_2

$$V_2 = \frac{\frac{P_2 - jQ_2}{V_2} - \sum\limits_{\substack{j=1 \\ j\neq 2}}^{n} Y_{2j}V_j}{Y_{22}}, \quad i = 1,\ldots\ldots,j \neq i$$

$$V_2 = \frac{1}{Y_{22}}\left[\frac{1.2}{V_2} - \{(-4)(1.0) + (-5)(1)\}\right]$$

The updated bus 2 voltage is given as

$$V_2^{(1)} = \frac{1}{9}\left[\frac{1.2}{1} + 4 + 5\right] = 1.1333 \text{ p.u}$$

We continue the iterative process and update bus 3 voltage.

$$\sum_{\substack{j=1 \\ j\neq 3}}^{3} Y_{3j}V_j = Y_{31}V_1 + Y_{32}V_2$$

We update the bus 3 voltage, V_3

$$V_3 = \frac{1}{Y_{33}}\left[\frac{-1.5}{V_3} - [Y_{31}V_1 + Y_{32}V_2]\right]$$

$$V_3 = \frac{1}{15}\left[\frac{-1.5}{1.0} - \{(-10)(1) + (-5)(1.1333)\}\right]$$

$$V_3^{(1)} = \frac{1}{15}[-1.5 + 10 + 5.666] = \frac{14.1666}{15} = 0.9444 \text{ p.u}$$

We continue the approximation by calculating the mismatch at bus 2 and bus 3.

The mismatch at bus 2 is

$$P_{2(\text{Calculated})} = \sum_{\substack{j=0 \\ j\neq2}}^{3} V_2 I_{2j}$$

$$P_{2(\text{Calculated})} = V_2 I_{20} + V_2 I_{21} + V_2 I_{23}$$

$$P_{2(\text{Calculated})} = 0 + 1.1333 \left(\frac{V_2 - V_1}{0.25} \right) + 1.333 \left(\frac{V_2 - V_3}{0.2} \right)$$

$$P_{2(\text{Calculated})} = 0 + 1.1333 \left(\frac{1.1333 - 1.0}{0.25} \right) + 1.1333 \left(\frac{1.1333 - 0.944}{0.2} \right) = 1.6769$$

$$\Delta P_2 = P_{2(\text{Scheduled})} - P_{2(\text{Calculated})} = 1.2 - 1.6769$$

$$\Delta P_2 = -0.4769 \text{ p.u}$$

The mismatch at bus 3 is

$$P_{3(\text{Calculated})} = \sum_{\substack{j=0 \\ j\neq3}}^{3} V_3 I_{3j}$$

$$P_{3(\text{Calculated})} = V_3 I_{30} + V_3 I_{31} + V_3 I_{32}$$

$$P_{3(\text{Calculated})} = 0 + 0.944 \left(\frac{V_3 - V_1}{0.1} \right) + 0.944 \left(\frac{V_3 - V_2}{0.2} \right)$$

$$P_{3(\text{Calculated})} = 0.944 \left(\frac{0.944 - 1.0}{0.1} \right) + 0.944 \left(\frac{0.944 - 1.1333}{0.2} \right) = -1.4221$$

$$\Delta P_3 = P_{3(\text{Scheduled})} - P_{3(\text{Calculated})} = -1.5 - (-1.4221)$$

$$\Delta P_3 = -0.0779 \text{ p.u}$$

The process is continued until error reduces to a satisfactory value. The result is obtained after seven iterations with c_p of 1×10^{-4}. The results are given in Table 10.1.

TABLE 10.1   Example 10.4 Results

Bus	p.u Voltage	p.u Power Mismatch
2	1.078	0.63×10^{-4}
3	0.917	0.28×10^{-4}

The power supplied by bus 1, the swing bus, is equal to the total load minus total generation by all other buses plus the losses.

$$P_1 = V_1 I_1 = V_1 \sum_{j=1}^{3} Y_{1j} V_j = V_1(Y_{11}V_1 + Y_{12}V_2 + Y_{13}V_3)$$

The bus voltage of the swing bus is $1\angle 0$ and the p.u power injected by the bus is $P_1 = 0.522$ p.u.

The total power loss of the transmission lines is 0.223 p.u.

10.11 THE GAUSS–SEIDEL Z_{BUS} ALGORITHM

In the Gauss–Seidel Z_{Bus} algorithm method,[6] the power problem can be expressed as

$V_1 = 1 \angle 0$ defines the swing bus voltage.

$$V_{Bus} = Z_{Bus} \, I_{Bus}$$

The Z_{Bus} defines the power flow through the transmission system.

$$S_k = V_k \cdot I_k^*$$

The Z_{Bus} defines the injection model.

The Gauss–Seidel Z_{Bus} algorithm can be summarized in Figure 10.15.

Example 10.5   Consider the system shown in Figure 10.16. Find the bus voltages using the Gauss–Seidel Z_{Bus} algorithm.

Bus 1 is the swing bus and its voltage is set to $V_1 = 1 \angle 0$.

Without the fictitious line to ground, the Z_{Bus} is not defined. The swing bus is grounded and power drawn by this bus will be accounted for when the problem is solved. The tie to ground is selected in the same order as the line impedances. The injected power is $S_2 = -\frac{1}{2}$, $S_3 = -1$, and $S_4 = -\frac{1}{2}$. All data are expressed in per unit.

Solution

The Z_{Bus} with respect to the ground bus is

$$Z_{Bus} = \begin{bmatrix} 0.01 & 0.01 & 0.01 & 0.01 \\ 0.01 & 0.0186 & 0.0157 & 0.0114 \\ 0.01 & 0.0157 & 0.0271 & 0.0143 \\ 0.01 & 0.0114 & 0.0143 & 0.0186 \end{bmatrix}$$

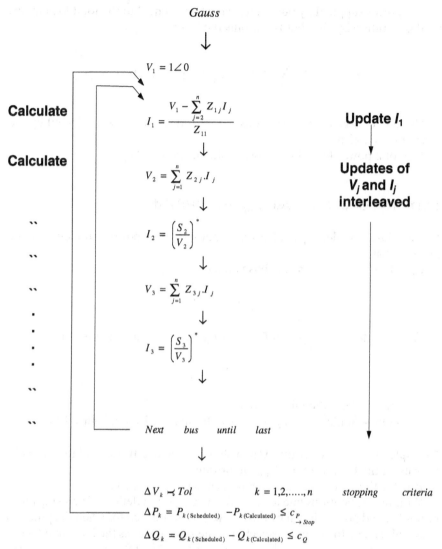

Figure 10.15 The Gauss–Seidel Z_{Bus} algorithm.

The power flow problem can be expressed as

$$V_1 = 1 \angle 0$$

$$V_{\text{Bus}} = Z_{\text{Bus}} I_{\text{Bus}}$$

For a DC system, we will only have active power flow.

$$S_k = V_k \cdot I_k^* = V_k I_k = P_k, \quad Q_k = 0$$

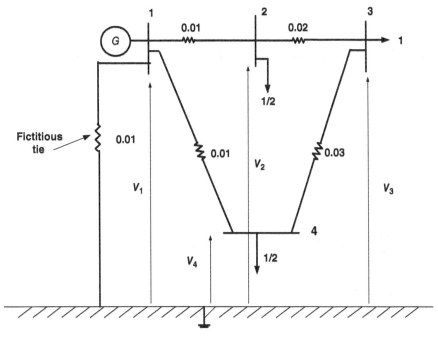

Figure 10.16   The one-line diagram of Example 10.5.

The busload and generation can be expressed as
$$-0.5 = V_2 I_2$$
$$-1.0 = V_3 I_3$$
$$-0.5 = V_4 I_4$$
The first step is to compute the gauss zero iteration (0) values for bus voltages and current as given below.

$$V_{Bus}^{(0)} = \begin{bmatrix} 1 \\ 0 \\ 0 \\ 0 \end{bmatrix}, I_{Bus}^{(0)} = \begin{bmatrix} 0 \\ 0 \\ 0 \\ 0 \end{bmatrix}$$

For the first row of $V_{Bus} = Z_{Bus}I_{Bus}$, we have the following:
Update I_1

$$I_1 = \frac{V_1 - \sum\limits_{j=2}^{n} Z_{1j}I_j}{Z_{11}} = \frac{1-0}{0.01} = 100 \qquad n = 4$$

Update V_2

$$V_2 = \sum_{j=1}^{n} Z_{2j}I_j = (0.01)(100) + (0.0186)(0) + (0.0157)(0) + (0.0114)(0) = 1$$

Update I_2

$$I_2 = \left(\frac{S_2}{V_2}\right) = \left(\frac{-0.5}{1}\right) = -0.5$$

Update V_3

$$V_3 = \sum_{j=1}^{n} Z_{3j}I_j = (0.01)(100) + (0.0157)(-0.5) + (0.0271)(0) + (0.0143)(0)$$

$$= 1 - 0.0078 = 0.9922$$

Update I_3

$$I_3 = \left(\frac{S_3}{V_3}\right) = \left(\frac{-1}{0.9922}\right) = -1.0079$$

Update V_4

$$V_4 = \sum_{j=1}^{n} Z_{4j}I_j = (0.01)(100) + (0.0114)(-0.5) + (0.0143)(-1.0079)$$

$$+ (0.0186)(0) = 1 - 0.0057 - 0.0144 = 0.9799$$

Update I_4

$$I_4 = \left(\frac{S_4}{V_4}\right) = \left(\frac{-0.5}{0.9799}\right) = -0.5102$$

From the above calculation, we have

$$V_{\text{Bus}}^{(1)} = \begin{bmatrix} 1 \\ 1 \\ 0.9922 \\ 0.9799 \end{bmatrix}, \quad I_{\text{Bus}}^{(1)} = \begin{bmatrix} 100 \\ -0.5 \\ -1.0079 \\ -0.5102 \end{bmatrix}$$

Update I_1

$$I_1 = \frac{1 - (0.01)(-0.5) - (0.01)(-1.0079) - (0.01)(-0.5102)}{0.01}$$

$$= \frac{1.02019}{0.01} = 102.019$$

Update V_2

$$V_2 = (0.01)(102.019) + (0.0186)(-0.5) + (0.0157)(-1.0079)$$
$$+ (0.0114)(-0.5102) = 0.9892$$

Update I_2

$$I_2 = \left(\frac{S_2}{V_2}\right) = \left(\frac{-0.5}{0.9892}\right) = -0.5055$$

Update V_3

$$V_3 = (0.01)(102.019) + (0.0157)(-0.5055) + (0.0271)(-1.0079)$$
$$+ (0.0143)(-0.5102) = 0.9776$$

Update I_3

$$I_3 = \left(\frac{S_3}{V_3}\right) = \left(\frac{-1}{0.9776}\right) = -1.023$$

Update V_4

$$V_4 = \sum_{j=1}^{n} Z_{4j} I_j = (0.01)(102.019) + (0.0114)(-0.5055) + (0.0143)(-1.023)$$
$$+ (0.0186)(-0.5102) = 0.9903$$

Update I_4

$$I_4 = \left(\frac{S_4}{V_4}\right) = \left(\frac{-0.5}{0.9903}\right) = -0.5049$$

The mismatch at each bus k is given as

$$P_{(\text{Calculated})k} = V_k I_k$$

$$\Delta P_k = P_{(\text{Scheduled})k} - P_{(\text{Calculated})k}$$

TABLE 10.2   Example 10.5 Results

Bus	p.u Voltage	p.u Power Mismatch
2	0.99	0.1306×10^{-5}
3	0.98	0.2465×10^{-5}
4	0.99	0.6635×10^{-5}

For bus 2

$$P_{2(\text{Calculated})} = V_2 I_2$$

$$\Delta P_2 = P_{2(\text{Scheduled})} - P_{2(\text{Calculated})}$$

For bus 3

$$P_{3(\text{Calculated})} = V_3 I_3$$

$$\Delta P_3 = P_{3(\text{Scheduled})} - P_{3(\text{Calculated})}$$

The process is continued until error reduces to a satisfactory value. The results are obtained after four iterations with c_p of 1×10^{-4}. The results are provided in Table 10.2.

If the error is more than that for satisfactory performance, the next iteration is followed.

The MATLAB simulation testbed for the above problem is given below:

```
%Power Flow: Gauss-Seidel method
clc; clear all;
tolerance= 1e-4;
N=4; % no. of buses
Y=[1/.01+1/.01+1/.01 -1/.01 0 -1/.01;
-1/.01 1/.01+1/.02 -1/.02 0;
0 -1/.02 1/.02+1/.03 -1/.03;
-1/.01 0 -1/.03 1/.01+1/.03];
Z=inv(Y)
P_sch=[1 -.5 -1 -.5]'
I=[0 0 0 0]';
V=[1 0 0 0]';
iteration=0;
while (iteration <= 999)
iteration=iteration+1;
VZ=0;
for n=2:N
VZ=VZ+Z(1,n)*I(n);
end
I(1)=(V(1)-VZ)/Z(1,1);
for m=2:N
```

```
V(m)=Z(m,:)*I;
I(m)=P_sch(m)/V(m);
end
P_calc = V.*(Y*V);
mismatch=[P_sch(2:N)-P_calc(2:N)];
if (norm(mismatch,'inf') < tolerance)
break;
end
end
P_calc(1)=P_calc(1)-V(1)^2/.01; % Subtracting the power of the fic-
titious branch
iteration
for i = 1:N
fprintf(1, 'Bus %d:\n', i);
fprintf(1, 'Voltage = %f p.u\n',V(i));
fprintf(1, 'Injected P = %f p.u\n', P_calc(i));
end
```

The power supplied by bus 1, the swing bus, is equal to the total load minus total generation by all other buses plus the losses.

$$P_1 = V_1 I_1 = V_1 \sum_{j=1}^{3} Y_{1j} V_j = V_1 (Y_{11} V_1 + Y_{12} V_2 + Y_{13} V_3)$$

The bus voltage of swing bus is $1\angle 0$ and the p.u power is $P_1 = 102.033$.

The power loss in the fictitious line is $= \dfrac{V_1^2}{z_{\text{fictitious, p.u}}} = \dfrac{1^2}{0.01} = 100$ p.u.

Therefore, the actual power injected by bus 1 is $P_1 = 102.033 - 100 = 2.033$ p.u.

The total power loss of the transmission lines is 0.033 p.u.

10.12 COMPARISON OF THE Y_{BUS} AND Z_{BUS} POWER FLOW SOLUTION METHODS

For a power grid with n number of buses, there are n nonlinear complex equations. The complete complex Y_{Bus} matrix or Z_{Bus} matrix has about $2n^2$ elements. If all elements were to be stored, for a system with 500 buses, the core requirements would be 500,000 words. However, because both Y_{Bus} and Z_{Bus} are symmetric matrices, the upper triangle is stored. Therefore, the storage requirement for a 500-bus system is 250,000 words. The Z_{Bus} matrix model is full because the elements of Z_{Bus} matrix are generally non-zero. The computation and storage requirement is astronomical for very large problems. On the other hand, the Y_{Bus} matrix is a sparse matrix because each element Y_{kj} will be non-zero only if there is a direct connection, that is, a transmission line or a transformer is located between buses k and j. The zero elements are not stored. Therefore, for large power system problems, by exploiting this natural sparsity, the storage and computation time are reduced substantially. The Y_{Bus} should possess strict diagonal dominance. This condition may not be satisfied for some practical power grids. The power grid with long-distance extra-high voltage (EHV) lines, series, and shunt compensation, an abnormally high impedance, or very low series impedances, and cable circuits with high charging capacitances may need a large number of iterations. These systems have a Y_{Bus} with weak diagonal elements. For a power flow solution, because the swing bus voltage is known, the corresponding row and column are deleted from Y_{Bus}. For power grids with the least diagonally dominant Y_{Bus}, the swing bus can be located on that bus, thus ensuring convergence. The Z_{Bus} matrix method is not usually sensitive to the choice of a swing bus. The disadvantage of the Z_{Bus} matrix method is the need to obtain, store, and iterate the non-sparse Z_{Bus}. The advantage of the Y_{Bus} matrix method is that both the storage requirements for the network and the computation per iteration are small and roughly proportional to the number of buses. However, the disadvantages of the Y_{Bus} matrix are a slow convergence property and that at times it may not converge. On the other hand, the Z_{Bus} method has a fast convergence property and always will converge.

10.13   THE SYNCHRONOUS AND ASYNCHRONOUS OPERATION OF MICROGRIDS

Figure 10.17 depicts a typical microgrid connected to a local power grid. Depending on the size of the microgrid generation sources, the local network

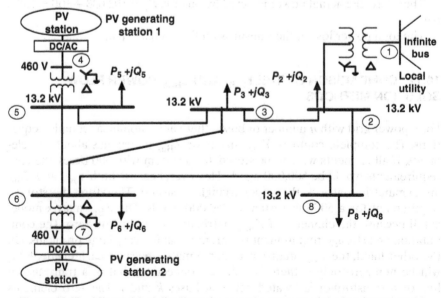

Figure 10.17   Microgrid of distributed generation as part of the local power grid.

can be designed to operate in a 480 V to 20 kV voltage class. It is clear that in this microgrid, bus 1 must be designated as a swing bus because the amount of power available at the local power grid is many times larger than the PV microgrid system of Figure 10.17.

The operation of the microgrid of Figure 10.17 can take two forms: (1) The microgrid can operate as part of the interconnected system, and (2) the microgrid can operate as a standalone once it is separated from the local network. When the microgrid is operating as part of the local power grid, the load and frequency control and voltage control are the responsibility of the local power-grid control center. As we discussed in Chapter 4, when a microgrid of PV or wind is connected to the local power grid, the entire system operates at a single frequency. The voltage of the power grid bus is also controlled by the local control center. However, the microgrid load buses can still have low voltages if adequate reactive power, VAr, support is not provided. However, the microgrid of PV or wind can be designed to operate asynchronously. Figure 10.18 depicts such a design.

The microgrid distributed generation system of Figure 10.18 is designed to operate both as a synchronous and an asynchronous system. This distributed generation system has a variable-speed wind doubly fed induction generator (DFIG) and a gas turbine synchronous generator. When this microgrid is operating as part of the local power grid, the frequency control and voltage control are the responsibility of the local power-grid control center. The local power grid operator monitors and controls the system frequency as discussed in Chapter 4. The system operator also controls the local power-grid bus voltages.

The gas turbine generator can be operated as a P-V bus: the bus voltage and active power injected into the microgrid are fixed and the reactive power and phase angle are computed via a load flow analysis. The DFIG generating station will be modeled as a P-V bus that injects only active power at a constant voltage.

When the distributed generation section of a microgrid is separated from the local power grid, the gas turbine unit is responsible for both voltage control and load frequency control. In this case, the gas turbine unit should be modeled as a swing bus for power flow analysis. To ensure stable operation, the local load control of the distributed generation system is essential. If the load control is also provided, this independent microgrid is termed a smart microgrid, because it can remain stable with control over its loads.

10.14 AN ADVANCED POWER FLOW SOLUTION METHOD: THE NEWTON–RAPHSON ALGORITHM

To formulate the Newton–Raphson algorithm,[8] the main equations of a power flow problem are restated below:

$$V_1 = 1 \angle 0 \qquad (10.38)$$

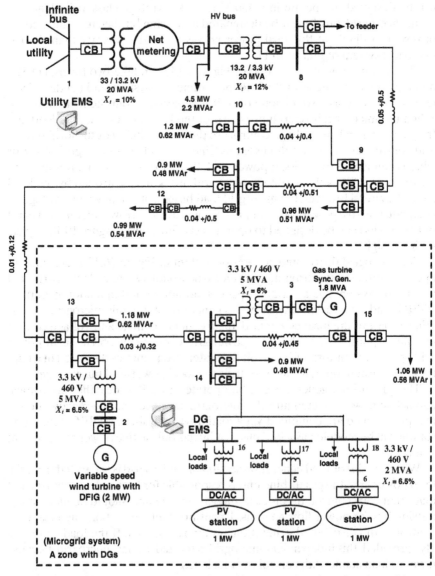

Figure 10.18 The synchronous and asynchronous operation of a microgrid.

The matrix bus admittance describes the flow of net-injected current through the transmission systems as given blow:

$$I_{Bus} = Y_{Bus} \, V_{Bus} \tag{10.39}$$

The bus injection is described as

$$S_k = V_k \cdot I_k^* \tag{10.40}$$

where S_k is the net-injected complex power at bus k, V_k is the complex voltage of bus k, and I_k is the net-injected current at bus k.

We can substitute for bus net-injected current from Equation (10.39) in Equation (10.40) to obtain the residue form of the equation for each bus k as

$$S_k - V_k \sum_{j=1}^{n} Y_{kj}^* V_j^* = 0 \qquad k = 1, \ldots \ldots, n \qquad (10.41)$$

For $k = 1$, we have

$$S_1 - V_1 \sum_{j=1}^{n} Y_{1j}^* V_j^* = 0 \qquad f_1(X) = 0 \qquad (10.42)$$

And $k = 2$

$$S_2 - V_2 \sum_{j=1}^{n} Y_{2j}^* V_j^* = 0 \qquad f_2(X) = 0 \qquad (10.43)$$

And $k = n$

$$S_n - V_n \sum_{j=1}^{n} Y_{nj}^* V_j^* = 0 \qquad f_n(X) = 0 \qquad (10.44)$$

The above can be expressed in a compact form as

$$F(x) = \begin{bmatrix} f_1(x) \\ f_2(x) \\ . \\ . \\ . \\ f_n(x) \end{bmatrix} = 0 \qquad (10.45)$$

where $[X]^T = [x_1, x_2, x_3, \ldots \ldots, x_n]$.

In the above equation, the variable vector X represents the bus voltage vector.

Expanding row 1 of $F(x) = 0$ in a Taylor series, about a guess solution $X^{(o)}$, we have

$$f_1(X) = f_1(x_1, x_2 \ldots x_n) = f_1\left(x_1^{(0)}, x_2^{(0)} \ldots x_n^{(0)}\right) + \left.\frac{\partial f_1}{\partial x_1}\right|_{X^{(0)}} \Delta x_1 + \left.\frac{\partial f_1}{\partial x_2}\right|_{X^{(0)}} \Delta x_2$$

$$+ \ldots\ldots \left.\frac{\partial f_n}{\partial x_n}\right|_{X^{(0)}} + \text{higher order terms.} \tag{10.46}$$

Equation (10.46) can be expressed in compact form as

$$f_1(X) = f_1(X^{(0)}) + \sum_{j=1}^{n} \left.\frac{\partial f_1}{\partial x_j}\right|_{X^{(0)}} \Delta x_j = 0 \tag{10.47}$$

$$f_2(X) = f_2(X^{(0)}) + \sum_{j=1}^{n} \left.\frac{\partial f_2}{\partial x_j}\right|_{X^{(0)}} \Delta x_j = 0 \tag{10.48}$$

$$f_n(X) = f_n(X^{(0)}) + \sum_{j=1}^{n} \left.\frac{\partial f_n}{\partial x_j}\right|_{X^{(0)}} \Delta x_j = 0 \tag{10.49}$$

where $\Delta x_j = x_j - x_j^{(0)}$. In compact matrix notation,

$$F(X) = F\left(X^{(0)}\right) + \begin{bmatrix} \dfrac{\partial f_1}{\partial x_1} \dfrac{\partial f_1}{\partial x_2} & \cdots & \dfrac{\partial f_1}{\partial x_n} \\ \dfrac{\partial f_2}{\partial x_1} \dfrac{\partial f_2}{\partial x_2} & \cdots & \dfrac{\partial f_2}{\partial x_n} \\ & \cdot & \\ \dfrac{\partial f_n}{\partial x_1} \dfrac{\partial f_n}{\partial x_2} & \cdots & \dfrac{\partial f_n}{\partial x_n} \end{bmatrix}_{X^{(0)}} \begin{bmatrix} \Delta x_1 \\ \Delta x_2 \\ \cdot \\ \cdot \\ \Delta x_n \end{bmatrix} = 0 \tag{10.50}$$

The matrix of Equation (10.50) is referred to as the Jacobian matrix. The above equation can be rewritten as

$$F\left(X^{(0)}\right) + [J]\big|_{X^{(0)}} [\Delta X] = 0 \tag{10.51}$$

$$f_k(X^0) = S_k - V_k \sum_{j=1}^{n} Y_{kj}^* V_j^*$$

and

$$[X^0] = [V_1^{(0)} \ldots\ldots V_i^{(0)} \ldots\ldots V_n^{(0)}] \tag{10.52}$$

where S_k is the scheduled net-injected (generator bus or load) complex power into bus k (positive for generation). Therefore, the term $F(X^{(0)})$ represents the power mismatch at each bus. When the term $F(X^{(0)})$ is very small the power flow solution has been obtained.

$$S_{k(\text{Calculated})} = V_k \sum_{j=1}^{n} Y_{kj}^* V_j^* \qquad (10.53)$$

Equation (10.53) represents the calculated power flowing away from bus k to all the other buses j.

Equation (10.51) can be solved for ΔX

$$[\Delta X] = -[J]_{X^{(0)}}^{-1} \, F(X^{(0)}) \qquad (10.54)$$

$$F_k(X^0) = \Delta S_k = \Delta P_k + j\Delta Q_k = S_{k(\text{Scheduled})} - S_{k(\text{Calculated})} \qquad (10.55)$$

$$\Delta P_k = P_{k(\text{Scheduled})} - P_{k(\text{Calculated})}$$

$$\Delta Q_k = Q_{k(\text{Scheduled})} - Q_{k(\text{Calculated})} \qquad (10.56)$$

10.14.1 The Newton–Raphson Algorithm

The solution steps are as follows:

1. Write load flow equations in residual form.
   Form $F(X) = 0$
2. Guess a solution vector, that is, $X^{(0)}$ and evaluate $F(X^{(0)})$
3. Calculate J at $X^{(0)}$

$$J_{X^{(0)}} = \begin{vmatrix} \dfrac{\partial f_1}{\partial x_1} \dfrac{\partial f_1}{\partial x_2} & \cdots\cdots & \dfrac{\partial f_1}{\partial x_n} \\[2mm] \dfrac{\partial f_2}{\partial x_1} \dfrac{\partial f_2}{\partial x_2} & \cdots\cdots & \dfrac{\partial f_2}{\partial x_n} \\[2mm] \cdot & & \\[2mm] \dfrac{\partial f_n}{\partial x_1} \dfrac{\partial f_n}{\partial x_2} & \cdots\cdots & \dfrac{\partial f_n}{\partial x_n} \end{vmatrix}_{X^{(0)}} = J(X^{(0)})$$

(or in general $X^{(i)}$)

The ΔX is computed from the equation below.

Initial computation: Use an initial guess for X.

Step 1. $[\Delta X] = -[J]_{X^{(i)}}^{-1} \, F(X^{(i)})$.

Step 2. Update $X^{(i+1)} = X^{(i)} + \Delta X$.

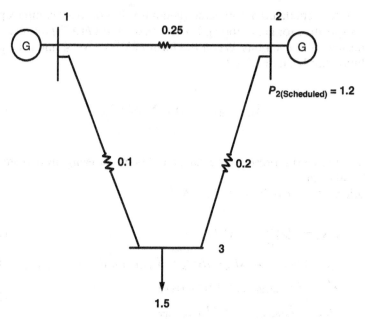

Figure 10.19 The one-line diagram of Example 10.6.

Step 3. Calculate $F(X)$ at $X^{(i+1)}$.

Step 4. Check if $F(X^{(i+1)}) < 10^{-6}$, then the solution has converged. Store the solution vector $X^{(i+1)}$.

Step 5. Update i to $i + 1$.

If not, go to Step 1.

Example 10.6 Consider the three-bus power system given in Figure 10.19. The system data are as follows:

a. Bus 1 is a swing bus with $V_1 = 1\angle 0$.
b. Scheduled injected power at bus 2 is 1.2 p.u.
c. Scheduled load (negative injection) at bus 3 is 1.5 p.u.

Solution

The bus admittance matrix of the system in Example 10.6 is

$$Y_{\text{Bus}} = \begin{bmatrix} 14 & -4 & -10 \\ -4 & 9 & -5 \\ -10 & -5 & 15 \end{bmatrix}$$

The bus powers can be expressed as nonlinear functions of bus voltages in residue form as

$$P_1 (V_1, V_2, V_3) - V_1(Y_{11}V_1 + Y_{12}V_2 + Y_{13}V_3) = 0$$

$$P_2(V_1, V_2, V_3) - V_2(Y_{21}V_1 + Y_{22}V_2 + Y_{23}V_3) = 0$$

$$P_3(V_1, V_2, V_3) - V_3(Y_{31}V_1 + Y_{32}V_2 + Y_{33}V_3) = 0$$

Using a Taylor series expansion about a guess solution, that is, $V_2^{(0)}, V_2^{(0)}$ and $V_1^{(0)} = 1 \angle 0$

$$
\begin{bmatrix} \Delta P_1 \\ \Delta P_2 \\ \Delta P_3 \end{bmatrix} = -
\begin{bmatrix}
\dfrac{\partial P_1}{\partial V_1} \dfrac{\partial P_1}{\partial V_2} \dfrac{\partial P_1}{\partial V_3} \\[2mm]
\dfrac{\partial P_2}{\partial V_1} \dfrac{\partial P_2}{\partial V_2} \dfrac{\partial P_2}{\partial V_3} \\[2mm]
\dfrac{\partial P_3}{\partial V_1} \dfrac{\partial P_3}{\partial V_2} \dfrac{\partial P_3}{\partial V_3}
\end{bmatrix}_{V^{(0)}}
\begin{bmatrix} \Delta V_1 \\ \Delta V_2 \\ \Delta V_3 \end{bmatrix}
$$

In compact notation, we have

$$[\Delta V] = -[J]^{-1} \times [\Delta P]$$

where

$$P_{2(\text{Calculated})} = V_2^{(0)}(Y_{21}V_1^{(0)} + Y_{22}V_2^{(0)} + Y_{23}V_3^{(0)})$$

$$\Delta P_2 = P_{2(\text{Scheduled})} - P_{2(\text{Calculated})} = 0$$

$$P_{3(\text{Calculated})} = V_3^{(0)}(Y_{31}V_1^{(0)} + Y_{32}V_2^{(0)} + Y_{33}V_3^{(0)})$$

$$\Delta P_3 = P_{3(\text{Scheduled})} - P_{3(\text{Calculated})} = 0$$

since $V_1 = 1 \angle 0$ (slack bus).

Therefore $\Delta V_1 = 0.0$. Therefore, only bus 2 and bus 3 voltages are to be calculated. The Jacobian matrix is a 2×2 matrix as shown below.

$$
\begin{bmatrix} \Delta P_2 \\ \Delta P_3 \end{bmatrix} =
\begin{bmatrix}
\dfrac{\partial P_2}{\partial V_2} \dfrac{\partial P_2}{\partial V_3} \\[2mm]
\dfrac{\partial P_3}{\partial V_2} \dfrac{\partial P_3}{\partial V_3}
\end{bmatrix}_{V^{(0)}}
\begin{bmatrix} \Delta V_2 \\ \Delta V_3 \end{bmatrix}
$$

From the above equation, we can calculate ΔV_2 and ΔV_3.

$$\begin{bmatrix} \Delta V_2 \\ \Delta V_3 \end{bmatrix} = -[J]^{-1}_{|V^{(0)}} \begin{bmatrix} \Delta P_2 \\ \Delta P_3 \end{bmatrix}$$

The above solution can be restated as

$$\Delta V_2 = V_{2(new)} - V_{2(old)}$$

$$\Delta V_3 = V_{3(new)} - V_{3(old)}$$

$$\begin{bmatrix} V_2 \\ V_3 \end{bmatrix}_{|new} = \begin{bmatrix} V_2 \\ V_3 \end{bmatrix}_{|old} + [J]^{-1}_{|V^{(0)}} \begin{bmatrix} \Delta P_2 \\ \Delta P_3 \end{bmatrix}$$

The elements of the Jacobian matrix are

$$\frac{\partial P_2}{\partial V_2} = Y_{21} V_1 + 2 Y_{22} V_2 + Y_{23} V_3$$

$$\frac{\partial P_2}{\partial V_3} = Y_{23} V_2$$

$$\frac{\partial P_3}{\partial V_2} = Y_{32} V_3$$

$$\frac{\partial P_3}{\partial V_3} = Y_{31} V_1 + Y_{32} V_2 + 2 Y_{33} V_3$$

Assuming that $V_2^{(0)} = 1$, $V_3^{(0)} = 1$, the elements of the Jacobian matrix are

$$\frac{\partial P_2}{\partial V_2} = -4 + 2 \times 9 - 5 = 9$$

$$\frac{\partial P_2}{\partial V_3} = -5$$

$$\frac{\partial P_3}{\partial V_2} = -5$$

$$\frac{\partial P_3}{\partial V_3} = -10 - 5 + 2 \times 15 = 15$$

Therefore, the Jacobian matrix is given below as

$$[J] = \begin{bmatrix} 9 & -5 \\ -5 & 15 \end{bmatrix}$$

And the $[J]^{-1}$ is

$$[J]^{-1} = \frac{1}{110} \begin{bmatrix} 15 & 5 \\ 5 & 9 \end{bmatrix}$$

The mismatch power at each bus k is

$$P_{2(\text{Calculated})} = V_2^{(0)}\left(Y_{21}V_1^{(0)} + Y_{22}V_2^{(0)} + Y_{23}V_3^{(0)}\right)$$

$$P_{2(\text{Calculated})} = 1.0(-4.(1) + 9.(1) - 5.(1)) = 0$$

$$\Delta P_2 = P_{2(\text{Scheduled})} - P_{2(\text{Calculated})}$$

$$\Delta P_2 = 1.2 - 0 = 1.2$$

$$P_{3(\text{Calculated})} = V_3^{(0)}\left(Y_{31}V_1^{(0)} + Y_{32}V_2^{(0)} + Y_{33}V_3^{(0)}\right)$$

$$P_{3(\text{Calculated})} = 1.0\,(-10.(1) - 5.(1) + 15.(1)) = 0$$

$$\Delta P_3 = P_{3(\text{Scheduled})} - P_{3(\text{Calculated})}$$

$$\Delta P_3 = -1.5 - 0 = -1.5$$

$$\begin{bmatrix} V_2 \\ V_3 \end{bmatrix} = \begin{bmatrix} 1.0 \\ 1.0 \end{bmatrix} + \frac{1}{110} \begin{bmatrix} 15 & 5 \\ 5 & 9 \end{bmatrix} \begin{bmatrix} 1.2 \\ -1.5 \end{bmatrix}$$

$$\begin{bmatrix} V_2 \\ V_3 \end{bmatrix} = \begin{bmatrix} 1.095 \\ 0.932 \end{bmatrix}$$

This new value is now used in the next iteration. The iteration is continued until error does not go below the satisfactory level. The MATLAB simulation testbed for solving the above problem is given below.

```
%Power Flow: Newton Raphson
clc; clear all;
mis_match=0.0001;
Y_bus=[1/.25+1/.1 -1/.25 -1/.1;
-1/.25 1/.25+1/.2 -1/.2;
-1/.1 -1/.2 1/.1+1/.2];
```

```
P_sch = [1; 1.2; -1.5];
N = 3; % no. of buses
% allocate storage for Jacobian
J = zeros(N-1,N-1);
% initial mispatch
V = [1 1 1]';
P_calc = V.*(Y_bus*V);
mismatch = [P_sch(2:N)-P_calc(2:N)];
iteration=0;
% Newton-Raphson iteration
while (iteration<10)
iteration=iteration+1;
% calculate Jacobian
for i = 2:N
for j = 2:N
if (i = j)
J(i-1,j-1)=Y_bus(i,:)*V+Y_bus(i,i)*V(i);
else
J(i-1,j-1)=Y_bus(i,j)*V(i);
end
end
end
% calculate correction
correction =inv(J)*mismatch;
V(2:N) = V(2:N)+correction(1:(N-1));
% calculate mismatch and stop iterating
% if the solution has converged
P_calc = V.*(Y_bus*V);
mismatch = [P_sch(2:N)-P_calc(2:N)];
if (norm(mismatch,'inf') < mis_match)
break;
end
end
iteration
% output solution data
for i = 1:N
fprintf(1, 'Bus %d:\n', i);
fprintf(1, ' Voltage = %f p.u\n', abs(V(i)));
fprintf(1, ' Injected P = %f p.u \n', P_calc(i));
end
```

The result is obtained after three iterations for $c_p = 1 \times 10^{-4}$.

Table 10.3 depicts the bus voltage and power mismatch of Example 10.3. The power supplied by bus 1, the swing bus, is equal to the total load minus total generation by all other bus plus the losses.

The bus voltage of swing bus is $1\angle 0$ and the p.u power is $P_1 = 0.522$.

The total power loss of the transmission lines is 0.223 p.u.

TABLE 10.3   Example 10.6 Results

Bus	p.u Voltage	p.u Power Mismatch
2	1.08	0.3036×10^{-6}
3	0.92	0.6994×10^{-6}

10.15 GENERAL FORMULATION OF THE NEWTON–RAPHSON ALGORITHM

In the above discussions, we presented the basic concepts of Newton–Raphson algorithm. In the following, we present a general formulation of the Newton–Raphson method in calculating the element of the Jacobian matrix. Let us start from a basic equation again.

$$I_{\text{Bus}} = Y_{\text{Bus}} \ V_{\text{Bus}}$$

For each bus k, $I_k = \sum_{j=1}^{n} Y_{kj} V_j$, where n is the number of buses and Y_{kj} is the element of Y_{Bus} matrix. For each bus k, we can also write:

$$P_k + jQ_k = V_k I_k^*$$

where P_k and Q_k are real and the reactive power is entering node k (* is a complex conjugate and $j = \sqrt{-1}$).
  Let

$$P_k + jQ_k = V_k \sum_{j=1}^{n} Y_{kj}^* V_j^* \qquad (10.57)$$

$$v_k = V_k \cdot e^{j\theta_k} = e_k + j \cdot f_k; \theta_k = \tan^{-1} \frac{f_k}{e_k} \qquad (10.58)$$

$$Y_{kj} = Y_{kj} \cdot e^{j\alpha_{kj}} = G_{kj} + j \cdot B_{kj}; \alpha_{kj} = \tan^{-1} \frac{B_{kj}}{G_{kj}} \qquad (10.59)$$

Using Equations (10.58) and (10.59) in (10.57), we have

$$P_k + jQ_k = V_k e^{j\theta_k} \sum_{j=1}^{n} Y_{kj} e^{-j\alpha_{kj}} V_j e^{-j\theta_j} \qquad (10.60)$$

Using the Taylor series expansion, express the power flow problem as

$$\Delta P_k = \sum_{j=1}^{n} H_{kj}\Delta\theta_j + \sum_{j=1}^{n} N_{kj}\frac{\Delta V_j}{V_j} \qquad (10.61)$$

$$\Delta Q_k = \sum_{j=1}^{n} J_{kj}\Delta\theta_j + \sum_{j=1}^{n} L_{kj}\frac{\Delta V_j}{V_j} \qquad (10.62)$$

And in compact form as

$$\begin{bmatrix} \Delta P \\ \Delta Q \end{bmatrix} = \begin{bmatrix} \dfrac{\partial P}{\partial\theta} & \dfrac{\partial P}{\partial V}V \\ \dfrac{\partial Q}{\partial\theta} & \dfrac{\partial Q}{\partial V}V \end{bmatrix} \cdot \begin{bmatrix} \Delta\theta \\ \dfrac{\Delta V}{V} \end{bmatrix} \qquad (10.63)$$

The basic load flow nonlinear equation in a complex domain is given by Equation (10.60). Students can take the derivative of Equation (10.60) and calculate for $j = k$ using the following diagonal elements of the Jacobian matrix:

$$H_{kk} = \frac{\partial P_k}{\partial\theta_k} = -Q_k - V_k^2 B_{kk}$$

$$J_{kk} = \frac{\partial Q_k}{\partial\theta_k} = P_k - V_k^2 G_{kk}$$

$$N_{kk} = \frac{\partial P_k}{\partial V_k}V_k = P_k + V_k^2 G_{kk} \qquad (10.64)$$

$$L_{kk} = \frac{\partial Q_k}{\partial V_k}V_k = Q_k - V_k^2 B_{kk}$$

So far, we have calculated (general equations) the diagonal elements of the Jacobian matrix. To calculate the off-diagonal elements of the Jacobian matrix, we first calculate the following:

$$I_j = a_j + j \cdot b_j = Y_{kj} \cdot V_j \qquad (10.65)$$
$$Y_{kj} \cdot V_j = (G_{kj} + j \cdot B_{kj}).(e_j + j \cdot f_j) \qquad (10.66)$$

$$Y_{kj} \cdot V_j = (G_{kj} \cdot e_j - B_{kj}f_j) + j(B_{kj}e_j + G_{kj}f_j)$$

$$Y_{kj} \cdot V_j = a_j + jb_j$$

where

$$a_j = G_{kj} \cdot e_j - B_{kj}f_j$$

and

$$b_j = B_{kj}e_j + G_{kj}f_j.$$

For $j \neq k$, we have the following off-diagonal elements for the Jacobian matrix:

$$H_{kj} = \frac{\partial P_k}{\partial \theta_j} = a_j f_k - b_j e_k$$

$$J_{kj} = \frac{\partial Q_k}{\partial \theta_j} = -a_j e_k - b_j f_k$$

$$N_{kj} = \frac{\partial P_k}{\partial V_j} V_j = a_j e_k + b_j f_k$$

$$L_{kj} = \frac{\partial Q_k}{\partial V_j} V_j = a_j f_k - b_j e_k \qquad (10.67)$$

$$\begin{bmatrix} \Delta P \\ \Delta Q \end{bmatrix} = \begin{bmatrix} H & N \\ J & L \end{bmatrix} \cdot \begin{bmatrix} \Delta\theta \\ \frac{\Delta V}{V} \end{bmatrix} \qquad (10.68)$$

Power flow program software incorporates the following steps:

1. Renumber the system buses by creating an internal bus numbering system by bus types.
2. A swing bus is selected as bus 1 and followed by all P-V bus types.
3. Sparsity programming is used to eliminate storage requirements.
4. The Jacobian matrix is factored into upper and lower triangular matrices.
5. The diagonal and off-diagonal elements are computed as summarized below.
   For $j = k$, the diagonal elements of the Jacobian matrix are

$$\left. \begin{aligned} H_{kk} &= -Q_k - B_{kk}V_k^2 \\ L_{kk} &= Q_k - B_{kk}V_k^2 \\ N_{kk} &= P_k + G_{kk}V_k^2 \\ J_{kk} &= P_k - G_{kk}V_k^2 \end{aligned} \right\} \; j = k \qquad (10.69)$$

For $j \neq k$, the off-diagonal elements of the Jacobian matrix are

$$\left. \begin{aligned} H_{kj} &= L_{kj} = a_j f_k - b_j e_k \\ N_{kj} &= -J_{kj} = a_j e_k + b_j f_k \end{aligned} \right\} \; j \neq k \qquad (10.70)$$

6. The $P_{k(\text{Calculated})}$ and $Q_{k(\text{Calculated})}$ for each bus k are computed.

$$P_{k(\text{Calculated})} = \sum_{j=1}^{n} [e_k(e_j G_{kj} - f_j B_{kj}) + f_k(e_j B_{kj} + f_j G_{kj})] \quad (10.71)$$

$$Q_{k(\text{Calculated})} = \sum_{j=1}^{n} [f_k(e_k G_{kj} - f_j B_{kj}) - e_k(e_j B_{kj} + f_j G_{kj})] \quad (10.72)$$

where $V_k = e_k + j \cdot f_k$; $Y_{kj}^* = G_{kj} - j \cdot B_{kj}$; $V_k^* == e_k - j \cdot f_k$.

7. The power mismatch for active and reactive powers are calculated as follows:

$$\Delta P_k^{(i)} = P_{k(\text{Scheduled})}^{(i)} - P_{k(\text{Calculated})}^{(i)} < c_P \quad (10.73)$$

$$\Delta Q_k^{(i)} = Q_{k(\text{Scheduled})}^{(i)} - Q_{k(\text{Calculated})}^{(i)} < c_Q \quad (10.74)$$

The Jacobian matrix is evaluated at each iterative approximation at the last computed solution.

$$V = V^{(i)} \text{ and } \theta = \theta^{(i)}$$

$$\Delta V^{(k)} = V^{(k+1)} - V^{(k)}$$

$$\Delta \theta^{(k)} = \theta^{(k+1)} - \theta^{(k)}$$

The power mismatch at each bus is calculated based on $P_{k(\text{Scheduled})}$ and $Q_{k(\text{Scheduled})}$ and the calculated power flow at the bus $P_{k,(\text{Calculated})}$, $Q_{k,(\text{Calculated})}$ The calculated active and reactive powers are based on the voltage profile of the network and Y_{Bus} model using $V^{(i)}$ and $\theta^{(i)}$ computed bus voltages.

For P-V buses, the ΔV_k is set equal to zero. Notice that for a P-V bus, the reactive power generation has to be calculated and checked for violation of Q-limits. In case of limit violations, the bus type has to be switched from P-V to PQ type to maintain the reactive generation within the specified limits.

10.16 THE DECOUPLED NEWTON–RAPHSON ALGORITHM

Numerical studies of many systems clearly indicate that the change in voltage magnitudes has little effect on power flows. In addition, the changes in voltage angles have little effect on reactive power flows. These observations may not be true for cables and short lines. However, for bulk power flows they are generally true. Using the above assumptions facilitates a substantial reduction in the amount of computer memory required for a large-scale power grid load flow analysis.

We can restate the above assumptions as

1. The partial of $\frac{\partial P}{\partial V}$ is assumed to be zero.
2. The partial of $\frac{\partial Q}{\partial \theta}$ is assumed to be zero.

With the above assumptions, we can decouple the $\Delta P - \Delta \theta$ equations and $\Delta Q - \Delta V$ equations of the load flow problems.

$$
\begin{bmatrix} \Delta P \\ \Delta Q \end{bmatrix} = \begin{bmatrix} \frac{\partial P}{\partial \theta} & 0 \\ 0 & \frac{\partial Q}{\partial V}V \end{bmatrix} \begin{bmatrix} \Delta \theta \\ \frac{\Delta V}{V} \end{bmatrix} \tag{10.75}
$$

Therefore, Equation (10.75) will become two decoupled and independent equations.

$$
[\Delta P] = \begin{bmatrix} \frac{\partial P}{\partial \theta} \end{bmatrix} \quad [\Delta \theta]
$$
$$
[\Delta Q] = \begin{bmatrix} \frac{\partial Q}{\partial V}V \end{bmatrix} \quad \begin{bmatrix} \frac{\Delta V}{V} \end{bmatrix} \tag{10.76}
$$

Therefore, Equation (10.75) can be rewritten as

$$
\begin{bmatrix} \Delta P \\ \Delta Q \end{bmatrix} = \begin{bmatrix} H & 0 \\ 0 & L \end{bmatrix} \begin{bmatrix} \Delta \theta \\ \frac{\Delta V}{V} \end{bmatrix} \tag{10.77}
$$

where

$$
[H] = \begin{bmatrix} \frac{\partial P}{\partial \theta} \end{bmatrix}
$$

and

$$
[L] = \begin{bmatrix} \frac{\partial Q}{\partial V}V \end{bmatrix}.
$$

Equation (10.77) presents the decoupled Newton–Raphson method.

10.17 THE FAST DECOUPLED LOAD FLOW ALGORITHM

The FDLF algorithm is a modified version of the Newton–Raphson algorithm. The FDLF algorithm[9] takes advantage of the weak coupling between the real and reactive powers and uses two constant matrices to approximate and decouple the Jacobian matrix.[8] We rewrite the H matrix as

$$[\Delta P] = [H][\Delta\theta]$$

$$[H] = \left[\frac{\partial P}{\partial\theta}\right] = [V]\,[B']\,[V]$$

$$[\Delta P] = [V]\,[B']\,[V][\Delta\theta]$$

Dividing both sides by V, we have

$$\left[\frac{\Delta P}{V}\right] = [B']\,[V][\Delta\theta]$$

To obtain a constant linear approximation with constant coefficient, we set the V in the above equation to 1 and we obtain:

$$\left[\frac{\Delta P}{V}\right] = [B'] \cdot [\Delta\theta] \qquad (10.78)$$

Similarly, we can observe that the L matrix in the decoupled Newton–Raphson can be written as

$$[\Delta Q] = [L]\left[\frac{\Delta V}{V}\right]$$

$$[L] = \left[\frac{\partial Q}{\partial V}V\right] = [V]\,[B]\,[V]$$

Then, we divide both sides by V and setting the second V to 1 p.u, we have

$$\left[\frac{\Delta Q}{V}\right] = [B''] \cdot [\Delta V] \qquad (10.79)$$

Equations (10.78) and (10.79) present the FDLF algorithm. The matrix B'' is the imaginary part of the Y matrix and B' is the same as B'' except for the line resistances and shunt elements that are neglected. The FDLF method reduces the memory requirement and converges in most problems to an acceptable solution. However, the number of iterations increases. When the power grid has short cables and has phase-shifting transformers, the FDLF can fail to converge. However, we should observe that even when the FDLF is constructed from approximation of the Jacobian matrix, the convergence of the power flow

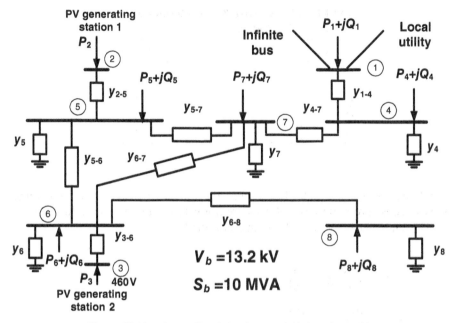

Figure 10.20   A one-line injection model of a microgrid.

problem is still checked based on the original power flow mismatch of active and reactive power. Therefore, if FDLF converges, then the solution is as accurate as the full Newton–Raphson method.

10.18 ANALYSIS OF A POWER FLOW PROBLEM

Let us consider the case of the decoupled Newton–Raphson method and analyze the injection model of Figure 10.20.

We will use the data in Tables 10.4 and 10.5 to show the linearized relationship of known variables in terms of the unknown variables. Bus 1, bus 2, and bus

TABLE 10.4   Active Power Injection as a Function of Bus Angle

Knowns	Unknowns
P_2	θ_2
P_3	θ_3
P_4	θ_4
P_5	θ_5
P_6	θ_6
P_7	θ_7
P_8	θ_8

TABLE 10.5   Reactive Power Injection as a Function of Bus Voltage

Knowns	Unknowns
Q_4	V_4
Q_5	V_5
Q_6	V_6
Q_7	V_7
Q_8	V_8

3 voltages are specified at a fixed value; therefore, for these buses the change in voltage is zero and these voltages are not listed as unknowns in Table 10.5. The same is true for the phase angle of bus 1.

$$\Delta P_2 = \frac{\partial P_2}{\partial \theta_2}\Delta \theta_2 + \frac{\partial P_2}{\partial \theta_5}\Delta \theta_5$$

$$\Delta P_3 = \frac{\partial P_3}{\partial \theta_3}\Delta \theta_3 + \frac{\partial P_3}{\partial \theta_6}\Delta \theta_6$$

$$\Delta P_4 = \frac{\partial P_4}{\partial \theta_4}\Delta \theta_4 + \frac{\partial P_4}{\partial \theta_7}\Delta \theta_7$$

$$\Delta P_5 = \frac{\partial P_5}{\partial \theta_2}\Delta \theta_2 + \frac{\partial P_5}{\partial \theta_5}\Delta \theta_5 + \frac{\partial P_5}{\partial \theta_6}\Delta \theta_6 + \frac{\partial P_5}{\partial \theta_7}\Delta \theta_7 \qquad (10.80)$$

$$\Delta P_6 = \frac{\partial P_6}{\partial \theta_3}\Delta \theta_3 + \frac{\partial P_6}{\partial \theta_5}\Delta \theta_5 + \frac{\partial P_6}{\partial \theta_6}\Delta \theta_6 + \frac{\partial P_6}{\partial \theta_7}\Delta \theta_7 + \frac{\partial P_6}{\partial \theta_8}\Delta \theta_8$$

$$\Delta P_7 = \frac{\partial P_7}{\partial \theta_4}\Delta \theta_4 + \frac{\partial P_7}{\partial \theta_5}\Delta \theta_5 + \frac{\partial P_7}{\partial \theta_6}\Delta \theta_6 + \frac{\partial P_7}{\partial \theta_7}\Delta \theta_7$$

$$\Delta P_8 = \frac{\partial P_8}{\partial \theta_6}\Delta \theta_6 + \frac{\partial P_8}{\partial \theta_8}\Delta \theta_8$$

If you carefully evaluate the sets of equations given in Equation (10.80), you will observe that an injected power at each bus is only the function of the bus under consideration and the buses they are connected to in the network. For example, bus 4 is connected to bus 1 and bus 7. The injection at bus 4 changes the flow between bus 1 and bus 7. However, the bus 1 change in angle is zero, because bus 1 is the swing bus. As we know, the swing bus injection is computed last when bus voltages are computed and the power flow problem has converged. Again, the same condition is true for all lines connecting buses in the network. For example, bus 7 is connected to buses 4, 5, and 6 and corresponding terms are given in the bus 7 injection model. As we have stated before, each active power mismatch is computed based on injected power and the calculated active power based on the system Y_{Bus} and bus voltages.

$$
\begin{bmatrix} \Delta P_2 \\ \Delta P_3 \\ \Delta P_4 \\ \Delta P_5 \\ \Delta P_6 \\ \Delta P_7 \\ \Delta P_8 \end{bmatrix} =
\begin{array}{ccccccc}
P_2 & P_3 & P_4 & P_5 & P_6 & P_7 & P_8 \\
\end{array}
\begin{bmatrix}
\dfrac{\partial P_2}{\partial \theta_2} & 0 & 0 & \dfrac{\partial P_2}{\partial \theta_5} & 0 & 0 & 0 \\[2mm]
0 & \dfrac{\partial P_3}{\partial \theta_3} & 0 & 0 & \dfrac{\partial P_3}{\partial \theta_6} & 0 & 0 \\[2mm]
0 & 0 & \dfrac{\partial P_4}{\partial \theta_4} & 0 & 0 & \dfrac{\partial P_4}{\partial \theta_7} & 0 \\[2mm]
\dfrac{\partial P_5}{\partial \theta_2} & 0 & 0 & \dfrac{\partial P_5}{\partial \theta_5} & \dfrac{\partial P_5}{\partial \theta_6} & \dfrac{\partial P_5}{\partial \theta_7} & 0 \\[2mm]
0 & \dfrac{\partial P_6}{\partial \theta_3} & 0 & \dfrac{\partial P_6}{\partial \theta_5} & \dfrac{\partial P_6}{\partial \theta_6} & \dfrac{\partial P_6}{\partial \theta_7} & \dfrac{\partial P_6}{\partial \theta_8} \\[2mm]
0 & 0 & \dfrac{\partial P_7}{\partial \theta_4} & \dfrac{\partial P_7}{\partial \theta_5} & \dfrac{\partial P_7}{\partial \theta_6} & \dfrac{\partial P_7}{\partial \theta_7} & 0 \\[2mm]
0 & 0 & 0 & 0 & \dfrac{\partial P_8}{\partial \theta_6} & 0 & \dfrac{\partial P_8}{\partial \theta_8}
\end{bmatrix}
\begin{bmatrix} \Delta \theta_2 \\ \Delta \theta_3 \\ \Delta \theta_4 \\ \Delta \theta_5 \\ \Delta \theta_6 \\ \Delta \theta_7 \\ \Delta \theta_8 \end{bmatrix}
$$

$$(10.81)$$

Equation (10.81) presents the matrix formulation of Equation (10.80). As we described earlier, the elements of the Jacobian matrix are computed by taking the partial derivative of the general load flow equation. For each iteration, we use the voltages computed from the last iteration to compute the matrix H of Equation (10.82).

$$
\begin{bmatrix} \Delta P_2 \\ \Delta P_3 \\ \Delta P_4 \\ \Delta P_5 \\ \Delta P_6 \\ \Delta P_7 \\ \Delta P_8 \end{bmatrix} =
\begin{array}{ccccccc}
P_2 & P_3 & P_4 & P_5 & P_6 & P_7 & P_8 \\
\end{array}
\begin{bmatrix}
H_{22} & 0 & 0 & H_{25} & 0 & 0 & 0 \\
0 & H_{33} & 0 & 0 & H_{36} & 0 & 0 \\
0 & 0 & H_{44} & 0 & 0 & H_{47} & 0 \\
H_{52} & 0 & 0 & H_{55} & H_{56} & H_{57} & 0 \\
0 & H_{63} & 0 & H_{65} & H_{66} & H_{67} & H_{68} \\
0 & 0 & H_{74} & H_{75} & H_{76} & H_{77} & 0 \\
0 & 0 & 0 & 0 & H_{86} & 0 & H_{88}
\end{bmatrix} \cdot
\begin{bmatrix} \Delta \theta_2 \\ \Delta \theta_3 \\ \Delta \theta_4 \\ \Delta \theta_5 \\ \Delta \theta_6 \\ \Delta \theta_7 \\ \Delta \theta_8 \end{bmatrix}
$$

$$(10.82)$$

$$[\Delta P] = [H][\Delta \theta] \qquad (10.83)$$

The linearized $\Delta Q - \Delta V$ equations are written using Table 10.5 and Figure 10.20. Again, we recognize that the bus voltage is a function of the reactive power from other buses. For example, the voltage at bus 4 is a function of reactive power at bus 4 and bus 7 only because bus voltage is fixed.

$$
\begin{bmatrix} \Delta Q_4 \\ \Delta Q_5 \\ \Delta Q_6 \\ \Delta Q_7 \\ \Delta Q_8 \end{bmatrix} =
\begin{array}{ccccc} Q_4 & Q_5 & Q_6 & Q_7 & Q_8 \end{array}
\begin{bmatrix}
\dfrac{\partial Q_4}{\partial V_4}V_4 & 0 & 0 & \dfrac{\partial Q_4}{\partial V_7}V_7 & 0 \\[2ex]
0 & \dfrac{\partial Q_5}{\partial V_5}V_5 & \dfrac{\partial Q_5}{\partial V_6}V_6 & \dfrac{\partial Q_5}{\partial V_7}V_7 & 0 \\[2ex]
0 & \dfrac{\partial Q_6}{\partial V_5}V_5 & \dfrac{\partial Q_6}{\partial V_6}V_6 & \dfrac{\partial Q_6}{\partial V_7}V_7 & \dfrac{\partial Q_6}{\partial V_8}V_8 \\[2ex]
\dfrac{\partial Q_7}{\partial V_4}V_4 & \dfrac{\partial Q_7}{\partial V_5}V_5 & \dfrac{\partial Q_7}{\partial V_6}V_6 & \dfrac{\partial Q_7}{\partial V_7}V_7 & 0 \\[2ex]
0 & 0 & \dfrac{\partial Q_8}{\partial V_6}V_6 & 0 & \dfrac{\partial Q_8}{\partial V_8}V_8
\end{bmatrix}
\cdot
\begin{bmatrix}
\dfrac{\Delta V_4}{V_4} \\[2ex]
\dfrac{\Delta V_5}{V_5} \\[2ex]
\dfrac{\Delta V_6}{V_6} \\[2ex]
\dfrac{\Delta V_7}{V_7} \\[2ex]
\dfrac{\Delta V_8}{V_8}
\end{bmatrix}
\qquad (10.84)
$$

$$
\begin{bmatrix} \Delta Q_4 \\ \Delta Q_5 \\ \Delta Q_6 \\ \Delta Q_7 \\ \Delta Q_8 \end{bmatrix} =
\begin{array}{ccccc} Q_4 & Q_5 & Q_6 & Q_7 & Q_8 \end{array}
\begin{bmatrix}
L_{44} & 0 & 0 & L_{47} & 0 \\
0 & L_{55} & L_{56} & L_{57} & 0 \\
0 & L_{65} & L_{66} & L_{67} & L_{68} \\
L_{74} & L_{75} & L_{76} & L_{77} & 0 \\
0 & 0 & L_{86} & 0 & L_{88}
\end{bmatrix}
\begin{bmatrix}
\dfrac{\Delta V_4}{V_4} \\[2ex]
\dfrac{\Delta V_5}{V_5} \\[2ex]
\dfrac{\Delta V_6}{V_6} \\[2ex]
\dfrac{\Delta V_7}{V_7} \\[2ex]
\dfrac{\Delta V_8}{V_8}
\end{bmatrix}
\qquad (10.85)
$$

$$[\Delta Q] = [L]\left[\frac{\Delta V}{V}\right]$$

Example 10.7   Consider the microgrid given in Figure 10.21. Assume the following data:

a. Transformers connected to the PV generating station are rated at 460 V Y-grounded/13.2 kV Δ and have 10% reactance and 10 MVA capacity. The transformer connected to the power grid is rated at 13.2/63 kV and has 10 MVA capacity and 10% reactance.

b. Assume the load on bus 4 is 1.5 MW, 0.9 p.f. lagging; on bus 5 is 2.5 MW, 0.9 p.f. lagging; on bus 6 is 1.0 MW, 0.95 p.f. lagging; on bus 7 is 2 MW, 0.95 p.f. leading; and on bus 8 is 1.0 MW, 0.9 p.f. lagging.

c. The transmission line has a resistance of 0.0685 Ω/mile, reactance of 0.40 Ω/mile, and half of line-charging admittance $(Y'/2)$ of 11×10^{-6} Ω^{-1}/mile.

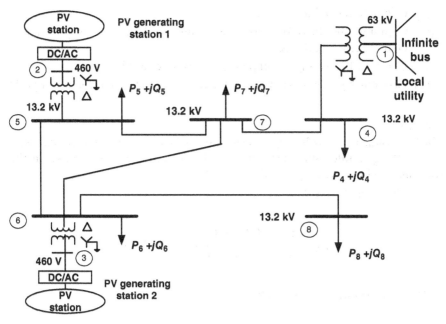

Figure 10.21   The photovoltaic microgrid of Example 10.7.

The line 4–7 is 5 miles, 5–6 is 3 miles, 5–7 is 2 miles, 6–7 is 2 miles, and 6–8 is 4 miles long.

Figure 10.22 depicts the transmission-line pie model.

d. Assume the PV generating station #1 is rated at 0.75 MW and PV generating station #2 is rated at 3 MW. Assume PV generating stations are operating at unity power factors.

Perform the following:

(i) Assume an S_b of 10 MVA and a voltage base of 460 V in PV generator #1 and compute the p.u model.

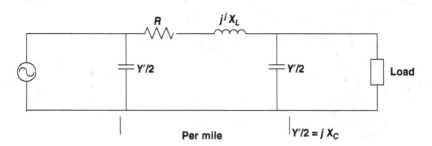

Figure 10.22   The transmission-line pie model.

(ii) Compute the Y bus model.

(iii) Compute the load bus voltages using the Newtown–Raphson and Gauss–Seidel methods.

(iv) Compute bus voltages and power flow of microgrid.

(v) How much green power is imported or exported to the local power grid?

Solution

The base value of the volt-amp is designated as $S_b = 10$ MVA; the voltage base on the PV generator side is specified as 460 V, and the voltage base on the transmission line side is specified as $V_b = 13.2$ kV.

The base impedance is

$$Z_b = \frac{V_b^2}{S_b} = \frac{(13.2 \times 10^3)^2}{10 \times 10^6} = 17.424 \ \Omega$$

The base admittance is given by

$$Y_b = \frac{1}{Z_b} = \frac{1}{17.424} = 0.057$$

The current injection model of the PV system is shown in Figure 10.23.

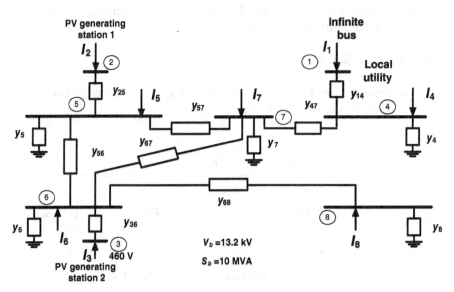

Figure 10.23   The current injection model of Example 10.7.

From transmission line data, the primitive impedance and admittance are calculated as follows:

$$Z_{1-4\text{p.u}} = Z_{2-5}\text{p.u} = Z_{3-6}\text{p.u} = j0.1, y_{14\text{p.u}} = y_{25\text{p.u}} = y_{36\text{p.u}} = \frac{1}{j0.1} = -j10$$

$$Z_{4-7} = 5\,(0.0685 + j0.4)\,/Z_b = 0.020 + j0.115, \quad y_{47\ \text{p.u}} = \frac{1}{0.020 + j0.115}$$

$$= 1.45 - j8.46 \ \text{p.u}\ \Omega$$

$$Z_{5-6} = 3\,(0.0685 + j0.4)\,/Z_b = 0.012 + j0.069 \quad y_{56\ \text{p.u}} = \frac{1}{0.012 + j0.069}$$

$$= 2.42 - j14.11 \ \text{p.u}\ \Omega$$

$$Z_{5-7} = Z_{6-7} = 2\,(0.0685 + j0.4)\,/Z_b = 0.008 + j0.046 \ \text{p.u}\ \Omega$$

$$y_{57} = y_{67\text{p.u}} = \frac{1}{0.008 + j0.046} = 3.62 - j21.16$$

$$Z_{6-8} = 4\,(0.0685 + j0.4)\,/Z_b = 0.016 + j0.092, \ \text{p.u}\ \Omega$$

$$y_{68} = \frac{1}{0.016 + j0.092} = 1.18 - 10.57 \tag{10.86}$$

The line-charging admittance is the same as the $\frac{Y'}{2}$ as shown in Figure 10.22. The line-charging admittances in per unit are as follows:

$$y'_{47} = 5 \times \frac{j11 \times 10^{-6}}{Y_b} = j9.58 \times 10^{-4}, \quad y'_{56\ \text{p.u}} = 3 \times \frac{j11 \times 10^{-6}}{Y_b} = j5.75 \times 10^{-4}$$

$$y'_{57} = y'_{67} = 2 \times \frac{j11 \times 10^{-6}}{Y_b} = j3.83 \times 10^{-4}, \quad y'_{68} = 4 \times \frac{j11 \times 10^{-6}}{Y_b} = j7.67 \times 10^{-4}$$

The p.u admittance matrix, Y_{Bus}, is calculated using Equation (10.18).

$$Y_{\text{Bus}} = \begin{bmatrix}
-j10 & 0 & 0 & j10 & 0 & 0 & 0 & 0 \\
0 & -j10 & 0 & 0 & j10 & 0 & 0 & 0 \\
0 & 0 & -j10 & 0 & 0 & j10 & 0 & 0 \\
j10 & 0 & 0 & 1.45 - j18.46 & 0 & 0 & -1.45 + j8.46 & 0 \\
0 & j10 & 0 & 0 & 6.04 - j45.26 & -2.42 + j14.11 & -3.62 + j21.16 & 0 \\
0 & 0 & j10 & 0 & -2.42 + j14.11 & 7.85 - j55.84 & -3.62 + j21.16 & -1.81 + j10.58 \\
0 & 0 & 0 & -1.45 + j8.46 & -3.62 + j21.16 & -3.62 + j21.16 & 8.69 - j50.78 & 0 \\
0 & 0 & 0 & 0 & 0 & -1.81 + j10.58 & 0 & 1.81 - j10.58
\end{bmatrix}$$

A MATLAB program was written to perform the load flow analysis. The PV generator buses are treated as PV buses in the load flow problem. The voltage at the PV generator buses are fixed at 1 p.u. Table 10.6 provides the scheduled powers of each bus.

TABLE 10.6   The Scheduled Active and Reactive Power of Each Bus

Bus	2	3	4	5	6	7	8
$P_{scheduled}$	0.075	0.3	−0.150	−0.250	−0.100	−0.200	−0.100
$Q_{scheduled}$	–	–	−0.073	−0.121	−0.033	0.066	−0.048

TABLE 10.7a   The Voltages and Power of Each Bus Using the Newton–Raphson Method

$$S_b = 10 \text{ MVA}, V_b = 13.2 \text{ kV}$$

			Generation		Load			
Bus #	Volts (p.u)	Angle (°)	MW (p.u)	MVAr (p.u)	MW (p.u)	MVAr (p.u)	ΔP	ΔQ
1	1.000	0.0	0.427	0.065	0	0	0	0
2	1.000	−4.1	0.075	0.094	0	0	0.005×10^{-10}	0
3	1.000	−2.5	0.300	0.085	0	0	0.021×10^{-10}	0
4	0.994	−2.5	0	0	0.150	0.073	0.189×10^{-10}	0.424×10^{-10}
5	0.991	−4.6	0	0	0.250	0.121	0.127×10^{-10}	0.002×10^{-10}
6	0.992	−4.3	0	0	0.100	0.033	0.079×10^{-10}	0.071×10^{-10}
7	0.992	−4.3	0	0	0.200	−0.066	0.233×10^{-10}	0.110×10^{-10}
8	0.986	−4.8	0	0	0.100	0.048	0.074×10^{-10}	0.009×10^{-10}

For this load flow problem, the tolerance of error chosen was 1×10^{-5}. The load flow problem was solved using the Newton–Raphson and Gauss–Seidel methods. Table 10.7 gives the results using the Newton–Raphson method, which converges in three iterations. The Gauss–Seidel converges in 392 iterations (Table 10.8). The Table 10.7a gives the results of power flow of Example 10.7 and Table 10.7b gives the power flow through transmission lines.

To maintain PV bus 2 and PV bus 3 at 1 p.u, the required reactive power at bus 2 is 0.094 p.u and at bus 3 is 0.085 p.u. The active power flow from the local

TABLE 10.7b   The Power Flow through Transmission Lines and Transformers Using the Newton–Raphson Method

$$P_{loss} \text{ (p.u)} = 0.002, Q_{loss} \text{ (p.u)} = 0.035$$

From Bus #	To Bus #	MW Flow (p.u)	MVAr Flow (p.u)
1	4	0.427	0.065
2	5	0.075	0.094
3	6	0.300	0.085
4	7	0.277	−0.025
5	6	−0.079	−0.006
5	7	−0.097	−0.021
6	7	0.021	−0.012
6	8	0.100	0.049

TABLE 10.8a   The Voltages and Power of Each Bus Using the Gauss–Seidel Method

			Generation		Load			
			$S_b = 10$ MVA, $V_b = 13.2$ kV					
Bus #	Volts (p.u)	Angle (°)	MW (p.u)	MVAr (p.u)	MW (p.u)	MVAr (p.u)	ΔP	ΔQ
1	1.000	0.0	0.427	0.065	0	0	0	0
2	1.000	−4.1	0.075	0.094	0	0	0.179×10^{-5}	0
3	1.000	−2.5	0.300	0.085	0	0	0.181×10^{-5}	0
4	0.994	−2.5	0	0	0.150	0.073	0.141×10^{-5}	0.018×10^{-5}
5	0.991	−4.6	0	0	0.250	0.121	0.783×10^{-5}	0.100×10^{-5}
6	0.992	−4.3	0	0	0.100	0.033	0.979×10^{-5}	0.132×10^{-5}
7	0.992	−4.3	0	0	0.200	−0.066	0.822×10^{-5}	0.137×10^{-5}
8	0.986	−4.8	0	0	0.100	0.048	0.188×10^{-5}	0.031×10^{-5}

power grid is 0.427 p.u. The reactive power flow from the local power grid is 0.065 p.u.

It can be seen from Tables 10.7 and 10.8 that the reactive power of 0.094 p.u and 0.085 p.u is needed from PV generating station connected to buses 2 and 3 to maintain a voltage of 1 p.u at these buses. If the PV system has a local storage system, the control of the inverter can be set to provide the reactive power. The reactive power can also be provided by placing the required VAr support at the AC side of the PV generating stations. The scheduled reactive VAr support at bus 2 is 0.094 p.u and at bus 3 is 0.085 p.u. The power flow problem was solved with scheduled VAr support reactive power injected into the PV generating station buses. The results are shown in Table 10.9. The program converged in three iterations.

It is seen that with the reactive power injected into the PV generator buses, the voltage of these buses are maintained at 1 p.u. From Table 10.9, we see that the active power supplied by bus 1 is positive 0.427 p.u. This means, the power is being imported (bought) from the local power grid.

TABLE 10.8b   The Power Flow through Transmission Lines and Transformers Using the Gauss–Seidel Method

		P_{loss} (p.u) = 0.002, Q_{loss} (p.u) = 0.035	
From Bus #	To Bus #	MW Flow (p.u)	MVAr Flow (p.u)
1	4	0.427	0.065
2	5	0.075	0.094
3	6	0.300	0.085
4	7	0.277	−0.025
5	6	−0.079	−0.006
5	7	−0.097	−0.021
6	7	0.021	−0.012
6	8	0.100	0.049

TABLE 10.9a   The Voltage and Power of Each Bus with Reactive Power Correction for the PV Generation Buses

			Generation		Load			
Bus #	Volts (p.u)	Angle (°)	MW (p.u)	MVAr (p.u)	MW (p.u)	MVAr (p.u)	ΔP	ΔQ
				$S_b = 10$ MVA, $V_b = 13.2$ kV				
1	1.000	0.0	0.427	0.065	0	0	0	0
2	1.000	−4.1	0.075	0.094	0	0	0.049×10^{-8}	0.003×10^{-8}
3	1.000	−2.5	0.300	0.085	0	0	0.196×10^{-8}	0.081×10^{-8}
4	0.994	−2.5	0	0	0.150	0.073	0.064×10^{-8}	0.086×10^{-8}
5	0.991	−4.6	0	0	0.250	0.121	0.230×10^{-8}	0.016×10^{-8}
6	0.992	−4.3	0	0	0.100	0.033	0.131×10^{-8}	0.148×10^{-8}
7	0.992	−4.3	0	0	0.200	−0.066	0.131×10^{-8}	0.042×10^{-8}
8	0.986	−4.8	0	0	0.100	0.048	0.146×10^{-8}	0.007×10^{-8}

The base volt-amp is 10 MVA ($S_b = 10$ MVA); therefore, the power imported from the local power grid is $P_1 = 0.427 \times S_b = 4.27$ MW. The power generated by the PV generator 1 connected to bus 2 is $P_2 = 0.075 \times S_b = 0.75$ MW.

The power generated by the PV generator 2 connected to bus 3 is $P_3 = 0.300 \times S_b = 3.00$ MW.

Example 10.8   For Example 10.7, perform the following:

(i) Compute the matrix of B'' and matrix B' of FDLF.

(ii) Compute the load bus voltages and the mismatch using the FDLF method.

TABLE 10.9b   The Power Through the Transmission Lines and Transformers with Reactive Power Correction for the PV Generation Buses

From Bus #	To Bus #	MW Flow (p.u)	MVAr Flow (p.u)
	P_{loss} (p.u) $= 0.002$, Q_{loss} (p.u) $= 0.035$		
1	4	0.427	0.065
2	5	0.075	0.094
3	6	0.300	0.085
4	7	0.277	−0.025
5	6	−0.079	−0.006
5	7	−0.097	−0.021
6	7	0.021	−0.012
6	8	0.100	0.049

Solution

(i) From Example 10.7, the bus number 1 is the swing bus and buses 2 and 3 are PV buses. Hence, out of eight buses, there are five PQ buses. The matrix B'' is 5×5 and is equal to the imaginary part of the last five rows and columns of the Y_{Bus} matrix:

$$B'' = \begin{bmatrix} -18.46 & 0 & 0 & 8.46 & 0 \\ 0 & -45.26 & 14.11 & 21.16 & 0 \\ 0 & 14.11 & -55.84 & 21.16 & 10.58 \\ 8.46 & 21.16 & 21.16 & -50.78 & 0 \\ 0 & 0 & 10.58 & 0 & -10.18 \end{bmatrix}$$

To calculate B', the resistance and shunt elements are neglected. The B' matrix is a 7×7 matrix. The diagonal elements of B' are

$$B'_{22} = -j10,$$
$$B'_{33p.u} = -j10,$$
$$B'_{44p.u} = -j10 - j8.71 = -j18.71,$$
$$B'_{55} = -j10 - j14.52 - j21.78 = -j46.30,$$
$$B'_{66} = -j10 - j14.52 - j21.78 - j10.89 = -j57.19,$$
$$B'_{77} = -j8.71 - j21.78 - j21.78 = -j52.27,$$
$$B'_{88} = -j10.89$$

The off-diagonal elements of B' are $B'_{kj} = \dfrac{1}{x_{kjp.u}}$, where $x_{kjp.u}$ is the per unit series reactance between buses k and j.

The matrix is symmetric. The upper triangular elements are as follows:

$$B'_{25} = B'_{36} = \frac{1}{x_{25,p.u}} = \frac{1}{x_{36,p.u}} = -\frac{1}{j0.1} = j10,$$

$$B'_{57} = \frac{1}{x_{57,p.u}} = -\frac{1}{j0.0459} = j21.78,$$

$$B'_{47} = \frac{1}{x_{47,p.u}} = -\frac{1}{j0.1148} = j8.71,$$

$$B'_{56} = \frac{1}{x_{56,p.u}} = -\frac{1}{j0.0689} = j14.52,$$

$$B'_{68} = \frac{1}{x_{68,p.u}} = -\frac{1}{j0.0918} = j10.89,$$

$$B'_{67} = \frac{1}{x_{67,p.u}} = -\frac{1}{j0.0459} = j21.78$$

The elements of the matrix are the imaginary part of the above.

$$
B' = \begin{array}{c} \\ \\ \\ \end{array}
\begin{array}{cccccccc}
2 & 3 & 4 & 5 & 6 & 7 & 8 \\
-10 & 0 & 0 & 10 & 0 & 0 & 0 \\
0 & -10 & 0 & 0 & 10 & 0 & 0 \\
0 & 0 & -18.71 & 0 & 0 & 8.71 & 0 \\
10 & 0 & 0 & -46.30 & 14.52 & 21.78 & 0 \\
0 & 10 & 0 & 14.52 & -57.19 & 21.78 & 10.89 \\
0 & 0 & 8.71 & 21.78 & 21.78 & -52.27 & 0 \\
0 & 0 & 0 & 0 & 10.89 & 0 & -10.89
\end{array}
\begin{array}{c}
2 \\ 3 \\ 4 \\ 5 \\ 6 \\ 7 \\ 8
\end{array}
$$

(ii) The problem was solved using the FDLF method with the tolerance of error chosen as 1×10^{-5}. The solution converged in six iterations. The results are given in Table 10.10.

In this chapter, we have used the bus admittance matrix and bus impedance matrix to formulate the power flow in power grids. Power flow problems must be solved during the system planning of large, interconnected power grids using the forecasted bus loads to ensure that planned generation and transmission systems are at acceptable bus load voltages so that the system lines, transformers, etc., are not overloaded. Power flow problems must also be solved during daily operations and for a PV or wind generating microgrid operating as an

TABLE 10.10a   The Voltages and Power of Each Bus Using the Fast Decoupled Method

			$S_b = 10$ MVA, $V_b = 13.2$ kV					
			Generation		Load			
Bus #	Volts (p.u)	Angle (°)	MW (p.u)	MVAr (p.u)	MW (p.u)	MVAr (p.u)	ΔP	ΔQ
1	1.000	0.0	0.427	0.065	0	0	0	0
2	1.000	−4.1	0.075	0.094	0	0	0.002×10^{-5}	0
3	1.000	−2.5	0.300	0.085	0	0	0.002×10^{-5}	0
4	0.994	−2.5	0	0	0.150	0.073	0.052×10^{-5}	0.086×10^{-5}
5	0.991	−4.6	0	0	0.250	0.121	0.063×10^{-5}	0.129×10^{-5}
6	0.992	−4.3	0	0	0.100	0.033	0.388×10^{-5}	0.295×10^{-5}
7	0.992	−4.3	0	0	0.200	−0.066	0.301×10^{-5}	0.504×10^{-5}
8	0.986	−4.8	0	0	0.100	0.048	0.190×10^{-5}	0.062×10^{-5}

TABLE 10.10b   The Power Flow Through Transmission Lines and Transformers Using the Fast Decoupled Method

P_{loss} (p.u) = 0.002, Q_{loss} (p.u) = 0.035			
From Bus #	To Bus #	MW Flow (p.u)	MVAr Flow (p.u)
1	4	0.427	0.065
2	5	0.075	0.094
3	6	0.300	0.085
4	7	0.277	−0.025
5	6	−0.079	−0.006
5	7	−0.097	−0.021
6	7	0.021	−0.012
6	8	0.100	0.049

isolated system using the forecasted bus loads and also to ensure that the microgrid can support its loads at acceptable bus load voltages and that cables, lines, transformers, etc., are not overloaded.

PROBLEMS

10.1 A three-phase generator rated 440 V, 20 kVA is connected by one cable with impedance of $1 + j0.012\ \Omega$ to a motor load rated 440 V, 15 kVA, 0.9 p.f. lagging. Assume the load voltage to be set at 5% above its rated value. Perform the following:

    (i) Give the three-phase circuit if the load is Y connected.
    (ii) Give the three-phase circuit if the load Δ connected.
    (iii) Give a one-line diagram.
    (iv) Compute the generator voltage.

10.2 A three-phase generator rated 440 V, 20 kVA is connected through one cable with impedance of $1 + j0.012\ \Omega$ to a Δ-connected motor load rated 440 V, 10 kVA, 0.9 p.f. lagging. Assume the generator voltage is to be controlled at its rated voltage and its phase angle is used as the reference angle. Perform the following:

    (i) What is the number of unknown variables?
    (ii) How many equations are needed to solve for bus voltage? Give the expressions.
    (iii) Compute the load bus voltage.

10.3 The radial feeder of Figure 10.24 is connected to a local power grid rated at 11.3 kV distribution. Assume the base voltage of 10 kVA and a voltage base of 11.3 kV. Perform the following:

    (i) Compute the per unit model.
    (ii) Write the number of equations that are needed to solve for the bus load voltages.

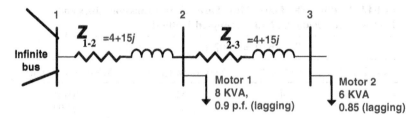

Figure 10.24   The radial feeder for Problem 10.3.

(iii) Use the Gauss–Seidel method and compute the bus voltages.

(iv) Compute the power at bus 1. Assume the power mismatch of 0.00001 per unit.

(v) Compute the total active and reactive power losses.

10.4 The radial feeder in Figure 10.24 is connected to a local power grid rated at 11.3 kV distribution. Assume a base voltage of 15 kVA and a voltage base of 11.3 kV. Perform the following:

(i) Compute the per unit model.

(ii) Write the number of equations that are needed to solve for the bus load voltages.

(iii) Compute the Y bus matrix.

(iv) Compute the matrices B' and B''.

(v) Compute the bus voltages. Assume the power mismatch of 0.00001 per unit.

(vi) Compute the power at bus 1.

(vii) Compute the total active and reactive power losses.

10.5 For the power grid in Figure 10.25, perform the following:

(i) Compute the bus admittance and bus impedance model for power flow studies.

(ii) Add a parallel line between bus 1 and bus 2 with the same impedance and compute the bus impedance model.

(iii) What is the driving point impedance of bus 1 before and after adding the line?

(iv) Remove the shunt element to ground and compute the bus admittance and bus impedance model.

10.6 Consider the power grid in Figure 10.26.
Assume the following data:

a. The transformers connected to the PV generating station are rated at 460 V Y-grounded/13.2 kV Δ and have 10% reactance and 8 MVA capacity. The transformer connected to the power grid is rated at 13.2/63 kV and has 8 MVA capacity and 10% reactance.

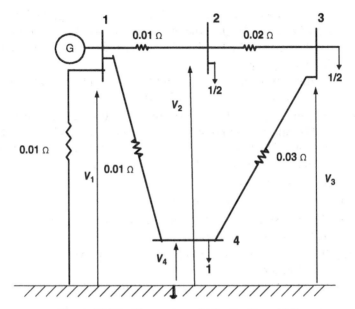

Figure 10.25 The power grid for Problem 10.5.

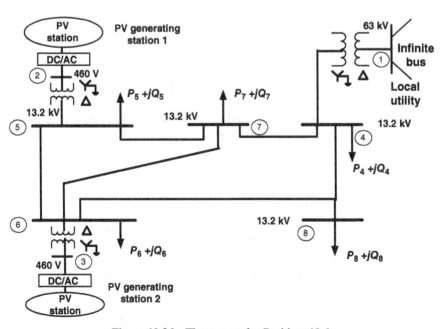

Figure 10.26 The system for Problem 10.6.

b. Assume the bus 5 load is 1.5 MW, 0.85 p.f. lagging; the bus 6 load is 1.2 MW, 0.9 p.f. lagging; the bus 7 load is 2.4 MW, 0.9 p.f. leading; the bus 4 load is 1.5 MW, 0.85 p.f. lagging; and the bus 8 load is 1.3 MW, 0.95 p.f. lagging.

c. Assume PV generating station #1 is rated at 0.95 MW and PV generating station #2 is rated at 3.5 MW.

d. Transmission line has a resistance of 0.0685 Ω/mile, reactance of 0.40 Ω/mile, and half of line-charging admittance ($Y'/2$) of 11×10^{-6} Ω^{-1}/mile. The line 4–7 is 4 miles, 4–8 is 2 miles, 5–6 is 4 miles, 5–7 is 1 mile, 6–7 is 3 miles, and 6–8 is 5 miles long.

Perform the following:

(i) Find the per unit Y_{Bus} matrix.

(ii) Write a MATLAB program and compute the load bus voltages using the FDLF method.

(iii) If line 6–7 is out of service, compute the power flow through each transformer.

(iv) If 500 kVAr is added to bus 5, compute the bus load voltages.

10.7 Consider a power grid where the system is modeled as $I_{Bus} = Y_{Bus} V_{Bus}$.

For each bus k we get $I_k = \sum_{m=1}^{n} Y_{km} V_m$, where m is the number of buses and Y_{km} is the element of Y_{Bus} matrix. For each bus k, we can also write:

$$P_k + jQ_k = V_k I_k^*$$

where P_k and Q_k are real and reactive power entering node k (* is a complex conjugate and $j = \sqrt{-1}$), respectively.

Let

$$P_k + jQ_k = V_k \sum_{m=1}^{n} Y_{km}^* V_m^*$$

$$V_k = V_k e^{j\theta}k = e_k + jf_k; \theta_k = \tan^{-1} \frac{f_k}{e_k}$$

$$Y_{km} = Y_{km} e^{j\alpha}km = G_{km} + jB_{km}; \alpha_{km} = \tan^{-1} \frac{B_{km}}{G_{km}}$$

Using the above equation in $P_k + jQ_k = V_k \sum_{m=1}^{n} Y_{km}^* V_m^*$, we obtain

$$P_k + jQ_k = V_k e^{j\theta_k} \sum_{m=1}^{n} Y_{km} e^{-j\alpha_{km}} V_m e^{-j\theta_m}$$

Then using Taylor series expansion, express the power flow problem as

$$\Delta P_k = \sum_{m=1}^{n} H_{km}\Delta\theta_m + \sum_{m=1}^{n} N_{km}\frac{\Delta V_m}{V_m}$$

$$\Delta Q_k = \sum_{m=1}^{n} J_{km}\Delta\theta_m + \sum_{m=1}^{n} L_{km}\frac{\Delta V_m}{V_m}$$

And in compact form as

$$\begin{bmatrix} \Delta P \\ \Delta Q \end{bmatrix} = \begin{bmatrix} H & N \\ J & L \end{bmatrix} \cdot \begin{bmatrix} \Delta\theta \\ \frac{\Delta V}{V} \end{bmatrix}$$

$$\begin{bmatrix} \Delta P \\ \Delta Q \end{bmatrix} = \begin{bmatrix} \frac{\partial P}{\partial\theta} & \frac{\partial P}{\partial V}V \\ \frac{\partial Q}{\partial\theta} & \frac{\partial Q}{\partial V}V \end{bmatrix} \cdot \begin{bmatrix} \Delta\theta \\ \frac{\Delta V}{V} \end{bmatrix}$$

Show that for $m = k$, we have

$$H_{kk} = \frac{\partial P_k}{\partial\theta_k} = -Q_k - V_k^2 B_{kk}$$

$$J_{kk} = \frac{\partial Q_k}{\partial\theta_k} = P_k - V_k^2 G_{kk}$$

10.8 For the system given in Problem 10.7 show that for $m = k$, we have the following expressions

$$N_{kk} = \frac{\partial P_k}{\partial V_k}V_k = P_k + V_k^2 G_{kk}$$

$$L_{kk} = \frac{\partial Q_k}{\partial V_k}V_k = Q_k - V_k^2 B_{kk}$$

10.9 Consider Problem 10.7, first let

$$I_m = a_m + jb_m = Y_{km}V_m$$

and $Y_{km}V_m = (G_{km} + jB_{km})(e_m + jf_m)$

$$Y_{km}V_m = (G_{km}e_m - B_{km}f_m) + j(B_{km}e_m + G_{km}f_m)$$

then using the following:

$$a_m = G_{km} \cdot e_m - B_{km} f_m$$
$$b_m = B_{km} \cdot e_m + G_{km} f_m$$

derive the following expressions:

$$H_{km} = \frac{\partial P_k}{\partial \theta_m} = a_m f_k - b_m e_k$$

$$J_{km} = \frac{\partial Q_k}{\partial \theta_m} = -a_m e_k - b_m f_k$$

10.10 For Problem 10.7, show that

For $m \neq k$, we have the following off-diagonal elements for the Jacobian matrix:

$$N_{km} = \frac{\partial P_k}{\partial V_m} V_m = a_m e_k + b_m f_k$$

$$L_{km} = \frac{\partial Q_k}{\partial V_m} V_m = a_m f_k - b_m e_k$$

10.11 For Problem 10.7, show that the off-diagonal elements of the Jacobian matrix for $m \neq k$ are:

$$\left.\begin{array}{l} H_{km} = L_{km} = a_m f_k - b_m e_k \\ N_{km} = -j_{km} = a_m e_k + b_m f_k \end{array}\right\} \text{Off-diagonal elements of Jacobian matrix}$$

And the diagonal elements of the Jacobian matrix for $m = k$ are:

$$\left.\begin{array}{l} H_{kk} = -Q_k - B_{kk} V_k^2 \\ L_{kk} = Q_k - B_{kk} V_k^2 \\ N_{kk} = P_k + G_{kk} V_k^2 \\ J_{kk} = P_k - G_{kk} V_k^2 \end{array}\right\} \text{Diagonal elements of Jacobian matrix}$$

10.12 For Problem 10.7, compute $P_{k(\text{Calculated})}$ and $Q_{k(\text{Calculated})}$ for each bus k as expressed below.

$$P_k = \sum_{m=1}^{n} [e_k(e_m G_{km} - f_m B_{km}) + f_k(e_m B_{km} + f_m G_{km})]$$

$$Q_k = \sum_{m=1}^{n} [f_k(e_k G_{km} - f_m B_{km}) - e_k(e_m B_{km} + f_m G_{km})]$$

where $V_k = e_k + jf_k$; $Y_{km}^* = G_{km} - jB_{km}$; $V_m^* = e_m - jf_m$

10.13 Assume that the power balance equation for a power system network can be written as

$$S = P + jQ = [V]^T [I^*] = [V]^T [YV]^*$$

where S is the complex power injection vector, P is the real power injection vector, Q is the reactive power injection vector, I is the current injection vector, V is the bus voltage vector, and $Y_{kj} = G_{kj} + jB_{kj}$ is the system admittance matrix.

Assume that in polar coordinate system, the complex voltage can be written as

$$V_k = V_k(\cos \theta_k + j \sin \theta_k)$$

Show that the calculated real and reactive powers can be expressed as

$$P_{k(\text{Calculated})} = V_k \sum_{j=1}^{n} V_j(G_{kj} \cos \theta_{kj} + B_{kj} \sin \theta_{ij})$$

$$Q_{k(\text{Calculated})} = V_k \sum_{j=1}^{n} V_j(G_{kj} \sin \theta_{kj} - B_{kj} \cos \theta_{ij})$$

where $\theta_{kj} \cong \theta_k - \theta_j$.

And in the Cartesian coordinate system the calculated real and reactive powers can be expressed as

$$P_{k(\text{Calculated})} = e_k \sum_{j=1}^{n} (G_{kj}e_j - B_{kj}f_j) + f_k \sum_{j=1}^{n} (G_{kj}f_j + B_{ik}e_j)$$

$$Q_{k(\text{Calculated})} = f_k \sum_{j=1}^{n} (G_{kj}e_j - B_{kj}f_j) - e_k \sum_{j=1}^{n} (G_{kj}f_j + B_{kj}e_j)$$

where the bus voltage is given by

$$V_k = e_k + jf_k$$

10.14 Consider the feeder given in Figure 10.27.

System data: $V_1 = 1\angle 0$; $Z_{25} = 5 + j10\ \Omega$, $Z_{34} = 2 + j8\ \Omega$; $Z_{23} = 5.41 + j3.34\ \Omega$.

$S_2 = 3$ MVA, p.f. = 0.75 lagging; $S_3 = 3$ MVA, p.f. = 0.8 lagging; $S_4 = 4$ MVA, p.f. = 0.9 lagging; $S_5 = 2$ MVA, p.f. = 0.9 lagging.

$T_1 = 63/20$ kV, 10% reactance, 20 MVA; T_2 is the same as T_1.

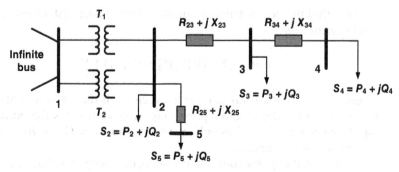

Figure 10.27   The feeder of Problem 10.14.

Perform the following:

(i) Calculate the per-unit equivalent circuit model. Use S_b equal to the MVA rating of transformer T_1.

(ii) Let $V_1 = V_1 \angle 0$, $V_K = V_{RK} + j V_{IK}$, $K = 2, 3, 4, 5$. Develop an equation in a Cartesian coordinate for $V_{RK} = f (V_{RK}, V_{IK})$, ΔP_K and ΔQ_K, $K = 2, 3, 4, 5$ for each bus. Use an iterative Gauss–Seidel approximation technique and compute bus voltages and system losses. Use five iterations and assume initial voltages are $V_K = 1 \angle 0$, $K = 1, 2, 3, 4, 5$. Put your results in a table.

(iii) Compute the Y_{Bus} B' and B'' matrices and use the FDLF technique to compute bus voltages. Make a table and compare your results with (ii) above. Use five iterations.

(iv) Use the FDLF technique and compute bus voltages after correcting the power factor of each bus to unity. Make a table and compare your results with (iii) above.

10.15 Consider the five-bus system given in Figure 10.28.
Table 10.11 presents the load and generation schedules.
Perform the following:

(i) Calculate $Y_{Bus} = Y_r + jY_i$ (internal Y bus), that is, compute the Y_r and Y_i matrices.

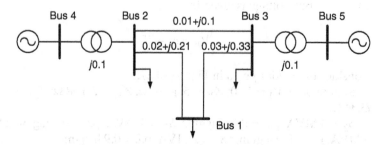

Figure 10.28   The system for Problem 10.15.

TABLE 10.11   Bus Data for a Five-Bus System

Bus #	Bus Type	V_N (kV)	V (p.u)	θ (rad)	P_G (MW)	P_L (MW)	Q_L (MVAr)
1	PQ	138	–	–	–	160	80
2	PQ	138	–	–	–	200	100
3	PQ	138	–	–	–	370	130
4	PV	1	1.05	–	500	–	–
5	Swing	4	1.05	0.0	–	–	–

(ii) Compute the Jacobian matrix $[H, N, J, L]$.

(iii) Compute ΔP and ΔQ.

(iv) Solve the Newton–Raphson power flow for the five-bus system of Figure 10.28.

10.16 Consider the power grid given in Figure 10.29.

  a. Bus 1 is swing bus $V_1 = 1\angle 0$.

  b. Bus 2 is $P - V$ bus çV_2ç = 1.05 and $P_{2(\text{Scheduled})} = 0.9$ p.u.

  c. Bus 2 Q_2 is assumed to be $-1 \le Q_2 \le 2$.

  Compute the bus voltages using the decoupled Newton–Raphson method.

10.17 Consider the power grid given in Figure 10.30.

  Consider bus 1 as a swing bus with $1.05\angle 0$ and bus 2 as $P - V$ with $P = 0.20$ and çV_2ç= 0.96.

  Compute the following:

  (i) The B', B'', and Y_{Bus} matrices.

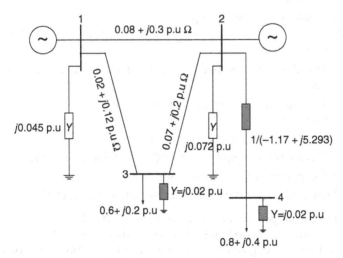

Figure 10.29   The system for Problem 7.16.

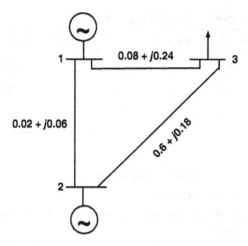

Figure 10.30 The power grid for Problem 10.17. (All units in p.u.)

(ii) One iteration of an FDLF. Assume the following starting voltages: $V_1 = 1.0\angle0$, $V_2 = 0.9\angle2°$, $V_3 = 0.9\angle{-1.2°}$. System losses are based on the computed voltages after the first iteration.

10.18 Consider the power system given in Figure 10.30. Assume that bus 1 has a gas turbine generator. It has a load of 1 p.u; the bus voltage is fixed at 1 p.u and is the swing bus. Bus 2 has a number of PV generators with total injected power of 1.5 p.u into the bus. The transmission data are given in per unit as specified. Bus 3 has a number of loads with a total connected load of 2.0 p.u. Perform the following by writing a MATLAB simulation testbed. The maximum mismatch tolerance is 0.001. Perform the following:

(i) Compute the system Y_{Bus}.

(ii) Use the Gauss–Seidel Y_{Bus} method and compute the bus 2 and 3 voltages.

(iii) Use the Gauss–Seidel Z_{Bus} method and compute the bus 2 and 3 voltages.

(iv) Use the Newton–Raphson method and compute the bus 2 and 3 voltages.

(v) Make a table and compare the above methods.

(vi) Determine the power provided by a swing generator.

(vii) Determine the total power losses.

Figure 10.31 presents one-line diagram of Problem 10.18.

10.19 Consider the power system given in Figure 10.32.

Assume $V_1 = 1\angle0°$p.u (swing bus). Also, assume transmission line impedances are given in per unit on 440 V, 100 MVA base ($S_b = $ 100MVA for the entire system)

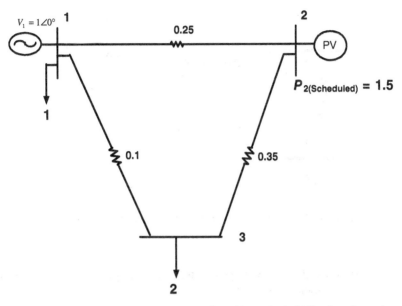

Figure 10.31   The one-line diagram of Problem 10.18. (All values in p.u.)

Assume the generation and load schedules are as follows:
a. Bus 1: Load #1: 4 MVA, p.f. = 0.85 lagging
b. Bus 2: G2: 2 MW, p.f. = 0.95 lagging
    Load #2: 4 MVA, p.f. = 0.90 leading
c. Bus 3: G3: 1 MW, p.f. = 0.95 leading
    Load #3: 2 MVA, p.f. = 0.90 leading

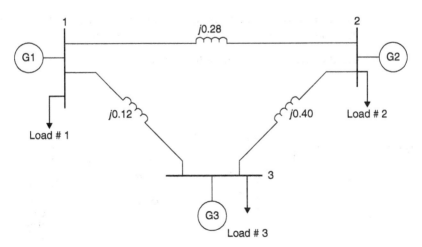

Figure 10.32   The power grid of Problem 10.19.

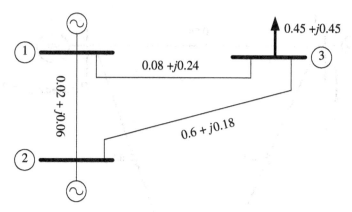

Figure 10.33   The system for Problem 10.20. (All values in p.u.)

Compute the bus voltages using the decoupled Newton–Raphson and Gauss–Seidel methods.

10.20   Consider the system in Figure 10.33.
Assume that generator's internal reactance in p.u is 0.8. Assuming the load voltage is 1 p.u, compute the following:
(i) The Y_{Bus} model for short-circuit studies.
(ii) The Z_{Bus} model for short-circuit studies

10.21   For the radial power system given in Figure 10.34, compute the following:

(i) For the load of 500 W, which bus should be considered as an infinite bus?
(ii) Assume the load voltage is to be maintained at its rated value. Which bus should be considered as an infinite bus?

10.22   Consider the microgrid power system given in Figure 10.35.

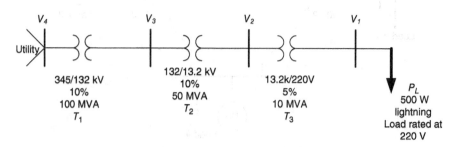

Figure 10.34   The system of Problem 10.21.

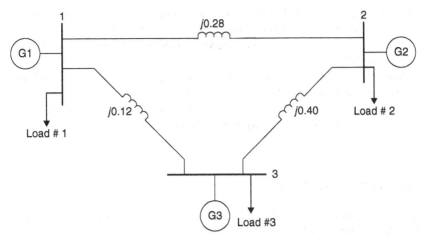

Figure 10.35   The system of Problem 10.22. (All values are in p.u.)

The transmission line impedances are given in per unit on 100 MVA base (S_b = 100 MVA for the entire system). The generation and load schedules are as follows:

a. Bus 1: Load #1: 4 MVA, p.f. = 0.85 lagging
b. Bus 2: G2: 2 MW, p.f. = 0.95 lagging, $X = 10\%$
   Load #2: 4 MVA, p.f. = 0.90 leading
c. Bus 3: G3: 1 MW, p.f. = 0.95 leading, $X = 25\%$
   Load #3: 2 MVA, p.f. = 0.90 leading

(i) Compute bus load voltages using the FDLF method.
(ii) Compute the load impedance of each load.

REFERENCES

1. Gross, A.C. (1986) *Power System Analysis*, Wiley, Hoboken, NJ.
2. Grainger, J. and Stevenson, W.D. (2008) *Power Systems Analysis*, McGraw-Hill, New York.
3. Dommel, H. and Tinny, W. (1968) Optimal power flow solution. *IEEE Transactions on Power Apparatus and Systems*, PAS-87(10), 1866–1876.
4. Duncan Glover, J. and Sarma, M.S. (2002) *Power System Analysis and Design*, Brooks/Cole Thomson Learning, Pacific Grove, CA.
5. Weedy, R.M. (1970) *Electric Power Systems*, Wiley, Hoboken, NJ.
6. Heydt, G.T. (1986) *Computer Analysis Methods for Power Systems*, Macmillan, New York.
7. Stagg, G. and El-Abiad, A. (1968) *Computer Methods in Power System Analysis*, McGraw-Hill, New York.

8. Tinney, W.F. and Hart, C.E. (1967) Power flow solution by Newton's method. *IEEE Transactions on Power Apparatus and Systems*, PAS-86, 1449–1456.

9. Stott, B. (1974) Review of load-flow calculation methods. *Proceedings of the IEEE*, 62(7), 916–929.

10. Chabert, J-C. and Barbin, E. (1999) *A History of Algorithms: From the Pebble to the Microchip*, Springer, New York/Heidelberg/Berlin.

11. Khayyám, O. (1997) *Rubáiyát of Omar Khayyám: A Critical Edition* (FitzGerald, E., Trans., Decker,C., ed.), University of Virginia Press, Charlottesville, VA.

ADDITIONAL RESOURCES

Anderson, P.M. and Fouad, A.A. (1977) *Power System Control and Stability*, 1st edn, Iowa State University Press, Ames, IA.

Bergen, A. and Vittal, V. (2000) *Power Systems Analysis*, Prentice Hall, Englewood Cliffs, NJ.

Bohn, R., Caramanis, M., and Schweppe, F. (1984) Optimal pricing in electrical networks over space and time. *Rand Journal on Economics*, 18(3), 360–376.

Elgerd, O.I. (1982) *Electric Energy System Theory: An Introduction*. 2d edn, McGraw-Hill, New York.

El-Hawary, M.E. (1983) *Electric Power Systems: Design and Analysis*, Reston Publishing, Reston, VA.

Energy Information Administration, Official Energy Statistics from the US Government. Available at http://www.eia.doe.gov. Accessed October 7, 2010.

IEEE Brown Book. (1980) *IEEE Recommended Practice for Power System Analysis*, Wiley-Interscience, New York.

Institute for Electric Energy. Available at http://www.edisonfoundation.net/IEE. Accessed October 7, 2010.

Institute of Electrical and Electronics Engineers. (1999) *1451.1-1999—IEEE Standard for a Smart Transducer Interface for Sensors and Actuators–Network Capable Application Processor (NCAP) Information Model*. IEEE, Piscataway, NJ.

Masters, G.M. (2004) *Renewable and efficient electric power systems*, Wiley, New York.

Midwest ISO. Available at http://toinfo.oasis.mapp.org/oasisinfo/MMTA_Transition_Plan_V2_3.pdf. Accessed January 12, 2010.

North American Electric Reliability Corp. (NERC) (2008). Long-Term Reliability Assessment 2008–2017. Available at http://www.nerc.com/files/LTRA2008.pdf. Accessed October 8, 2010.

Nourai, A. and Schafer, C. (2009) Changing the electricity game. *IEEE Power and Energy Magazine*, 7(4), 42–47.

Phadke, A.G. (1993) Computer applications in power. *IEEE Power and Energy Society*, 1(2), 10–15.

Sauer, P. and Pai, M.A. (1998) *Power Systems Dynamics and Stability*, Prentice Hall, Englewood Cliffs, NJ.

Schweppe, F.C. and Wildes. J. (1970) Power system static-state estimation, part I: exact model. *IEEE Transactions on Power Apparatus and Systems*, PAS-89(1), 120–125.

Shahidehpour, M. and Yamin, H. (2002) *Market Operations in Electric Power Systems: Forecasting, Scheduling, and Risk Management*, Wiley/IEEE, New York/Piscataway, NJ.

Wood, A.J. and Wollenberg, B.F. (1996) *Power Generation, Operation, and Control*, Wiley, New York.

CHAPTER 11

POWER GRID AND MICROGRID FAULT STUDIES

11.1 INTRODUCTION

A fault in a power grid is any condition that results in abnormal operation. When energized parts of the system are accidently connected to the ground, two phase conductors are connected together, or a conductor is broken, the result is a faulted power grid. As an example, when a transmission line is accidentally grounded due to weather conditions, such as lightening from an electrical storm, the result is a flashover of the insulation and a flow of high fault currents.

When a fault or short circuit occurs in a power grid, all synchronous generators contribute current directly to that fault until protective equipment acts to isolate the fault as quickly as possible. If the fault current is not isolated, the protective system of the power grid will trip (switch-off) the generators, and as a result, the balance between system loads and power generation is lost and the power grid is unstable. Most blackouts are the consequence of an unstable power grid. The power grid must be designed to operate successfully for the isolation of faults at the highest levels of current that can be anticipated for power grid operation. If the fault current exceeds the ability of breakers to extinguish the high fault current and to protect the grid, the result could be a catastrophic failure, fire, and permanent damage to significant portions of the power grid infrastructure. Therefore, before microgrids of distributed generation are connected to a local power grid, the fault current contribution must be calculated and mitigating measures must be taken prior to connection.

In fault studies of power grids, it is assumed that the power grids remain balanced except for the faulted point. Therefore, when a fault occurs, the power grid *must remain* balanced. As soon as a fault occurs, the faulted part of the

Design of Smart Power Grid Renewable Energy Systems, Second Edition. Ali Keyhani.
© 2017 John Wiley & Sons, Inc. Published 2017 by John Wiley & Sons, Inc.
Companion website: www.wiley.com/go/smartpowergrid2e

power grid must be quickly isolated and removed from service. Therefore, in fault studies of power grids, we are working with an "if–then condition": if a point in a power grid is faulted, then we want to calculate the fault current and protect the power grid's equipment by isolating the faulted part of the system.[1–7]

The study of a faulted power grid is the topic of this chapter. In fault studies, we learn how a balanced three-phase network can be modeled as positive, negative, and zero sequence networks.

Most faults are single line-to-ground or double lines that are faulted and then grounded. For any unbalanced fault current calculation that involves a ground, we use positive, negative, and zero sequence networks. Balanced three-phase faults are also used to size the circuit breakers. For balanced faults, we use the positive sequence network.

Solved examples and homework problems for symmetrical components; positive, negative, and zero sequence networks; and modeling balanced three-phase faults: single line, double line-to-ground faults, and double line-to-line faults are given at the end of the chapter.

11.2 POWER GRID FAULT CURRENT CALCULATION

Figure 11.1 depicts a power grid and its circuit breakers. The elements of a power grid include generators, transformers, and transmission lines—all of

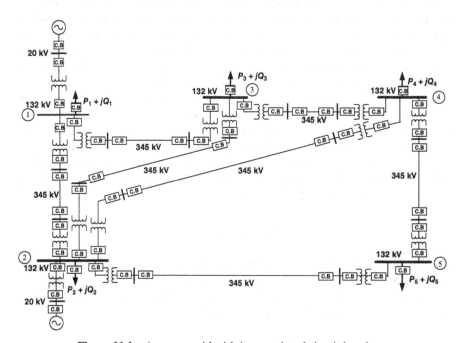

Figure 11.1 A power grid with its associated circuit breakers.

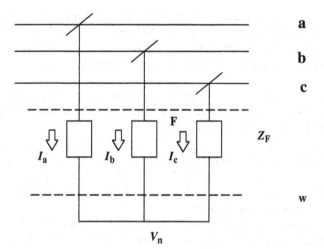

Figure 11.2   A balanced three-phase fault.

which must be protected so that if a fault occurs the fault currents can be isolated by the circuit breakers. For example, if a fault occurs because of a storm, a line between bus 3 and bus 4 is faulted, both circuit breakers will open by a control action issued from a ground relay fault current detection system. The main objective of the short-circuit study is to determine the power interruption capability of a circuit breaker at each switching location. To determine the required interrupting capability of breakers, a three-phase fault is assumed at each bus to calculate the maximum fault current. All breakers connected to the bus must be able to interrupt the fault current. For a protective relay system to issue control commands, the voltages and currents for balanced and unbalanced faults at many locations in the power grid must be determined.

To compute the short-circuit current flow through a power grid due to three-phase balanced and unbalanced faults, the power grid system must be modeled to reflect the intended study. The types of faults are a three-phase balanced fault and an unbalanced fault. Figure 11.2 depicts a balanced fault at a bus of a power grid where the three phases a, b, and c are shown. We know that in a balanced power grid, the sum of phase a, phase b, and phase c currents adds up to zero. Therefore, the neutral current is zero.

$$I_n = I_a + I_b + I_c$$

where I_n is the neutral current. If the system is unbalanced, the neutral current will flow through the neutral conductor. However, if a balanced fault occurs in a balanced power grid, the neutral point where three phases are connected is at zero potential and neutral current will not flow.

$$I_n = 0$$

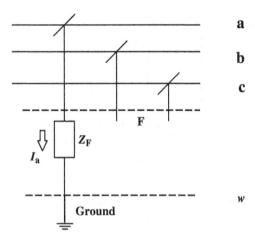

Figure 11.3   An unbalanced fault: one line to ground.

The dashed lines marked by "w" in Figure 11.2 delineate the section of a three-phase system that has a fault.

Figure 11.3 depicts a single line-to-ground fault. For an if–then study of a single line-to-ground fault, the phase designation is arbitrary. In single line-to-ground fault studies, it is customary to designate the faulted phase at phase a, with the two other phases operating as normal. The dashed lines marked by "w" depicted in Figure 11.3 delineate the faulted phase. Because phase a is faulted, the ground current flow is equal to the fault current of phase a.

Figure 11.4 depicts a double line-to-ground fault. For an if–then study of a double line-to-ground fault, again, the phase designation is arbitrary. In double

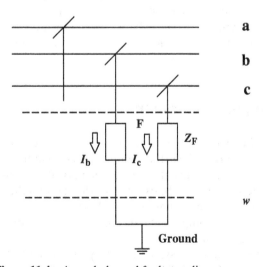

Figure 11.4   An unbalanced fault: two lines to ground.

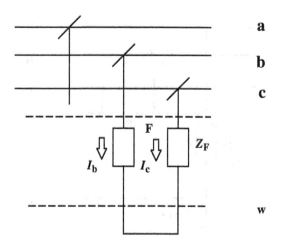

Figure 11.5   An unbalanced fault: line to line.

line-to-ground fault studies, it is customary to designate the faulted phases as phases b and c, with phase a operating as normal. The dashed lines marked by "w" in Figure 11.4 delineate the section and the phases with faults in a three-phase system. Because phases b and c are faulted to ground, the ground current flow is equal to the sum of the fault currents of phases b and c.

Figure 11.5 depicts faulted double lines. For an if–then study of double lines, again, the phase designation is arbitrary. In double-line fault studies, it is customary to designate the faulted phases on phases b and c, with phase a operating as normal. The dashed lines marked by "w" in Figure 11.5 delineate the section and the phases with faults in a three-phase system. Because phase a and the faulted phases are not connected to ground, the ground current will not flow. The fault current is equal to phase b current. The phase c current is the negative of phase b. To calculate fault currents, we must model the system using symmetrical components. We will study this topic in Section 11.3.

11.3 SYMMETRICAL COMPONENTS

In 1918, Charles Legeyt Fortescue[8] described how a set of three unbalanced phasors could be expressed as the sum of three symmetrical sets of balanced phasors. These three sets are defined as a positive sequence, a negative sequence, and a zero sequence. In essence, Fortescue's method converts three unbalanced phases into three independent balanced sets. Although in practice, power grids are not completely balanced, a balanced approximation is an acceptable engineering solution.

In a balanced system, it is assumed that every element of the power grid is balanced. Generators are designed to generate a balanced set of voltages:

all phase voltages have the same magnitude and are 120 degrees apart. Similarly, the three-phase transmission lines are balanced: each line has the same impedance and admittance. However, in practice, the spacing of the lines may not be the same; hence, unbalanced inductances and therefore line impedances will result. However, the lines are transposed as they are hung on transmission line towers or poles, that is, the positions of phases a, b, and c are changed every few miles. This results in approximately balanced spacing and balanced inductance per mile. Three-phase transformers and loads are also balanced. The transposition of three-phase transmission lines was discussed in Chapter 2.

The loads in secondary distribution systems are not balanced. However, as a first approximation, we can assume the loads are also balanced because power grids are designed as balanced systems. Therefore, as we discussed in the previous chapter, the balanced three-phase system can be analyzed by analyzing phase a with respect to ground. However, if a line in the system has been damaged during a storm and falls on the ground, the system will become unbalanced. The unbalanced systems are analyzed using the symmetrical components of the power grid. For example, on line-to-ground faults, we compute the ground current and then set the protection system to isolate the faults by opening the circuit breakers on both ends of the line. However, we need to isolate the faulted line in a very short time—within a few cycles—to prevent damage to any of the power grid equipment. This requirement dictates that the system must remain balanced during the faults; only the unbalanced part of the power grid is the faulted segment. This is a key concept in the calculation of fault current. First, we need to develop an impedance model of a balanced system and model the different fault types. Once that is done, we will better understand how to construct a model of faulted systems.

The basic concepts of symmetrical components can be introduced by presenting the three-phase systems in terms of positive, negative, and zero sequences.[1-7] Normally, the designation of "1" or "+" is used to present the positive sequence variables of voltage, current, and impedance. The balanced three-phase voltages can be expressed as shown in Figure 11.6.

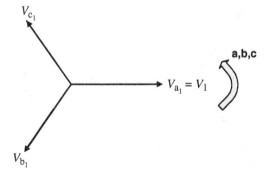

Figure 11.6 Three-phase voltage presented in terms of positive sequence quantities.

Figure 11.6 depicts a positive sequence voltage with phase a as the reference followed by phase b and then phase c. The phase voltages a, b, and c are given as follows:

$$V_a = V_{a_1} = V \angle 0° = V_1 \angle 0° \tag{11.1}$$
$$V_b = V_{b_1} = V_1 \angle 240° \tag{11.2}$$
$$V_c = V_{c_1} = V_1 \angle 120° \tag{11.3}$$

Let $a = 1 \angle 120°$ and $a^2 = 1 \angle 240°$, then we have

$$V_{a_1} = V_1 \quad V_{b_1} = a^2 V_1 \quad V_{c_1} = a V_1 \tag{11.4}$$

Rearranging the above in matrix form, we have

$$\begin{bmatrix} V_{a_1} \\ V_{b_1} \\ V_{c_1} \end{bmatrix} = \begin{bmatrix} 1 \\ a^2 \\ a \end{bmatrix} V_1 \tag{11.5}$$

The negative sequence voltage is presented by designating "2" or a negative sign ("−") as a subscript or superscript. The negative sequence is depicted in Figure 11.7.

From Figure 11.7, we can write the following:

$$V_{a_2} = V_2 \tag{11.6}$$
$$V_{b_2} = a V_2 \tag{11.7}$$
$$V_{c_2} = a^2 V_2 \tag{11.8}$$

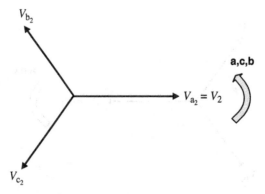

Figure 11.7   Three-phase voltage presented in terms of negative sequence quantities.

When the above is expressed in matrix form, we have

$$\begin{bmatrix} V_{a_2} \\ V_{b_2} \\ V_{c_2} \end{bmatrix} = \begin{bmatrix} 1 \\ a \\ a^2 \end{bmatrix} V_2 \qquad (11.9)$$

The zero sequence voltages are presented by a set of voltages that are in phase. The zero sequence voltages are designated by "0." The zero sequence voltages are

$$V_{a_0} = V_0 \qquad (11.10)$$
$$V_{b_0} = V_0 \qquad (11.11)$$
$$V_{c_0} = V_0 \qquad (11.12)$$

Using the above presentation, a set of three-phase voltages can be expressed in terms of its sequence voltages. In general, a set of unbalanced voltages, V_a, V_b, V_c, can be written as

$$V_a = V_{a_0} + V_{a_1} + V_{a_2} \Rightarrow V_a = V_0 + V_1 + V_2 \qquad (11.13)$$
$$V_b = V_{b_0} + V_{b_1} + V_{b_2} \Rightarrow V_b = V_0 + a^2 V_1 + a V_2 \qquad (11.14)$$
$$V_c = V_{c_0} + V_{c_1} + V_{c_2} \Rightarrow V_c = V_0 + a V_1 + a^2 V_2 \qquad (11.15)$$

The above equations can be expressed in compact matrix form as

$$\begin{bmatrix} V_a \\ V_b \\ V_c \end{bmatrix} = \begin{bmatrix} 1 & 1 & 1 \\ 1 & a^2 & a \\ 1 & a & a^2 \end{bmatrix} \begin{bmatrix} V_0 \\ V_1 \\ V_2 \end{bmatrix} \qquad (11.16)$$

We can designate the matrix transformation by T and present as

$$T = \begin{bmatrix} 1 & 1 & 1 \\ 1 & a^2 & a \\ 1 & a & a^2 \end{bmatrix} \qquad (11.17)$$

Therefore, the three-phase voltages can be expressed in terms of its sequence variables as given below:

$$[V_{abc}] = [T][V_{012}] \qquad (11.18)$$

We can multiply the above by inverse of matrix T.

$$[T]^{-1}[V_{abc}] = [T]^{-1}[T][V_{012}]$$

$$[V_{012}] = [T]^{-1}[V_{abc}]$$

$$\text{(11.19)}$$

$$[T]^{-1} = \frac{1}{3}\begin{bmatrix} 1 & 1 & 1 \\ 1 & a & a^2 \\ 1 & a^2 & a \end{bmatrix}$$

Therefore, the sequence voltages are

$$V_0 = \frac{1}{3}(V_a + V_b + V_c)$$
$$V_1 = \frac{1}{3}(V_a + aV_b + a^2V_c)$$
$$V_2 = \frac{1}{3}(V_a + a^2V_b + aV_c)$$

$$\text{(11.20)}$$

Similarly, we can compute the symmetrical components of current in terms of three-phase currents:

$$[I_{abc}] = [T][I_{012}]$$

$$[I_{012}] = [T]^{-1}[I_{abc}]$$

$$\text{(11.21)}$$

Example 11.1 Consider a balanced, Y-connected, 460-V generator. Compute the positive, negative, and zero sequence voltages.

Solution:
Let us assume that phase a is selected as the reference phase. The phase a, b, and c voltages are

$$V_{a_n} = \frac{460}{\sqrt{3}}\angle 0° = 265.9\angle 0°$$

$$V_{b_n} = 265.9\angle 240° = 265.9\,a^2$$

$$V_{c_n} = 265.9\angle 120° = 265.9\,a$$

The sequence voltages can be computed as

$$\begin{bmatrix} V_0 \\ V_1 \\ V_2 \end{bmatrix} = \frac{1}{3}\begin{bmatrix} 1 & 1 & 1 \\ 1 & a & a^2 \\ 1 & a^2 & a \end{bmatrix}\begin{bmatrix} V_{a_n} \\ V_{b_n} \\ V_{c_n} \end{bmatrix}$$

$$\text{(11.22)}$$

$$V_0 = \frac{1}{3}(265.9 + 265.9\,a^2 + 265.9\,a)$$

$$= \frac{265.9}{3}(1 + a + a^2) = 0 \Rightarrow V_0 = 0$$

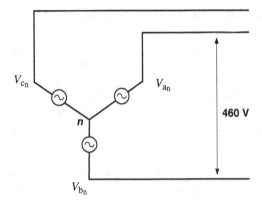

Figure 11.8   The three-phase generator of Example 11.1.

Since $(1 + a + a^2) = 0$

$$V_1 = \tfrac{1}{3}(265.9 + 265.9a^3 + 265.9a^3)$$

Since $a^3 = 1\angle0°, V_1 = 265.9\angle0°$.
Since $a^4 = a = 1\angle120°$, for the negative sequence voltages, we have

$$V_2 = \tfrac{1}{3}(265.9 + 265.9\,a^4 + 265.9\,a^2)$$
$$V_2 = \tfrac{1}{3}[265.9(1 + a + a^2)] = 0$$

Figure 11.8 depicts the three-phase generator voltage of Example 11.1.
We can conclude that for three-phase balanced voltages, the positive voltage sequence is the only source of power to the grid.
The transformation of the three-phase system to a symmetrical component can be used to show the relationship between the two systems. The three-phase power expressed in terms of phase a, phase b, and phase c can be expressed as

$$S_{3\varphi} = \left[V_a I_a^* + V_b I_b^* + V_c I_c^* \right] \tag{11.23}$$

In matrix notation, we have

$$S_{3\varphi} = [V_{abc}]^T [I_{abc}]^*$$

We can use the symmetrical transformation for both voltage and current and obtain the power in a symmetrical system.

$$S_{3\varphi} = [V_{012}]^T [T]^T [T]^* [I_{012}]^*$$
$$S_{3\varphi} = 3 \left[V_0 I_0^* + V_1 I_1^* + V_2 I_2^* \right] = 3 [S_{012}] \tag{11.24}$$

11.4 SEQUENCE NETWORKS FOR POWER GENERATORS

Figure 11.9 depicts an impedance model of a synchronous generator. The impedance Z_n is the grounding impedance. Its function is to limit the ground current fault if a ground fault occurs in the generator. The model depicts the steady-state operation of the generator. In this model, the shaft speed, ω_m, and the field current, I_f, are constant. The generator supplies balanced three-phase voltages. Using the equivalent model depicted in Figure 11.9, we can write the following:

$$E_a = (R_a + jX_s + Z_n)I_a + (jX_m + Z_n)I_b + (jX_m + Z_n)I_c + V_a$$
$$E_b = (R_a + jX_s + Z_n)I_b + (jX_m + Z_n)I_a + (jX_m + Z_n)I_c + V_b$$
$$E_c = (R_a + jX_s + Z_n)I_c + (jX_m + Z_n)I_a + (jX_m + Z_n)I_b + V_c \qquad (11.25)$$

Let us assume that the generator is supplying balanced three-phase voltages as depicted in Figure 11.10. The supply voltage of each phase can be expressed eas

$$E_a = E$$
$$E_b = a^2 E$$
$$E_c = aE \qquad (11.26)$$

We can rewrite the Z_s and Z_m as

$$Z_s = R_a + jX_s + Z_n$$
$$Z_m = jX_m + Z_n$$

Then the set of equations given by Equation (11.25) can be written as

$$\begin{bmatrix} E_a \\ E_b \\ E_c \end{bmatrix} = \begin{bmatrix} Z_s & Z_m & Z_m \\ Z_m & Z_s & Z_m \\ Z_m & Z_m & Z_s \end{bmatrix} \begin{bmatrix} I_a \\ I_b \\ I_c \end{bmatrix} + \begin{bmatrix} V_a \\ V_b \\ V_c \end{bmatrix} \qquad (11.27)$$

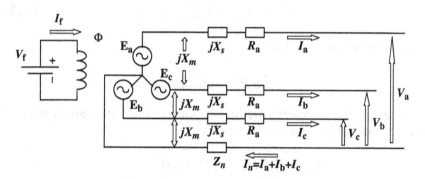

Figure 11.9   A synchronous generator reactance model.

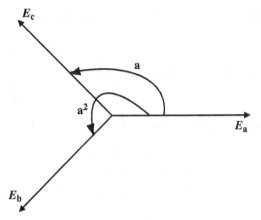

Figure 11.10   A balanced three-phase generator.

The set of equations presented by Equation (11.27) can be expressed in a matrix form

$$[E_{abc}] = [Z_{abc}][I_{abc}] + [V_{abc}] \tag{11.28}$$

Substituting $[I_{abc}] = [T_s][I_{012}]$ in Equation (11.28) and then pre-multiplying by $[T_s]^{-1}$, we have

$$[T_s]^{-1}[E_{abc}] = [T_s]^{-1}[Z_{abc}][T_s][I_{012}] + [T_s]^{-1}[V_{abc}] \tag{11.29}$$

Recall the transformation from *abc* to 012 as given below.

$$[T_s]^{-1} \cdot [E_{abc}] = [T_s]^{-1} \cdot [T_s] \cdot [E_{012}]$$
$$[E_{012}] = [T_s]^{-1} \cdot [E_{abc}]$$
$$[T_s]^{-1} = \frac{1}{3}\begin{bmatrix} 1 & 1 & 1 \\ 1 & a & a^2 \\ 1 & a^2 & a \end{bmatrix}$$

Because the generator is supplying balanced three-phase voltages, the right-hand side of Equation (11.29) can be written as the generator sequence voltages as given by Equation (11.30).

$$\begin{bmatrix} E_0 \\ E_1 \\ E_2 \end{bmatrix} = \frac{1}{3}\begin{bmatrix} 1 & 1 & 1 \\ 1 & a & a^2 \\ 1 & a^2 & a \end{bmatrix}\begin{bmatrix} E \\ a^2 E \\ aE \end{bmatrix} = \frac{E}{3}\begin{bmatrix} 0 \\ 3 \\ 0 \end{bmatrix} = \begin{bmatrix} 0 \\ E \\ 0 \end{bmatrix} \tag{11.30}$$

We can rewrite the expression $[T]^{-1}[Z_{abc}][T] = [Z_{012}]$ as expressed by Equation (11.31).

$$[T]^{-1}[Z_{abc}][T] = [Z_{012}] = \begin{bmatrix} Z_0 & 0 & 0 \\ 0 & Z_1 & 0 \\ 0 & 0 & Z_2 \end{bmatrix} \tag{11.31}$$

where $Z_0 = Z_{0,\text{Gen}} + 3Z_n$ and $Z_{0,\text{Gen}} = R_a + j(X_s + 2X_m)$ and

$$\begin{aligned} Z_0 &= R_a + j(X_s + 2X_m) + 3Z_n \\ Z_1 &= Z_s - Z_m = R_a + j(X_s - X_m) \\ Z_2 &= Z_s - Z_m = R_a + j(X_s - X_m) \end{aligned} \tag{11.32}$$

Now Equation (11.29) can be written as

$$\begin{bmatrix} 0 \\ E \\ 0 \end{bmatrix} = \begin{bmatrix} Z_0 & 0 & 0 \\ 0 & Z_1 & 0 \\ 0 & 0 & Z_2 \end{bmatrix} \begin{bmatrix} I_0 \\ I_1 \\ I_2 \end{bmatrix} + \begin{bmatrix} V_0 \\ V_1 \\ V_2 \end{bmatrix} \tag{11.33}$$

The above system of equations will result in the following sequence of zero, positive, and negative networks.

$$\begin{aligned} Z_0 I_0 + V_0 &= 0 \\ Z_1 I_1 + V_1 &= E \\ Z_2 I_2 + V_2 &= 0 \end{aligned}$$

Therefore, when the three-phase generator supplies balanced three-phase voltages, only the positive sequence network is excited by positive sequence voltage, that is the same as phase a of the three-phase system. Figure 11.11 depicts the positive, zero, and negative impedance sequences of a generator.

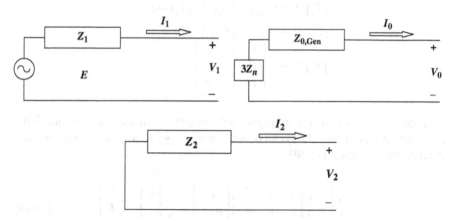

Figure 11.11 The positive, zero, and negative sequences of a generator.

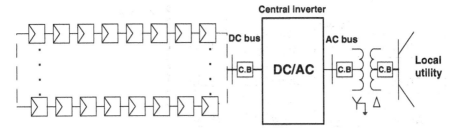

Figure 11.12   A central DC/AC inverter for a large-scale photovoltaic power configuration.

11.5 THE MODELING OF A PHOTOVOLTAIC-GENERATING STATION

In Chapter 5, we studied the modeling of photovoltaic (PV)-generating modules. We learned that the PV-module input impedance is purely resistive; it is a function of the input irradiance and temperature of the module. In Chapter 3, we studied the operation of DC/AC inverters. We modeled inverters using ideal power switches. Because inverters are operating at zero frequency on the DC side and at the power grid frequency on the AC side, the disturbance on the AC side has a limited propagation to the DC side. In fact, the DC/AC inverters block the propagation of disturbance. At the same time, because there is no inertia energy and the stored energy in PV arrays is limited to irradiance energy, the fault current contribution of PV is quite low. Therefore, the input impedance of PV-generating stations is high. Figures 11.12 and 11.13 depict PV-generating stations. For estimating the fault current contributions of PV-generating stations, we need to model the input impedance of PV-generating stations from measured operating data. The modeling of PV-generating stations is currently being investigated. Students can find updated models for PV- and wind-generation systems from the National Renewable Energy Laboratory (NREL) website, www.nrel.gov. However, we can safely use a high impedance model using the model developed in Chapter 5 for the PV module operating at maximum PV-generating station voltage and current.

11.6 SEQUENCE NETWORKS FOR BALANCED THREE-PHASE TRANSMISSION LINES

Figure 11.14 depicts the balanced three-phase network model of a transmission line. The voltage equations expressing the voltage drop across the lines can be expressed as given by Equation (11.34).

$$
\begin{aligned}
V_a &= jX_sI_a + jX_mI_b + jX_mI_c + V'_a \\
V_b &= jX_sI_b + jX_mI_a + jX_mI_c + V'_b \\
V_c &= jX_sI_c + jX_mI_a + jX_mI_b + V'_c
\end{aligned}
$$

$$(11.34)$$

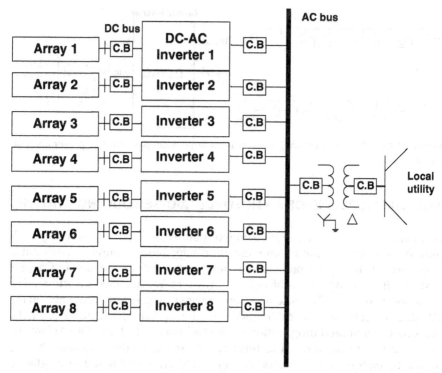

Figure 11.13   The general structure of photovoltaic arrays with inverters.

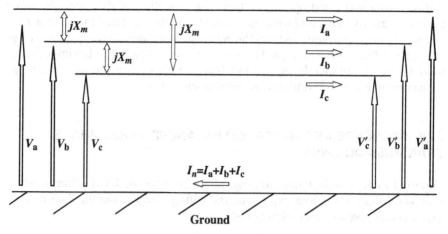

Figure 11.14   A balanced three-phase transmission reactance model.

The set of equations can be expressed in compact matrix as given below.

$$
\begin{bmatrix} V_a \\ V_b \\ V_c \end{bmatrix} - \begin{bmatrix} V'_a \\ V'_b \\ V'_c \end{bmatrix} = j \begin{bmatrix} X_s & X_m & X_m \\ X_m & X_s & X_m \\ X_m & X_m & X_s \end{bmatrix} \begin{bmatrix} I_a \\ I_b \\ I_c \end{bmatrix}
\tag{11.35}
$$

$$
[V_{abc}] - [V'_{abc}] = [Z_{abc}][I_{abc}]
\tag{11.36}
$$

$$
[V_{abc}] = [T_s][V_{012}]
\tag{11.37}
$$

$$
[I_{abc}] = [T_s][I_{012}]
\tag{11.38}
$$

Replace V_{abc} with Equation (11.37) and I_{abc} with Equation (11.38) to obtain

$$
[T_s][V_{012}] - [T_s][V'_{012}] = [Z_{abc}][T_s][I_{012}]
\tag{11.39}
$$

Multiplying Equation (11.39) by $[T_s]^{-1}$, we get

$$
[T_s]^{-1}[T_s][V_{012}] - [T_s]^{-1}[T_s][V'_{012}] = [T_s]^{-1}[Z_{abc}][T_s][I_{012}]
$$

Since $[T_s]^{-1}[T_s] = $ Identity matrix
we obtain

$$
[V_{012}] - [V'_{012}] = [Z_{abc}][I_{012}]
\tag{11.40}
$$

$$
[Z_{012}] = [T_s]^{-1}[Z_{abc}][T_s]
$$

$$
Z_{012} = \frac{1}{3} \begin{bmatrix} 1 & 1 & 1 \\ 1 & a & a^2 \\ 1 & a^2 & a \end{bmatrix} j \begin{bmatrix} X_s & X_m & X_m \\ X_m & X_s & X_m \\ X_m & X_m & X_s \end{bmatrix} \begin{bmatrix} 1 & 1 & 1 \\ 1 & a^2 & a \\ 1 & a & a^2 \end{bmatrix}
\tag{11.41}
$$

Equation (11.40) simplifies as

$$
Z_{012} = \frac{j}{3} \begin{bmatrix} X_s + 2X_m & X_s + 2X_m & X_s + 2X_m \\ X_s - X_m & aX_s + (1+a^2)X_m & a^2X_s + (1+a)X_m \\ X_s - X_m & a^2X_s + (1+a)X_m & aX_s + (1+a^2)X_m \end{bmatrix} \begin{bmatrix} 1 & 1 & 1 \\ 1 & a^2 & a \\ 1 & a & a^2 \end{bmatrix}
$$

After further simplification, we obtain

$$
Z_{012} = j \begin{bmatrix} X_s + 2X_m & 0 & 0 \\ 0 & X_s - X_m & 0 \\ 0 & 0 & X_s - X_m \end{bmatrix}
\tag{11.42}
$$

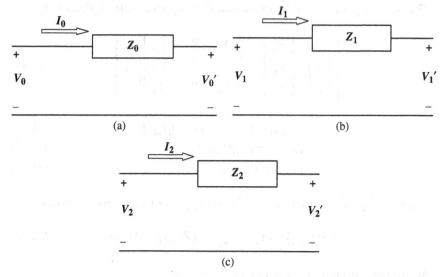

Figure 11.15 The sequence networks for a transmission line's (a) zero, (b) positive, and (c) negative circuit models.

Therefore, the symmetrical sequence network model of a transmission line can be expressed as

$$\begin{bmatrix} V_0 \\ V_1 \\ V_2 \end{bmatrix} - \begin{bmatrix} V'_0 \\ V'_1 \\ V'_2 \end{bmatrix} = j \begin{bmatrix} X_s + 2X_m & 0 & 0 \\ 0 & X_s - X_m & 0 \\ 0 & 0 & X_s - X_m \end{bmatrix} \begin{bmatrix} I_0 \\ I_1 \\ I_2 \end{bmatrix} \quad (11.43)$$

And zero, positive, and negative sequence model networks are

$$Z0 = \text{Zero sequence impedance} = j(X_s + 2X_m) \quad (11.44)$$
$$Z1 = \text{Positive sequence impedance} = j(X_s - X_m) \quad (11.45)$$
$$Z2 = \text{Negative sequence impedance} = j(X_s - X_m) \quad (11.46)$$

Figure 11.15 depicts the sequence networks for a transmission line's zero, positive, and negative circuit models.

11.7 GROUND CURRENT FLOW IN BALANCED THREE-PHASE TRANSFORMERS

The neutral point of a power grid is often connected to earth ground. Often ground and neutral are the same electrical point if there is no ground impedance between neutral and the earth ground point. The conductor that

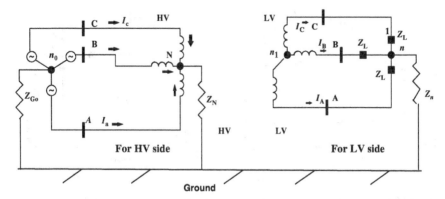

Figure 11.16 The Y–Y-connected transformer with high voltage side grounded and low voltage side not grounded.

connects the power grid's neutral point and ground will not carry load current; its function is to detect the ground fault current. If a power grid is faulted and the faulted bus is connected to ground, then the first question of interest is how ground current fault will flow through the power grid.

The transformers' high and low voltage sides are not electrically connected. The voltages induced in either side of the transformers are due to magnetic coupling of respective windings. Consider a Y–Y-connected transformer depicted in Figure 11.16. Let us assume the three-phase voltages supplied to the three-phase transformers are not balanced.

Let us assume that the transformer is grounded on the high voltage side and it is not grounded on the low voltage side as shown in Figure 11.16. The question is whether ground current can flow in the transformer when it is grounded on the generator side (high voltage side) and its low voltage side is not grounded. To answer this question, we need to remember that in transformers, voltage is induced in the low voltage side by magnetic induction and resulting currents that are flowing must obey the Kirchhoff current law. This means the current must return to its generating source. Therefore, the ground current cannot flow in either the high voltage side or the low voltage side. If the ground current flows in the low voltage side, it must return to neutral on the low voltage side. However, if the low voltage side is not grounded, it cannot complete the flow path. Therefore, if there is no path for the ground current to flow, then the low voltage side neutral will be at a value that will satisfy the following:

$$V_n = V_{an} + V_{bn} + V_{cn} \qquad (11.47)$$

By the same reasoning, the ground current cannot flow on the high voltage side. If there is ground current flowing on the high voltage side, the magnetic coupling must induce the three-phase voltages on the low voltage side and have the low-side phase current flowing, which will add up and result in ground current flow on the low voltage side. However, the low voltage side is not grounded; therefore, the low-side voltages will add up as given in Equation (11.47).

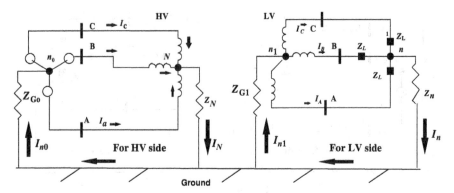

Figure 11.17 The Y–Y-connected transformer with the high voltage side and the low voltage side grounded.

Now consider a case where both high and low voltage sides are grounded, and ground current can flow on both sides as shown in Figure 11.17.

11.8 ZERO SEQUENCE NETWORK

11.8.1 Transformers

We have concluded that for a Y–Y transformer with grounded neutral on one side and ungrounded neutral on the other side, the ground current cannot flow, because there is no ground path on the ungrounded side as shown in Figure 11.18a. We have also stated that ground current can flow if the neutral point of both sides of a Y–Y-connected transformer is grounded as shown in Figure 11.18b. The arrows in Figure 11.18b show the direction of the current flow and resulting ground current, I_n. Figure 11.18b displays a transformer that has been supplied with unbalanced voltages on one side that results in the current I_n because for the unbalanced three-phase currents, $I_n = I_a + I_b + I_c$. As shown in Figure 11.18b, the ground current I_n flows through the ground conductor; the arrow is pointing downward. On the other side of the transformer in Figure 11.18b, the unbalanced voltages result in a ground current flowing out of the ground so that $I_n = I_a + I_b + I_c$. Of course, the relationship of ground currents on both sides of the transformer is governed by the turn ratio of the transformer.

Figure 11.18c depicts a grounded Y–Δ transformer. In this case, the ground current flows on the grounded Y side because we have a circulating current on the Δ side of the transformer. Again, we should remember that if unbalanced three-phase voltages are applied to the grounded Y side, we have created a ground path for ground current to flow as shown in Figure 11.18c. Figures 11.18d and 11.18e depict the conditions of an ungrounded Y–Δ connection and Δ–Δ connection.

Symbols   Connection diagrams   Zero-sequence
                                                      equivalent circuits

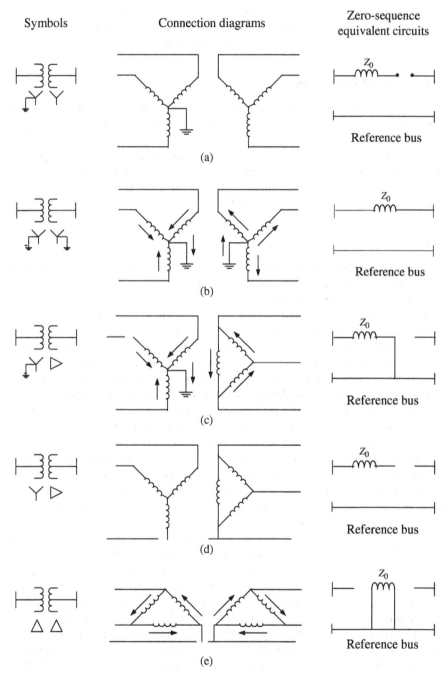

Figure 11.18 Zero sequence equivalent circuits of three-phase (a) Y–Y-connected transformer banks with one side grounded, (b) Y–Y-connected transformer banks with both sides grounded, (c) grounded y–δ-connected transformer banks, (d) ungrounded y–δ-connected transformer banks, and (e) Δ–Δ-connected transformer banks.

Ground current flows if a path is established for it to flow; however, in a transformer, we must keep in mind Faraday's law of induction, the Kirchhoff current law, and the current laws for ground current flow.

11.8.2 Load Connections

Figure 11.19a depicts the zero sequence of a Y-connected load when the load is not grounded. As seen from the zero sequence of the ungrounded load, no ground current flows.

In Figure 11.19c, three times the value of the ground impedance Z_n appears in the zero sequence network for the grounded load because $I_n = 3\,I_0$. Finally, Figure 11.19d depicts the Δ-connected load. In this case, the zero sequence current circulates as shown in Figure 11.19d.

11.8.3 Power Grid

As we have discussed in this chapter, when a fault occurs in a power grid that involves the ground, the ground current will flow. The ground current is three times the zero sequence current. Again, we must remember that in the calculation of fault current, it is assumed the power grid remains balanced, except for the faulted part. Finally, we study the if–then condition to calculate the faulted currents and set the protective relay to open the associated circuit breakers to isolate the fault before the power grid becomes unbalanced. Power grids are always designed and operated as balanced three-phase systems. To compute the ground fault current, we need to use a zero sequence network. In the preceding sections, we presented each element of a power grid zero sequence network. Now, we need to learn how to construct a zero sequence network for power grids to compute the unbalanced fault currents.

Figure 11.20 depicts a one-line diagram of a power microgrid with the designation of how the transformers in the power grid are connected. As we can see in Figure 11.20, the generator connected to bus 6 is a gas turbine unit; its neutral point is grounded through impedance Z_G. Bus 6 is connected through an ungrounded Y-Δ transformer to bus 5.

Figure 11.21 depicts the zero sequence network of Figure 11.20. In Figure 11.21, we have replaced the gas turbine generator with its zero sequence impedance of gas turbine generator, Z_{G0}. However, we have placed Z_{G0} in series with three times the Z_G, because the ground current flowing into the ground reference is equal to three times the zero sequence current.

The transformer T_1 is Y-Δ-connected and the Y side is not grounded. Therefore, this transformer is not connected to ground through its neutral point and breaks the ground connection in the Y-Δ transformer. Therefore, this transformer is shown as an open circuit in the zero sequence network depicted in Figure 11.21.

The three-bus power grid transmission network is connected to a PV-generating station and its local power grid. The zero sequence network of the transmission system is shown in Figure 11.21. Through bus 2, the transformer

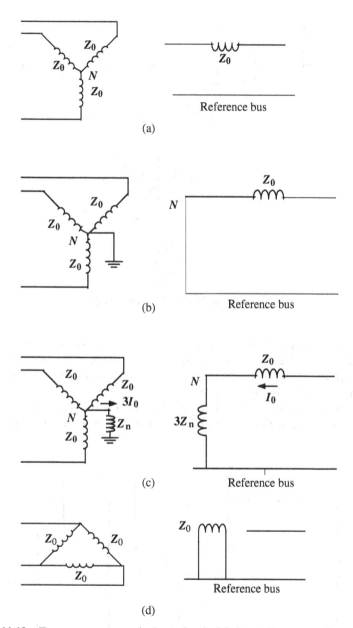

Figure 11.19 Zero sequence equivalent circuit (a) for a Y-connected ungrounded load, (b) for a grounded Y-connected load, (c) for a Y-connected load with grounding impedance, and (d) for a Δ-connected load.

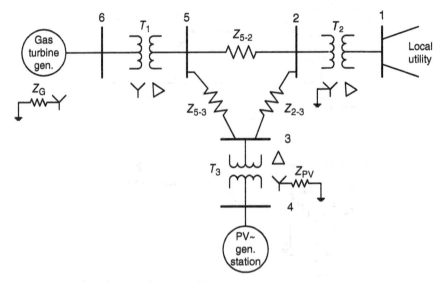

Figure 11.20   A balanced three-phase microgrid.

T_2 is connected to bus 1, and then to the local power grid bus. The transformer T_2 is a grounded Y–Δ. Because grounded Y–Δ is connected to the ground on the Y side, bus 2 is connected to the ground reference through its zero sequence impedance as shown in Figure 11.21.

Transformer T_3 connects bus 3 to bus 4 through a grounded Δ–Y transformer to a PV-generating station. Ground impedance, Z_{PV}, is inserted in series from

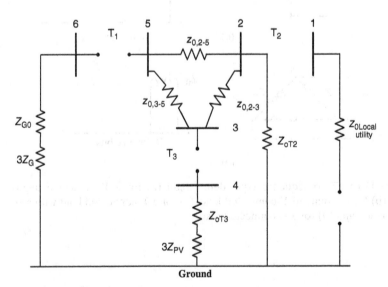

Figure 11.21   The zero sequence network of Figure 11.20.

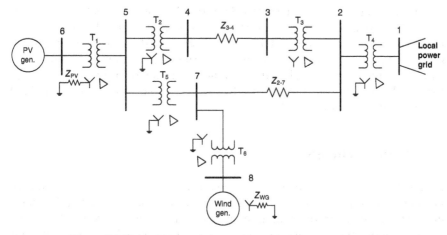

Figure 11.22   A balanced three-phase wind-power microgrid.

the neutral point of the Y side of the transformer to ground. Again, because the grounded Y is connected to bus 4, bus 4 is connected to the ground reference as shown in Figure 11.21. However, bus 3 is shown as an open circuit. On the Δ side of the transformer T_2, the zero sequence current (ground current) does not flow and an open circuit reflects this condition as shown in Figure 11.21.

As another illustration, the one-line diagram is given in Figure 11.22. Figure 11.23 depicts the zero sequence network of the microgrid given in Figure 11.22.

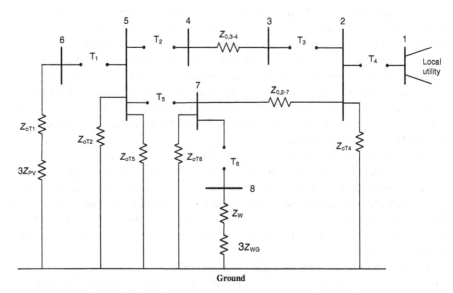

Figure 11.23   The zero sequence network of Figure 11.22.

To construct zero sequence networks correctly, close observation of ground connections is essential. In addition, close attention should be paid to how the transformers are connected. Basically, ungrounded transformers break the ground current path as shown by the transformer connections in Figures 11.22 and 11.23. However, particular attention should be paid to grounded Y–Y transformers. In these transformers, the path of zero sequence currents is maintained as can be seen in Figure 11.18b.

11.9 FAULT STUDIES

For a balanced three-phase fault, only the positive sequence network is excited. For unbalanced faults, all three sequence networks may be excited. If a fault involves a connection to ground, the ground current will flow. The back emf voltage behind the generator reactance is assumed equal in magnitude and phase angle; normally, it is assumed to be equal to 1 per unit.

For a fault study, normally, all shunt elements including loads and line charging are neglected. Loads may be represented with constant load impedance models. All tap changing transformers are assumed to be at their nominal tap settings and balanced transmission lines are assumed. The negative and positive sequence networks are assumed equal; coupling between adjacent circuits only is taken into account in a zero sequence network.

Power grid generators are maintained by a constant voltage behind the transient reactance. Normally, three reactances are used depending on when the faults are cleared. The subtransient reactance, X'', of a generator is estimated from tests performed on the generator and from recordings of the generator current response. The subtransient reactance is estimated from the slope of the current response during the first quarter cycle. After the subtransient reactance, the transient reactance, X', is estimated from the slope of current response from the first half cycle and synchronous reactance, X, is estimated from the steady-state current. The value of the subtransient reactance is less than the transient reactance. The transient reactance is less than the steady state reactance. The lower value of reactance results in higher short-circuit current. The selection of reactance is a function of circuit breakers' timing. If the breakers are fast acting and can interrupt higher fault current, then the lower value of reactances is used in the fault calculation.

The study of unbalanced faults requires the modeling of the power grid using the symmetrical analysis of a power grid. Therefore, we use the sequence network of generators, transformers, transmission lines, and loads. Based on the type of unbalanced faults, one line-to-ground, two lines-to-ground, line-to-line faults, etc., the symmetrical models of the power grid are constructed. These sequence network models are used to construct the unbalanced fault models of the power grid.

The objectives of balanced fault studies are to determine the required circuit breaker short capacity in kVA or MVA. The objectives of unbalanced fault studies are to determine how to set the protective relay systems. It is important

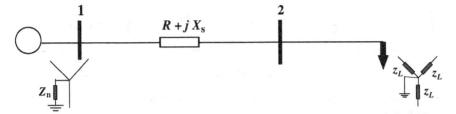

Figure 11.24   A one-line diagram of Example 11.2.

that close attention be paid to the representation of large motors in fault studies. The back emf of the motors can contribute substantially to the fault currents. Although the short-circuit capacity (SCC) of breakers in power grids are computed based on balanced faults, in some system single line-to-ground faults, the fault currents may be higher than balanced three-phase faults. Finally, a fault can happen anywhere in a power grid. For example, faults may occur in the middle of a transmission line. For these cases, we place a bus at that location in the if–then condition of the fault current calculations.

Example 11.2   Consider the power grid given in Figure 11.24.

Assume the generator positive, negative, and zero sequence impedances are $Z_{gen,1}$, $Z_{gen,2}$, and $Z_{gen,0}$, the generator is grounded through the ground impedance Z_n. Also, assume the transmission line model as depicted in Figure 11.25.

Perform the following:

(i) If the supply generator is unbalanced and supplies three-phase unbalanced voltages, determine the positive, negative, and zero sequence networks for the one-line diagram in Figure 11.24.

(ii) If the supply generator is balanced and supplies three-phase balanced voltages, determine the positive, negative, and zero sequence networks for the one-line diagram in Figure 11.24.

Solution

(i) The positive, negative, and zero sequence networks of the transmission line of Example 11.2 is presented in Figure 11.26.

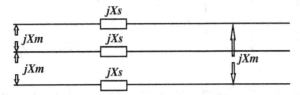

Figure 11.25   The balanced three-phase transmission line of Figure 11.24.

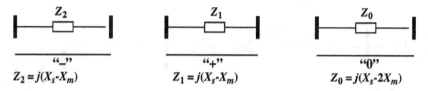

Figure 11.26   The negative, positive, and zero sequence model of the transmission line of Example 11.2.

If the transmission line has the total length of "L," the distributed line impedance and line charging capacitance are depicted in Figure 11.27a and its lumped model is presented in Figure 11.27b.

$$z_1 = L \times z$$
$$Y' = L \times y \qquad\qquad (11.48)$$

The zero, positive, and negative sequence networks for Example 11.2 are depicted in Figure 11.28.

(i) If the supply generator is balanced, the zero, positive, and negative sequence networks are as depicted in Figure 11.29.

11.9.1 Balanced Three-Phase Fault Analysis

For a balanced three-phase fault study, only the positive network model must be constructed. Figure 11.30 depicts a one-line diagram of a three-bus power grid. Figure 11.31 depicts the positive sequence network model for balanced fault studies.

Figure 11.31 depicts the complete model of a power system including the shunt elements and loads. In the figure, the load is represented by its equivalent impedance model

$$z_{\text{load}} = \frac{V_{\text{load}}^2}{P_{\text{load}} - jQ_{\text{load}}} \qquad\qquad (11.49)$$

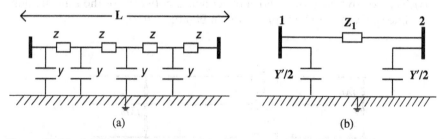

Figure 11.27   (a) The distributed model of the transmission line and (b) the lumped model of the transmission line.

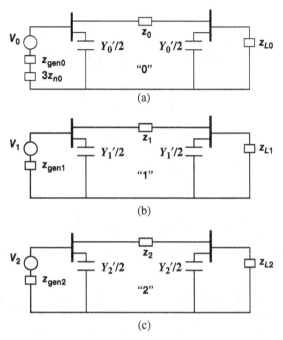

Figure 11.28 (a) The zero sequence, (b) positive sequence, and (c) negative sequence for Example 11.2 when the power supply generator voltages are unbalanced.

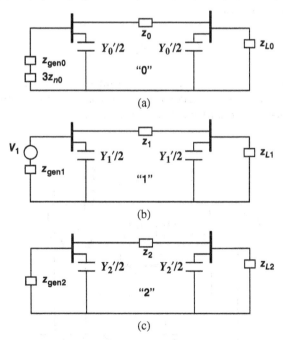

Figure 11.29 (a) The zero sequence, (b) positive sequence, and (c) negative sequence for Example 11.2 when the power supply generator voltages are balanced.

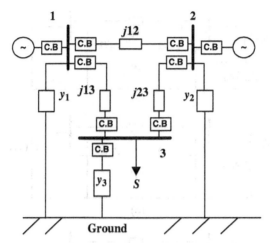

Figure 11.30 A one-line diagram of a balanced three-bus power grid.

The model given in Figure 11.31 can be simplified by removing the shunt elements and load. This will result in a higher value for the fault currents.

In the design of power grids, the voltage calculation and power flow studies are performed before the short-circuit currents are calculated. Then, the calculated bus load voltages are used to determine the circuit breakers' interrupting capacities. Therefore, the pre-fault voltages are calculated from power flow studies and are known.

$$E_{\text{Bus}(0)} = Z_{\text{Bus}}I_{\text{Bus}(0)} \tag{11.50}$$

In Equation (11.50), $E_{\text{Bus}(0)}$ is the bus voltage vector before the fault and Z_{Bus} is the bus impedance matrix model with respect to the ground bus. $I_{\text{Bus}(0)}$ is the generator-injected currents before the fault.

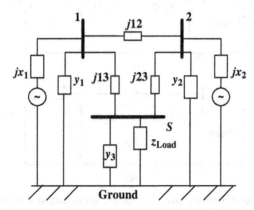

Figure 11.31 A positive sequence network model for balanced fault studies.

During the fault, the faulted network variables are designated by "F" and the bus voltage during fault can be expressed as

$$E_{Bus(F)} = E_{Bus(0)} - Z_{Bus(F)}I_{Bus(F)} \tag{11.51}$$

where $E_{Bus(F)}$ is the bus voltage vector during the fault, $E_{Bus(0)}$ is the bus voltage before the fault, and $Z_{Bus(F)}$ is the bus impedance matrix, and $I_{Bus(F)}$ is the bus fault current during the fault.

For the bus system of Figure 11.30 with a fault at bus 3, we have

$$\begin{bmatrix} E_{1(F)} \\ E_{2(F)} \\ E_{3(F)} \end{bmatrix} = \begin{bmatrix} E_{1(0)} \\ E_{2(0)} \\ E_{3(0)} \end{bmatrix} - \begin{bmatrix} Z_{11} & Z_{12} & Z_{13} \\ Z_{21} & Z_{22} & Z_{23} \\ Z_{31} & Z_{32} & Z_{33} \end{bmatrix} \begin{bmatrix} 0 \\ 0 \\ I_{3(F)} \end{bmatrix} \tag{11.52}$$

Therefore, the bus voltage at each bus during the fault can be expressed as

$$E_{1(F)} = E_{1(0)} - Z_{13}I_{3(F)}$$
$$E_{2(F)} = E_{2(0)} - Z_{23}I_{3(F)}$$
$$E_{3(F)} = E_{3(0)} - Z_{33}I_{3(F)} \tag{11.53}$$

If the fault has impedance Z_f, then the voltage across the fault impedance is given as

$$E_{3(F)} = Z_f I_{3(F)} \tag{11.54}$$

Substituting Equation (11.54) in Equation (11.53), we have

$$Z_f I_{3(F)} = E_{3(0)} - Z_{33}I_{3(F)} \tag{11.55}$$

Therefore, the fault current at bus 3 for a balanced three-phase fault can be calculated as

$$I_{3(F)} = \frac{E_{3(0)}}{Z_{33} + Z_f} \tag{11.56}$$

In Equation (11.56), $E_{3(0)}$ is pre-fault voltage and Z_{33} is the Thevenin impedance of bus 3 with respect to the ground bus.

For the general case with a balanced three-phase fault on a bus "i," we have

$$\begin{bmatrix} E_{1(F)} \\ E_{2(F)} \\ \vdots \\ E_{i(F)} \\ \vdots \\ E_{n(F)} \end{bmatrix} = \begin{bmatrix} E_{1(0)} \\ E_{2(0)} \\ \vdots \\ E_{i(0)} \\ \vdots \\ E_{n(0)} \end{bmatrix} - \begin{bmatrix} Z_{11} & Z_{12} \ldots Z_{1i} \ldots Z_{1n} \\ \vdots \\ \vdots \\ Z_{i1} & Z_{i2} \ldots Z_{ii} \ldots Z_{in} \\ \vdots \\ Z_{n1} & Z_{n2} \ldots Z_{nP} \ldots Z_{nn} \end{bmatrix} \begin{bmatrix} 0 \\ 0 \\ \vdots \\ I_{i(F)} \\ 0 \\ \vdots \\ 0 \end{bmatrix} \tag{11.57}$$

From the above, we can express the fault at bus i as

$$E_{i(F)} = E_{i(0)} - Z_{ii}I_{P(F)} \tag{11.58}$$

And, $E_{i(F)} = Z_f I_{i(F)}$
Therefore, the fault current at bus i is

$$I_{i(F)} = \frac{E_{i(0)}}{Z_{ii} + Z_f} \tag{11.59}$$

In Equation (11.59), $E_{i(0)}$ is pre-fault voltage and Z_{ii} is the Thevenin impedance of bus "i" with respect to the ground bus.

$$Z_{Bus} = \begin{matrix} & 1 & & i & & n \\ \begin{bmatrix} Z_{11} & \cdots & Z_{i1} & \cdots & Z_{1n} \\ \cdot & \cdots & \cdot & \cdots & \cdot \\ Z_{i1} & \cdots & Z_{ii} & \cdots & Z_{in} \\ \cdot & \cdots & \cdot & \cdots & \cdot \\ Z_{n1} & \cdots & Z_{ni} & \cdots & Z_{nn} \end{bmatrix} & \begin{matrix} 1 \\ \cdot \\ i \\ \cdot \\ n \end{matrix} \end{matrix}$$

The balanced three phase fault is depicted in Figure 11.32. The elements of a Z_{Bus} positive sequence matrix are computed from the positive sequence network.

We can generalize an algorithm for a balanced three-phase fault calculation as

Step 1. Build the Z_{Bus} (matrix) for a positive impedance network.
Step 2. Obtain the pre-fault bus voltage from load flow.
Step 3. Compute: $I_{i(F)} = \frac{E_{i(0)}}{Z_{ii}+Z_f}$.

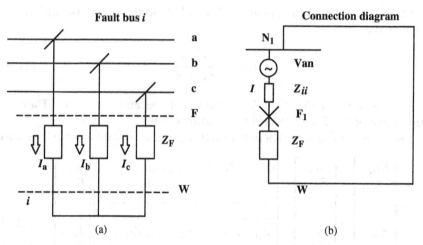

Figure 11.32 depicts Fault bus i and the Connection diagram.

Figure 11.32 The balanced three-phase (a) fault and (b) the Thevenin equivalent circuit.

Using the above algorithm, we can calculate the balance fault current for each bus "i."

The SCC for breakers connected to each bus is computed based on the balanced three-phase fault at each bus. The SCC is defined as

$$SCC = \left|V_{\text{prefault(p.u)}}\right| \cdot \left|I_{\text{fault(p.u)}}\right| \tag{11.60}$$

If the pre-fault voltage and fault current are given in per unit (p.u), the SCC is also given in p.u of MVA. To obtain SCC in MVA, the p.u MVA is multiplied by S_b.

$$SCC = S_b \left|V_{\text{prefault(p.u)}}\right| \cdot \left|I_{\text{fault(p.u)}}\right| \tag{11.61}$$

$$SCC = \sqrt{3} \cdot \left|V_{\text{prefault, line-line}}\right| \cdot \left|I_{\text{fault, line}}\right| \text{VA} \tag{11.62}$$

In Equation (11.62), the voltage is in volts and current is in amperes. Therefore, at each bus, the fault level of the bus is determined by its Thevenin impedance to the ground. To explain this concept clearly, consider the system given in Figure 11.33.

To calculate the fault current at bus 3, the Thevenin equivalent circuit is given in Figure 11.34. V_{th} is the voltage at bus 3 before the fault.

$Z_{\text{th}} = Z_{33}$ is obtained from the Z_{bus} impedance matrix; the impedance between bus 3 and ground by looking into the system at bus 3.

$$I_f = \frac{V_{\text{th}}}{Z_{\text{th}} + Z_f} \tag{11.63}$$

If the fault impedance is zero, that is we have a bolted fault, then Equation (11.63) can be expressed as

$$I_f = \frac{V_{\text{th}}}{Z_{\text{th}}} \tag{11.64}$$

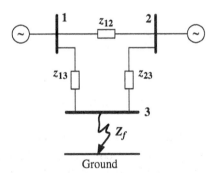

Figure 11.33 A fault at bus 3.

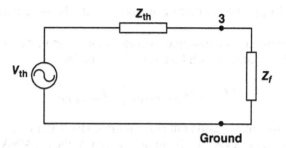

Figure 11.34   The Thevenin equivalent circuit for the fault current calculation at bus 3.

The power grid system is designed such that the voltage at each bus of the system is around 1 per unit. Therefore,

$$V_{th} \cong 1 \text{ p.u}$$

As a result, the fault current is

$$I_f = \frac{1}{Z_{th}} \tag{11.65}$$

Therefore, the Thevenin can be computed as

$$Z_{th} = \frac{1}{I_f} \tag{11.66}$$

Because the SCC in per unit is expressed as

$$SCC = |V_{prefault}| |I_{fault}| \text{ p.u}$$

and,

$$V_{prefault} \cong 1 \text{ p.u;}$$

therefore,

$$SCC = I_{f(fault)} \text{p.u} \tag{11.67}$$

and

$$Z_{th} = \frac{1}{SCC} \tag{11.68}$$

Power companies calculate the SCC for each bus of their system. When a microgrid of a green energy system is connected to the local power grid, the SCC is provided to assess the fault current contribution from the local power grid to the microgrid. Therefore, SCC specifies the input Thevenin impedance that is needed for computing the interrupting capacities of the microgrid circuit breakers.

Example 11.3   Consider a microgrid as part of interconnected power grid. Assume the following data:

Local power grid short-circuit capacity = 320 MVA

PV-generating station #1: PV arrays = 2 MVA, internal impedance = highly resistive, 50% of its rating

Gas turbine station: combined heat and power (CHP) units = 10 MVA, internal reactance = 4%. Units are Y connected and grounded.

Transformers = 460 V Y grounded/13.2 kV Δ, 10% reactance, 10 MVA capacity.

Power grid transformer: 20 MVA, 63 kV/13.2 kV, 7% reactance.

Bus 4 load = 1.5 MW, power factor (p.f.) = 0.85 lagging; bus 5 load = 5.5 MW, p.f. = 0.9 lagging; bus 6 load = 4.0 MW, p.f. = 0.95 leading; bus 7 load = 5 MW, p.f. = 0.95 lagging; bus 8 load = 1.0 MW, p.f. = 0.9 lagging.

Transmission line: resistance = 0.0685 Ω/mile, reactance = 0.40 Ω/mile, and half of line charging admittance ($Y'/2$) of 11×10^{-6} Ω^{-1}/mile. Line 4–7 = 5 miles, 4–8 = 1 mile, 5–6 = 3 miles, 5–7 = 2 miles, 6–7 = 2 miles, 6–8 = 4 miles.

Perform the following:

(i) Develop a per unit equivalent model for balanced three-phase fault studies based on a 20-MVA base.

(ii) If the bus load voltages are at 1 per unit, compute the per impedance model of each load.

(iii) For three-phase faults, compute the SCC of each distribution network bus.

(iv) To increase the security of the system two identical transformers are used at each distribution network and the interconnection to the local power grid. Compute the SCC of each bus.

Solution

(i) The base value of the volt-amp is selected as S_b = 20 MVA. The voltage base selected on the PV generator and gas turbine side is selected to be 460 V. The voltage base on the transmission lines side is therefore V_b = 13.2 kV.

The p.u SCC of the local power grid is given by

$$\frac{\text{SCC}}{S_b} = \frac{320}{20} = 16$$

Therefore, the internal p.u impedance of the local power grid is

$$Z_{th} = \frac{1}{SCC_{p.u}} = \frac{1}{16} = j0.063$$

The internal p.u impedance of the PV-generating station at 20 MVA base is given by

$$Z_{p.u(new)} = Z_{p.u(old)} \times \frac{VA_{b(new)}}{VA_{b(old)}} \times \left(\frac{V_{b(old)}}{V_{b(new)}}\right)^2$$

$$= 0.5 \times \frac{20 \times 10^6}{2 \times 10^6} \times \left(\frac{460}{460}\right)^2 = 5$$

The p.u impedance of the gas turbine is

$$z = j0.4 \times \frac{20 \times 10^6}{10 \times 10^6} \times \left(\frac{460}{460}\right)^2 = j0.8$$

The base impedance transmission system is

$$Z_b = \frac{V_b^2}{S_b} = \frac{(13.2 \times 10^3)^2}{20 \times 10^6} = 8.712 \ \Omega$$

The base admittance given by

$$Y_b = \frac{1}{Z_b} = \frac{1}{8.712} = 0.115$$

The p.u impedance of the local power grid transformer is 7% based on 20 MVA.

(ii) The loads represented by their equivalent impedance are calculated from the equation

$$z_{load} = \frac{V_{load}^2}{P_{load} - jQ_{load}}$$

(iii) The per unit impedance of a line between buses i and j is given by

$$z_{i-j,p.u} = \frac{z_{i-j}}{Z_b}$$

Using the above equation, the transmission line p.u parameters are calculated and listed in Table 11.2a.

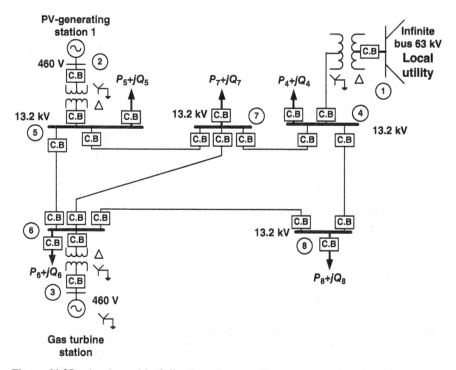

Figure 11.35 A microgrid of distributed generation connected to a local power network.

Figure 11.35 depicts a microgrid of distributed generation connected to a local power network.

Figure 11.36 depicts the equivalent impedance of the transmission line model for Example 11.3.

Figure 11.37 depicts the impedance model for short-circuit studies.

Table 11.1 presents the load of each bus and its equivalent load impedance of network of Figure 11.37. Table 11.2 presents the transmission line parameters of Figure 11.37.

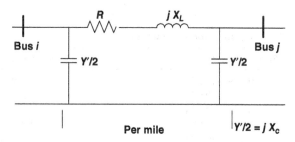

Figure 11.36 The equivalent impedance of the transmission line model for Example 11.3.

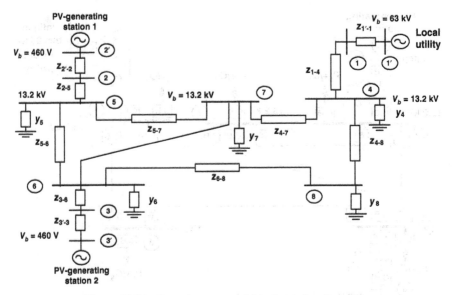

Figure 11.37   Impedance model for short-circuit studies.

TABLE 11.1   The Load of Each Bus and Its Equivalent Load Impedance

Bus	Load	Complex Power (p.u)	Equivalent Load Impedance (p.u)
4	1.5 MW at 0.85 p.f. (lagging) = 1.5 + j0.92	0.075 + j0.046	9.69 + j5.94
5	5.5 MW at 0.9 p.f. (lagging) = 5.5 + j2.66	0.275 + j0.133	2.95 + j1.43
6	4.0 MW at 0.95 p.f. (leading) = 4.0 − j1.31	0.20 − j0.066	4.51 − j1.49
7	5 MW at 0.95 p.f. (lagging) = 5.0 + j1.64	0.25 + j0.082	3.61 + j1.18
8	1.0 MW at 0.9 p.f. (lagging) 1.0 + j0.48	0.05 + j0.024	16.25 + j7.80

TABLE 11.2a   Transformer Per Unit Impedance

Line	Series Impedance (p.u)
1–4	j0.07
2–5	j0.2
3–6	j0.2

TABLE 11.2b   Transmission Line Parameters

Line	Series Impedance (p.u)	Line Charging Admittance (p.u)
4–7	0.039 + j0.229	$j479 \times 10^{-6}$
4–8	0.008 + j0.046	$j96 \times 10^{-6}$
5–6	0.024 + j0.138	$j287 \times 10^{-6}$
5–7	0.016 + j0.092	$j192 \times 10^{-6}$
6–7	0.016 + j0.092	$j192 \times 10^{-6}$
6–8	0.031 + j0.184	$j383 \times 10^{-6}$

The bus admittance matrix is calculated using the Y_{Bus} algorithm. For short-circuit studies, it is industry practice to omit the line charging and the load impedances for calculating the balanced fault current for each bus. However, we can include the load impedance by using the bus load voltages from power flow calculation.

$$z_{\text{load}} = \frac{V_{\text{load}}}{I_{\text{load}}} = \frac{V_{\text{load}}}{(V_{\text{load}}/I_{\text{load}})^*} = \frac{V_{\text{load}}^2}{S_{\text{load}}^*}$$

However, because in a designed power grid the load bus voltages are around 1 per unit with tolerance of 5%, we can use 1 per unit for load voltages and compute the load impedance for use in short-circuit studies.

The Y_{Bus} matrix for short-circuit studies will include the internal impedance of the generator buses. For Example 11.3, the Y_{Bus} matrix will be 8×8 with the voltage sources replaced by their internal impedance to find Thevenin's equivalent impedance.

$$Y_{\text{Bus}} = \begin{bmatrix} & 1 & 2 & 3 & 4 & 5 & 6 & 7 & 8 & \\ -j30.2 & 0 & 0 & j14.3 & 0 & 0 & 0 & 0 & 1 \\ 0 & 0.2-j5 & 0 & 0 & j5.0 & 0 & 0 & 0 & 2 \\ 0 & 0 & -j6.25 & 0 & 0 & j5.0 & 0 & 0 & 3 \\ j14.3 & 0 & 0 & 4.35-j39.68 & 0 & 0 & -0.72+j4.23 & -3.6+j21.16 & 4 \\ 0 & j5.0 & 0 & 0 & 3.02-j22.63 & -1.2+j7.1 & -1.8+j10.58 & 0 & 5 \\ 0 & 0 & j5.0 & 0 & -1.2+j7.1 & 3.93-j27.92 & -1.8+j10.58 & -0.91+j5.29 & 6 \\ 0 & 0 & 0 & -0.7+j4.2 & -1.8+j10.6 & -1.8+j10.58 & 4.35-j25.39 & 0 & 7 \\ 0 & 0 & 0 & -3.6+j21.2 & 0 & -0.9+j5.29 & 0 & 4.53-j26.45 & 8 \end{bmatrix}$$

The bus admittance matrix, Y_{Bus} is inverted to get the Z_{Bus} matrix. The diagonal elements of the Z_{Bus} matrix are Thevenin's equivalent impedance of the bus looking back into the system. That is Z_{ii} is the Thevenin impedance of bus i with respect to ground.

$$\text{SCC}_{\text{p.u}} = \frac{1}{Z_{ii}} \qquad i = 1, 2, ..., 8$$

where Z_{ii} is the diagonal element of Z_{Bus} matrix. The SCC of each bus is given in Table 11.3.

(iv) The SCC of each bus when two transformers are used for each generation bus is given in Table 11.4.

From the data in Tables 11.3 and 11.4 we find that when we parallel two transformers, we decrease the net Thevenin impedance of each bus and increase the short-circuit current. Therefore, when we make a change in a power grid, we must perform a short-circuit study of the grid to ensure that we have not exceeded the SCC of the breakers.

TABLE 11.3 The Short-Circuit Capacity (SCC) of Each Bus

Bus	SCC (p.u)
1	16.71
2	2.21
3	3.42
4	8.41
5	3.92
6	4.82
7	4.57
8	6.45

TABLE 11.4 The Short-Circuit Capacity (SCC) of Each Bus with Two Transformers

Bus	SCC (p.u)
1	16.81
2	3.05
3	4.31
4	11.18
5	4.36
6	5.51
7	5.20
8	7.91

Example 11.4   Consider the power grid as given in Figure 11.38. Assume a base of 100 MVA and compute the following:

(i) The SCC of bus 5 (415 V) when all the transformers and the generator G_1 are in service.

(ii) The SCC of bus 5 (415 V) when all transformers are in service, but generator G_1 is not in service.

Solution
The base volt-amp is given as $S_b = 100$ MVA.
   The SCC of local power grid bus is given as SCC = 10,000 MVA.
   Therefore, p.u SCC is given as

$$SSC_{p.u} = \frac{SSC}{S_b}$$

$$SSC_{p.u} = \frac{10,000}{100} \, 100 \text{ p.u}$$

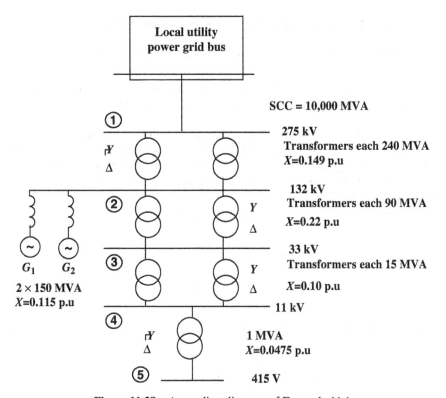

Figure 11.38   A one-line diagram of Example 11.4.

From Equation (11.68), the internal impedance of the power grid is

$$X_{th} = \frac{1}{100} = 0.01 \text{ p.u}$$

The p.u impedance of the transformers and generators at a new base is given by

$$Z_{p.u(new)} = Z_{p.u(old)} \times \frac{VA_{b(new)}}{VA_{b(old)}} \times \left(\frac{V_{b(old)}}{V_{b(new)}}\right)^2$$

Therefore, the p.u impedance of the transformers and generators at 100 MVA is as listed below.

- For a 240 MVA transformer, $X_{Tr240} = 0.149 \left(\frac{100}{240}\right) = 0.062$ p.u.
- For a 90 MVA transformer, $X_{Tr90} = 0.22 \left(\frac{100}{90}\right) = 0.244$ p.u.
- For a 15 MVA transformer, $X_{Tr15} = 0.1 \left(\frac{100}{15}\right) = 0.67$ p.u.

- For a 1 MVA transformer, $X_{Tr1} = 0.0475 \left(\frac{100}{1}\right) = 4.75$ p.u.
- For the generators G_1 and $G_2 = X_G = 0.115 \left(\frac{100}{150}\right) = 0.0766$ p.u.

(i) The equivalent circuit diagram for fault analysis when all the generators and transformers are in service is given in Figure 11.39.

Figure 11.39a shows the equivalent circuit of Figure 11.38 under fault conditions. The parallel impedances of Figure 11.39a have been reduced to single equivalent impedances in Figure 11.39b, and in Figure 11.39c the series impedances are combined together to form only three equivalent impedances. In Figure 11.39d a simplified equivalent circuit is shown.

The p.u fault current from Equation (11.68) is given by

$$I_f = \frac{1}{Z_{th}}$$
$$= \frac{1}{j5.227} = 0.1913 \text{ p.u A}$$

It should be noted that the p.u fault current and the p.u SCC are equal at 1 p.u voltage.

$$SCC = SCC_{p.u} \times S_b$$
$$SCC = 0.1913 \times 100 = 19.13 \text{ MVA}$$

The fault current at bus 5 can also be calculated by finding the Thevenin impedance at bus 5 obtained from the Z_{Bus}^+ matrix as shown below.

$$Z_{Bus}^+ = \begin{bmatrix} j0.009 & j0.005 & j0.005 & j0.005 & j0.005 \\ j0.005 & j0.020 & j0.020 & j0.020 & j0.020 \\ j0.005 & j0.020 & j0.142 & j0.142 & j0.142 \\ j0.005 & j0.020 & j0.142 & j0.477 & j0.477 \\ j0.005 & j0.020 & j0.142 & j0.477 & j5.227 \end{bmatrix}$$

The positive sequence Thevenin impedance for bus 5 is given by $Z_{Th,5}^+ = Z_{55}^+ = j5.227$.

This value obtained from the Z_{Bus}^+ is the same as the value calculated from the circuit simplification as shown in Figure 11.39. The fault current is calculated as shown before.

(ii) The equivalent circuit for fault analysis with generator G_1 out of service is given in Figure 11.40.

The simplification of the faulted network of Figure 11.40a is shown in Figures 11.40b to 11.40d following the same steps as explained in Part (i).

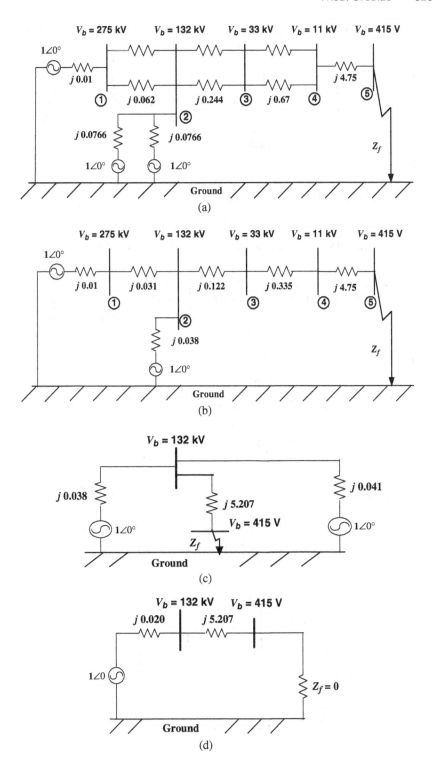

Figure 11.39 (a–d) The impedance diagram of Example 11.4.

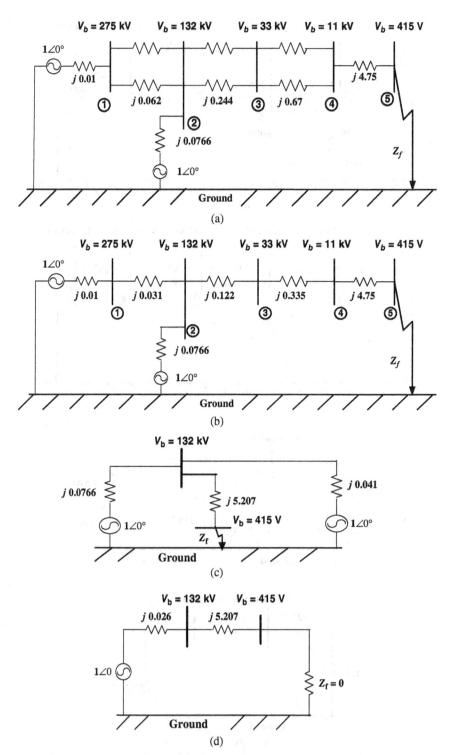

Figure 11.40 (a–d) The impedance diagram of Example 11.4, Part (ii).

The p.u fault current from Equation (11.68) is given by

$$I_f = \frac{1}{Z_{th}}$$

$$Z_{th} = \frac{1}{j5.233} = 0.1911 \text{ p.u}$$

It should be noted that the p.u fault current and the p.u. SCC are equal at 1 p.u voltage.

$$SCC = SCC_{p.u} \times S_b$$
$$SCC = 0.1911 \times 100 = 19.11 \text{ MVA}$$

The fault current at bus 5 can also be calculated by finding the Thevenin impedance at bus 5 obtained from the Z^+_{Bus} matrix as shown below.

$$Z^+_{Bus} = \begin{bmatrix} j0.009 & j0.007 & j0.007 & j0.007 & j0.007 \\ j0.007 & j0.027 & j0.027 & j0.027 & j0.027 \\ j0.007 & j0.027 & j0.149 & j0.149 & j0.149 \\ j0.007 & j0.027 & j0.149 & j0.484 & j0.484 \\ j0.007 & j0.027 & j0.149 & j0.484 & j5.233 \end{bmatrix}$$

The positive sequence Thevenin impedance for bus 5 is given by $Z^+_{Th,5} = Z^+_{55} = j5.233$.

This value obtained from the Z^+_{Bus} is the same as the value calculated from the circuit simplification as shown above. The fault current is calculated using the equivalent Thevenin impedance at bus 5 obtained from the Z^+_{Bus} matrix.

11.9.2 Unbalanced Faults

Depending on the type of unbalanced faults, we need the positive, negative, and zero sequence networks. When a fault involves ground, the ground current will flow on part of the network if low impedance paths exist for the flow of ground current. The unbalanced fault currents are used to set the protection system relays. The most common fault is the single line-to-ground fault. For safety and for the protection of the power grid, a grounded grid is designed. When the ground current flow is detected, the relay system identifies the types of faults and isolates the faulted part of the systems by opening the circuit breakers as required.

Power grids are designed to operate as balanced three-phase systems. Again, as we discussed before, in an unbalanced fault analysis, the only part of the system that is unbalanced is the faulted part. For example, when one of the three-phase lines is faulted due to a passing storm, the faulted line is removed very quickly so that the power grid during fault still remains balanced.

11.9.3 Single Line-to-Ground Faults

To analyze the single line-to-ground fault, let us assume that a phase conductor of a transmission line of a power grid is broken due to a heavy storm and it has fallen on a tree and has an impedance Z_f. This is a typical single line-to-ground fault condition. It is customary to designate the faulted phase as phase a. Assume the faulted point is designated as bus i. Because phase b and phase c are not faulted, the following condition holds at bus i.

$$I_{fb} = 0, \; I_{fc} = 0 \text{ and } V_{fa} = Z_f \cdot I_{fa}$$

where Z_f is the fault impedance to ground at phase a of bus i.

Recall from the symmetrical transformation of phase currents a, b, and c to zero, positive, and negative currents, respectively.

$$\begin{bmatrix} I_{fa}^0 \\ I_{fa}^+ \\ I_{fa}^- \end{bmatrix} = \frac{1}{3} \begin{bmatrix} 1 & 1 & 1 \\ 1 & a & a^2 \\ 1 & a^2 & a \end{bmatrix} \begin{bmatrix} I_{fa} \\ I_{fb} \\ I_{fc} \end{bmatrix} \tag{11.69}$$

When the conditions of a single line-to-ground fault are substituted for current, we have

$$\begin{bmatrix} I_{fa}^0 \\ I_{fa}^+ \\ I_{fa}^- \end{bmatrix} = \frac{1}{3} \begin{bmatrix} 1 & 1 & 1 \\ 1 & a & a^2 \\ 1 & a^2 & a \end{bmatrix} \begin{bmatrix} I_{fa} \\ 0 \\ 0 \end{bmatrix} \tag{11.70}$$

The above matrix simplifies to the following:

$$I_{fa}^0 = I_{fa}^+ = I_{fa}^- \tag{11.71}$$

The above equation clearly states that the zero, positive, and negative sequence network must be connected in series for a single line-to-ground fault calculation.

Again, we should remember that we are interested in an if–then condition at bus i. That is before bus i is faulted, the Thevenin voltage at bus i is the open-circuit voltage of bus i. Before the fault, the power grid is balanced and during the time of the fault, it is assumed it remains at the same voltage because the fault is isolated very quickly. The impedance in bus i is the Thevenin impedance. Therefore, we need to compute the Thevenin impedance by looking into bus i for its zero, positive, and negative sequence networks. Figure 11.41b depicts the sequence network connection for the single line to ground.

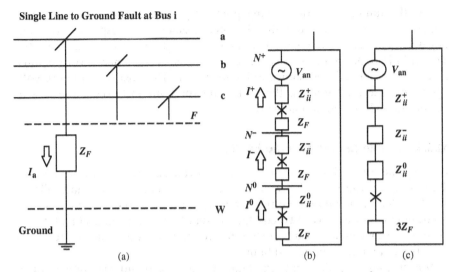

Single Line to Ground Fault at Bus i

Figure 11.41 (a) A single line-to-ground fault, (b) the connection of sequence networks for a single line-to-ground fault calculation, and (c) the simplified connection of sequence networks for a single line-to-ground fault calculation.

To compute the single line-to-ground fault, the positive, negative, and zero sequence networks of the power grid are constructed, and then Z_{Bus}^+, Z_{Bus}^-, and Z_{Bus}^0 are computed as given in Equations (11.72)–(11.74).

$$
Z_{Bus}^+ =
\begin{matrix}
 & 1 & & i & & n & \\
 & \begin{bmatrix} Z_{11}^+ & \cdots & Z_{i1}^+ & \cdots & Z_{1n}^+ \\ \cdot & \cdots & \cdot & \cdots & \cdot \\ Z_{i1}^+ & \cdots & Z_{ii}^+ & \cdots & Z_{in}^+ \\ \cdot & \cdots & \cdot & \cdots & \cdot \\ Z_{n1}^+ & \cdots & Z_{ni}^+ & \cdots & Z_{nn}^+ \end{bmatrix} & \begin{matrix} 1 \\ \cdot \\ i \\ \cdot \\ n, \end{matrix}
\end{matrix}
\tag{11.72}
$$

$$
Z_{Bus}^- =
\begin{matrix}
 & 1 & & i & & n & \\
 & \begin{bmatrix} Z_{11}^- & \cdots & Z_{i1}^- & \cdots & Z_{1n}^- \\ \cdot & \cdots & \cdot & \cdots & \cdot \\ Z_{i1}^- & \cdots & Z_{ii}^- & \cdots & Z_{in}^- \\ \cdot & \cdots & \cdot & \cdots & \cdot \\ Z_{n1}^- & \cdots & Z_{ni}^- & \cdots & Z_{nn}^- \end{bmatrix} & \begin{matrix} 1 \\ \cdot \\ i \\ \cdot \\ n, \end{matrix}
\end{matrix}
\tag{11.73}
$$

$$
Z_{Bus}^0 =
\begin{matrix}
 & 1 & & i & & n & \\
 & \begin{bmatrix} Z_{11}^0 & \cdots & Z_{i1}^0 & \cdots & Z_{1n}^0 \\ \cdot & \cdots & \cdot & \cdots & \cdot \\ Z_{i1}^0 & \cdots & Z_{ii}^0 & \cdots & Z_{in}^0 \\ \cdot & \cdots & \cdot & \cdots & \cdot \\ Z_{n1}^0 & \cdots & Z_{ni}^0 & \cdots & Z_{nn}^0 \end{bmatrix} & \begin{matrix} 1 \\ \cdot \\ i \\ \cdot \\ n \end{matrix}
\end{matrix}
\tag{11.74}
$$

For the faulted bus, the driving point impedance, that is, the Thevenin impedance is selected from the diagonal elements of respective positive, negative, and zero sequence networks. For a single line-to-ground fault, the Thevenin impedance of positive, negative, and zero sequence networks must be connected in series as shown in Figure 11.41b. The positive sequence voltage before the fault remains the same after the fault and excites the flow of current though the faulted network.

11.9.4 Double Line-to-Ground Faults

In a typical double line-to-ground fault, you can envision the following incident. Suppose the insulation of a one-phase conductor breaks and falls on another phase conductor, then the conductor falls on a tree. Because the tree is grounded, and we have two conductors that have made an accidental connection, we have created a double line-to-ground fault condition. Figure 11.42a depicts a double line-to-ground fault.

It is customary to designate the faulted phases as b and c as shown in Figure 11.42a. The condition of the fault indicates that phase a is not faulted, $I_{fa} = 0$. We can calculate the sequence currents during the fault from the symmetrical transformation.

$$\begin{bmatrix} I_{fa}^0 \\ I_{fa}^+ \\ I_{fa}^- \end{bmatrix} = \frac{1}{3} \begin{bmatrix} 1 & 1 & 1 \\ 1 & a & a^2 \\ 1 & a^2 & a \end{bmatrix} \begin{bmatrix} 0 \\ I_{fb} \\ I_{fc} \end{bmatrix} \tag{11.75}$$

Therefore, the zero sequence current is given as

$$I_{fb} + I_{fc} = 3 I_{fa}^0 \tag{11.76}$$

Because the power grid is balanced and phase a, b, and c voltages at bus i remain balanced:

$$V_{fb} = V_{fc} = (I_{fb} + I_{fc}) Z_{FG} \tag{11.77}$$

Substituting for phase b and c currents $I_{fb} + I_{fc} = 3 I_{fa}^0$, we have

$$V_{fb} = V_{fc} = 3 \cdot Z_{FG} \cdot I_{fa}^0$$

The above analysis clearly indicates that for computing double line-to-ground fault, we need to use the sequence network connections given in Figure 11.42b.

Similar to the single line-to-ground fault in that the fault involves the ground connection, for the calculation of the double line-to-ground fault current, we need the Thevenin impedance from the fault point and the Thevenin voltage before the fault. Therefore, we need to construct the positive, negative, and zero

Double line to ground fault at bus i

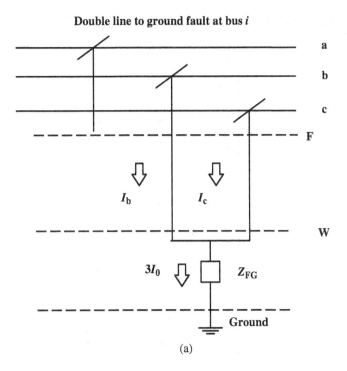

(a)

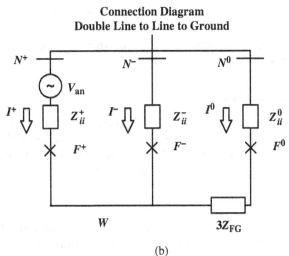

(b)

Figure 11.42 (a) A double line-to-ground fault and (b) the sequence network connection for a double line-to-ground fault.

sequence impedance matrices as given by Z_{Bus}^+, Z_{Bus}^-, and Z_{Bus}^0 in Equations (11.78)–(11.80).

$$
Z_{\text{Bus}}^+ =
\begin{array}{c}
\quad\ 1 \qquad\quad i \qquad\quad n \\
\begin{bmatrix}
Z_{11}^+ & \cdots & Z_{i1}^+ & \cdots & Z_{1n}^+ \\
\cdot & \cdots & \cdot & \cdots & \cdot \\
Z_{i1}^+ & \cdots & Z_{ii}^+ & \cdots & Z_{in}^+ \\
\cdot & \cdots & \cdot & \cdots & \cdot \\
Z_{n1}^+ & \cdots & Z_{ni}^+ & \cdots & Z_{nn}^+
\end{bmatrix}
\begin{array}{c} 1 \\ \cdot \\ i \\ \cdot \\ n, \end{array}
\end{array}
\tag{11.78}
$$

$$
Z_{\text{Bus}}^- =
\begin{array}{c}
\quad\ 1 \qquad\quad i \qquad\quad n \\
\begin{bmatrix}
Z_{11}^- & \cdots & Z_{i1}^- & \cdots & Z_{1n}^- \\
\cdot & \cdots & \cdot & \cdots & \cdot \\
Z_{i1}^- & \cdots & Z_{ii}^- & \cdots & Z_{in}^- \\
\cdot & \cdots & \cdot & \cdots & \cdot \\
Z_{n1}^- & \cdots & Z_{ni}^- & \cdots & Z_{nn}^-
\end{bmatrix}
\begin{array}{c} 1 \\ \cdot \\ i \\ \cdot \\ n, \end{array}
\end{array}
\tag{11.79}
$$

$$
Z_{\text{Bus}}^0 =
\begin{array}{c}
\quad\ 1 \qquad\quad i \qquad\quad n \\
\begin{bmatrix}
Z_{11}^0 & \cdots & Z_{i1}^0 & \cdots & Z_{1n}^0 \\
\cdot & \cdots & \cdot & \cdots & \cdot \\
Z_{i1}^0 & \cdots & Z_{ii}^0 & \cdots & Z_{in}^0 \\
\cdot & \cdots & \cdot & \cdots & \cdot \\
Z_{n1}^0 & \cdots & Z_{ni}^0 & \cdots & Z_{nn}^0
\end{bmatrix}
\begin{array}{c} 1 \\ \cdot \\ i \\ \cdot \\ n \end{array}
\end{array}
\tag{11.80}
$$

Again, the diagonal elements of the Z_{Bus} matrices represent the corresponding Thevenin impedance.

11.9.5   Line-to-Line Faults

Figure 11.43a depicts a line-to-line fault. This type of fault can occur when a phasor conductor falls on another phasor conductor due to insulation failure caused by high wind. It is customary to designate the faulted phases on phase b and phase c. Therefore, the fault point can be expressed as

$$
I_{\text{fa}} = 0, \ I_{\text{fc}} = -I_{\text{fb}}
\tag{11.81}
$$

The power grid before the fault and after the fault remains balanced, because the fault is removed very quickly by opening the appropriative circuit breakers.

$$
V_{\text{fb}} - V_{\text{fc}} = I_{\text{fb}} \cdot Z_{\text{f}}
\tag{11.82}
$$

$$
\begin{bmatrix}
I_{\text{fa}}^0 \\
I_{\text{fa}}^+ \\
I_{\text{fa}}^-
\end{bmatrix}
= \frac{1}{3}
\begin{bmatrix}
1 & 1 & 1 \\
1 & a & a^2 \\
1 & a^2 & a
\end{bmatrix}
\begin{bmatrix}
0 \\
I_{\text{fb}} \\
-I_{\text{fb}}
\end{bmatrix}
\tag{11.83}
$$

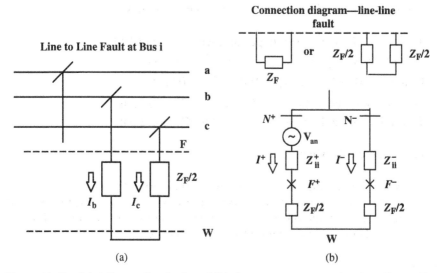

Figure 11.43   (a) A line-to-line fault and (b) the sequence connection for a line-to-line fault.

From the above, we have

$$I_{fa}^0 = 0 \tag{11.84}$$

$$I_{fa}^+ = -I_{fa}^- \tag{11.85}$$

The above results clearly indicate that we need to use the sequence network connection given by Figure 11.43b to calculate the line-to-line fault.

To compute the Thevenin impedance of the faulted bus, we need to construct positive and negative sequence impedance matrices as given by Z_{Bus}^+ and Z_{Bus}^- in Equations (11.86) and (11.87). The diagonal elements of the Z_{Bus} matrix are equal to the Thevenin impedance. The Thevenin voltage is equal to the faulted bus before the fault.

$$
Z_{Bus}^+ =
\begin{array}{c}
\begin{array}{ccccc} 1 & & i & & n \end{array} \\
\left[
\begin{array}{ccccc}
Z_{11}^+ & \cdots & Z_{i1}^+ & \cdots & Z_{1n}^+ \\
\vdots & \cdots & \vdots & \cdots & \vdots \\
Z_{i1}^+ & \cdots & Z_{ii}^+ & \cdots & Z_{in}^+ \\
\vdots & \cdots & \vdots & \cdots & \vdots \\
Z_{n1}^+ & \cdots & Z_{ni}^+ & \cdots & Z_{nn}^+
\end{array}
\right]
\begin{array}{c} 1 \\ \cdot \\ i \\ \cdot \\ n, \end{array}
\end{array}
\tag{11.86}
$$

$$
Z_{Bus}^- =
\begin{array}{c}
\begin{array}{ccccc} 1 & & i & & n \end{array} \\
\left[
\begin{array}{ccccc}
Z_{11}^- & \cdots & Z_{i1}^- & \cdots & Z_{1n}^- \\
\vdots & \cdots & \vdots & \cdots & \vdots \\
Z_{i1}^- & \cdots & Z_{ii}^- & \cdots & Z_{in}^- \\
\vdots & \cdots & \vdots & \cdots & \vdots \\
Z_{n1}^- & \cdots & Z_{ni}^- & \cdots & Z_{nn}^-
\end{array}
\right]
\begin{array}{c} 1 \\ \cdot \\ i \\ \cdot \\ n \end{array}
\end{array}
\tag{11.87}
$$

Example 11.5   Consider the power system given below.

The system data are given in per unit with a base of 100 MVA.
Generator A: $X_G^+ = 0.25$, $X_G^- = 0.15$, $X_G^0 = 0.03$ p.u
Generator B: $X_G^+ = 0.2$, $X_G^- = 0.12$, $X_G^0 = 0.02$ p.u
Transmission line C–D: $Z^+ = Z^- = j0.08$, $Z^0 = j0.14$ p.u
Transmission line D–E: $Z^+ = Z^- = j0.06$, $Z^0 = j0.12$ p.u

Perform the following:

(i) Assume generator A is Y connected and ungrounded and generator B is Y connected and grounded. Compute the single line-to-ground fault at bus D. Determine the sequence networks and show the flow of line-to-ground current in the faulted system at bus D.

Solution
The sequence networks of Example 11.5 are calculated from the equivalent circuit in Figure 11.46:

$$I = \frac{V}{Z} = \frac{1\angle 0°}{j0.39} = 2.56\angle -90°$$

$$I_{aCD}^0 = 0$$

$$I_{aCD}^+ = 2.56\angle -90° \times \frac{j0.26}{j0.26 + j0.33} = 1.12\angle -90°$$

$$I_{aCD}^- = 2.56\angle -90° \times \frac{j0.18}{j0.18 + j0.23} = 1.12\angle -90°$$

The actual line currents in line C–D are as follows:

$$\begin{bmatrix} I_{aCD} \\ I_{bCD} \\ I_{cCD} \end{bmatrix} = [T_s] \begin{bmatrix} I_{aCD}^0 \\ I_{aCD}^+ \\ I_{aCD}^- \end{bmatrix} = \begin{bmatrix} 1 & 1 & 1 \\ 1 & a^2 & a \\ 1 & a & a^2 \end{bmatrix} \begin{bmatrix} 0 \\ 1.12\angle -90° \\ 1.12\angle -90° \end{bmatrix} = \begin{bmatrix} -j2.24 \\ j1.12 \\ j1.12 \end{bmatrix}$$

The line currents in line E–D are as follows:

$$I_{aED}^0 = I = 2.56\angle -90°$$

$$I_{aED}^+ = I - I_{aCD}^+ = 1.44\angle -90°$$

$$I_{aED}^- = I - I_{aCD}^- = 1.44\angle -90°$$

$$\begin{bmatrix} I_{aED} \\ I_{bED} \\ I_{cED} \end{bmatrix} = [T_s] \begin{bmatrix} I_{aED}^0 \\ I_{aED}^+ \\ I_{aED}^- \end{bmatrix} = \begin{bmatrix} 1 & 1 & 1 \\ 1 & a^2 & a \\ 1 & a & a^2 \end{bmatrix} \begin{bmatrix} 2.56\angle -90° \\ 1.44\angle -90° \\ 1.44\angle -90° \end{bmatrix} = \begin{bmatrix} -j5.44 \\ -j1.12 \\ -j1.12 \end{bmatrix}$$

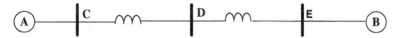

Figure 11.44   A one-line diagram of Example 11.5.

Alternatively, the Z_{th} for bus D can also be computed from the Z_{Bus}^+, Z_{Bus}^-, and Z_{Bus}^0, which are shown below.

$$Z_{Bus}^+ = \begin{bmatrix} j0.14 & j0.11 & j0.08 \\ j0.11 & j0.15 & j0.11 \\ j0.08 & j0.11 & j0.13 \end{bmatrix} \begin{matrix} C \\ D \\ E \end{matrix}, \quad Z_{Bus}^- = \begin{bmatrix} j0.10 & j0.07 & j0.04 \\ j0.07 & j0.10 & j0.07 \\ j0.04 & j0.07 & j0.08 \end{bmatrix} \begin{matrix} C \\ D \\ E \end{matrix},$$

$$Z_{Bus}^0 = \begin{bmatrix} j0.28 & j0.14 & j0.02 \\ j0.14 & j0.14 & j0.02 \\ j0.02 & j0.02 & j0.02 \end{bmatrix} \begin{matrix} C \\ D \\ E \end{matrix}$$

The positive sequence Thevenin impedance for bus D is given by $Z_{Th,D}^+ = Z_{DD}^+ = j0.15$.

The negative sequence Thevenin impedance for bus D is given by $Z_{Th,D}^- = Z_{DD}^- = j0.10$.

The zero sequence Thevenin impedance for bus D is given by $Z_{Th,D}^0 = Z_{DD}^0 = j0.14$.

Figure 11.44 depicts a one-line diagram of Example 11.5.

The above values are the same as those shown in Figure 11.46e and the corresponding sequence currents can be calculated as shown before.

Figure 11.45 presents a three-phase diagram of a single line-to-ground fault at bus D.

The equivalent circuit for a single line-to-ground fault at bus D is shown in Figure 11.46.

Figure 11.46 depicts network connection for Example 11.5: (a) positive sequence network, (b) negative sequence network, (c) zero sequence network, (d) connection diagram of sequence networks for single line-to-ground fault at bus D, and (e) simplified circuit of (d).

Figure 11.47 depicts the current flow in the sequence networks of Example 11.5.

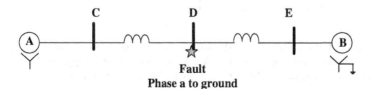

Fault
Phase a to ground

Figure 11.45   A three-phase diagram of a single line-to-ground fault at bus D.

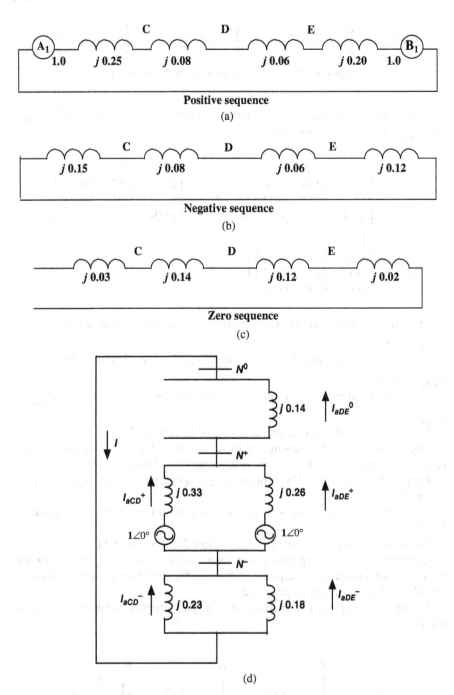

Figure 11.46 For Example 11.5: (a) positive sequence network, (b) negative sequence network, (c) zero sequence network, (d) connection diagram of sequence networks for single line-to-ground fault at bus D, and (e) simplified circuit of (d).

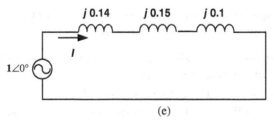

(e)

Figure 11.46   (*Continued*)

Example 11.6   Consider the power grid given in Figure 11.48. The sequence network data based on 100 MVA are

$$ZG^+ = ZG^- = j0.1 \text{ p.u}, \ ZG^0 = j0.05 \text{ p.u}$$
$$Z_{12}^+ = Z_{12}^- = j0.3 \text{ p.u}, \ Z_{12}^0 = j0.6 \text{ p.u}$$
$$Z_{23}^+ = Z_{23}^- = j0.4 \text{ p.u}, \ Z_{23}^0 = j0.5 \text{ p.u}$$
$$Z_{\text{Trans}}+ = Z_{\text{Trans}}^- = Z_{\text{Trans}}^0 = j0.08 \text{ p.u}$$

Assume the load is given as

$$S_{\text{Load}} = 1 + j0.5 \text{ p.u}$$

And the load voltage is

$$V_{\text{L}} = 0.9\angle -4.0 \text{ p.u}$$

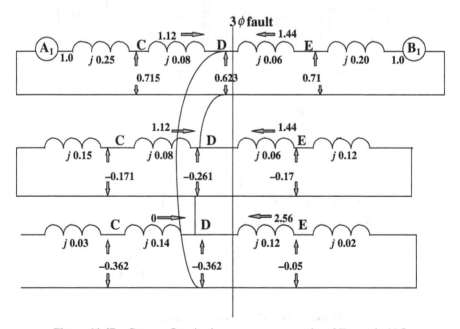

Figure 11.47   Current flow in the sequence networks of Example 11.5.

The ground impedance in the generator is equal to $j0.01$ p.u. For a single line-to-ground fault at bus 2, compute the following:

(i) The fault currents flowing from bus 1 to bus 2 (faulted bus) and from bus 3 when the load is ignored.

(ii) The same as Part (i), but take the load into consideration.

(iii) The same as Part (i), but assume the generator is not grounded.

Solution

(i) The sequence circuits for a single line-to-ground fault at bus 2 without considering the loads is shown in Figure 11.49.

From Figure 11.49, the sequence currents are calculated as follows:

$$I = \frac{V}{Z} = \frac{1\angle 0°}{j1.11} = 0.9\angle -90°$$

$$I_{a12}^{0} = 0.9\angle -90° \times \frac{j0.58}{j0.58 + j0.65} = 0.42\angle -90°$$

$$I_{a12}^{+} = I_{a12}^{-} = I = 0.9\angle -90°$$

The actual currents in line 1–2 are given by

$$\begin{bmatrix} I_{a12} \\ I_{b12} \\ I_{c12} \end{bmatrix} = [T_s] \begin{bmatrix} I_{a12}^{0} \\ I_{a12}^{+} \\ I_{a12}^{-} \end{bmatrix} = \begin{bmatrix} 1 & 1 & 1 \\ 1 & a^2 & a \\ 1 & a & a^2 \end{bmatrix} \begin{bmatrix} 0.42\angle -90° \\ 0.9\angle -90° \\ 0.9\angle -90° \end{bmatrix} = \begin{bmatrix} -j2.22 \\ j0.48 \\ j0.48 \end{bmatrix}$$

The sequence currents in line 3–2 are given by

$$I_{a32}^{0} = I - I_{a12}^{0} = 0.48\angle -90°$$

$$I_{a32}^{+} = I_{a32}^{-} = 0$$

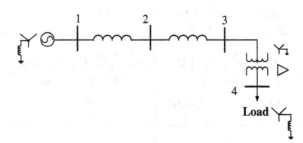

Figure 11.48   The system for Example 11.6.

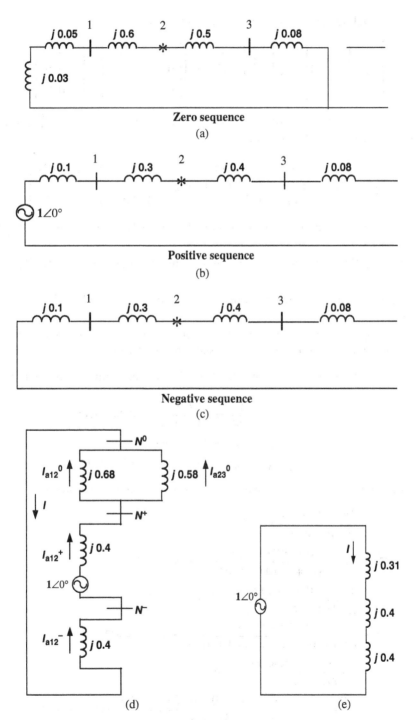

Figure 11.49 For Example 8.6(i): (a) zero sequence network, (b) positive sequence network, (c) negative sequence network, (d) connection diagram for single line-to-ground fault at bus 2, and (e) simplified circuit of (d).

The actual line currents in line 3–2 are given by

$$\begin{bmatrix} I_{a32} \\ I_{b32} \\ I_{c32} \end{bmatrix} = [T_s] \begin{bmatrix} I^0_{a32} \\ I^+_{a32} \\ I^-_{a32} \end{bmatrix} = \begin{bmatrix} 1 & 1 & 1 \\ 1 & a^2 & a \\ 1 & a & a^2 \end{bmatrix} \begin{bmatrix} 0.48\angle -90° \\ 0 \\ 0 \end{bmatrix} = \begin{bmatrix} -j0.48 \\ -j0.48 \\ -j0.48 \end{bmatrix}$$

Alternatively, the Z_{th} for bus 2 can also be computed from the Z^0_{Bus}, Z^-_{Bus}, and Z^0_{Bus}, which are shown below.

$$Z^+_{\text{Bus}} = Z^-_{\text{Bus}} = \begin{bmatrix} j0.1 & j0.1 & j0.1 \\ j0.1 & j0.4 & j0.4 \\ j0.1 & j0.4 & j0.8 \end{bmatrix}, \quad Z^0_{\text{Bus}} = \begin{bmatrix} j0.07 & j0.04 & j0.01 \\ j0.04 & j0.31 & j0.04 \\ j0.01 & j0.04 & j0.07 \end{bmatrix}$$

The positive sequence Thevenin impedance for bus 2 is given by $Z^+_{\text{Th},2} = Z^+_{22} = j0.4$.

The negative sequence Thevenin impedance for bus 2 is given by $Z^-_{\text{Th},2} = Z^-_{22} = j0.4$.

The zero sequence Thevenin impedance for bus 2 is given by $Z^0_{\text{Th},2} = Z^0_{22} = j0.31$.

From Figure 11.49e we see that the values of Thevenin impedance are the same as that calculated from the Z_{Bus} sequence matrices.

(ii) Taking the load into consideration, the Δ equivalent load is given by

$$\bar{Z}_Y = \frac{V^2}{S^*} = \frac{(0.9/\sqrt{3})^2}{1 - j0.5} = 0.24\angle 26.6°$$

$$\bar{Z}_\Delta = 3\bar{Z}_Y = 0.72\angle 26.6°$$

$$Z_{\text{load00}} = Z_{\text{load11}} = Z_{\text{load22}} = 0.72\angle 26.6°$$

From Figure 11.50, the sequence currents are calculated as follows:

$$Z^0_{\text{th}} = \frac{(j0.68)(j0.58)}{(j0.68 + j0.58)} = j0.31$$

$$Z^+_{\text{th}} = Z^-_{\text{th}} = \frac{(j0.4)(0.64 + j0.8)}{(j0.4) + (0.64 + j0.8)} = 0.06 + j0.3$$

$$V^+_{\text{th}} = \frac{1\angle 0° \times (0.64 + j0.8)}{(j0.4) + (0.64 + j0.8)} = 0.75\angle -10.58°$$

$$I = \frac{0.75\angle -10.58°}{0.06 + j0.3 + 0.06 + j0.3 + j0.31} = 0.82\angle -93°$$

$$I^0_{12} = \frac{0.82\angle -93° \times j0.58}{j0.58 + j0.68} = 0.38\angle -93°$$

$$I^+_{12} = \frac{0.82\angle -93° \times (0.64 + j0.8) - 0.75\angle -10.88°}{j0.4 + 0.64 + j0.8} = 0.32\angle -167°$$

$$I^-_{12} = \frac{0.82\angle -93° \times (0.64 + j0.8)}{j0.4 + 0.64 + j0.8} = 0.62\angle -103°$$

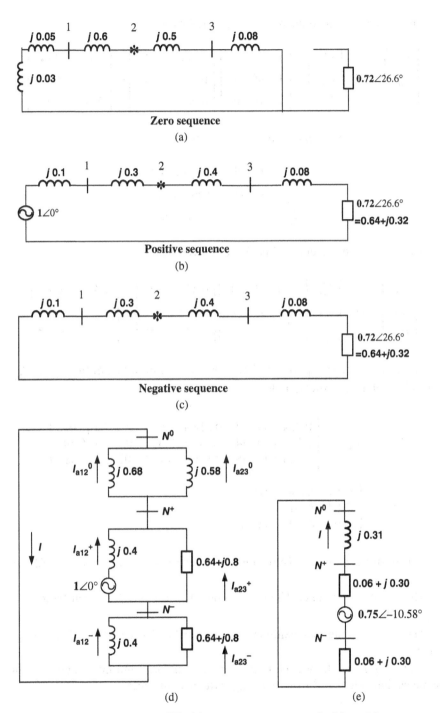

Figure 11.50 For Example 11.6(ii): (a) zero sequence network, (b) positive sequence network, (c) negative sequence network, (d) connection diagram for single line-to-ground fault at bus 2, and (e) the simplified circuit of (d).

The actual line currents in line 1–2 are as follows:

$$\begin{bmatrix} I_{a12} \\ I_{b12} \\ I_{c12} \end{bmatrix} = [T_s] \begin{bmatrix} I^0_{a12} \\ I^+_{a12} \\ I^-_{a12} \end{bmatrix} = \begin{bmatrix} 1 & 1 & 1 \\ 1 & a^2 & a \\ 1 & a & a^2 \end{bmatrix} \begin{bmatrix} 0.38\angle -93° \\ 0.32\angle -167° \\ 0.62\angle -103° \end{bmatrix} = \begin{bmatrix} 1.16\angle -114° \\ 0.67\angle 8° \\ 0.32\angle -142° \end{bmatrix}$$

The sequence currents in line 3–2 are as follows:

$$I^0_{32} = I - I^0_{12} = 0.44\angle -93°$$
$$I^+_{32} = I - I^+_{12} = 0.79\angle -70°$$
$$I^-_{32} = I - I^-_{12} = 0.24\angle -65°$$

The actual line currents in line 3–2 are given by

$$\begin{bmatrix} I_{a32} \\ I_{b32} \\ I_{c32} \end{bmatrix} = [T_s] \begin{bmatrix} I^0_{a32} \\ I^+_{a32} \\ I^-_{a32} \end{bmatrix} = \begin{bmatrix} 1 & 1 & 1 \\ 1 & a^2 & a \\ 1 & a & a^2 \end{bmatrix} \begin{bmatrix} 0.44\angle -93° \\ 0.79\angle -70° \\ 0.24\angle -65° \end{bmatrix} = \begin{bmatrix} 1.44\angle -106° \\ 0.35\angle -145° \\ 0.62\angle -8.97° \end{bmatrix}$$

Alternatively, the Z_{th} for bus 2 can also be computed from the Z^+_{Bus}, Z^-_{Bus}, and Z^0_{Bus}, which are shown below.

$$Z^+_{Bus} = Z^-_{Bus} = \begin{bmatrix} 0.004 + j0.094 & 0.014 + j0.074 & 0.028 + j0.048 \\ 0.014 + j0.074 & 0.06 + j0.30 & 0.111 + j0.192 \\ 0.028 + j0.048 & 0.111 + j0.192 & 0222 + j0.385 \end{bmatrix},$$

$$Z^0_{Bus} = \begin{bmatrix} j0.07 & j0.04 & j0.01 \\ j0.04 & j0.31 & j0.04 \\ j0.01 & j0.04 & j0.07 \end{bmatrix}$$

The positive sequence Thevenin impedance for bus 2 is given by $Z^+_{Th,2} = Z^+_{22} = 0.06 + j0.3$.

The negative sequence Thevenin impedance for bus 2 is given by $Z^-_{Th,2} = Z^-_{22} = 0.06 + j0.3$.

The zero sequence Thevenin impedance for bus 2 is given by $Z^0_{Th,2} = Z^0_{22} = j0.31$.

From Figure 11.50e we see that the values of Thevenin impedance are the same as that calculated from the Z_{Bus} sequence matrices.

(iii) The sequence circuits, at no load when the generator is not grounded, are given in Figure 11.51.

The sequence currents, calculated from Figure 11.51, are as follows:

$$I = \frac{V}{Z} = \frac{1\angle 0°}{j1.38} = 0.72\angle -90°$$

$$I^0_{a12} = 0$$

$$I^+_{a12} = I^-_{a12} = I = 0.72\angle -90°$$

The actual line currents in line 1–2 are as follows:

$$\begin{bmatrix} I_{a12} \\ I_{b12} \\ I_{c12} \end{bmatrix} = [T_s] \begin{bmatrix} I^0_{a12} \\ I^+_{a12} \\ I^-_{a12} \end{bmatrix} = \begin{bmatrix} 1 & 1 & 1 \\ 1 & a^2 & a \\ 1 & a & a^2 \end{bmatrix} \begin{bmatrix} 0 \\ 0.72\angle -90° \\ 0.72\angle -90° \end{bmatrix} = \begin{bmatrix} -j1.44 \\ j0.72 \\ j0.72 \end{bmatrix}$$

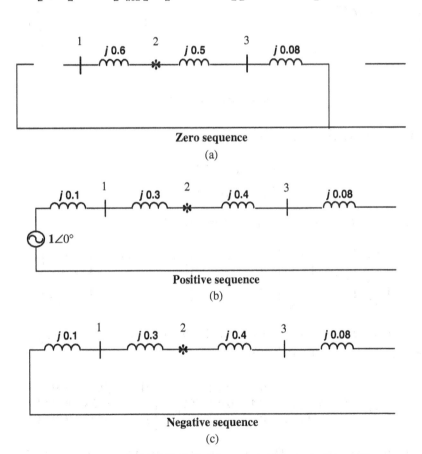

Zero sequence
(a)

Positive sequence
(b)

Negative sequence
(c)

Figure 11.51   For Example 11.6(iii): (a) zero sequence network, (b) positive sequence network, (c) negative sequence network, (d) connection diagram for single line-to-ground fault at bus 2, and (e) the simplified circuit of (d).

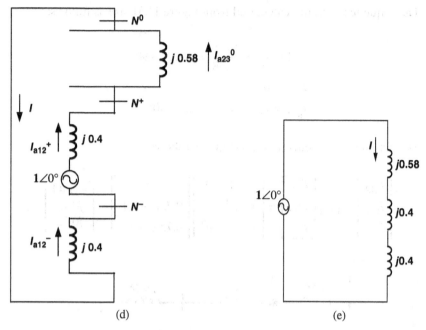

(d)   (e)

Figure 11.51 (*Continued*)

The sequence currents in line 3–2 are as follows:

$$I^0_{a32} = I = 0.72\angle - 90°$$
$$I^+_{a32} = I^-_{a32} = 0$$

The actual line currents in line 3–2 are given by:

$$
\begin{bmatrix} I_{a32} \\ I_{b32} \\ I_{c32} \end{bmatrix} = [T_s]
\begin{bmatrix} I^0_{a32} \\ I^+_{a32} \\ I^-_{a32} \end{bmatrix} =
\begin{bmatrix} 1 & 1 & 1 \\ 1 & a^2 & a \\ 1 & a & a^2 \end{bmatrix}
\begin{bmatrix} 0.72\angle - 90° \\ 0 \\ 0 \end{bmatrix} =
\begin{bmatrix} -j0.72 \\ -j0.72 \\ -j0.72 \end{bmatrix}
$$

Alternatively, the Z_{th} for bus 2 can also be computed from the Z^+_{Bus}, Z^-_{Bus}, and Z^0_{Bus}, which are shown below.

$$
Z^+_{Bus} = Z^-_{Bus} =
\begin{bmatrix} j0.1 & j0.1 & j0.1 \\ j0.1 & j0.4 & j0.4 \\ j0.1 & j0.4 & j0.8 \end{bmatrix}, \;
Z^0_{Bus} =
\begin{bmatrix} j1.18 & j0.58 & j0.08 \\ j0.58 & j0.58 & j0.08 \\ j0.08 & j0.08 & j0.08 \end{bmatrix}
$$

The positive sequence Thevenin impedance for bus 2 is given by $Z^+_{Th,2} = Z^+_{22} = j0.4$.

The negative sequence Thevenin impedance for bus 2 is given by $Z^-_{Th,2} = Z^-_{22} = j0.4$.

The zero sequence Thevenin impedance for bus 2 is given by $Z^0_{Th,2} = Z^0_{22} = j0.58$.

From Figure 11.51e we see that the values of Thevenin impedance are the same as that calculated from the Z_{Bus} sequence matrices.

Example 11.7   For Example 11.6, assume a double line fault at bus 2 and compute the currents in each line without considering the loads.

Solution
The equivalent sequence circuits for a double line fault are shown in Figure 11.52.

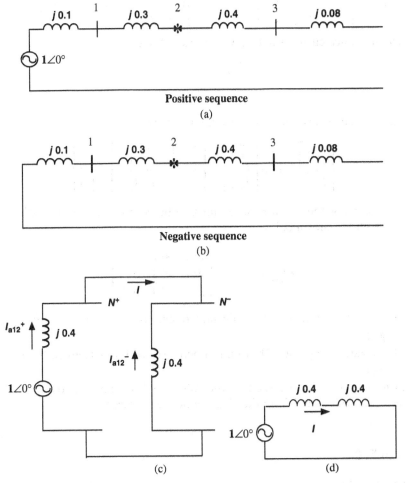

Figure 11.52   For Example 11.7: (a) positive sequence network, (b) negative sequence network, (c) connection diagram for double line fault at bus 2, and (d) the simplified circuit of (c).

The sequence currents for the double line fault at bus 2, calculated from Figure 11.52, are as follows:

$$I = \frac{V}{Z} = \frac{1\angle 0°}{j0.8} = 1.25\angle -90°$$

$$I^0_{a12} = 0$$

$$I^+_{a12} = I = 1.25\angle -90°$$

$$I^-_{a12} = -I = 1.25\angle 90°$$

The actual line currents in line 1–2 are given by

$$\begin{bmatrix} I_{a12} \\ I_{b12} \\ I_{c12} \end{bmatrix} = [T_s] \begin{bmatrix} I^0_{a12} \\ I^+_{a12} \\ I^-_{a12} \end{bmatrix} = \begin{bmatrix} 1 & 1 & 1 \\ 1 & a^2 & a \\ 1 & a & a^2 \end{bmatrix} \begin{bmatrix} 0 \\ 1.25\angle -90° \\ 1.25\angle 90° \end{bmatrix} = \begin{bmatrix} 0 \\ 2.17\angle 180° \\ 2.17\angle 0° \end{bmatrix}$$

The sequence currents in line 3–2 are as follows:

$$I^0_{a32} = 0$$

$$I^+_{a32} = -I^-_{a32} = 0$$

The sequence currents in line 3–2 are given by

$$\begin{bmatrix} I_{a32} \\ I_{b32} \\ I_{c32} \end{bmatrix} = [T_s] \begin{bmatrix} I^0_{a32} \\ I^+_{a32} \\ I^-_{a32} \end{bmatrix} = \begin{bmatrix} 1 & 1 & 1 \\ 1 & a^2 & a \\ 1 & a & a^2 \end{bmatrix} \begin{bmatrix} 0 \\ 0 \\ 0 \end{bmatrix} = \begin{bmatrix} 0 \\ 0 \\ 0 \end{bmatrix}$$

Alternatively, the Z_{th} for bus 2 can also be computed from the Z^+_{Bus} and Z^-_{Bus}, which are shown below.

$$Z^+_{Bus} = Z^-_{Bus} = \begin{bmatrix} j0.1 & j0.1 & j0.1 \\ j0.1 & j0.4 & j0.4 \\ j0.1 & j0.4 & j0.8 \end{bmatrix}$$

The positive sequence Thevenin impedance for bus 2 is given by $Z^+_{Th,2} = Z^+_{22} = j0.4$.

The negative sequence Thevenin impedance for bus 2 is given by $Z^-_{Th,2} = Z^-_{22} = j0.4$.

From Figure 11.52d we see that the values of Thevenin impedance are the same as that calculated from the Z_{Bus} sequence matrices.

PROBLEMS

11.1 Consider a typical power system given in Figure 11.53.
All reactances are in p.u. on a base of 100 MVA.

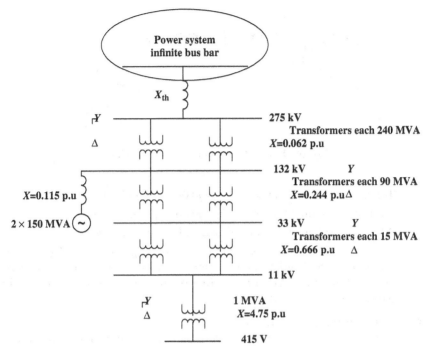

Figure 11.53   A typical power grid system.

$X_{th} = 0.01$ when the maximum number of generators is in service.
$X_{th} = 0.015$ when the minimum number of generators is in service.
Compute the following:

(i) The SCC of the 415 V bus when all transformers are in service, but generator G_1 is not in service. Assume that the maximum numbers of generators are in service.

(ii) The SCC of the 415 V bus when all the transformers and G_1 are in service. Assume one generator is in service.

11.2 Consider the power system given in Figure 11.54.
Generator A:

$$X''_{G(1)} = 0.25, \ X''_{G(2)} = 0.15, \ X''_{G(0)} = 0.03 \text{ p.u.}$$

Generator B:

$$X''_{G(1)} = 0.2, \ X''_{G(2)} = 0.12, \ X''_{G(0)} = 0.02 \text{ p.u.}$$

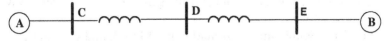

Figure 11.54   A one-line diagram for Problem 11.2.

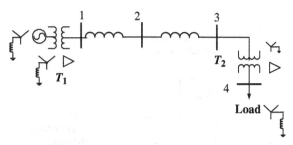

Figure 11.55   The system for Problem 11.3.

Transmission line C–D: $Z_1 = Z_2 = j0.08$, $Z_0 = j0.14$ p.u.
Transmission line D–E: $Z_1 = Z_2 = j0.06$, $Z_0 = j0.12$ p.u.
All values are given in per unit with a base of 100 MVA.

Assume generator A is Y connected and ungrounded and generator B is Y connected and grounded, compute the single line-to-ground fault at bus D and the current and actual phase voltages (i.e., V_a, V_b, V_c in p.u) of buses C, D, and E.

11.3   Consider the power grid given below.

Figure 11.55 presents the one-diagram of a power grid for Problem 11.3.

The sequence network data are

Generator: 20 kV, 100 MVA, 10% positive reactance, negative sequence network = positive sequence reactance, zero sequence reactance = 8% based on the generator's rating

Transmission line length: bus 1–2 = 50 km, reactance = 0.5 Ω/km
                             bus 2–3 = 100 km, reactance = 0.7 Ω/km

Transformer T_1 = 20 kV/138 kV, 8% reactance, 150 MVA

Transformer T_2 = 138 kV/13.8 kV, 10% reactance, 200 MVA

Transmission line sequence impedances: $Z_1 = Z_2 = j0.06$ p.u, $Z_0 = j0.12$ in p.u on a 100 MVA base

Transformers' sequence impedances: positive = negative = zero

Load: S_{Load} = 50 MVA, p.f. = 0.95 lagging

Generator ground impedance = $j0.01$ p.u based on its rating

Perform the following:

(i) Compute the per unit model for positive, negative, and zero sequence networks based on 100 MVA.

11.4   For Problem 11.3, perform the following:

(i) Compute the load voltage if the generator is set at 5% above its own rating

(ii) For a double line-to-ground fault at bus 3, find the fault currents flowing from bus 1 and bus 2 to bus 3 (faulted bus) when the load is ignored

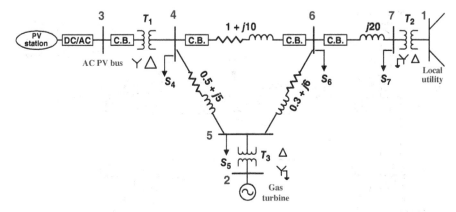

Figure 11.56   A one-line diagram for Problem 11.6.

11.5 For Problem 11.3, perform the following:
  (i) Compute the single line-to-ground fault at bus 2, but take the load into consideration
  (ii) The same as Part (i), but assume the generator is not grounded

11.6 Consider the microgrid given in Figure 11.56.
  The impedances of the transmission lines are given in the one-line diagram (Figure 11.56).
  The system data are
  PV-generating station: 2 MW, 460 V AC; positive, negative, and zero sequence impedances = 10%.
  Gas-turbine-generating station: positive sequence impedance = 10 MVA, 3.2 kV, 10% reactance.
  Sequence impedance: negative sequence = positive sequence, zero sequence = $\frac{1}{2}$ positive sequence.
  Transformers' sequence impedances: positive = negative = zero
  Transformer T_1: 10 MVA, 460 V/13.2 kV, 7% reactance
  Transformer T_2: 25 MVA, 13.2 kV/69 kV, 9% reactance
  Transformer T_3: 20 MVA, 13.2 kV/3.2 kV, 8% reactance
  Loads: S_4 = 4 MW, p.f. = 0.9 lagging; S_5 = 8 MW, p.f. = 0.9 lagging; S_6: 10 MVA, p.f. = 0.9 leading; S_7 = 5 MVA, p.f. = 0.85 lagging
  Local power grid: positive, negative, and zero sequence internal reactance = 10 Ω; negligible internal reactance
  Perform the following:
  (i) Per unit impedance model for positive, negative, and zero sequences
  (ii) Ignore the loads and compute a single line-to-ground fault at bus 4
  (iii) Compute the load voltages; use the load impedance models and compute the single line-to-ground fault at bus 4

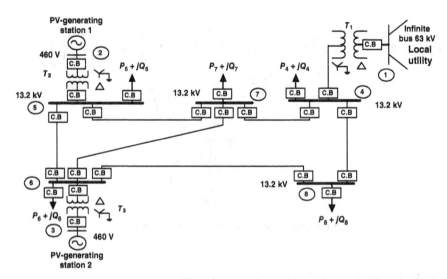

Figure 11.57 A microgrid of distributed generation connected to a local power network.

11.7 Consider the microgrid given in Figure 11.57.

The system data are

Local power grid SCC = 1600 MVA

PV-generating stations: ungrounded, approximate internal impedance (resistive only) = 50% for positive, negative, and zero sequence, 100 MVA

Transformers: 460 V Y grounded/13.2 kV Δ, 10% reactance, 10 MVA

Local power grid transformer: 20 MVA, 63 kV/13.2 kV, 7% reactance

Transmission line: resistance = 0.0685 Ω/mile, reactance = 0.40 Ω/mile, half of line charging admittance $(Y'/2) = 11$ Ω/mile. Line 4–7 = 10 miles, 4–8 = 7 miles, 5–6 = 12 miles, 5–7 = 7 miles, 6–7 = 6 miles, 6–8 = 8 miles

Transmission line sequence impedance: positive = negative, zero sequence = 2 × positive sequence impedance

Perform the following:

(i) Per unit equivalent model for positive, negative, and zero sequence impedances based on a 20-MVA base

(ii) For a three-phase fault on bus 4, compute the bus 4 SCC

11.8 For Problem 11.7, for a single line-to-ground fault on bus 1, compute the ground fault current.

11.9 For Problem 11.7, to increase the security of the system two identical transformers are used at each distribution network and at the interconnection to the local power grid. Compute the SCC of bus 4.

11.10 For Problem 11.7, assume that transformer T1 is grounded Y–Y. Compute the line-to-ground fault current at bus 4.

11.11 For Problem 11.7, assume that transformer T1 is grounded Y–Y. Compute the line-to-ground fault current at bus 4.

REFERENCES

1. Gross, A.C. (1986) *Power System Analysis*, Wiley, New York.
2. El-Hawary, M.E. (1983) *Electric Power Systems: Design and Analysis*, Reston Publishing, Reston, VA.
3. IEEE Brown Book (1980) *IEEE Recommended Practice For Power System Analysis*, Wiley-Interscience, New York.
4. Grainger, J. and Stevenson, W.D. (2008) *Power Systems Analysis*, McGraw-Hill, New York.
5. Duncan Glover, J. and Sarma, M.S. (2002) *Power System Analysis and Design*, Brooks/Cole Thomson Learning, Pacific Grove, CA.
6. Stagg, G.W. and El-Abiad, A.H. (1968) *Computer Methods in Power System Analysis*, McGraw-Hill, New York.
7. Bergen, A. and Vittal, V. (2000) *Power Systems Analysis*, Prentice Hall, Englewood Cliffs, NJ.
8. Fortescue, C.L. (1918) Method of symmetrical co-ordinates applied to the solution of polyphase networks. *AIEE Transactions*, 37(part II), 1027–1140.

INDEX

Note: Page numbers followed by f refer to figures; page numbers followed by t refer to tables.

Design of Smart Power Grid Renewable Energy Systems, Second Edition. Ali Keyhani.
© 2017 John Wiley & Sons, Inc. Published 2017 by John Wiley & Sons, Inc.
Companion website: www.wiley.com/go/smartpowergrid2e